AF411181

Behavior of Metallic and Composite Structures (Second Volume)

Behavior of Metallic and Composite Structures (Second Volume)

Editors

Tomasz Sadowski
Holm Altenbach

MDPI • Basel • Beijing • Wuhan • Barcelona • Belgrade • Manchester • Tokyo • Cluj • Tianjin

Editors
Tomasz Sadowski
Department of Solid Mechanics
Lublin University of Technology
Lublin
Poland

Holm Altenbach
Institute for Mechanics
Otto-von-Guericke-Universität Magdeburg
Magdeburg
Germany

Editorial Office
MDPI
St. Alban-Anlage 66
4052 Basel, Switzerland

This is a reprint of articles from the Special Issue published online in the open access journal *Materials* (ISSN 1996-1944) (available at: www.mdpi.com/journal/materials/special_issues/ behavior_metallic_composite).

For citation purposes, cite each article independently as indicated on the article page online and as indicated below:

LastName, A.A.; LastName, B.B.; LastName, C.C. Article Title. *Journal Name* **Year**, *Volume Number*, Page Range.

ISBN 978-3-0365-1492-5 (Hbk)
ISBN 978-3-0365-1491-8 (PDF)

Contents

Review

Interactions between Dislocations and Boundaries during Deformation

Hongjiang Pan [1,2,*], **Yue He** [3] **and Xiaodan Zhang** [2,*]

[1]   Faculty of Materials Science and Engineering, Kunming University of Science and Technology, Kunming 650093, China
[2]   Department of Mechanical Engineering, Technical University of Denmark, 2800 Kongens Lyngby, Denmark
[3]   Department of Chemical and Materials Engineering, University of Auckland, Private Bag 92019, Auckland 1142, New Zealand; yue_he_1225@outlook.com
*   Correspondence: hj_pan@outlook.com (H.P.); xzha@mek.dtu.dk (X.Z.); Tel.: +45-5274-3486 (H.P.)

Abstract: The interactions between dislocations (dislocations and deformation twins) and boundaries (grain boundaries, twin boundaries and phase interfaces) during deformation at ambient temperatures are reviewed with focuses on interaction behaviors, boundary resistances and energies during the interactions, transmission mechanisms, grain size effects and other primary influencing factors. The structure of boundaries, interactions between dislocations and boundaries in coarse-grained, ultrafine-grained and nano-grained metals during deformation at ambient temperatures are summarized, and the advantages and drawbacks of different in-situ techniques are briefly discussed based on experimental and simulation results. The latest studies as well as fundamental concepts are presented with the aim that this paper can serve as a reference in the interactions between dislocations and boundaries during deformation.

Keywords: dislocation–boundary interaction; dislocation–interface interaction; deformation twin–boundary interaction; size effect; boundary structure; boundary strengthening; characterization techniques

Citation: Pan, H.; He, Y.; Zhang, X. Interactions between Dislocations and Boundaries during Deformation. *Materials* **2021**, *14*, 1012. https://doi.org/10.3390/ma14041012

Academic Editor: Tomasz Sadowski

Received: 21 December 2020
Accepted: 15 February 2021
Published: 21 February 2021

1. Introduction

The mechanical properties of metals at ambient temperatures (from about 0 to 200 °C [1–3]) are mainly affected by their microstructures, except for the intrinsic properties determined by their chemical compositions [4,5]. As plastic deformation of metals proceeds via the dislocation motion, including nucleation, multiplication and slip of full dislocations with an edge/screw/mixed characteristics as well as nucleation and propagation of partial dislocations/deformation twins [6–9]. Compared with the dislocation–dislocation interaction, interactions between dislocations and boundaries, such as grain boundaries, twin boundaries or phase interfaces, play a more important role in the engineering of the mechanical properties such as strength [10–18].

It is well known that the grain refinement by thermomechanical processing is an effective way for the improvement of strength. Generally, the boundary strengthening is based on the impediment of dislocation motions by all sorts of boundaries [16,18–22]. The concept of grain boundary engineering (GBE) has been developed based on the knowledge of the grain boundary structure and energy in the past twenty years [23–25]. These developments have motivated scholars to discover the nature of dislocation–boundary interactions at different length scales. For decades, the progress has been made in the studies of boundary structures experimentally [26,27], as well as in the field of interactions between dislocations and boundaries, such as the processes and products of interactions [12–14,16,21,28]. Different influencing factors have been found and several models on the interaction mechanisms have been established [11,13,29–32].

In general, the boundary is an obstacle to dislocation motion and twin propagation during the deformation of metals. Dislocations or twins may either transmit across boundaries or be blocked by boundaries. The transmitted dislocation may be a rotated or a changed dislocation of the original incident dislocation, or it may be a new dislocation induced by the stress concentration derived from the pile-ups of incident dislocations.

The resistance of boundaries to dislocation motion during the interactions can be divided into three aspects [33]:

- The long-range elastic stress field between the dislocations and the boundaries, i.e., the image force;
- The dislocation accommodations to the boundaries when impinging on boundaries;
- The resistance of the adjacent crystal due to different orientations, Bravais lattices and lattice parameters.

All sorts of dislocation–boundary interactions show some similar behaviors and characteristics. This is due to the similarity between the dislocation–boundary and the deformation twin-boundary interactions: the essence of a deformation twin-boundary interaction is the dislocations of the incident deformation twin boundary interacting with the obstacle boundary, where the dislocation–boundary interaction is one of the essential parts during the whole process. However, different dislocations and boundaries have different interaction features, including interaction behaviors, dominating factors and transmission mechanisms.

Taken the transmission behavior as an example, it is reasonable to assume that the transmission behavior during the dislocation–boundary interactions can only occur when the resolved shear stress reaches an appropriate level and the transmission process must follow the rule of minimizing boundary energy [29,34,35]. Besides these, there are many other factors, which have significant influences on the transmission behavior. These factors include the misorientation [14,16,19,32], the geometry [20,30,36], the energy barrier [37], the dislocation type [31], the dimension [21], the Bravais lattice [13,38,39], the stacking fault energy (SFE) [31]. In order to predict the transmission behaviors of dislocation–boundary interactions in metals with a coarse-grained microstructure, numerous transmission mechanisms have been proposed. Most of them are established based on a single factor regardless of the others. However, the transmission behavior is influenced by a combined effect of many factors and none of them can be considered to be absolutely independent. The accuracy of models varies from case to case and should be restricted by the specific conditions. For instance, according to the in-situ transmission electron microscopy (TEM) observations on dislocation–twin boundary interactions, even in the case of a direct transmission without any residual dislocation, an absorption process can be found before the occurrence of any dislocation transmission [40]. Moreover, a recent study shows that the strain rate plays a significant role in the dislocation–grain boundary interactions [17]: the dislocations can transmit across a high angle grain boundary at a low strain rate but cannot at a high strain rate. This is ascribed to the different absorption processes at low and high strain rates, respectively. These findings indicate that the strain rate is not a negligible factor during the dislocation–grain boundary interaction. Even in the case where the strain rate is fixed, the observation of dislocation-boundary interactions is also related to the specimen size and the characterization techniques. Thus, a comprehensive transmission mechanism that can be applied in any circumstances is hard to obtain.

The dislocation–boundary interaction behaviors are significantly determined by the grain size [18,41]. This is because the dislocations are subjected to different stress states of image forces in metals with different grain sizes at micro- and nano-scales. For ultrafine-grained or nano-grained metals, a dislocation may be subjected to almost equivalent image forces derived from several boundaries due to the small grain boundary spacing. In this case, the resistance against dislocation motion in nano-grained metals is stronger than the one in coarse-grained metals, which is verified by a larger dislocation curvature in the grain interior [18,42,43]. The strong image force also changes the dislocation–boundary interaction behaviors in nano-grained metals through the dislocation behaviors: the dislocations nucleated initially at boundaries are stored and constrained in the vicinities of boundaries rather than glide into the grain interior. The vicinity of the boundaries is believed to be the favored places for plastic deformation. This phenomenon is different from that in a coarse grain, where the plastic deformation occurs throughout the grain with the slip of dislocations mostly in the grain interior. Moreover, the dislocation–boundary interactions change the boundary structure and generate severe distortions [41,44,45]. In the case of the deformation–twin boundary interaction, the tip of deformation twin subjects to a strong image force when approaching the boundary [46]. This makes the straight coherent twin boundary distorted and incoherent. Thus, the deformation twin is unable to grow larger and unlikely to transmit across the boundary.

For metals with a micro-scale boundary spacing, when a dislocation approaches a boundary, the image force derived from this boundary can be considered to be the dominant contributor of the resistance. In this case, the characteristics of the dislocations and boundaries, such as the types of dislocations and boundaries, geometrical conditions and energies during the interactions, are significant factors to the interaction between dislocations and boundaries. Besides, for the different types of dislocation–boundary interactions, the dominant factor may change. For example, the dominant factors of a dislocation-coherent twin boundary interaction are the dislocation types, the geometrical conditions and the SFE owing to the definite structure of coherent twin boundary [31,39,40]. However, for the twin–twin interaction, the Bravais lattice and the chemical composition are the primary factors due to the different twin boundary structures, slip systems and twin variants [13,38,39].

In the case of the dislocation–phase interface interaction, the phase interface is supposed to be a stronger barrier to dislocation motion compared with the grain boundary and twin boundary [21,30,47,48]. This is because the dislocation transmission across a phase interface must not only overcome the image force and energy barrier of the boundary, but also accommodate the different Bravais lattice, orientation and lattice parameters of the other phase. Thus, the primary factors of a dislocation–phase interface interaction are different from other sorts of dislocation–boundary interactions. The atomic bonding seems to be the most important factor: the resistance of a non-metallic compound interface is much stronger than the one between two different metallic phases. Additionally, according to the phase shape, size and interface configurations, such as particles and lamellae, their interactions between dislocations and interfaces are different. In terms of the non-metallic compound interface, the size of particle and lamella is the key factor [21,22]:

- The dislocations can only transmit across the non-metallic compound particle interface when the particle size is smaller than a critical value, which is always several nanometers;
- The slip are able to transmit across the non-metallic compound lamellar interface by slip steps only if the lamellar thickness is fine enough, otherwise the fracture of lamellae may occur due to the stress concentration of dislocation pile-ups.

With regard to the interfaces between two different metallic phases (particles or lamellae at the micro-scale), the alignment or deviation of slip systems of different phases is considered as a remarkable factor to predict the dislocation transmission behavior [47–49]. These indicate that the characteristics of two different phases, the morphology and dimension of phase interfaces play a more important role in this type of interaction.

In recent years, new discoveries on interactions between dislocations and boundaries are mostly obtained by in-situ scanning electron microscopy (SEM)/TEM mechanical testing. For example, at the nano-/atomic-scale, the detailed processes and products of interactions are investigated by TEM. Multiple phenomena are capable to observe, including the nucleation, slip, pile-up (dissociation or absorption), transmission (deflection or emission) and reflection of individual dislocations [16,33,50], the formation of steps and facets [12], the generation of local distortions and reconstructions [39] in the boundaries. While the macroscopic mechanical properties can be interpreted by the studies conducted at or above the micro-scale. For instance, to investigate the effect of grain boundaries and twin boundaries on slip traces and mechanical properties, a typical method is to perform in-situ SEM investigations on bi-crystal pillars [15,17,20,31,51]. Since the studies at different length scales have large distinctions in experimental conditions, specimen dimensions, deformation strains and strain rates, resolutions and research interests, each approach has achieved breakthroughs in its own specialized fields. Therefore, it is important and necessary to verify the analysis and conclusion by methods at different length scales to avoid one-sided understanding.

In this paper, based on experimental and simulation results, the structure of boundaries (the grain boundary, the dislocation boundary, the twin boundary and the multiphase interface), the interactions between dislocations and boundaries in coarse-grained, ultrafine-grained and nanocrystalline metals (dislocation–grain boundary interactions, dislocation–twin boundary interactions, dislocation–phase interface interactions, deformation twin-grain boundary interactions and deformation twin–twin boundary interactions), and the available characterization methods are briefly reviewed and discussed as follows:

1. Summary of boundary structure.
2. Transmission mechanisms of dislocation–boundary interactions in coarse-grained metals.
3. Interaction behaviors and influencing factors between dislocations and boundaries in coarse-grained metals.
4. Effect of grain size: dislocation-boundary interaction behaviors and influencing factors in ultrafine-grained and nano-grained metals.
5. Applications and characterization techniques.
6. Summary and outlook.

2. Structure of Boundary

A boundary is a broad concept referring to the structure which separates crystals with different orientations (grain boundary or twin boundary), Bravais lattices and compositions (phase interface) [26]. During plastic deformation of metals at ambient temperatures, dislocations or deformation twins nucleate due to applied stresses. To a great extent, their interactions with the existing grain boundaries, twin boundaries and phase interfaces determine the mechanical properties of metals. Since the intrinsic structure of boundaries has a remarkable effect on these interactions, it is indispensable to understand the different types of boundaries before any specific narration.

Due to the thorough consideration of structure, composition and interfacial energy, it is reasonable to classify the boundaries in metals into two primary categories. The first category is that two crystals on either side of a boundary have the same phase but only differ by crystallographic orientations. This category includes the grain boundary, the twin boundary and the stacking fault. On the contrary, the other category is an interface that differentiates two crystals by composition or Bravais lattice. Generally, the complexity level and interfacial energy of a boundary have similar variation tendency. Both increase with the increase of misorientation, defect density, heterogeneity of Bravais lattices and composition across the boundary. Therefore, twin boundaries and grain boundaries with perfect CSL have simple and definite structures and relatively low interfacial energies. While crystals with poor matching and multiple defects at boundaries may generate large lattice distortions, which may form high interfacial-energy boundaries with complex geometries.

Hereby, the structures of grain boundaries and twin boundaries introduced by thermomechanical processing, as well as the interfaces formed in the matrix of alloy are briefly outlined in the following sections. To simplify the description, the word "boundary" only represents grain boundary and twin boundary. While the word "interface" in the following chapters mainly refers to the phase interface formed in the matrix of metals.

2.1. Grain Boundary

The structures of grain boundaries depend on the crystal Bravais lattices, interatomic interactions, point defects and segregations [52]. Although the real grain boundaries do not extend infinitely and may contain defects and facets, we only reviewed the grain boundary structures without any segregations in this paper.

The grain boundary is only about two or three atom layers in thickness but accommodates crystals with the same phase but different orientations by dislocations. The simplest grain boundary structure is a straight and non-defective low angle grain boundary: a periodic array of dislocations [53–55], as sketched in Figure 1. Notably, the low angle grain boundary may be constructed by two types of dislocation series (each series terminates atomic planes at one side of the boundary) [56,57]. For the high angle grain boundary, the structure may become more complicated.

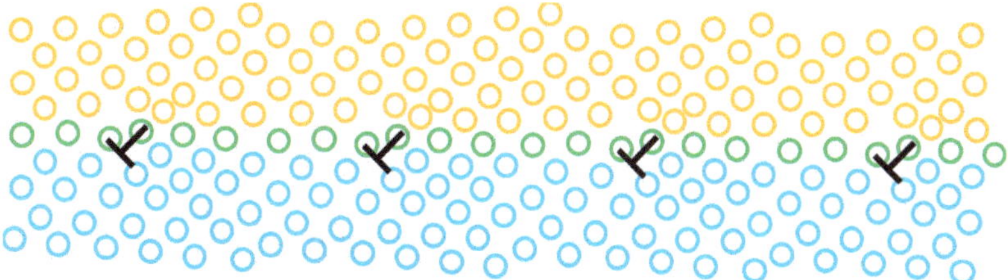

Figure 1. An illustration of a low angle grain boundary.

Generally, there are five degrees of freedom for grain boundaries. The adjacent crystals can either rotate around three perpendicular axes to form a misorientation, or the grain boundary plane can rotate around two vertical axes with respect to the crystals [58]. However, the structures of grain boundaries have some regular patterns determined by minimizing the grain boundary energy. For example, the grain boundary plane is always the densest-packed plane and exhibits a good fitting between crystals [59–62]. In most cases, when two lattices form a grain boundary, one lattice can be obtained by rotating another lattice around a certain axis with an angle.

Grain boundaries can be roughly classified into symmetrical and asymmetrical boundaries according to their grain boundary planes and misorientations. If two crystals share an equivalent crystallographic plane at the grain boundary and show a mirror symmetry on each side of the boundary, it is a symmetrical boundary and vice versa [52]. The symmetrical tilt grain boundary is always used to interpret low angle tilt grain boundaries [63]. For the high angle tilt grain boundaries, this model only works well when the dislocations are aligned in an array with a constant spacing in the grain boundary [64].

The twist grain boundary is a type of asymmetrical grain boundary, where two crystals rotate around the common grain boundary plane normally and form a grain boundary, as shown in Figure 2. In the right illustration, the twist boundary is parallel to the plane of the figure, and one lattice is indicated by circles and the other by crosses. It can be seen that the coincident positions of circles and crosses are able to constitute a superstructure.

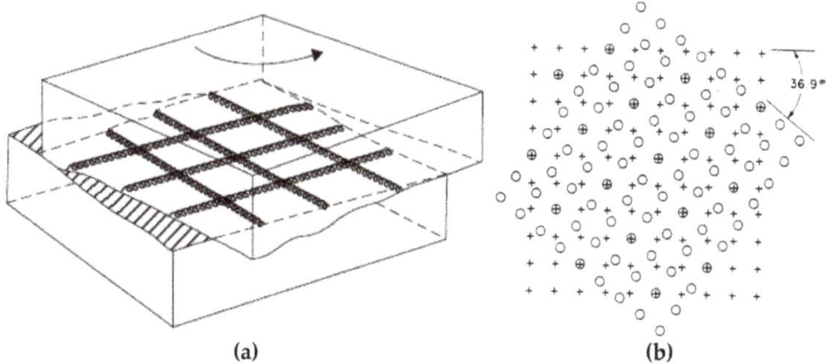

Figure 2. Illustrations of twist grain boundaries. (**a**) Two crystals rotate around the common grain boundary plane; (**b**) Coincident positions of two crystals. (Reprinted from References [26,58], with permission from Elsevier).

For the analytical quantification of grain boundaries with all tilt angles or orientations, the coincidence site lattice (CSL) and displacement shift complete lattice (DSC) models have been developed from the concept of coincident positions [65]. If two lattices with different orientations are overlapped with each other, some lattice points can be found to coincide periodically. The coincident points define a superstructure: a coincidence site lattice. The number of lattice points in the unit cell of a CSL is expressed as Σ, which characterizes the unit cell volume of CSL lattice compared with the crystal unit cell. For example, Figure 3 shows the $\Sigma = 5$ grain boundary and its CSL lattice. The Σ value is always odd and there are two special boundaries among them. The $\Sigma = 1$ boundary represents a perfect crystal without a grain boundary. Additionally, a twin boundary can be expressed as an $\Sigma = 3$ boundary in the CSL model.

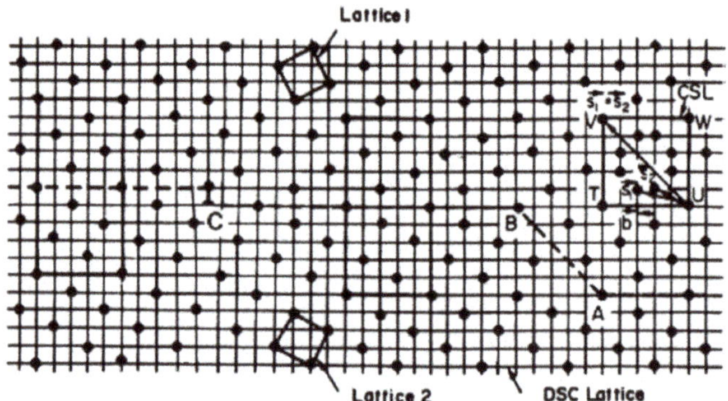

Figure 3. An illustration of CSL lattice and DSC lattice in a grain boundary with CSL lattices of $\Sigma = 5$ by a 36.9° rotation around [001] direction. (Reprinted from Reference [65], with permission from Elsevier).

The DSC model is developed from the CSL model and provides a more specific way to quantify grain boundaries. As shown in Figure 3, the CSL lattice can be retained by shifting a large displacement (CSL vector) to another coincidence point. However, there is another effective way to retain the CSL lattice: just preserving the coincidence instead of shifting a large displacement. When two lattices are overlapped with each other, a coarsest sub-mesh can be established. If all lattice points of one crystal shift a displacement of this sub-mesh vector along any direction, the coincidence still exists although the coincide points have changed. This sub-mesh defines a sub-lattice called the DSC lattice.

As mentioned above, a low angle tilt boundary consists of a periodic array of dislocations. The translation vectors of the DSC lattice are possible Burgers vectors for such grain boundary dislocations [66]. In other words, the DSC lattice dislocation is a sort of grain boundary dislocation. A low angle tilt grain boundary can be treated as small deviation from the $\Sigma = 1$ grain boundary (perfect single crystal). Geometrically, any practical grain boundary can be regarded as a small deviation from the perfect CSL lattice.

It is reasonable to assume that the grain boundaries with perfect CSL and low Σ are expected to be low interfacial-energy boundaries due to the high atomic density of the planes. The experimental results of symmetrical tilt boundaries validate the above assumption [26]. Figure 4 shows the dependence of the experimental grain boundary energies on the misorientation angle for the <110> symmetrical tilt boundaries in face-centered cubic (fcc) metals (Cu and Au). The visible drops of {111} and {113} planes are associated with the low-energy twin boundaries. It indicates that the grain boundaries with perfect CSL and low-indexed boundary planes (such as {111}, {100}, {110} planes) exhibit low energies. On the contrary, the other grain boundaries deviating from these special orientations show much higher boundary energies. These grain boundaries possess lattice distortion due to the elastic stresses between atoms or defects such as dislocations, facets and ledges at the grain boundaries.

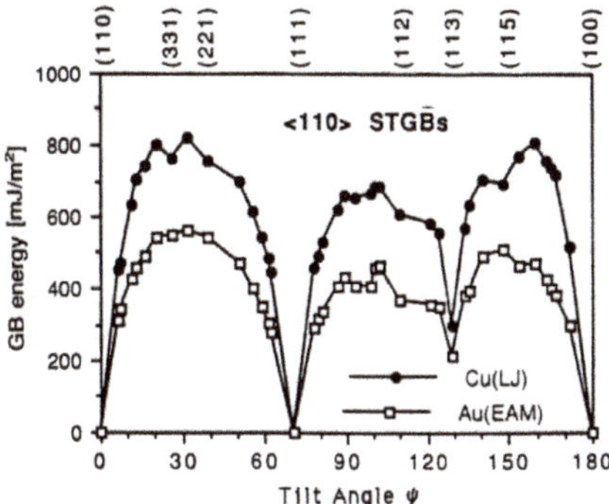

Figure 4. The correlation between grain boundary energies and misorientation angles of <110> symmetrical tilt boundaries in fcc metals (Cu and Au). (Reprinted from Reference [26], with permission from Elsevier).

Figure 4 demonstrates a clear correlation between the grain boundary energy and the atomic arrangement at the grain boundary. The actual grain boundary structure is more complicated than the above models. Figure 5 illustrates the lattice points of grain boundaries with perfect $\Sigma = 5, 17, 37$ and 1. The $\Sigma = 17$ and 37 grain boundaries can be regarded as an ordered sequence interspersed by $\Sigma = 5$ and 1 grain boundaries, i.e., a mixture of basic structural units labeled A and B. In most cases, it has been clarified that a boundary structure can be described as a periodic array of structural units. In this case, the interfacial energies of grain boundaries are expected to be lower than those predicted by the CSL model. According to this theory, a local elastic distortion region exists at a boundary deviating from the perfect CSL model, i.e., the unrelaxed type. This leads to the crystal planes in the vicinity of the boundary have larger spacing than the bulk value [67]. This phenomenon is termed as the volume expansion [68,69]. The volume expansion has been manifested by experiments in dominating the grain boundary energy for all types of grain boundaries in fcc and body-centered cubic (bcc) metals, which shows a proportional relationship between them [70].

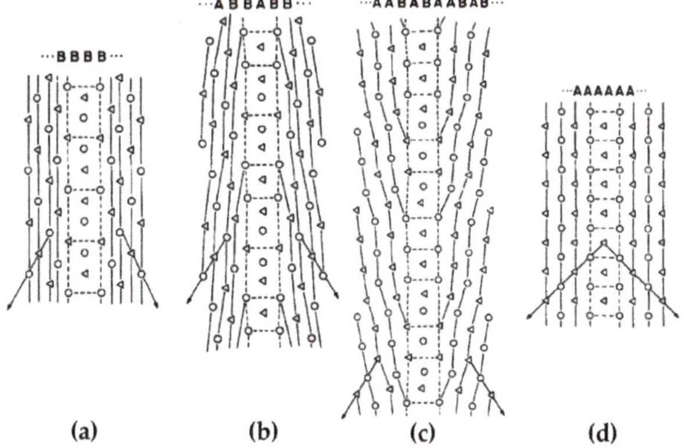

Figure 5. Illustrations of perfect $\Sigma = 5, 17, 37$ and 1 grain boundaries from (**a–d**), respectively. (Reprinted from Reference [26], with permission from Elsevier)

2.2. Dislocation Boundary Introduced during Deformation

Dislocation boundaries form by dislocation accumulation during deformation of fcc and bcc single crystals and polycrystalline metals (e.g., Cu, Al, Ni and Fe), which is also called the grain subdivision. These dislocation boundaries separate relatively clean regions with crystallographic misorientation angles [71]. It is commonly observed in medium to high SFE metals, which are dominated by dislocation slip during plastic deformation at ambient temperatures such as cold rolling, torsion, drawing, etc.

During deformation, the dislocation slip follow certain active slip patterns, and tend to accumulate and form morphologies aligned as lines or boundaries, leaving other regions of the matrix with a relatively low density of dislocations. These dislocation boundaries divide the matrix into small cell-like regions with different orientations called "cell blocks". As shown in Figure 6, two types of dislocation boundaries are distinguished by morphologies, which are called "extended boundaries" and "cell boundaries". These two types of dislocation boundaries come into being by different slip mechanisms. The cell boundaries are formed mostly by mutual trapping of dislocations, which are termed as "incidental dislocation boundaries (IDBs)". Whereas for extended boundaries, they are formed because the interaction of dislocations originated from different active slip patterns, which are termed as "geometrically necessary boundaries (GNBs)" [72–74].

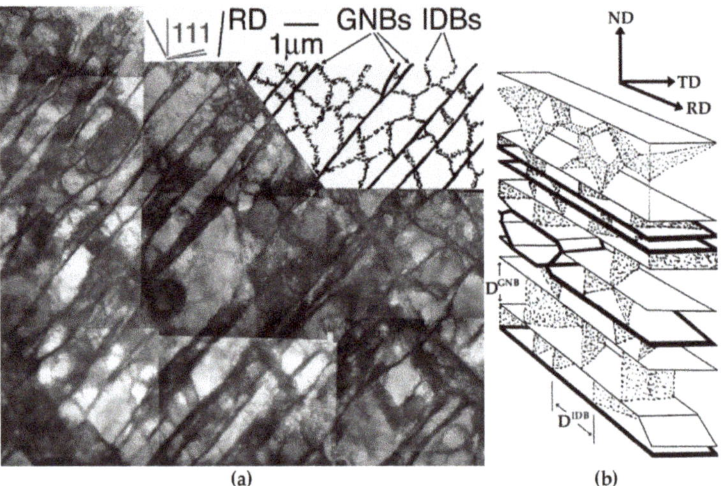

Figure 6. TEM image and the illustration of IDBs and GNBs. (**a**) A TEM image of pure Ni cold rolled to a reduction of 20%; (**b**) An illustration of IDBs and GNBs. (Reprinted from References [73,74], with permission from Elsevier).

The formation of different dislocation boundaries depends on the grain crystallographic orientations and deformation methods, and can be divided into three types [75,76]. The first type is totally composed of IDBs without GNBs and shows a cell structure, which forms in the grains with <001> parallel to the tensile direction. The other two types are cell-block structures and are composed of IDBs and GNBs. Among them, one type has straight and parallel GNBs and aligned approximately with the slip planes, and the other type has two sets of GNBs and deviate substantially from the slip planes. Figure 7 shows the TEM images of a GNB aligned approximately with the slip planes, which indicates that the nature of a dislocation boundary is a dislocation network due to the dislocation interactions [77,78].

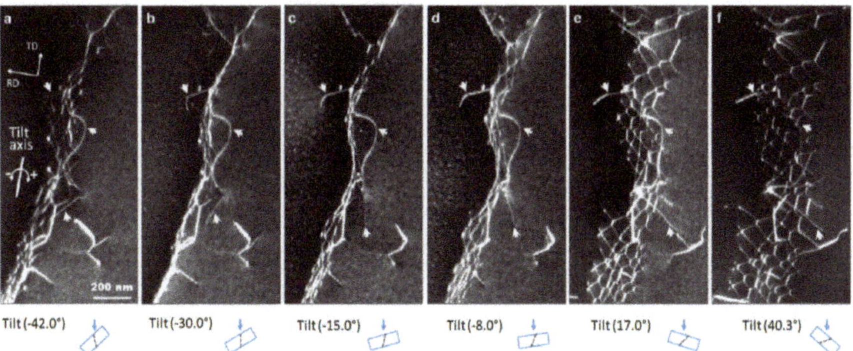

Figure 7. TEM images of a GNB observed at different tilting angles to incident electron beams. (**a–f**) show the images with tilting angles of −42.0°, −30°, −15°, −8°, 17° and 40.3°, respectively (Reprinted from Reference [77], with permission from Chuanshi Hong).

Since the IDBs and GNBs have distinguished morphologies and are generated by different mechanisms, their average misorientation angles and average boundary spacing during deformation are remarkably different [79], as shown in Figure 8. The variations of average misorientation angles and average boundary spacing of GNBs change much faster than those of IDBs with the increase of strain.

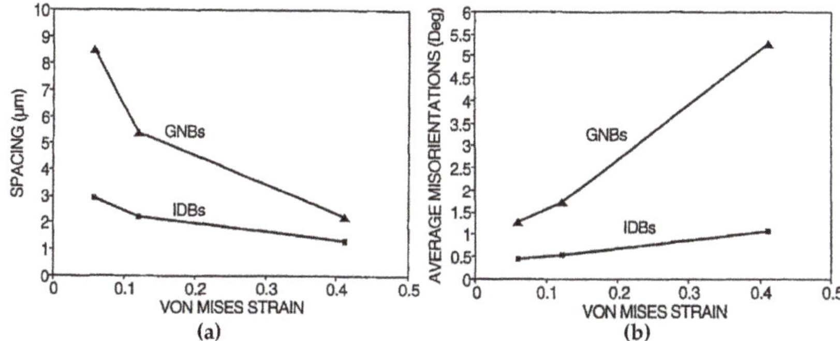

Figure 8. The variations of (**a**) boundary spacing and (**b**) misorientation angle of IDBs and GNBs in dependence of strain in cold-rolled pure Al, respectively. (Reprinted from Reference [79], with permission from Elsevier).

When the deformation strain increases, both the morphologies and dislocation boundary parameters of metals evolve with it. In most cases, at low to medium strains, GNBs of metals are paralleled along certain planes, which may be a slip plane or a plane related to slip planes [79]. At relatively high strains (e.g., a true strain of 5.0), dislocation boundaries appear to be lamellar-structured and paralleled to the rolling plane [80]. With further increase of strain, the boundary spacing of both GNBs and IDBs decreases and misorientation angles across boundaries increase. In the case of severe plastic deformation (SPD), which introduced much higher strains than common production methods, further structural refinements can be achieved [71].

2.3. Twin Boundary

The interface between twin and matrix can be distinguished as coherent and incoherent ones, as shown in Figure 9a. The two straight parallel sides of twin 1 and 2 are coherent twin boundaries, which have low interfacial free energy. On the other hand, the short end boundary of twin 2 is an incoherent boundary, which has high misfit and high boundary free energy. The measured boundary free energy of incoherent twin boundaries in Cu is about 25 times as high as the value of a coherent one [58].

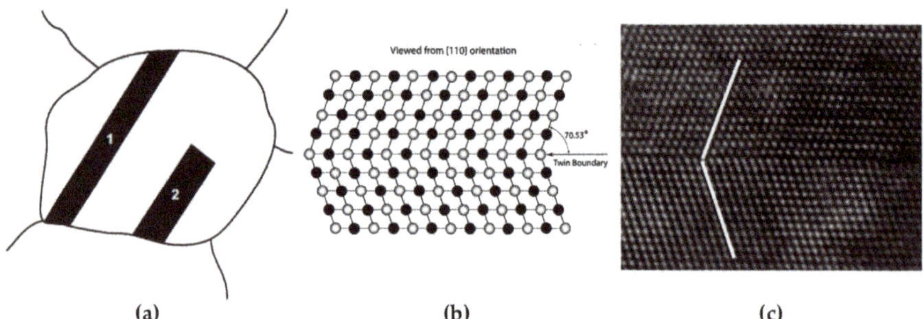

Figure 9. (**a**) A schematic diagram of coherent and incoherent twin boundaries; (**b**) A mirror symmetry of atoms in a coherent twin boundary in fcc metals viewed from the [110] direction; (**c**) A typical HREM image of a twin in a fcc metal. (Reprinted from References [27,58], with permission from Elsevier).

The twin boundary plane has atoms aligned along a coherent boundary (twin plane) with mirror symmetry. The atoms of twin plane belong to both lattices of two crystals without any lattice misfit. As mentioned in Section 2.1, a twin boundary can be regarded as a perfect $\Sigma = 3$ boundary with low interfacial energy. Figure 9b,c shows the illustration and the high-resolution transmission electron microscopy (HREM) image of a coherent twin boundary from the view of [110] direction in a fcc metal, whose twin boundary plane is the $(1\bar{1}1)$ close-packed plane and atoms show mirror symmetry arrangement across the boundary [27].

For metals with different Bravais lattices [81], they have different twin planes and growing directions. The twin plane and growing direction are {112} and <111> in bcc metals, {111} and <112> for fcc metals, and {1012} and <1011> for hexagonal close-packed (hcp) metals, respectively.

Twinning may occur during annealing or deformation, which are called "annealing twins" or "deformation twins", respectively. Generally speaking, metals with high SFEs have a little chance to form annealing twins due to the close relationship between twins and stacking faults, and annealing twins tend to occur in metals with low SFEs. Thus, it is easy to find more annealing twins in brass than in pure copper due to a relatively lower SFE [58].

The formation of a deformation twin requires relatively high stress compared with dislocation slip because a twin is a planar defect while a dislocation is a line defect [81]. It requires not only the activation of certain dislocations, but also additional surface energy to form an interface. Therefore, the dislocation slip is always the dominating mechanism during deformation at ambient temperatures. Whereas twin occurs in some certain conditions including low deformation temperatures, specific orientations, a limited number of slip systems and high strain rates—where a high critical shear stress can be obtained. For example, in hcp metals, the favorite deformation twin is due to the limited number of slip systems. For bcc and fcc metals, which has various slip systems at ambient temperatures, the deformation twin is likely to form at low deformation temperatures and high strain rates.

The formation of a twin is closely related to stacking faults. For example, the fcc single crystal has a stacking order of ABCABC . . . along {111} planes. If the stacking order from a certain plane is reversed, which turn to be ABCACBACBA . . . , the two parts of the crystal form a mirror-symmetrical structure, i.e., a twin. Since the nature of twin boundary is a stacking fault, its interfacial energy is expected to be relatively low.

Generally, deformation twins are considered to form by the successive slip of Shockley partial dislocations. In fcc metals, there are three equivalent Shockley partials on {111} slip planes, which are b1 = $a/6[2\bar{1}\bar{1}]$, b2 = $a/6[\bar{1}2\bar{1}]$, b3 = $a/6[\bar{1}\bar{1}2]$. In fcc single crystals, the stacking sequence of atoms in successive close-packed planes can be demonstrated as ABCABCABCABC. If a partial dislocation glides along the slip plane, a whole layer of atoms will move along the slip plane and generate a stacking fault. Indeed, the slip of any partial dislocation may lead to the same shift in stacking sequence, i.e., A → B, B → C, C → A, although the three Burgers vectors have different orientations. This microstructural transition by the stacking fault mechanism can also be found in the fcc-hcp martensitic transition of rare-gas solids [82,83].

Figure 10 illustrates the step-by-step formation process of a deformation twin by the slip of partial dislocations on successive slip planes. Since the slip of partial dislocation results in the same change in stacking positions, the twin can be obtained by the slip of a series of partial dislocations with the same Burgers vector (Figure 10a) or different Burgers vectors (Figure 10b). These two approaches bring about different macroscopic strains in the long range although both of them form the same stacking sequence. The partials with the identical Burgers vector generate shear strains in the same direction, which may produce a relatively large macroscopic strain. This type of twins often exists in the matrix of coarse grains, whose morphologies are plates with straight twin boundaries. In contrast, the second approach may not form a large macroscopic strain since their shear directions vary with each other. This situation is common in nano-grained fcc metals, which shows no obvious plate-like morphology.

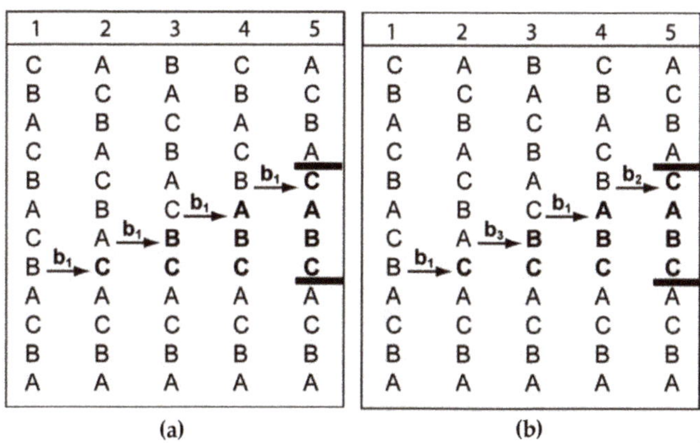

Figure 10. The step-by-step process of producing a four-layer deformation twin by the slip of (**a**) identical and (**b**) different partial dislocations on successive slip planes [27].

2.4. Interface between Different Phases

The interface seeks to maximize atomic matching and to minimize elastic strain for the minimization of interfacial free energy. As a result, the atoms of crystals at both sides of an interface tend to align along the close-packed planes or close-packed directions when an interface is formed [26]. Although the lowest interfacial free energy is achieved at the interface, a perfect matching is not always reached. Therefore, the interfacial energy increases with the increase of defect densities such as misfits, steps and ledges and the discrepancy of atomic bonding at interfaces.

In general, interfaces can be divided into three types according to the degree of atomic matching. "Coherency" is the terminology to describe the extent, and these three types are fully coherent, partly coherent and incoherent interfaces [84].

a. Fully coherent interface

This type of interface has complete matching between the atoms of two crystals and definite planes along the interface. It means that the lattice is continuous across the interface without disconnection or mismatch. A fully coherent interface is commonly observed between two phases with similar chemical composition but different Bravais lattice [85] (phase transformation) or two phases with both similar atoms size and Bravais lattice but different composition [86].

b. Partly coherent interface

Partly coherent interface refers to an interface with a similar arrangement of atoms between two crystal structures, which are accommodated by periodic misfit dislocations and/or steps at the interface. This probably happens when the two crystals have the same Bravais lattice, but different compositions and lattice parameters.

c. Incoherent interface

Incoherent interface means the atoms on each side of the interface show poor matching with each other. When the crystals have large differences in lattice parameters or the atomic bonding (such as covalent and metallic bonding), they seem to form an incoherent interface. Although it shows a tendency of aligning the close-packed planes of two crystals along the interface plane in order to minimize the interfacial energy, the interface is sometimes a high-indexed plane.

An incoherent interface with an almost flat interface plane between the Zr matrix and the ZrN precipitate in a Zr-N alloy is shown in Figure 11. It shows the faceted morphology along the interface and its orientation relationship is $[450]_{Zr}//[10\bar{1}]_{ZrN}$ and $(002)_{Zr}//(\bar{1}31)_{ZrN}$, a low-indexed Zr plane adjacent to a high-indexed ZrN plane [87]. However, the incoherent interface is not necessarily atomically flat. In this case, the incoherent interface shows a serrated shape and is composed of steps, which is called "disconnections" [88–90]. Although the crystals at both sides of a certain step may share a close-packed plane (also a low-indexed crystal plane), the whole interface may lie parallel to any arbitrary crystal plane according to different densities of steps.

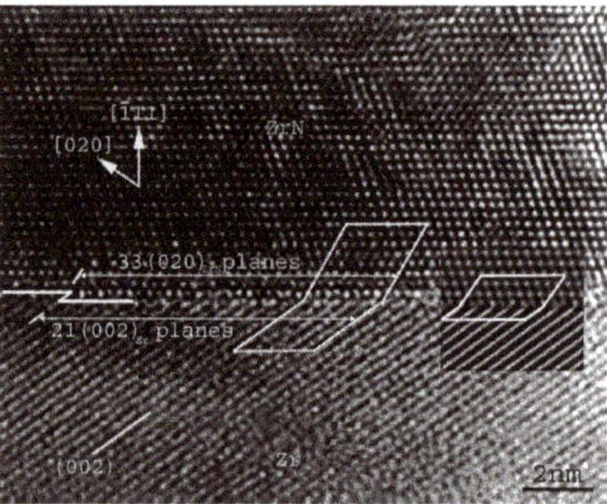

Figure 11. HREM image of an incoherent Zr-ZrN interface in a Zr-N alloy. (Reprinted from Reference [87], with permission from Elsevier).

3. Interactions between Dislocation and Boundary

Generally, the boundary is, more or less, a barrier for dislocation motion. Even in the weakest situation, i.e., a direct transmission across a boundary without any pile-up or residual dislocation at the boundary, the dislocations may be trapped in the boundary prior to being released from it, and the slip direction shall be changed [29,40].

The resistance of grain boundaries and twin boundaries mainly comes from the following three aspects [33]:

- When dislocations move towards the boundary, there exists a force against dislocation motion derived from the long-range elastic stress field between dislocations and the boundary, which is called the image force. The image force makes the group of following dislocations piled up in front of a boundary, and may increase due to the interaction between the leading dislocation and the boundary;
- Once impinging on the boundary, the incoming dislocation will face another resistance originated from the interaction between the dislocation and the boundary;
- When the dislocation is about to emit into the adjacent grain, the dislocation has to accommodate its orientation and Bravais lattice. This situation is more complicated for phase interface, whose resistance is stronger than those of grain boundaries and twin boundaries due to the dislocation accommodation to a different phase.

The dislocations may rotate, dissociate and reconstruct once impinging on the boundary. In some proper conditions, dislocation can overcome the impediment and transmit across the boundary. The dislocation transmissions of all sorts of boundaries are the result of stress concentration at boundaries induced by the accumulation of dislocations due to the increased applied stress. The transmission of dislocations across boundaries may dissociate, leave partial residual dislocations or form steps on the boundaries. Each type of dislocation–boundary interactions can be classified based on the different interaction behaviors, the dominating factors and the transmission mechanisms.

3.1. Dislocation–Grain Boundary Interactions in Coarse-Grained Metals
3.1.1. Basic

As the most common case, the dislocation–grain boundary interaction in coarse-grained metals has been investigated in relatively comprehensive and intensive studies for decades as detailed in the following. Numerous studies have reported on this interaction characterized by static and dynamic in-situ TEM as well as SEM methods, and obtained lots of quantitative and qualitative results [15–17,19,20,29,33–35,37,50,91]. It has been found that the dislocation–grain boundary interactions of bcc, fcc and hcp metals are similar in most aspects [29]. However, it is still not fully interpreted.

When a dislocation impinges on a grain boundary, the dislocation lines may rotate to be parallel to the grain boundary owing to the image force. Then, the dislocations may dissociate or glide along the grain boundary once impinging on the boundary, which further appears to be absorption, transmission or reflection [29,33,34,40,50,92,93], as shown in Figure 12. The above situations can happen at the same time, i.e., a primary impeded dislocation may dissociate, transmit and reflect at the grain boundary, as shown in Figure 13.

In specific, a dislocation may maintain itself or dissociate into several dislocation partials with smaller magnitudes of Burgers vectors when it is impeded by a grain boundary. Some decomposed dislocations can glide along the grain boundary called "glissile dislocations", while others reorganize with the local grain boundary dislocations to be a new periodic arrangement called "sessile dislocations" [33]. The glissile products can glide along the grain boundary and be impeded at steps or triple junctions, which may form cavitation due to the accumulated stresses. The sessile dislocation may change the local structure of the grain boundary. The absorption of dislocation is potential to emit dislocations with the increase of applied stress, which may transform into transmission or reflection.

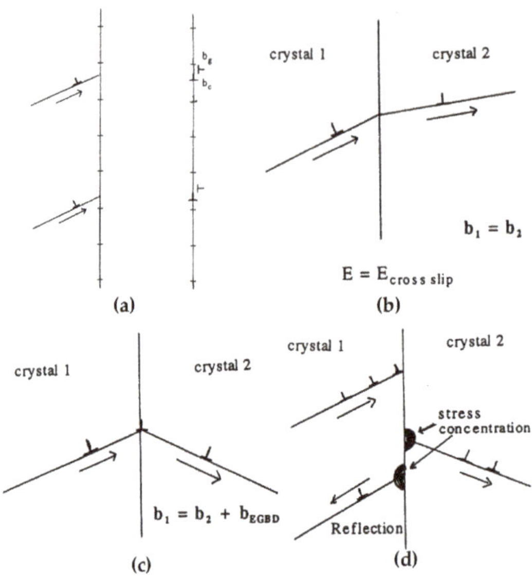

Figure 12. Three types of interactions between dislocations and grain boundaries. The schematic illustrations of (**a**) absorption; (**b**) transmission without residual dislocation; (**c**) transmission leaving residual dislocation; (**d**) reflection, respectively. (Reprinted from Reference [33], with permission from Elsevier).

Figure 13. TEM image of the dislocation–grain boundary interaction with direct and indirect transmission and reflection. (Reprinted from Reference [93], with permission from Elsevier).

Dislocation transmission can be considered as two types: direct and indirect. Direct transmission can be further divided into two modes. The first one refers to the situation that the slip planes of incoming and outgoing dislocation share the same line intersected on the grain boundary plane. This always happens when the boundary is a low angle symmetrical grain boundary sharing the same slip plane, so that the dislocations can glide across the boundary along the same slip plane by rotating a small angle deviated from the original slip direction. In this case, the Burgers vectors of incoming and outgoing dislocations do not change after crossing the grain boundary, and the energy required for transmission is equivalent to the cross-slip. Another type of direct transmission generates a residual dislocation at the grain boundary and the transmission energy is the formation energy of this residual dislocation. The magnitude $|b_{\text{residual}}|$ of the residual dislocation can be expressed as [34,94]:

$$b_{\text{residual}} = b_{\text{in}} - b_{\text{out}} \tag{1}$$

A criterion is found to be true in this mode that the magnitude of $|b_{\text{residual}}|$ should be minimized [40,93].

Indirect transmission has the same cause as reflection. Both of them occur when the dislocations are piled up at the grain boundary. The increasing stress concentration may activate dislocation sources at the vicinity of the intersection point, which emits dislocations to the other grain or back to the inner part of the same grain, i.e., indirect transmission or reflection.

Besides the dislocation transmission, the interaction between a dislocation and a grain boundary may result in the formation of a twin or a stacking fault ribbon on the other side of the grain boundary [35,91,93,95]. As shown in Figure 14, the incident dislocations dissociate at the grain boundary, where some partials are absorbed and reflected. Other partials emit from the grain boundary into the adjacent grain and form a stacking fault ribbon. This phenomenon often appears in hcp metals which have a high tendency to form deformation twins. When the orientation of adjacent grain is appropriate to nucleate deformation twins rather than activation of slip systems due to the stress concentration induced by dislocation–grain boundary interactions, such phenomenon will occur.

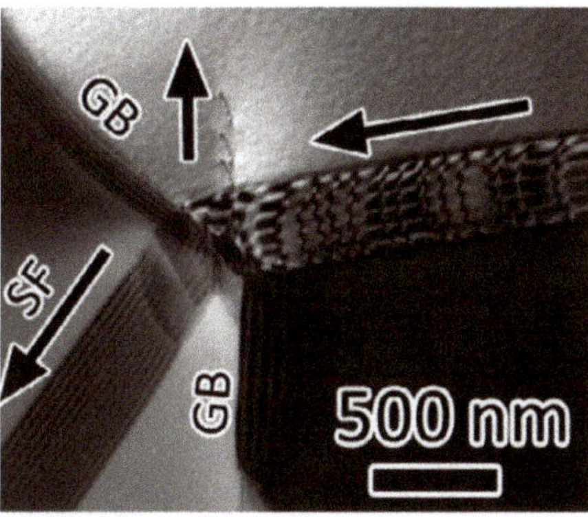

Figure 14. TEM image of twin formation induced by the dislocation–grain boundary interaction. (Reprinted from Reference [29], with permission from Elsevier).

3.1.2. Dislocation Transmission Mechanisms across a Grain Boundary

The dislocation transmission across a grain boundary in coarse-grained metals is somehow still a controversial topic, and several criteria have been proposed to elucidate this phenomenon [20,29,34,35,37,40,93]. These criteria are found, confirmed or established based on different types of standpoints, such as the misorientation, the geometrical conditions, the resolved shear stress, the grain boundary dislocation energy and the SFE.

Among these criteria, it is found that the dislocation transmission should follow the common criteria of minimizing the Burgers vector magnitude of residual dislocation partials on the grain boundary and of maximizing the resolved shear stress magnitude of the slip systems [29,34,35]. These two conditions must be met simultaneously, since the resolved shear stress is the power to overcome the resistance of grain boundary while the transmission requires small residual dislocation Burgers vector existing in the grain boundary due to the smallest increase in strain energy of grain boundaries. Additionally, some other factors are also reviewed in details as follows.

a. Misorientation

The misorientation of a grain boundary has an obvious effect on the interaction behavior between dislocations and the grain boundary. It can serve as a rough criteria to predict the transmission resistance of grain boundary [16,19,96]. This may be ascribed to the fact that the deviation between different slip planes of the incoming and outgoing dislocations across high angle grain boundaries is larger than that for low angle ones.

A low angle grain boundary can actually impede dislocation s to some extent, but much weaker than a high angle grain boundary. The high angle grain boundary is a strong barrier, which impedes the slip of dislocations and makes them pile-up against the boundary. Since a residual dislocation should form in a grain boundary if a dislocation transmits it by the direct mode, additional energy is required and the dislocation transmission becomes difficult. On the other hand, a low angle grain boundary can be taken as a periodic array of dislocations. The interaction between dislocations and low angle grain boundaries is the interaction of dislocations with different Burgers vectors. When passing through a low angle grain boundary, jogs are left in the grain boundary while kinks are formed in the transmitted dislocations. This implies an increase in the total energy during the interaction process.

b. Geometrical condition

The grain boundary misorientation is a rough criteria to estimate the resistance of a grain boundary to dislocation motion. However, it cannot determine whether the dislocation can or cannot transmit across a grain boundary.

It has been found that the dislocation transmission strongly depends on the geometrical condition of a grain boundary, which includes the orientations of slip systems and grain boundary parameters. The earliest criteria of the geometrical condition is established by Livingston and Chalmers (*LC*) [97,98], which is defined as:

$$LC = (e_{in} \cdot e_{out}) * (g_{in} \cdot g_{out}) + (e_{in} \cdot g_{out}) * (g_{in} \cdot e_{out}) \tag{2}$$

where e and g are the slip plane normal and slip directions, respectively. The subscripts refer to the incoming and outgoing slip systems, respectively. It is supposed to predict the dislocation transmission if the value of LC is minimized. However, this criteria has been considered not sufficient to account for the operative slip systems.

Clark et al. [34,94,97] combined geometrical and resolved shear stress conditions together and proposed another criteria, which is expressed as:

$$M_c = (l_{in} \cdot l_{out}) * (g_{in} \cdot g_{out}) \tag{3}$$

here l is the line traces of two slip planes intersected on the grain boundary plane. The slip system with the maximum value of M_c has the largest possibility to transmit across a grain boundary.

So far, several transmission factors have been widely used in order to predict the ease of dislocation transmission [20,29,35], such as m', *LRB* (Lee–Robertson–Birnbaum) and the Schmid factor. The m' and *LRB* are developed from the M criteria and based on the observation of dislocation transmission by TEM [29,35], which can be expressed as:

$$m' = \cos \psi \cos \kappa \tag{4}$$

$$LRB = \cos \theta \cos \kappa \tag{5}$$

where ψ is the angle between the slip plane normal of the two crystals, κ is the angle between the Burgers vectors of incoming and outgoing dislocations and θ is the angle between the line traces of two slip planes intersected on the grain boundary plane [35]. Higher values of m' or *LRB* indicate larger probabilities of dislocation transmission, whose maximum value is 1 referring to the easiest transmission condition [20].

Among them, m' can be evaluated easily by an orientation map of electron backscatter diffraction (EBSD), but it is hard to evaluate LRB depending on a non-destructive experiment [35]. Additionally, it remains controversial which is better between m' and LRB [20]. The Schmid factor is based on the macro-scale stress state, which is not so appropriate for the dislocation transmission dominated by the local micro-scale stress state [29]. Thus, m' is considered as the most convenient one to predict the dislocation transmission.

To better predict the dislocation transmission by simulation, Bieler et al. [35] proposed the following modified models of m' and LRB combining the accumulated shear γ:

$$m'_\gamma = \sum_{\text{in}}\sum_{\text{out}} m'(\gamma_{\text{in}}\gamma_{\text{out}}) / \sum_{\text{in}}\sum_{\text{out}} (\gamma_{\text{in}}\gamma_{\text{out}}) \tag{6}$$

$$LRB_\gamma = \sum_{\text{in}}\sum_{\text{out}} LRB(\gamma_{\text{in}}\gamma_{\text{out}}) / \sum_{\text{in}}\sum_{\text{out}} (\gamma_{\text{in}}\gamma_{\text{out}}) \tag{7}$$

The sums go over all possible combinations of slip systems in the two grains along the grain boundary. Furthermore, the criteria combining the m' and LRB together can be represented as:

$$s_\gamma = \sum_{\text{in}}\sum_{\text{out}} \cos\psi \cos\theta \cos\kappa (\gamma_{\text{in}}\gamma_{\text{out}}) / \sum_{\text{in}}\sum_{\text{out}} (\gamma_{\text{in}}\gamma_{\text{out}}) \tag{8}$$

Similarly, the m' can be modified using the Schmid factor m on each slip system:

$$m'_m = \sum_{\text{in}}\sum_{\text{out}} m'(m_{\text{in}}m_{\text{out}}) / \sum_{\text{in}}\sum_{\text{out}} (m_{\text{in}}m_{\text{out}}) \tag{9}$$

c. Energy

Among all the criteria of dislocation transmission, energy is the dominant factor that determines whether a dislocation is able to transmit across grain boundary or not. This viewpoint is generally accepted although some aspects are still ambiguous and require further study.

In the case of leaving residual partials on a grain boundary during dislocation transmission, the accumulation of residual dislocations leads to the increase of grain boundary energy, which is expressed as the strain energy density [29,34,40,93]. This energy is determined by the magnitude of Burgers vector of the residual dislocations and should be minimized during dislocation transmission, as shown by Equation (1). The strain energy density dominates the probability and behavior of dislocation transmission. For instance, the dislocation can be piled up if it may generate a residual dislocation with a large Burgers vector, even though it satisfies the condition of maximum resolved shear stress. On the other hand, according to the atomistic simulation, a dislocation is also possible to transmit across a grain boundary easily if the dislocation interaction changes the local misorientation and reduces the grain boundary energy [96]. Moreover, it is found that the accumulative strain energy density with the increase of strain may activate more dislocation sources and change the Burgers vectors of emission dislocations [93].

The effect of grain boundary energy on the dislocation transmission resistance across a grain boundary is still controversial in the literatures. However, there are tendencies verified by experiments that the resistance of a boundary to dislocation motion is related to the complexity level and the interfacial energy of boundary structure. Dislocations may pile up at impenetrable boundaries with complex structure and high interfacial energy by absorption or dissociation without any transmission into the adjacent grain, such as high angle grain boundaries. Whereas transmission is easy to occur on some specific activated slip systems across the boundaries with simple and definite structure, such as low angle boundaries, low Σ boundaries and coherent twin boundaries [35].

The above understandings are the most-accepted concepts, however, an opposite conclusion has been reported by Sangid et al. [37]. They proposed a methodology to measure energy barriers for dislocation transmission and use it in simulation, which is found to be consistent with experimental observations of dislocation transmission and the *LRB* criterion. These results indicate that a grain boundary with a lower interfacial energy offers a stronger barrier against dislocation transmission. For instance, the Σ3 grain boundary (twin boundary) with low interfacial energy has a much larger resistance to dislocation motion than the other ones, while the Σ19 grain boundary has a high interfacial energy shows easy resistance of dislocation transmission.

3.1.3. Other Influencing Factors

a.   Dislocation type

Most dislocation transmission mechanisms are established regardless of the incident dislocation type; however, some studies show that this is something significant during the dislocation–grain boundary interaction. According to the detailed investigations of dislocation–twin boundary interactions, the type of incident dislocation dominate the transmission behavior. This evidence will be reviewed in the next chapter. Additionally, different responses are reported in the interactions of high angle grain boundaries with edge and screw dislocations in ultrafine-grained Al [43]. The incident screw dislocations are absorbed by high angle grain boundaries and exert weak internal stresses to the grain boundary. While the incident edge dislocations seem to be piled up in front of high angle grain boundary and forms high internal stress to the grain boundary. These observations are further confirmed by unloading experiments: the screw dislocations are still accommodated in the grain boundary instead of moving back to the sources during unloading, whereas the edge dislocations are able to move back to the sources indicating a different response to the grain boundary. These phenomena may imply that the research on dislocation–grain boundary interactions should not only focus on the characteristics of a grain boundary, but also concern about the incident dislocation types.

b.   Strain and strain rate

Some recent research shows that the practical dislocation transmission across a grain boundary during deformation could be more complicated than what is known. It was found that the successive dislocation transmissions may destroy or interrupt the original structure of a grain boundary and form local misorientations, which results in activating more dislocation sources and changing the interaction behavior at the intersection point [29]. For example, the dislocation emission from a grain boundary may change from partial dislocations to perfect ones with increasing strain [93]. This change during dislocation transmission is always not considered in most studies but important for the transmission mechanism.

On the other hand, it is interesting that, under some situations, the individual high angle grain boundaries show weak influence on the stress–strain curve of pillar samples, whose curves are similar to the ones of single crystal samples whenever the slip transfer occurs or not [17,20]. It seems that these results are just the opposite to the common understanding that a high grain boundary is a strong barrier to dislocation motion. However, there are two main reasons to elucidate this conflictive phenomenon. The macroscopic mechanical properties are the average performances of both matrix and grain boundaries. It is hard to obtain a remarkable difference in macroscopic performances between samples of single crystal and bi-crystal that have only a single grain boundary. Moreover, it can be explained by the strain rate sensitivity (SRS) of the grain boundary [15,17]. The strain rate dependence of copper pillars containing a penetrable high angle grain boundary via in-situ compression tests at different strain rate are analyzed [17] and it is found that the grain boundary act as different roles in dislocation transmission when the strain rate changes. It is easy for a dislocation to transmit across a grain boundary at low strain rates, whereas pile-ups and dislocation multiplication always occur in front of a grain boundary at high strain rates. When low strain rates are applied, a similar yield stress is obtained in both single crystal and bi-crystal samples. However, the bi-crystal sample has higher yield stress at high strain rates owing to the difficulty of dislocation transmission. Therefore, due to no grain boundary interaction, the SRS of the single crystal sample is almost constant, which is distinguished with the SRS of bi-crystal samples. The difference of SRS between single crystal and bi-crystal samples is ascribed to the reorientation of dislocations at the grain boundary during transmission, which is hard to occur at high strain rate. Thus, a high SRS is found in bi-crystal samples as a result of dislocation–grain boundary interaction, while single crystal samples have a relatively low SRS due to dislocation–dislocation interactions and formation of dislocation networks.

This research indicates there are a large number of factors influencing the process of dislocation–grain boundary interactions in coarse-grained metals, which includes energy during interaction [34,35,37,93], the type and orientation (geometrical factors) of boundaries and dislocations [16,19,20,35,43] and strain and strain rate [15,17,93]. These factors may affect each other and their correlations are complicated. Since most researches focused only on one factor regardless of other important factors, this may lead to one-sided and incomplete conclusions. For example, the grain boundary interfacial energy has been considered as one of the dominating grain boundary characteristics influencing the dislocation–grain boundary interaction. However, this factor cannot be studied independently from the dislocation types, because even the same grain boundary may exhibit a varied response to different types of incident dislocations. Any isolated study will draw contradictory conclusions with the others, as described in Section 3.1.2.

3.2. Dislocation–Grain Boundary Interactions in Ultrafine-Grained and Nano-Grained Metals

Most researches of dislocation–grain boundary interactions narrated in the above sections are conducted in conventional metals with large grains above several micrometers. In these coarse-grained metals, the dislocation–grain boundary interaction plays a significant role in their mechanical properties. Owing to the impediment of grain boundary to dislocation motion, the plastic deformation of metals with relatively fine grain size requires larger applied stresses, which is known as the Hall–Petch relation [99,100]. In fact, the required stress for dislocation transmission across a grain boundary is much higher than the yield stress of a metal, which can reach four times the value of yield stress [50]. This is because the yield stress can be considered as an average of the contributions from grain boundaries and interior matrix.

When the grain size is smaller than 1 µm, i.e., ultrafine-grained and nano-grained metals, whose grain sizes are considered as 250–1000 nm and 1–250 nm, respectively [41]. The deformation mechanisms and dislocation–grain boundary interactions of metals with grain sizes at these length scales are different from those of conventional coarse-grained metals.

For ultrafine-grained and nano-grained metal, grain boundary mediated mechanisms, such as grain boundary sliding and rotation, have a large influence on the deformation mechanisms [41–43]. The limited distance between the adjacent grain boundaries generates a strong image force field which suppresses dislocation motions. The large resistance results in a confined number of dislocation pile-ups and requires larger stress to increase this number. With the decrease of the grain size to nano-scale, the number of pile-up eventually reduces to one [41]. Since the Hall–Petch relation is established according to dislocation pile-ups in front of grain boundaries, it does not work in some nano-grained metals and even shows an "Inverse Hall–Petch" phenomenon due to the activation of grain boundary sliding and rotation.

A model is proposed to interpret the dislocation–grain boundary interaction in nano-grained metals, which is called "core and mantle" [41]. The core refers to the grain interior with a relatively homogeneous stress state and area free of dislocations, while the mantle is the grain boundary region. Initially, the dislocations are likely to nucleate at grain boundaries and move inside the mantle region rather than emit into the grain interior, as shown in Figure 15a. This plastic deformation of the mantle region is much more severe than the core and leads to a fast increase of dislocation density. Due to the breakdown of dislocation pile-ups, the hardening contributor of the core is much lower compared with that of the mantle region, where dislocations are accumulated and cross-slip systems are active ultimately.

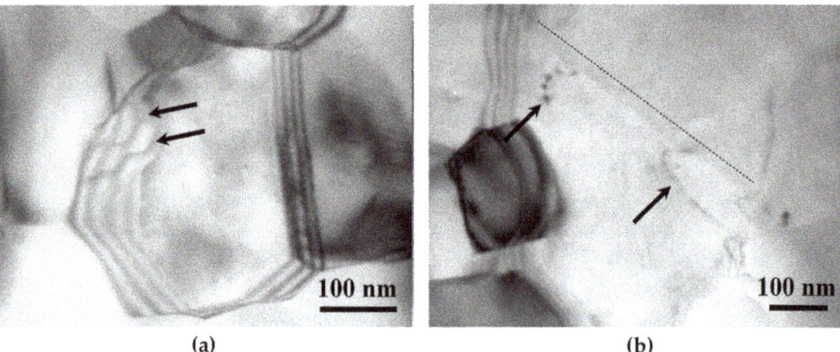

Figure 15. The dislocation–grain boundary interaction in an ultrafine-grained Al at different strains. (a) The dislocations move inside a grain boundary at a low strain; (b) The dislocations emit into the grain interior at a relatively high strain. (Reprinted from Reference [42], with permission from Elsevier).

At high strains, dislocations from grain boundaries is able to emit into the grain interior and glide forwards until absorbed at the opposite grain boundary, as shown in Figure 15b. At the same time, the grain boundary sliding and rotation are able to activate. The emitted dislocations in ultrafine-grained and nano-grained metals always show a curvature configuration than those in coarse-grained metals. This is due to the strong image force derived from grain boundaries close to each other. The process of dislocation absorption is accompanied by an increased number of grain-boundary dislocations, rearrangement of grain-boundary dislocations, and a change in the grain-boundary structure. Further strain increases will activate more dislocation sources and the emission process is continuous. Additionally, the dislocation transmission is also observed in ultrafine grains [43]. These indicate that the dislocation–grain boundary interactions in ultrafine-grained and nano-grained metal have some similar behaviors with those of coarse-grained metals when the incident dislocation impinges on grain boundaries.

3.3. Dislocation–Twin Boundary Interactions in Coarse-Grained Metals

3.3.1. Basic

There are two types of twin boundaries: coherent and incoherent ones. For the incoherent twin boundaries in coarse-grained metals, their interactions with dislocations show no different behaviors compared with grain boundaries [12,18]. For instance, the dislocations are able to transmit across incoherent twin boundaries due to applied stresses, and leave residual dislocations or form steps on the boundaries. This indicates that the incoherent twin boundary is not an impenetrable boundary to dislocation motion. Additionally, the incoherent twin boundary does not have a remarkable effect on the mechanical properties of metals, since the volume fraction of the coherent twin boundary is much larger than that of incoherent one in most cases. For example, the spacing of incoherent twin boundaries in nanotwinned Ag is at least an order of magnitude larger than the spacing of coherent twin boundaries, which shows a weak influence on the mechanical properties. Thus, most of the researches focus on the coherent twin boundaries. The following contents are engaged in these researches and the term "twin boundary" refers to the coherent twin boundary.

Since the coherent twin boundary is an $\Sigma = 3$ grain boundary, the dislocation–twin boundary interaction shows some similarities with the dislocation–grain boundary interaction. However, compared with the complicated and varied grain boundaries, more intensive and detailed studies can be done regardless of other influencing factors derived from twin boundaries and only focus on factors such as dislocation types and the relationship between the slip planes and the twin plane [27,29], owing to its definite mirror symmetry structure. When the dislocation approaches towards the coherent twin boundary, it tends to be parallel to the cross-line between the slip plane and the twin boundary before impingement. Then, it may be absorbed by the twin boundary through accommodation, reconstruction or dissociation into glissile and sessile partials. In some proper conditions, direct and indirect dislocation transmissions across twin boundaries are able to occur due to the increased applied stresses.

Figure 16 shows an example of dislocation pile-ups before transmission across a coherent twin boundary. It indicates that dislocations are primarily impeded by the coherent twin boundary until the stress reaches a critical value [101], below which dislocations are absorbed in the coherent twin boundary, and above this value the dislocation can transmit into the twinned region. In fcc metals, the critical stresses are estimated to be different for screw and non-screw dislocation, which are about 400 MPa [101–103] and 1 GPa [104,105], respectively. The critical stress value is found to be influenced by factors such as the SFE, the shear modulus and the Shockley partial Burgers vector.

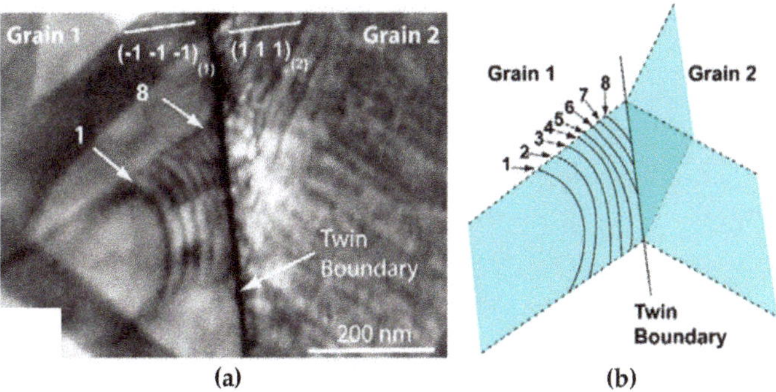

Figure 16. The process of dislocation transmission across a coherent twin boundary. (**a**) TEM image and (**b**) sketch showing dislocation pile-ups in front of a twin boundary, and the dislocation transmission by direct and indirect modes. (Reprinted from Reference [101], with permission from Elsevier).

When a screw dislocation glides parallel to the coherent twin boundary, direct transmission of dislocation may occur by means of cross-slip without leaving any residual dislocation on the twin boundary, as shown in Figure 17. In this case, the coherent twin boundary is the common intersection line of the slip systems on either side of the twin boundary [40]. Although no dislocation pile-ups form in front of the coherent twin boundary, the dislocations cannot transmit across directly, but are absorbed firstly by the coherent twin boundary and then emit into the neighbor twin interior.

In other cases, dislocations may dissociate into glissile and sessile partials when impinging on the coherent twin boundary. The sessile partials may form steps on the twin boundary while the glissile partials shall glide in the twin boundary leading to the migration of the twin boundary and resulting in thinning or thickening of the twin [27,29]. This sort of direct dislocation transmission may generate residual dislocations to form steps on the coherent twin boundaries, which is a cyclic and continuous process with the increase of strain [12,106]. For example, Figure 18 shows the TEM images of three dislocations transmitting across a coherent twin boundary and leaving a sharp step with three layers of atoms. The transmission process proceeds as the following sequences. The incoming dislocation glides to the coherent twin boundary and dissociates as: full dislocation → Frank dislocation + glissile Shockley partial. Then, the glissile Shockley partial glides away along the twin boundary, while the Frank dislocation further dissociates as: Frank dislocation → full dislocation + sessile Shockley partial. Finally, the full dislocation emits to the other side of the coherent twin boundary leaving a step of sessile Shockley partial.

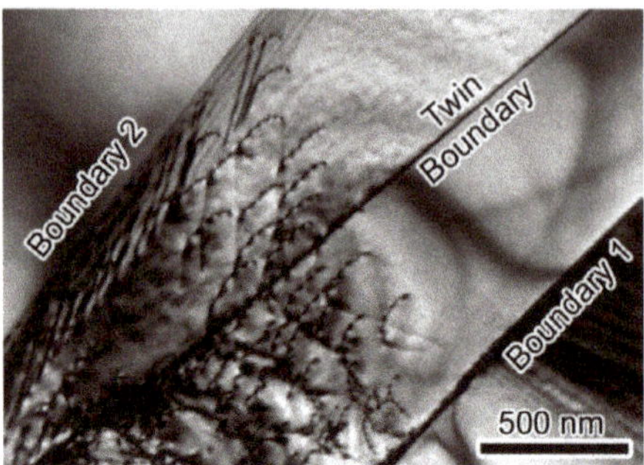

Figure 17. TEM image of direct dislocation transmission without leaving residual dislocations on the coherent twin boundary. (Reprinted from Reference [29], with permission from Elsevier).

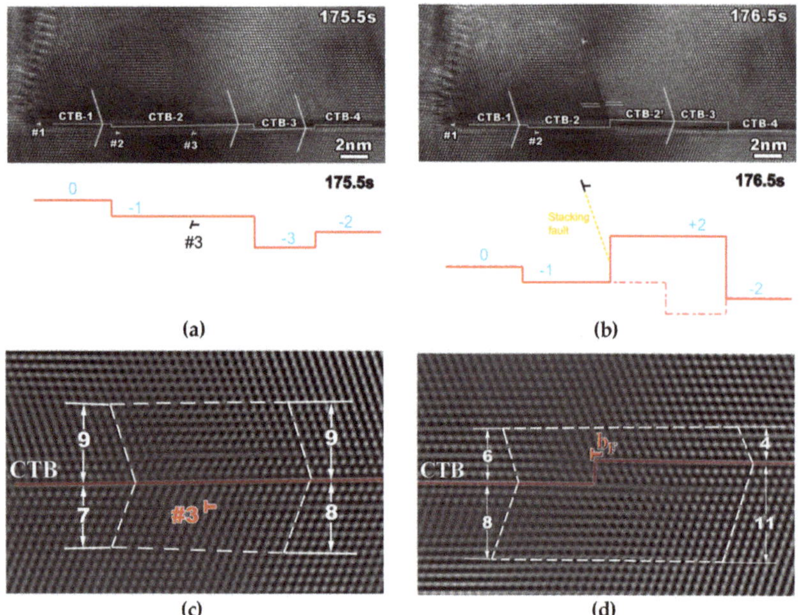

Figure 18. HREM images and illustrations of dislocation transmission leaving residual dislocations and forming steps on the coherent twin boundary. (**a,c**) Images and illustration before transmission; (**b,d**) Images and illustration after transmission. (Reprinted from Reference [106], with permission from Elsevier).

In TiAl alloy [39], when an incoming full dislocation with a Burgers vector of 1/2 <011> transmits across the coherent twin boundary, a residual sharp step dislocation is in the coherent twin boundary. This step dislocation may have the same Burgers vector as the full dislocation (1/2<011>). In some other cases of this study, the incoming full dislocation can either dissociate to a glissile $1/6[\bar{1}1\bar{2}]$ twinning dislocation and a sessile $1/3[1\bar{1}\bar{1}]$ Frank dislocation, or dissociate to a glissile $1/6[1\bar{1}2]$ twinning dislocation and a sessile $1/6[211]$ Shockley dislocation. These phenomena indicate that the behavior and products of dislocation–twin boundary interaction in the coarse-grained metals may be different from those of the dislocation–grain boundary interaction due to some influencing factors, which are going to be discussed in the following sections.

3.3.2. Dislocation Transmission Mechanisms across a Coherent Twin Boundary

Since the coherent twin boundary structure in one metal is the same, the different responses to impinged dislocations are almost due to different types of incident dislocations and different relationships between the slip planes and twin planes. Furthermore, the SFE plays a significant role in the dislocation–twin boundary interaction.

a.   Dislocation type

The dislocation–twin boundary interaction strongly depends on the types of incident dislocations [107]. The discrepancy of twin boundary resistance against dislocation motion may be extremely large [105]. For example, in fcc metals, the coherent twin boundary is a strong barrier to the non-screw dislocations, however, screw dislocation can transmit across a coherent twin boundary directly by cross-slip without leaving residual steps on the twin boundary, analogue to the Friedel–Escaig mechanism for cross-slip [40,108]. When non-screw dislocation impinges on a coherent twin boundary, it is more favorable to dissociate into partial dislocations on the interfacial plane rather than transmit to the other side of the twin boundary. A large number of non-screw dislocation pile-ups are found at the coherent twin boundary after deformation, and the transmission of non-screw dislocation requires high stress and a large accumulation number of dislocations [31]. The critical stress required to transmit across a coherent twin boundary can be as high as 1 GPa when the incident dislocation has a non-screw characteristics [104,105], while the pile-ups of screw dislocations are scarce which exert weak back stresses on the active dislocation sources [31]. The critical stress for screw dislocation transmitting across coherent twin boundary is estimated to be between 300 and 400 MPa [101–103].

In recent studies [31,105], the dislocation transmission mechanism in fcc metals is classified specifically into three categories according to the types of dislocation impinging on the coherent twin boundary: the cross-slip mode, the hard mode and the soft mode.

- The cross-slip mode denotes the transmission of perfect screw dislocations whose Burgers vector is parallel to the coherent twin boundary plane. In this mode, the screw dislocations are usually derived from grain boundaries, which are always activated favorably by applying tension/compression parallel to the coherent twin boundary. Then, they may transmit across a twin boundary by cross-slip without leaving residual dislocations and are ultimately impeded by the grain boundary at the opposite side;
- For the hard mode, the Burgers vector of incoming dislocations is inclined to the coherent twin boundary plane. This may require residual dislocations left in the twin boundary and a higher critical stress to transmit across a coherent twin boundary;
- Dislocations with a slip plane parallel to the twin boundary are defined as the soft mode, which occurs when the maximum shear stress is parallel to the coherent twin plane. The dislocations may glide on the coherent twin boundary leading to migration of the twin boundary by twinning or detwinning.

b.   Geometrical condition

Additionally to the dislocation type, the relationship between the slip plane and the twin plane also affects the behavior and products of the interaction [39,109]. The dislocation–twin boundary interaction can be different due to different slip systems or deviation between slip planes and twin planes even the dislocation type is the same. For example, two variants of $\{111\}\frac{a}{2}<011>$ of slip systems in 304 stainless steel, which are $(\bar{1}1\bar{1})\frac{a}{2}[110]$ and $(1\bar{1}\bar{1})\frac{a}{2}[101]$, are reported to show different responses when impinging on the coherent twin boundary, respectively [29,93]. The $(\bar{1}1\bar{1})\frac{a}{2}[110]$ dislocations can transmit across the twin boundary while the $(1\bar{1}\bar{1})\frac{a}{2}[101]$ dislocations just glide along the twin boundary rather than transmit it.

It is supposed that the dislocation types and geometrical conditions have a combined effect on the dislocation–twin boundary interaction [29]. In fcc metals, Zhu et al. [27] summarized four possible types of dislocation–twin boundary interactions: the 30° Shockley partial, the 90° Shockley partial, the screw perfect dislocation, and the 60° perfect dislocation. The screw perfect dislocation can transmit across the twin boundary by the cross-slip mode. The other types of interactions can either transmit across twin boundaries by the cross-slip mode or hard mode under different conditions. The alternative behaviors are considered to be determined by the SFE of metals, which is summarized in the next chapter.

c.   Stacking fault energy

The effects of SFE on the dislocation–twin boundary interaction are considered in two aspects [31]. The decrease of SFE may hinder the dislocation transmission across a coherent twin boundary by the cross-slip mode and increase the resistance of dislocation transmission by the hard mode. This is because the critical stresses required for nucleation, dissociation and constriction of perfect and partial dislocations are strongly related to the SFE, which further affects the dislocation interaction during transmission. High SFE encourages the constriction of Shockley partials into the screw type and facilities the cross-slip mode transmission. Whereas the low SFE increases the critical stress for dislocation emissions by the hard mode transmission, which leads to the increase of twin boundary resistance against non-screw dislocations. The detailed explanations can be found in the study by Liebig et al. [31].

The above assumptions are supported by the experiments on Cu and CuZn30 alloys. The Cu sample with a relatively high SFE activates more slip planes due to the favorable cross-slip mode dislocation transmission across the coherent twin boundary during compression than those in the $CuZn_{30}$ sample. Besides, the slip transfer steps across the coherent twin boundary show a smooth morphology in the Cu sample. While in the $CuZn_{30}$ sample, the slip transfer exhibits discrete and well-defined features and the serrated stress-strain response.

3.4. Dislocation–Twin Boundary Interactions in Nanotwinned Metals

Since the twin boundary is a barrier to dislocation motion, the nanotwinned boundaries confine the slip of non-screw dislocations perpendicular to the twin plane due to the limited spacing between two adjacent twin boundaries. Thus, the dislocation–twin boundary interaction in nanotwinned metals is different from that in coarse-twinned metals. Besides, the dislocation–boundary interactions in nanotwinned metals are also diverse from that in nano-grained ones. This is ascribed to the spacing between grain boundary is typically on the micro-scale in nanotwinned metals, while the spacing between twin layers is much smaller, which is on the nano-scale [105]. In this case, the twin spacing governs the strengthening of metals whereas the micro-scale grain boundary contributes much less strength, i.e., the bulk strength increases with the decrease of twin spacing [18].

Owing to the strong suppression from nanotwinned layers, dislocations are found to accumulate and glide in the vicinity of twin boundaries, which is likely the favored place for plastic deformation [44]. More specifically, the dislocations are favored to glide on the coherent twin boundaries by the soft mode, which results in twinning or detwinning due to migration of twin boundaries [105]. At the same time, dislocations tend to glide and accumulate in directions parallel to the twin boundaries rather than perpendicular to them. The dislocations glide towards twin boundaries may form Lomer–Cottrell (L-C) locks and resist the transmission of dislocations across a twin boundary [110]. Additionally, different from the large-spaced twin boundaries at the micro-scales, the progressive transmission of dislocations across the nano-scale coherent twin boundary by the hard mode may result in the severe distortion and loss of coherency on the twin boundary due to the formation of steps by residual sessile dislocations [45,111]. These changes of coherent twin boundaries lead to an increase of resistance to dislocation penetration and ultimately evolve into a distorted grain boundary with a high energy [44,46].

A recent study of nanotwinned Ag has found a possible reason responsible for the high strength of nanotwinned metals [18]. In fact, in some studies investigated by in-situ SEM tests [15,31,51,112], there is a ubiquitous phenomenon remaining unresolved: The strengthening contributed from dislocation transmission of individual twin boundary by the cross-slip mode is unexpectedly small, which is not consistent with the high strength obtained in nanotwinned metals [113]. Kini et al. proposes a possible explanation on this topic by testing nanotwinned Ag specimens with different twin spacing from 18 nm to 1800 nm. It is found that the strength increases dramatically when the twin spacing is smaller than 100 nm. This dramatic strength increase is found to come from the local curvature of the partial dislocations between two adjacent coherent twin boundaries, as shown in Figure 19. Since the cross-slip mode transmission requires the constrictions of partial dislocations into the perfect screw dislocations when transmitting across the coherent twin boundaries, the reduction of twin spacing leads to the increase of local curvature of partial dislocations, which further increase the resistance of dislocation transmission by the cross-slip mode. This is different from the observations in the twinned metals with a micro-scale spacing: the strengthening is contributed from the dislocation absorptions and the emissions during transmission across the coherent twin boundaries.

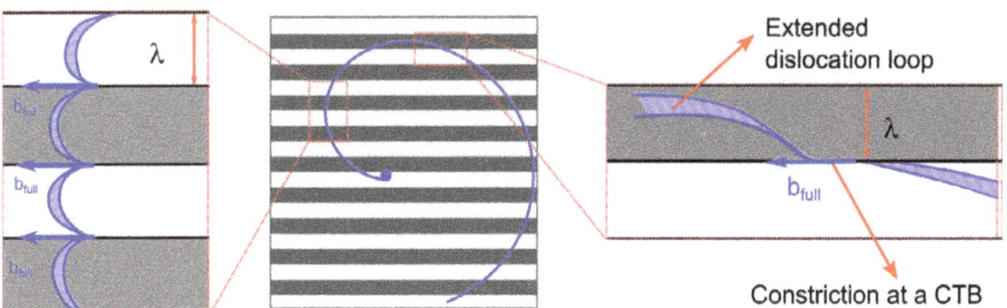

Figure 19. Schematic of the local curvature of partial dislocations between two adjacent coherent twin boundaries during dislocation transmission by the cross-slip mode. (Reprinted from Reference [18], with permission from Elsevier).

3.5. Dislocation–Phase Interface Interactions

The phase interface is supposed to have a much stronger capability on impeding dislocation motion than those of grain boundaries and twin boundaries due to the fact that dislocations must not only overcome the image force and accommodate the complex interface structure, but also accommodate different Bravais lattice, orientation and lattice parameters of the other phase. This stronger impediment makes the interaction between dislocations and phase interfaces different from the other kinds of interactions mentioned above.

Among all the influencing factors, the atomic bonding is likely to be the strongest one. The interface of non-metallic compound phases is a much stronger obstacle against dislocation motion compared with the interface between two different metallic phases, which is due to the complex incoherent interfaces with the poor matching of atoms and the high-indexed planes. Additionally, the interactions between dislocations and phase interfaces may be affected by the phase morphologies and dimensions, as well as the alignment or deviation of slip systems in different phases. In the following sections, the dislocation–phase interface interactions of two morphologies: particle and lamella, are reviewed.

3.5.1. Particle

a.   Particle at the nano-scale

The dislocation interaction with particle interface is governed by the particle size. There is a critical size determining this interaction, which is about 2–7 nm for non-metallic compounds [21]. When the size of the particle is smaller than the critical value, the dislocations are able to penetrate across the phase interface and glide through the whole crystal, which is called shearing. If the size is larger than the critical value, the dislocations may pass around the particle and form dislocation loops, which is termed as looping, as shown in Figure 20. No matter looping or shearing, both interactions generate new defects by consuming the deformation energy and increase the strength of matrix.

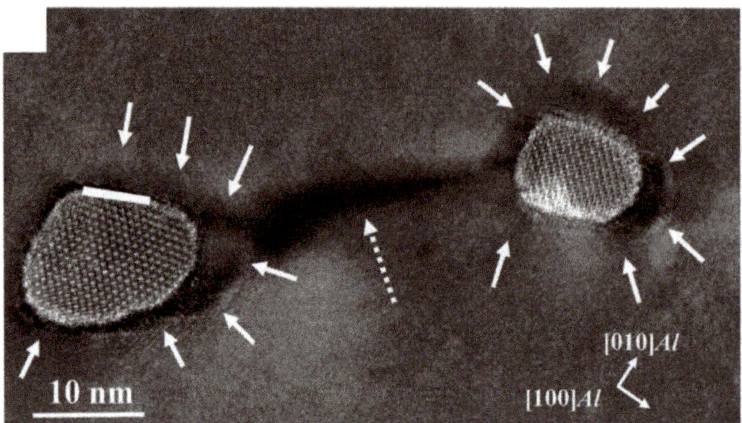

Figure 20. TEM image of looping interactions around particles. (Reprinted from Reference [21], with permission from Elsevier).

The process of shearing has three steps with the increase of applied stress [11]. Initially, the dislocations pile up in front of an interface, and the follow-up dislocations make the ahead dislocation curved along the interface. Then, due to the stress concentration, several dislocations reconstruct to be dislocations or superdislocations accommodating to the Bravais lattice, orientation and lattice parameters of the particle, and eventually glide through the whole particle. The shearing may lead to a large displacement or superlattice intrinsic stacking fault in the particle. As shown in Figure 21a–c, when the particle is partially sheared by dislocations, some regions become blurred due to the local displacements induced by dislocation shearing, compared with the crystal lattice before deformation [21,114]. When the particles are sheared by the arrays of dislocations, the displacements in the particles can be easily observed, as the steps shown in Figure 21d. Compared with the dislocation mobility in the metallic matrix, the mobility in the non-metallic particles is low due to the strong covalent bonds and high melting point of the particles [115,116].

b.   Particle at the micro-scale

For particles with the sizes above 1 μm, the dislocation is more likely to be impeded in front of the interface rather than penetrating it [48]. The dislocations can only transmit the phase interface between two different metallic phases, since few dislocations are able to penetrate the phase interface of non-metallic compound particles with a size larger than 10 nm. At this dimension, the slip transfer across the phase interface between two different metallic phases is dominated by the dislocation transmission instead of shearing.

Since the critical resolved shear stress to transmit across a phase interface is much higher than grain boundary, the stress state plays a more important role in dislocations-phase interface interactions: the Schmid factor seems to be one of the possible reasons for the dislocation transmission but not sufficient to predict it [48]. At the same time, the alignment or deviation of slip systems of different phases appears to be another important criteria for the dislocation transmission.

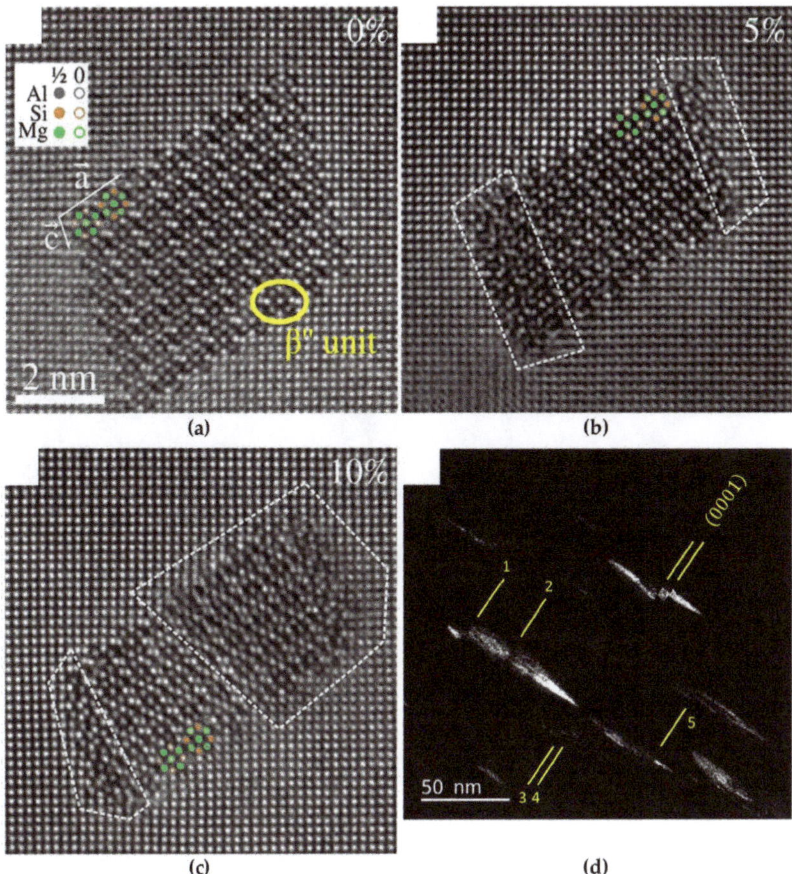

Figure 21. TEM images of shearing interactions. (**a–c**) Images of a particle partially sheared by dislocations at compressive engineering strains of 0%, 5% and 10%, respectively; (**d**) Image of particles sheared by arrays of dislocations. (Reprinted from References [114,117], with permission from Springer Nature and Elsevier).

3.5.2. Lamella

The dislocation–phase interface interaction of a lamella depends on different atomic bondings on both sides of an interface. The dislocation–phase interface interaction between two metallic phases is similar for both particles and lamellae. Their dislocation transmission is strongly influenced by the alignment or deviation of slip systems of different phases. Whereas the slip transfer across the non-metallic interface always leads to the yielding of non-metallic phases by fracture or deformation.

a. Interface between two different metallic phases

The resistance of lamellar interface between two different metallic phases can be diverse due to the alignment or deviation of slip systems between two metallic phases and the annihilation speed of the residual interfacial dislocations produced by dislocation transmission. Dislocations are easy to transmit across the interface when the two phases have a small misorientation between slip systems. For example, the dislocation transmission across the α–β lamellar interface of Ti alloy shows distinct morphologies when the incident dislocations belong to different slip systems [47,49,118]. As shown in Figure 22a, the β lamellae (white laths) can be sheared by the prism slip. However, no shearing of β laths occurs for the basal slip (Figure 22b). This phenomenon was further investigated by Savage et al. [47] through the TEM observations of dislocations in front of the α–β lamellar interface. They found that few pile-ups of prism dislocations were observed in front of the α–β interface. In contrast, the basal dislocations with large distortions are accumulated in front of the α–β interface. The reason is due to the prism slip in the α matrix have a smaller deviation with the slip in the β phase compared with the situation for basal slip. Additionally, the fast annihilation of residual interfacial dislocations during the prism dislocation transmission was considered to further facilitate this process. Thus, the transmission resistance to basal slip is relatively stronger than that of prism slip.

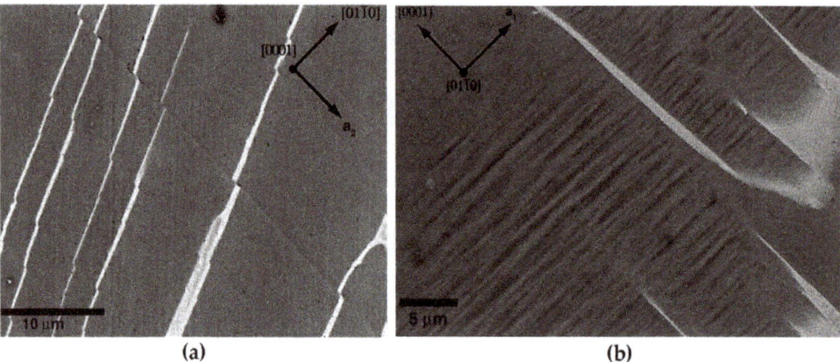

(a)   (b)

Figure 22. SEM images of interaction between slip and lamellar interfaces. (**a**) The prism slip; (**b**) The basal slip. (Reprinted from Reference [49], with permission from Elsevier).

b.   Non-metallic compound lamellar interface

The dislocation interactions with lamellar interfaces between metals and non-metallic compounds are complicated. Since the dislocation–phase interface interaction of high-carbon pearlite steel has been comprehensively studied, it is taken as an example here.

In a macroscopic view, the deformation process of pearlite occurs as following:

- Elastic deformation occurs in both ferrite and cementite phases;
- Plastic deformation occurs in the ferrite lamellae in advance from the dislocation sources at the ferrite–cementite interfaces [119–121] and then proceeds into the cementite lamellae [122,123];
- The dislocations slip in each lamella interior independently and then the slip transfers appear by the yielding of cementite lamellae.

This process proceeds with a reorientation of pearlite colonies and a reduction of lamellar spacing [124–127]. It is supposed that the ferrite–cementite interfaces serve as the dislocation sources and strong barriers against dislocation motions, which is confirmed by TEM observations [128–130]. Thus, at the first stage of plastic deformation, it is assumed that dislocations in ferrite lamellae nucleate firstly at the interface, and then dislocations in cementite lamellae may nucleate at the interface.

In contrast with other dislocation–boundary interactions, the resistance of a phase interface against dislocation motion is much stronger. The impediment of the interface is strong enough to hinder the dislocations moving perpendicular towards the lamellae, which make them piled up or pinned at the interface, so that dislocations are constrained to move parallel to the lamellae. In this case, the dislocation may rotate its moving direction parallel to the interface leaving "superkinks" at the interface and continue to propagate without any increase in the total dislocation length [28]. Therefore, the dislocations glide independently in their own lamellar interiors before slip transfer.

The slip transfer of a ferrite–cementite interface is also distinguished from those of other interfaces mentioned above [10]. The slip transfer occurs always from ferrite to cementite rather than from cementite to ferrite [122,131], which is induced by stress concentration at sites of dislocation pile-ups. The slip transfer occurs along slip systems of 1/2{110}<111> and 1/2{112}<111> in ferrite [10,30,132,133]. However, for cementite, the slip transfer leads to the yielding of cementite lamellae by fracture or bending by slip steps, as shown in Figure 23. For coarse cementite lamellae, the slip transfer is always by a fracture due to its brittleness [22], as shown in Figure 24. While both fracture and bending by slip steps may occur for fine cementite lamellae due to its deformability.

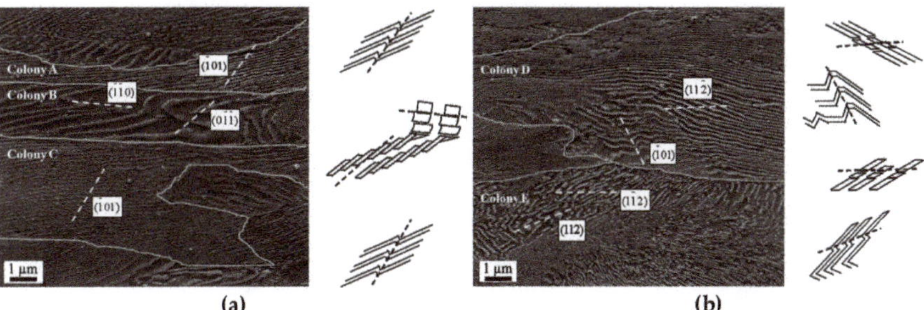

Figure 23. SEM micrographs showing the slip transfer of pearlite by fracture or bending of cementite lamellae. (**a**) {110} slip transfers; (**b**) {112} slip transfers. (Reprinted from Reference [10], with permission from Elsevier).

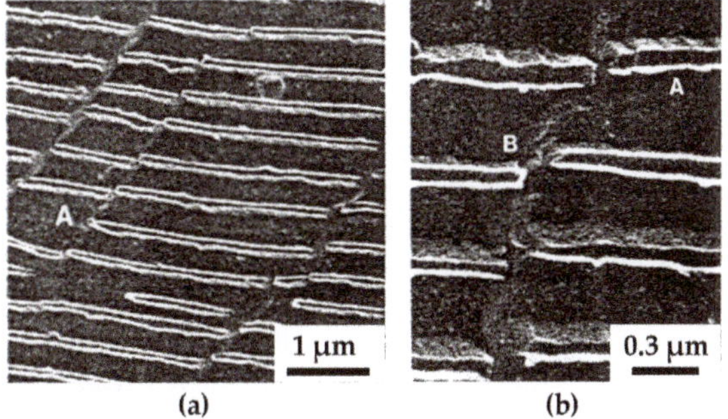

Figure 24. SEM micrographs showing the fracture of cementite lamellae caused by slip transfer. (**a**) and (**b**) are the SEM images at different magnifications. (Reprinted from Reference [22], with permission from Elsevier).

The mechanism of dislocation transmission across the ferrite–cementite interface is still a dubious topic. Several models have been proposed and simulations have been done to predict this probability, but no persuasive conclusion has been drawn. There is few solid evidence of the dislocation transmission across a ferrite–cementite interface by means of direct transmission, although the slip transfer has been observed in the pearlite. The feasibility of dislocation transmission is researched by using *LRB* criteria [132,133], dislocation transmission mechanism m' (mentioned in Section 3.1) [30] and simulation [122,131,134]. According to *LRB* criteria, two {110}<111> slip systems and one {112}<111> slip system of ferrite are possible to transmit across the ferrite–cementite interface. The dislocation transmission mechanism m' of grain boundary can be also applied.

Furthermore, most scholars consider that the ferrite–cementite interface is a strong barrier of dislocation motion, especially for dislocations with the slip direction perpendicular to the interface. At the early stage of deformation, dislocations in ferrite lamellae pile up in front of the interface and slip transfer occurs eventually owing to the increase of stress concentration. However, some researchers argue that dislocation transmission across the ferrite–cementite interface is relatively easy at the early stage of deformation.

The subsequent dislocation pile-ups are due to the changes of cementite structure induced by dislocation transmission. Zhao et al. [123] claims that dislocations can easily transmit from a ferrite lamella to a single crystalline cementite lamella across a flat ferrite–cementite interface. Then, the single crystalline cementite lamellae subdivide into polycrystals. The interface of polycrystalline cementite lamellae can effectively impede the dislocation motion. Therefore, the dislocation transmission mechanisms across the ferrite–cementite interface are proposed according to calculations and simulations, which are diverse and ambivalent requiring further studies and experimental verifications by advanced characterization methods such as high-resolution and in-situ TEM mechanical testing [135].

Dislocations-phase interface interactions play an important role for structural metals, in not only the strength [136,137], but also other mechanical properties such as ductility and toughness [138], wear resistance [139], fatigue properties [140–143] and micro-pitting [144], as well as the properties of heterogeneous structures [145–147]. All these need the systematic investigations with exquisitely designed experiments and/or simulations.

4. Interactions between Deformation Twin and Boundary

Since the twin nucleation and growth are driven by the motion of twin boundary dislocations, the deformation twin-boundary interaction is considered as the result of interaction between the incident twin boundary dislocations and the obstacle boundary [13]. Thus, it shares many similar characteristics with the dislocation–boundary interactions. However, the transmission of deformation twin across a boundary may be much harder than a single dislocation. This is ascribed to several possible reasons:

- Since the twin nucleation and growth are the result of partial dislocation motion, the dislocation slip must be active prior to the twin formations and the stress level for the twin-boundary interaction is higher compared with that for the dislocation–boundary interaction [148];
- Since the emitted twin keeps the same variant as the incident one, the resolved shear stress and slip system of the adjacent grains should be favorable for the twin nucleation of the same variant. In this case, both the crystals and twin pairs on each side of the boundary should have small misorientation.

On the other hand, for the twin–twin interaction, since the twin boundaries are fast channels for dislocation motions, the applied stress can be released by dislocation gliding on the twin boundaries from the deformation twin to the obstacle twin instead of forming twin transmission. Thus, the twin transmission can only occur under limited conditions.

4.1. Deformation Twin-Grain Boundary Interactions in Coarse-Grained Metals
4.1.1. Basic

Since hcp metals are favorable for the formation of deformation twins, they have been taken as one of the model metals to study the deformation twin-grain boundary interaction, especially pure Mg and Mg alloy. Similar results have been obtained in these studies [14,32,36,149–151] and several factors are engaged in the behavior of twin transmission across a grain boundary in coarse-grained metals, which has many characteristics in common with dislocation–grain boundary interaction.

The process of twin transmission in coarse-grained metals proceeds as the following sequences. The deformation twin propagation is primarily blocked by the grain boundary, which induces stress concentration in the vicinity. With the increase of stress concentration, the stress of the deformation twin relaxes in the neighboring grains and a new twin nucleates on the other side of the grain boundary due to local strain compatibility [151], as shown in Figure 25.

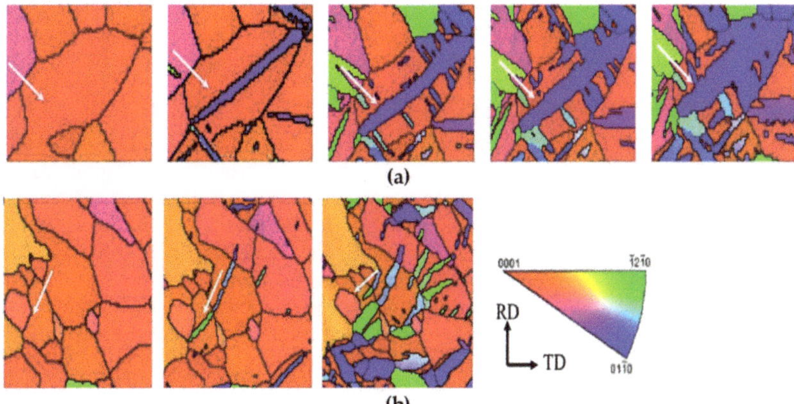

Figure 25. The process of twin transmission investigated by in-situ SEM mechanical testing with compressions strains of (**a**) 0, 1.6%, 3.1%, 4.2%, 5.4% and (**b**) 0%, 1%, 2%, respectively. (Reprinted from Reference [151], with permission from Elsevier).

4.1.2. Twin Transmission Mechanisms across a Grain Boundary

a. Misorientation

The misorientation has been considered as the primary factor determining the twin transmission in coarse-grained metals [14,32,36]: Deformation twins can transmit low angle grain boundaries but unlikely to do so when impinging on a high angle grain boundary. The impediment of twin propagation may cause high back stress and stress concentration in the vicinity, which generates local non-basal slip and secondary twinning. Twin transmission always occurs in the case of crossing a low angle grain boundary. The speed of twin transmission across low angle grain boundary is rapid, which is observed by in-situ SEM and in-situ optical microscopy (OM) mechanical testing [149,150].

Twin transmission determined by misorientation is probably due to the favorable twin nucleation conditions. Compared with high angle boundaries, low angle grain boundaries are composed of dislocation arrays, which are assumed to be favorable for activation of twin boundary dislocation and nucleation of deformation twins [14]. Furthermore, small misorientation facilitates twin nucleation due to a similar slip system and a sufficient resolved shear stress. With the increase of misorientation, the tendency of twin nucleation in the adjacent grain decreases due to the maximum resolved shear stress deviating away from the favorable slip systems [36].

b. Other influencing factors

Besides misorientation, the geometrical condition and chemical composition may also affect the behavior of twin transmission across grain boundaries in coarse-grained metals.

In most transmitted twin pairs, the angle between their twin plane normals tends to be minimized and the twin variant of incoming and outgoing twins should be maintained constant across the grain boundary [36]. These mechanisms are consistent with the criteria of misorientation. For low angle grain boundaries, the angle between twin plane normals of the incoming and outgoing twins is the misorientation of the grain boundary. The tendency of twin transmission increases with the reduction of misorientation owing to the increase of resolved shear stress. In this case, the easy slip system activated by the resolved shear stress must be the same as the incident twin. Thus, both the incident twin and emitting twin have the same variant.

The critical misorientation angles of twin transmission are slightly different for different metals. For instance, the highest transmission frequency in Mg is found to be the misorientation angle with a value below 15°–20° [32]. For pure Re, twin transmission is much easier to happen compared with Mg, leading to a relatively high critical misorientation angle [14,36]. This is ascribed to the low twin boundary energy and different deformation twinning system of Re.

4.2. Deformation Twin-Grain Boundary Interactions in Nano-Grained Metals

The deformation twin-grain boundary interaction in nano-grained metals shows distinct behaviors compared with that in coarse-grained metals. It is hard for a deformation twin to grow and intersect with grain boundary in nano-grained metals, let alone its transmission across a grain boundary [46]. Once the deformation twin nucleates at the grain boundary, the twin can grow into the grain interior due to the applied stress but is limited to a certain size. This is because the twin tip is imposed by a high image stress derived from grain boundaries nearby in nano-grained metals, which hinder the propagation of the deformation twin. At the same time, the increasing applied stress leads to nucleation of dislocations both from grain boundaries and twin boundaries. The progressive dislocation–twin boundary interactions result in severe distortion and loss of coherency on the twin boundary, which may eventually evolve into a distorted grain boundary with a high energy.

4.3. Twin–Twin Interactions

Although the coherent twin boundary has a definite structure, the twin–twin interaction is complicated and shows unique features compared with the deformation twin-grain boundary interaction. For example, in contrast to the dislocation gliding in the matrix, the twin boundaries can serve as the fast channels for dislocation motions due to no image force derived from the obstacle boundary [152]. Once the twin intersection is established, it is easy for dislocations to move from the twin boundaries of the incident twin to the twin boundaries of the obstacle twin. In this case, the applied stress can be released by this way rather than forming a twin transmission.

Furthermore, the twin–twin interaction behavior depends on the Bravais lattices of metals: The twin transmission is available to occur in bcc and fcc metals under suitable conditions [39,148], while the twin transmission is considered to be impossible for hcp metals, such as Mg [153,154]. These differences are reviewed in the following sections, separately.

4.3.1. Twin–Twin Interactions in bcc and fcc Metals

There are two types of twin–twin interactions in bcc and fcc metals [39,148]. Figure 26 shows an ordinary twin transmission in bcc and fcc structural metals. The process proceeds as follows:

- The incident twin impinges on one side of the obstacle twin boundary and the incident twin propagation along the original direction is blocked;
- The incident twin transmits the obstacle twin boundary and alters the propagate direction along the close-packed plane of the obstacle twin until it approaches the other side of twin boundary;
- The incident twin transmits the twin boundary again and continue to propagate along the original direction beyond the obstacle twin.

The other type of intersection may generate a local distortion as well as a low angle grain boundary at the position of the twin–twin intersection, and the twin boundaries disappear at the intersection position [39]. Additionally, the crystalline lattice of the twin and the matrix distort to some degree and a low angle grain boundary is formed due to the complex reaction.

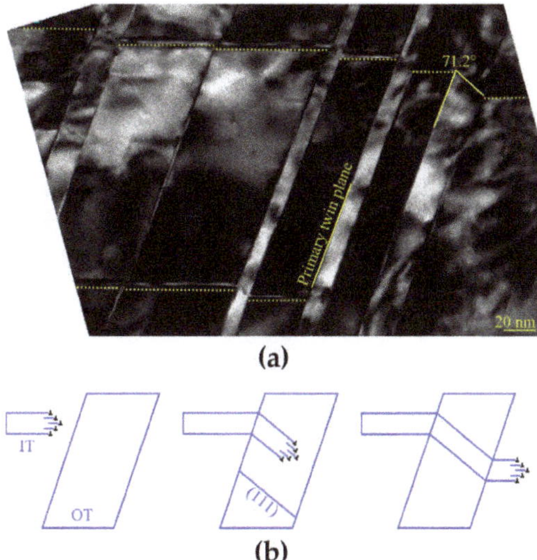

Figure 26. The ordinary twin transmission across twin boundaries in bcc and fcc metals. (**a**) The TEM image; (**b**) The schematic illustration showing the process of twin intersection. (Reprinted from Reference [39], with permission from Elsevier).

4.3.2. Twin–Twin Interactions in hcp Metals

a.　　Structure of the twin–twin intersection

The twin–twin intersection in the hcp metals can result in the large lattice deviation in the vicinity of twin–twin intersection and forms a unique structure, which consists of an incoherent interface and a step-like faceted structure [13,38]. The TEM images of a twin–twin intersection is shown in Figure 27. The intersection forms a complex step-like faceted structure, which is composed of an incoherent interface (twin–twin boundary marked as B-B or P-P) and several facets connected with the twin–twin boundary, marked as B-P or P-B. The B and P represent the basal and prismatic planes, respectively. The twin–twin boundary is often aligned as the basal plane of one twin nearly parallel to the basal plane of another twin with a small misorientation across the boundary (B-B boundary), or similarly, the prismatic plane of one twin is nearly parallel to the prismatic plane of another twin with a small misorientation across the boundary (P-P boundary). The structure of facets has the characteristics that the basal planes in the matrix almost align with the prismatic planes in the twin (B-P facet), or similarly, the basal planes in the twin align with the prismatic planes in the matrix (P-B facet). There are many facets in the vicinity of the twin–twin intersection due to the large lattice deviation, while the number of facets drops dramatically at the remote area from the twin boundary.

It is found these facets play a key role in the twin boundary migration, because the facets must move with the twin boundary propagation and interact with twin boundary dislocations [13]. They may facilitate the twin boundary dislocation motion and potential atom shuffling across the twin boundary, which improves the twin boundary mobility.

The reason for the formation of the faceted structure is that the actual twin plane of Mg metal has a small deviation from the theoretical ideal twinning system ($\{10\bar{1}2\}<10\bar{1}1>$) [38]. Compared with geometrically and energetically unfavorable twinning planes, the interfacial energy to accommodate this deviation by formation of small-scale facets along twin boundaries is lower [13].

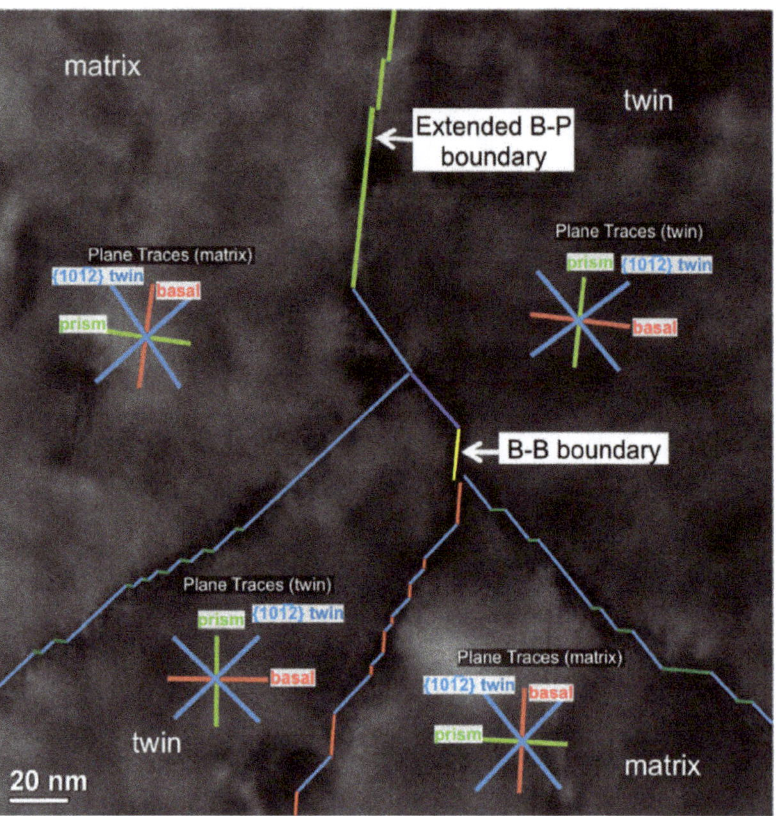

Figure 27. TEM image of a twin–twin intersection with a B-B boundary in Mg. (Reprinted from Reference [13], with permission from Elsevier).

b.   Behaviors during the twin–twin interaction

During the process of an incident twin impinging on an obstacle twin, a large number of dislocations are engaged [13]. Initially, the tip of the incident twin exhibits a sharp morphology when it is far away from the intersection point. Then, when the twin tip approaches the obstacle twin, it becomes blunted due to a high stress in front of the tip. Finally, a step forms at the intersection place of the obstacle twin boundary once the incident twin impinges on the obstacle twin.

After the twin–twin intersection, twins may grow thicker and coalesce together with the increase of applied stress [155]. As shown in Figure 28, the intersected twins grow thicker and the twins with the same variant may coalesce together once they meet with each other at low strains. At high strains, the twin variants with the highest Schmid factors may grow faster and larger than the others and take over the other twin variants eventually. The twin thickening is due to the twin boundary dislocation motion and twin boundary migration induced by the applied stress [153], as shown in Figure 29. The twin boundary dislocation glides from one twin boundary towards the other twin boundary with the stress increase. These dislocations may dissociate at the twin–twin intersection and glide on the new twin boundary leaving a tilt boundary behind, which results in the growth of the twin and the extension of the tilt boundary.

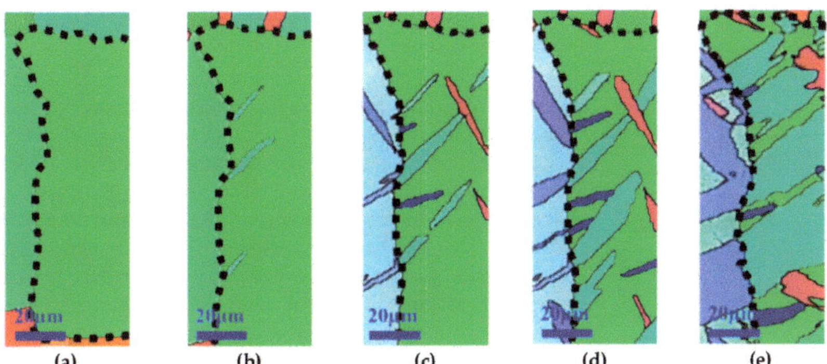

Figure 28. The twin–twin interaction in Mg at true strains of (**a**) 0%, (**b**) 1.2%, (**c**) 2.5%, (**d**) 5.5% and (**e**) 8%, respectively. (Reprinted from Reference [155], with permission from Elsevier).

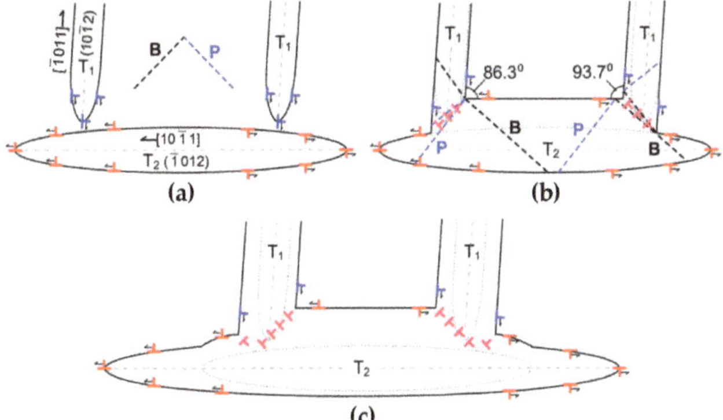

Figure 29. Schematic illustrations of a twin–twin intersection. (**a**) Before impingement; (**b**) Formation of a twin–twin intersection; (**c**) Twin growth after the twin–twin intersection. (Reprinted from Reference [153], with permission from Taylor & Francis).

c.   Interaction mechanisms

Notably, the twin transmission is unlikely to happen in the hcp metals in most cases [153,154]. It may be ascribed to two reasons. Since the twin formation requires the activation of proper twin boundary dislocations, the limited number of slip systems in hcp metals also means the limited number of twinning systems. In this case, the resolved shear stress derived from the twin–twin intersection always deviates from any possible twinning direction of the obstacle twin. Thus, it is unfavorable for twin nucleation even if the stress concentration increases at the twin–twin intersection. Instead, it activates a basal slip band at the intersection in the obstacle twin. Moreover, the dislocations tend to glide on the twin boundary, where dislocation motion is quicker than in the matrix without the suppression of image force [152]. The increase of applied stress can be relaxed by the twin thickening due to the motions of these dislocations.

On the other hand, the stress state dominates the twin thickening and coalesce after the twin–twin intersection. This is supported by the fact that the twin variants with the highest Schmid factors, or in other words, imposed by the maximum resolve shear stress, can grow fastest and consume the other twin variants [155].

5. Applications and Characterization Techniques

5.1. The Boundary Strengthening of Bulk Metals

For most bulk metals, tuning the microstructure to effectively impede dislocation motion is the typical method to improve strength. For example, the deformation and strengthening mechanisms of microstructural refinement by deformation at ambient temperatures is owing to the increase of grain and twin boundary densities as well as the increase of boundary misorientation according to the boundary strengthening [156–158] and dislocation strengthening [159]. Besides, the concept of precipitation strengthening is derived from the strong impediment of the phase interface to dislocation motion [160].

In addition to the usual methods to improve the mechanical properties of bulk metals, some approaches have been developed according to the other findings of dislocation–boundary interactions. For example, the concept of grain boundary engineering (GBE) by increasing the fraction of special or low Σ CSL boundaries of metals is based on the different structures and energies from random high angle boundaries [23–25]. On the other hand, the mechanical properties of Mg alloy has been improved by transition of deformation mechanism from twinning and <a> slip to <c + a> slip [161,162]. This is because the <c + a> slip system of Mg alloy has more independent slip directions than those of <a> slip system, which reduce the stress concentration induced by the dislocation–boundary interaction and increase its plasticity. Therefore, the knowledge of dislocation–boundary interaction is significant not only for fundamental science, but also for alloy/processing design strategies in the development of advanced bulk metals.

5.2. The Analytical Descriptions of Boundary Strengthening

Although massive researches show that there exists various kinds of dislocation–boundary interactions, it is still difficult to directly correlate these investigations to the mechanical properties of bulk samples due to the different kinds of restrictions such as the limitation of experimental techniques as summarized afterwards, the internal-microstructure and external-sample size, length-scale effect and the strain rate effect, etc. Hereby, the analytical descriptions of strengthening based on the dislocation–boundary interaction is briefly summarized in the following contents.

5.2.1. Hall–Petch Relation

The boundaries are barriers to dislocation motions. Consequently, in the coarse-grained polycrystalline metals, the strength increases as the grain size decreases [100]. The Hall–Petch relation is based on this rule and is expressed as the following:

$$\sigma_y = \sigma_0 + k_{HP} D_{av}^{-1/2} \tag{10}$$

where σ_y is the yield stress, D_{av} is the average grain size, σ_0 and k_{HP} are constants. σ_0 is the friction stress, which is the flow stress of an undeformed single crystal or approximately the yield stress of a very coarse-grained polycrystal without texture [99]. The Hall–Petch relation is the simplest analytical description of strengthening derived from dislocation–boundary interactions, which has a great impact on the other analytical descriptions.

Hansen [99] points out that the strengthening should not only come from grain boundaries but also dislocation accumulations in the grain interior, i.e., the dislocation boundaries (Section 2.2). Since the IDBs are assumed to be penetrable to slip and contribute via the forest hardening. Meanwhile the GNBs are assumed to be barriers to slip and contribute by the Hall–Petch strengthening [163,164]. The flow stress is the sum of friction stress, forest and Hall–Petch hardening, which can be expressed as:

$$\sigma_s = \sigma_0 + \left[k_{HP} \sqrt{f_{GNB}} + M\alpha\mu\sqrt{3b\theta_{IDB}(1 - f_{GNB})} \right] D_{av}^{-1/2} \tag{11}$$

where f_{GNB} is the fraction of GNBs, M is the Taylor factor, α is a constant, μ is the shear modulus, b is the Burgers vector and θ_{IDB} is the average misorientation angle of IDBs.

Since most GNBs and IDBs are high angle and low angle boundaries, respectively, the Equation (11) can be also expressed as:

$$\sigma_s = \sigma_0 + \left[k_{HP}\sqrt{f_{HAB}} + M\alpha\mu\sqrt{3b\theta_{LAB}(1 - f_{HAB})} \right] D_{av}^{-1/2} \tag{12}$$

here the subscripts "HAB" and "LAB" refer to the high angle and low angle boundary, respectively.

For the nanocrystalline metals with grain sizes smaller than 20 nm, usually, the dominant deformation mechanism switches from a dislocation-mediated process to the grain boundary sliding. However, Zhou et al. [100] has prevented grain boundary sliding and enhanced Hall–Petch strengthening in the nanocrystalline by relaxation and element segregation. They found that the Hall–Petch strengthening in nanocrystalline with grain size of 3 nm were different from coarse-grained polycrystals: the partial dislocation hardening was as important as full dislocation hardening. Therefore, they proposed a modified Hall–Petch relation as follow:

$$\sigma_y = \sigma_0 + \frac{k_{HP}}{\sqrt{D_{av}}} + \frac{k_{partial}}{D_{av}} \tag{13}$$

where $k_{partial}$ is a constant. The third term represents the contribution from partial dislocations, which is inversely proportional to grain size. Meanwhile, Zhang et al. [125,126] have also observed that the dislocation-based plasticity still exists in the sub-20 nm lamellar structures of pearlite steel wires, where the cementite lamellae dissolve at large strains and stabilize the ferrite lamellar boundaries for further deformation.

5.2.2. Thermal Activation Theory

The thermal activation theory is established based on the thermal and athermal nature of flow stress, where there are three types of barriers to dislocation motions: short range, long range, and drag components [165,166]. The short-range component originates from lattice and point defects in the grain interior, whose dimension is about 10 atomic diameters. Since the stress contributed by short range component is influenced by strain rate and temperature, it is absolutely a thermal stress σ_{TH}. The stress of long-range component is derived from forest dislocations and grain boundaries. This stress is not affected by strain rate and is almost independent of temperature, which is usually regard as an athermal stress σ_A. The stress contribution from drag component is considered to be small when the strain rate is smaller than 10^5–10^7 s^{-1}. In most cases, the flow stress can be expressed as [166]:

$$\sigma_s = \sigma_A + \sigma_{TH} \tag{14}$$

Since the athermal stress σ_A is contributed by forest dislocations and grain boundaries [167,168], it can be expressed as:

$$\sigma_A = M\alpha\mu b\sqrt{\rho} + k_{HP}D_{av}^{-1/2} \tag{15}$$

here ρ is the dislocation density. The two terms are contributions from forest and Hall–Petch strengthening, respectively. Visser and Ghonem [165] have taken grain boundaries as well as twin boundaries into account by considering the twin boundaries as a decrease of original grain size. The thermal stress is written as:

$$\sigma_A = \sigma_F\varepsilon^i + \sigma_T \tag{16}$$

where ε is the strain, σ_F and i are hardening parameters due to the forest strengthening. σ_T is the stress contributions of twin boundaries, which is expressed as:

$$\sigma_T = \sigma_0 + \beta F^j \tag{17}$$

where β and j are constant, F is the twin volume fraction.

5.2.3. Precipitation Strengthening

The analytical descriptions change according to the morphologies and the dimensions of phases [169]. In terms of the nano-scale particles (situation represented in the Section 3.5.1), if the particles have coherent interface, are small enough and can be sheared by the dislocations, the stress contribution is written as:

$$\sigma_{P1} = C_1 A_{\mathrm{eff}}^n \frac{R^n}{L + 2R} \tag{18}$$

here C_1 and n are constants, A_{eff} is the effective obstacle area, R is the radius of particle and L is the spacing between two adjacent phases. On the other hand, when the interfaces of particles are incoherent or the sizes of particles are relatively large, where the dislocations and the phase interfaces can only interact by looping (situation represented in the Section 3.5.1), the stress contribution is written as:

$$\sigma_{P2} = C_2 \frac{\mu b}{L} \tag{19}$$

where C_2 is constant. In the case of the phases with dimensions at micro-scale (situations represented in the Sections 3.5.1 and 3.5.2), e.g., the micro-scale particles and the lamellae, the stress contribution is written as:

$$\sigma_{P3} = \frac{C_3}{\sqrt{L}} \tag{20}$$

here C_3 is constant.

5.3. Characterization Techniques for Investigating Dislocation–Boundary Interaction

To investigate the strengthening mechanisms and the relationships between the mechanical properties of bulk metals and their dislocation–boundary interactions during deformation, the ideal characterization method is the dynamic observation of the deformation process and interaction behaviors at the full-length scales from nano to macro in three dimensions. Compared with ordinary static experiments, the dynamic observation by in-situ characterization techniques provides the real-time evolution and variation of structural data during the whole interaction process between dislocations and boundaries for further analysis and also provides an insight into the unknown field not obtainable in the post mortem results [12,13,16–18,114,151,155]. However, the ideal condition cannot be met by any single in-situ characterization technique due to the limitations on spatial resolution, states and dimensions of specimens, and deformation strains and strain rates. The in-situ techniques are compared below with a focus on SEM, TEM and XRD techniques.

The in-situ SEM technique can produce a high-quality image with a spatial resolution smaller than 1 nm and provide abundant information on microstructure, crystal orientations, phases, Schmid factors, and strains. It has been considered as one of the best methods for the investigation of the twinning behavior during deformation induced by the external stress [149,151,155]. Compared with the in-situ TEM technique, the in-situ SEM technique is capable for investigating the crystal orientation, microstructure and dynamic behavior at both large stains and large scales, which approaches the situation of macro-mechanical properties. However, it cannot give a detailed structure of a boundary as well as dislocation dissociation during twin-boundary interactions. These details need to be further verified by the TEM technique or simulation studies. On the other hand, it is difficult to observe individual dislocations interacting with boundaries by the SEM technique. Even so, multiple findings are obtained in observation of bulk metals coupled with mechanical testing, which are unlikely to be obtained by the TEM technique [15,18–20,170]. Therefore, it is promising to investigate dislocation–boundary interactions in bulk metals by in-situ SEM technique at large strains and large scales together with the post mortem observations by TEM.

Since it is difficult to observe individual dislocation interactions by other methods, the in-situ TEM technique has remarkable advantages in characterizing dislocation–boundary interactions due to its high spatial resolution. It facilitates the atomic-scale dynamic observation of dislocation–boundary interactions [12,16]. Additionally, details of dislocation motions during the twin–twin interaction can be revealed by the in-situ TEM technique [13,152]. Moreover, the TEM technique also provides abundant information including atomic number, crystal structure and orientation. Although it seems that the TEM technique meets all the needs for materials research, there are some drawbacks [50,171,172]:

- The specimen requires a thin foil with thickness typically below 300 nm, which cannot reflect a macroscopic performance of metals;
- The useful field of view is small, which is hard to reach a large quantity of statistics;
- Mechanical damage and relaxations of stored deformation may be induced during specimen preparation due to its large free surfaces;
- It is hard to control imaging and deformation conditions to obtain a qualified field of view during the in-situ experiment.

Consequently, it is hard to relate the data by in-situ TEM technique to the macroscopic mechanical properties of metals.

Compared with the in-situ SEM and TEM techniques, the in-situ XRD technique exhibits many advantages:

- Non-destructive measurements [173,174], which can be applied on bulk specimen tested by the conventional mechanical equipment rather than pillars or thin foils;
- A large detecting area up to macro-scale [173,175], which facilitates the research on average behaviors of multiple grains;
- The information obtained by penetrating several millimeters of bulk crystal volume instead of data from the surface (SEM) or the thin foil (TEM) [174], which is able to build 3D mapping and investigate 3D interactions between dislocations and boundaries [95,176,177];
- It is more sensitive to elastic strains, which is favorable to study the stress distribution before plastic deformation [178];
- It is capable to perform experiments at high temperatures (above 900 °C) [179].

Even though the in-situ XRD technique is powerful, some drawbacks have limited applying this method in the studies of interactions between dislocations and boundaries. At first, all information obtained by XRD is based on the analysis of Laue diffraction pattern, which is different from the straightforward results obtained by other microscopic techniques [174]. The most important reason for few research on interactions between dislocations and boundaries is the resolution of the in-situ XRD, which is only at the micro-scale [178,180,181].

6. Summary and Outlook

So far, there is still a gap between the knowledge of dislocation–boundary interactions and macroscopic mechanical properties due to many influencing factors and differences between experimental and normal testing conditions. Moreover, there are still uncovered interaction mechanisms which need intensive and comprehensive studies to better understand the nature of dislocation–boundary interactions.

6.1. Investigations of Unknown Interaction Mechanisms

6.1.1. Reveal New Mechanisms during Dislocation–Boundary Interaction

Although great progress has been achieved in understanding the dislocation–boundary interactions, new findings are emerging continuously. For example, in the dislocation–twin boundary interaction of nanotwinned Ag, the resistances of dislocation transmission across the nanotwin boundaries are relatively weak, which are not considered as the dominant reason for their high strength. The governing strengthening mechanism of nanotwinned Ag is that the small twin spacing of nanotwinned structure remarkably increases the local curvature of partial dislocations, which requires larger stresses to emerge the partial dislocations into the perfect screw dislocations before transmitting across twin boundaries [18]. Further researches need to reveal the unknown interaction mechanisms in particular for nanocrystalline and nanotwinned metals.

6.1.2. Discover the Origins of Boundary Resistance

The strengthening induced by dislocation–boundary interactions can originate from image forces, boundary absorptions and interactions, and accommodations to the adjacent grains. However, in most cases, the resistances of boundaries are regarded as a whole, and their strengthening contributions have not been clarified in detail. For example, Malyar et al. [17] found that the dislocation–grain boundary interaction shows different behaviors at different strain rates: the dislocations can easily transmit across a grain boundary at a low strain rate while are piled up in front of a grain boundary at high strain rates. This is due to varied difficulty levels of dislocation reorientation with strain rates when the dislocations are emitted into the adjacent grain. Since the resistances of boundaries at different stages of dislocation–boundary interactions may change according to the deformation status, it is important to distinguish their strengthening contributions from each other.

6.1.3. Validate Assumptions and Simulation Results by Experiments

For example, several models and simulations have been performed to explain the dislocation transmission mechanism across the ferrite–cementite interface in pearlitic steels [22,30,122,131–134]. However, the direct dislocation transmission across the interface has not been confirmed through experimental observations. This is ascribed to the fact that the observation of dislocations in cementite lamellae is relatively difficult by TEM. There are many cases remained at the stage of theory assumptions due to experimental difficulties, which are expected to be solved in future by advanced in-situ techniques.

6.2. Comprehensive Studies of Influencing Factors

Directly correlating the practical mechanicalproperties of metals with dislocation–boundary interaction models faces huge challenges. This is because there are numerous influencing factors of dislocation–boundary interactions and none of them can be considered as absolutely independent regardless of others. However, most researches focus on one factor. In this case, the obtained conclusions isolated to other influencing factors may be diverse from case to case and sometimes result in contradictions between different results. Moreover, deformation of polycrystalline metals is the dominant processing of structural metals. These conditions are different from most of the reported results in bi-crystals at relatively low strains and strains rate by TEM and SEM.

So far, the Hall–Petch relation is a model concerning the dislocation–boundary interactions that has been widely recognized as feasible to the practical mechanical properties of metallic materials [99,100]. However, even in this successful case, the grain size is the only factor considered in this relationship, regardless of important factors such as Bravais lattices, boundary structure, misorientation or geometrical condition, dislocation types, strain, strain rate and so on. Also, the other strengthening models on the dislocation–boundary interaction face the same situation as the Hall–Petch relation. However, the experimental results have verified the importance of influencing factors, which are not negligible in the dislocation–boundary interaction. For example, the dislocation types have been confirmed to show distinct behaviors when interacting with twin boundary [31,40,104,105,107] and grain boundary [43], respectively. Additionally, although the resistances to dislocation motions of low-energy and low Σ CSL boundaries are a controversial issue [35,37], they have been applied as grain boundary engineering (GBE) and improve the mechanical properties of metals [23–25].

On the other hand, compared with the dislocation–boundary interactions, there is few studies on the influencing factors of twin–boundary interactions. For example, the studies of twin–grain boundary interaction mainly focus on the grain boundary misorientation in hcp metals [14,32,36,149,150]. Whereas most researches of twin–twin interaction investigate what occurs during the interaction [13,38,39,148], but few of them involve investigations of influencing factors. Therefore, there is still a long way to go in the study of dislocation–boundary interactions. To better investigate the behaviors during dislocation–boundary interactions, more experimental data by combining different in-situ testing methods is needed to take cross-validations at different length scales, strains, and strain rates etc., as well as the correlation of these data to the mechanical properties of bulk metals.

Author Contributions: Conceptualization, H.P. and X.Z.; resources, H.P. and Y.H.; writing—original draft preparation, H.P.; writing—review & editing, X.Z. and Y.H.; supervision, X.Z.; funding acquisition, H.P. and X.Z. All authors have read and agreed to the published version of the manuscript.

Funding: This paper was funded by the National Natural Science Foundation of China (NSFC), grant number 51901091.

Institutional Review Board Statement: Not applicable.

Informed Consent Statement: Not applicable.

Data Availability Statement: No new data were created or analyzed in this study. Data sharing is not applicable to this article.

Acknowledgments: H. Pan acknowledges the support from China Scholarship Council, grant number 201908530046; The Basic Research Project of Yunnan Science and Technology Program, grant number 202001AU070081. X. Zhang acknowledges the support from the European Research Council (ERC) Under the European Union Horizon 2020 Research and Innovation Program, grant number 788567-M4D; Villum Fonden, grant number 00028216.

Conflicts of Interest: The authors declare that they have no conflicts of interest to this work. The founding sponsors had no role in the design of the study; in the collection, analyses, or interpretation of data; in the writing of the manuscript, and in the decision to publish the results.

References

1. Terentyev, D.; Malerba, L.; Bacon, D.J.; Osetsky, Y. The effect of temperature and strain rate on the interaction between an edge dislocation and an interstitial dislocation loop in α-iron. *J. Phys. Condens. Matter* **2007**, *19*, 456211. [CrossRef]
2. Chandra, S.; Samal, M.K.; Chavan, V.M.; Patel, R.J. Atomistic simulations of interaction of edge dislocation with twist grain boundaries in Al-effect of temperature and boundary misorientation. *Mater. Sci. Eng. A* **2015**, *646*, 25–32. [CrossRef]
3. Shim, J.H.; Kim, D.I.; Jung, W.S.; Cho, Y.W.; Hong, K.T.; Wirth, B.D. Atomistic study of temperature dependence of interaction between screw dislocation and nanosized bcc Cu precipitate in bcc Fe. *J. Appl. Phys.* **2008**, *104*, 1–5. [CrossRef]
4. Miracle, D.B.; Senkov, O.N. A critical review of high entropy alloys and related concepts. *Acta Mater.* **2017**, *122*, 448–511. [CrossRef]
5. Ovid'ko, I.A.; Valiev, R.Z.; Zhu, Y.T. Review on superior strength and enhanced ductility of metallic nanomaterials. *Prog. Mater. Sci.* **2018**, *94*, 462–540. [CrossRef]
6. Xu, D.S.; Chang, J.P.; Li, J.; Yang, R.; Li, D.; Yip, S. Dislocation slip or deformation twinning: Confining pressure makes a difference. *Mater. Sci. Eng. A* **2004**, *387–389*, 840–844. [CrossRef]
7. Koike, J.; Fujiyama, N.; Ando, D.; Sutou, Y. Roles of deformation twinning and dislocation slip in the fatigue failure mechanism of AZ31 Mg alloys. *Scr. Mater.* **2010**, *63*, 747–750. [CrossRef]
8. Ovid'ko, I.A.; Skiba, N.V. Crossover from dislocation slip to eformation twinning in nanostructured and oarse-grained metals. *Rev. Adv. Mater. Sci.* **2017**, *50*, 31–36.
9. Vasilev, E.; Merson, D.; Vinogradov, A. Kinetics of twinning and dislocation slip during cyclic deformation of ZK30 magnesium alloy. *KnE Eng.* **2018**, *3*, 156. [CrossRef]
10. Zhang, X.; Godfrey, A.; Hansen, N.; Huang, X.; Liu, W.; Liu, Q. Evolution of cementite morphology in pearlitic steel wire during wet wire drawing. *Mater. Charact.* **2010**, *61*, 65–72. [CrossRef]
11. Coujou, A.; Benyoucef, M.; Clement, N.; Priester, L. Interactions between dislocations and γ/γ′ interfaces in superalloys. *Interface Sci.* **1997**, *4*, 317–327. [CrossRef]
12. Li, N.; Wang, J.; Zhang, X.; Misra, A. In-situ TEM study of dislocation-twin boundaries interaction in nanotwinned Cu films. *Jom* **2011**, *63*, 62–66. [CrossRef]
13. Morrow, B.M.; Cerreta, E.K.; McCabe, R.J.; Tomé, C.N. Toward understanding twin-twin interactions in hcp metals: Utilizing multiscale techniques to characterize deformation mechanisms in magnesium. *Mater. Sci. Eng. A* **2014**, *613*, 365–371. [CrossRef]

14. Kacher, J.; Minor, A.M. Twin boundary interactions with grain boundaries investigated in pure rhenium. *Acta Mater.* **2014**, *81*, 1–8. [CrossRef]

15. Imrich, P.J.; Kirchlechner, C.; Motz, C.; Dehm, G. Differences in deformation behavior of bicrystalline Cu micropillars containing a twin boundary or a large-angle grain boundary. *Acta Mater.* **2014**, *73*, 240–250. [CrossRef]

16. Kondo, S.; Mitsuma, T.; Shibata, N.; Ikuhara, Y. Direct observation of individual dislocation interaction processes with grain boundaries. *Sci. Adv.* **2016**, *2*, 1–8. [CrossRef] [PubMed]

17. Malyar, N.V.; Dehm, G.; Kirchlechner, C. Strain rate dependence of the slip transfer through a penetrable high angle grain boundary in copper. *Scr. Mater.* **2017**, *138*, 88–91. [CrossRef]

18. Kini, M.K.; Dehm, G.; Kirchlechner, C. Size dependent strength, slip transfer and slip compatibility in nanotwinned silver. *Acta Mater.* **2020**, *184*, 120–131. [CrossRef]

19. Wang, H.; Boehlert, C.J.; Wang, Q.D.; Yin, D.D.; Ding, W.J. In-situ analysis of the slip activity during tensile deformation of cast and extruded Mg-10Gd-3Y-0.5Zr (wt.%) at 250°C. *Mater. Charact.* **2016**, *116*, 8–17. [CrossRef]

20. Weaver, J.S.; Li, N.; Mara, N.A.; Jones, D.R.; Cho, H.; Bronkhorst, C.A.; Fensin, S.J.; Gray, G.T. Slip transmission of high angle grain boundaries in body-centered cubic metals: Micropillar compression of pure Ta single and bi-crystals. *Acta Mater.* **2018**, *156*, 356–368. [CrossRef]

21. Teichmann, K.; Marioara, C.D.; Andersen, S.J.; Marthinsen, K. TEM study of β' precipitate interaction mechanisms with dislocations and β' interfaces with the aluminium matrix in Al–Mg–Si alloys. *Mater. Charact.* **2013**, *75*, 1–7. [CrossRef]

22. Smith, G.D. Dynamic studies of the tensile deformation and fracture of pearlite. *Acta Metall.* **1978**, *26*, 1405–1422.

23. Kalsar, R.; Kumar, L.; Suwas, S. Grain boundary engineering of medium Mn TWIP steels: A novel method to enhance the mechanical properties. *ISIJ Int.* **2018**, *58*, 1324–1331. [CrossRef]

24. Sinha, S.; Kim, D.I.; Fleury, E.; Suwas, S. Effect of grain boundary engineering on the microstructure and mechanical properties of copper containing austenitic stainless steel. *Mater. Sci. Eng. A* **2015**, *626*, 175–185. [CrossRef]

25. Watanabe, T.; Tsurekawa, S. Control of brittleness and development of desirable properties in polycrystalline systems by grain boundary engineering. *Acta Mater.* **1999**, *47*, 4171–4185. [CrossRef]

26. Howe, J.M. Structure, composition and energy of solid-solid interfaces. In *Physical Metallurgy: Fifth Edition*; Elsevier: Amsterdam, The Netherlands, 2014; pp. 1317–1451.

27. Zhu, Y.T.; Liao, X.Z.; Wu, X.L. Deformation twinning in nanocrystalline materials. *Prog. Mater. Sci.* **2012**, *57*, 1–62. [CrossRef]

28. Louchet, F.; Louchet, F.; Doisneau-Cottignies, B.; Bréchet, Y. Specific dislocation multiplication mechanisms and mechanical properties in nanoscaled multilayers: The example of pearlite. *Philos. Mag. A Phys. Condens. Matter, Struct. Defects Mech. Prop.* **2000**, *80*, 1605–1619.

29. Kacher, J.; Eftink, B.P.; Cui, B.; Robertson, I.M. Dislocation interactions with grain boundaries. *Curr. Opin. Solid State Mater. Sci.* **2014**, *18*, 227–243. [CrossRef]

30. Guziewski, M.; Coleman, S.P.; Weinberger, C.R. Atomistic investigation into interfacial effects on the plastic response and deformation mechanisms of the pearlitic microstructure. *Acta Mater.* **2019**, *180*, 287–300. [CrossRef]

31. Liebig, J.P.; Krauß, S.; Göken, M.; Merle, B. Influence of stacking fault energy and dislocation character on slip transfer at coherent twin boundaries studied by micropillar compression. *Acta Mater.* **2018**, *154*, 261–272. [CrossRef]

32. Fernández, A.; Jérusalem, A.; Gutiérrez-Urrutia, I.; Pérez-Prado, M.T. Three-dimensional investigation of grain boundary-twin interactions in a Mg AZ31 alloy by electron backscatter diffraction and continuum modeling. *Acta Mater.* **2013**, *61*, 7679–7692. [CrossRef]

33. Priester, L. "Dislocation-interface" interaction-stress accommodation processes at interfaces. *Mater. Sci. Eng. A* **2001**, *309–310*, 430–439. [CrossRef]

34. Lee, T.C.; Robertson, I.M.; Birnbaum, H.K. TEM in situ deformation study of the interaction of lattice dislocations with grain boundaries in metals. *Philos. Mag. A Phys. Condens. Matter Struct. Defects Mech. Prop.* **1990**, *62*, 131–153.

35. Bieler, T.R.; Eisenlohr, P.; Zhang, C.; Phukan, H.J.; Crimp, M.A. Grain boundaries and interfaces in slip transfer. *Curr. Opin. Solid State Mater. Sci.* **2014**, *18*, 212–226. [CrossRef]

36. Kacher, J.; Sabisch, J.E.; Minor, A.M. Statistical analysis of twin/grain boundary interactions in pure rhenium. *Acta Mater.* **2019**, *173*, 44–51. [CrossRef]

37. Sangid, M.D.; Ezaz, T.; Sehitoglu, H.; Robertson, I.M. Energy of slip transmission and nucleation at grain boundaries. *Acta Mater.* **2011**, *59*, 283–296. [CrossRef]

38. Sun, Q.; Ostapovets, A.; Zhang, X.; Tan, L.; Liu, Q. Investigation of twin-twin interaction in deformed magnesium alloy. *Philos. Mag.* **2018**, *98*, 741–751. [CrossRef]

39. Yang, G.; Ma, S.Y.; Du, K.; Xu, D.S.; Chen, S.; Qi, Y.; Ye, H.Q. Interactions between dislocations and twins in deformed titanium aluminide crystals. *J. Mater. Sci. Technol.* **2019**, *35*, 402–408. [CrossRef]

40. Kacher, J.; Robertson, I.M. In situ and tomographic analysis of dislocation/grain boundary interactions in titanium. *Philos. Mag.* **2014**, *94*, 814–829. [CrossRef]

41. Meyers, M.A.; Mishra, A.; Benson, D.J. Mechanical properties of nanocrystalline materials. *Prog. Mater. Sci.* **2006**, *51*, 427–556. [CrossRef]

42. Mompiou, F.; Legros, M.; Boé, A.; Coulombier, M.; Raskin, J.P.; Pardoen, T. Inter- and intragranular plasticity mechanisms in ultrafine-grained Al thin films: An in situ TEM study. *Acta Mater.* **2013**, *61*, 205–216. [CrossRef]

43. Mompiou, F.; Caillard, D.; Legros, M.; Mughrabi, H. In situ TEM observations of reverse dislocation motion upon unloading in tensile-deformed UFG aluminium. *Acta Mater.* **2012**, *60*, 3402–3414. [CrossRef]
44. Wu, X.L.; Ma, E. Accommodation of large plastic strains and defect accumulation in nanocrystalline Ni grains. *J. Mater. Res.* **2007**, *22*, 2241–2253. [CrossRef]
45. Wang, B.; Idrissi, H.; Galceran, M.; Colla, M.S.; Turner, S.; Hui, S.; Raskin, J.P.; Pardoen, T.; Godet, S.; Schryvers, D. Advanced TEM investigation of the plasticity mechanisms in nanocrystalline freestanding palladium films with nanoscale twins. *Int. J. Plast.* **2012**, *37*, 140–156. [CrossRef]
46. Zhang, X.Y.; Wu, X.L.; Zhu, A.W. Growth of deformation twins in roomerature rolled nanocrystalline nickel. *Appl. Phys. Lett.* **2009**, *94*, 121907.
47. Savage, M.F.; Tatalovich, J.; Mills, M.J. Anisotropy in the room-temperature deformation of α-β colonies in titanium alloys: Role of the α-β interface. *Philos. Mag.* **2004**, *84*, 1127–1154. [CrossRef]
48. Seal, J.R.; Crimp, M.A.; Bieler, T.R.; Boehlert, C.J. Analysis of slip transfer and deformation behavior across the α/β interface in Ti-5Al-2.5Sn (wt.%) with an equiaxed microstructure. *Mater. Sci. Eng. A* **2012**, *552*, 61–68. [CrossRef]
49. Savage, M.F.; Tatalovich, J.; Zupan, M.; Hemker, K.J.; Mills, M.J. Deformation mechanisms and microtensile behavior of single colony Ti-6242Si. *Mater. Sci. Eng. A* **2001**, *319–321*, 398–403. [CrossRef]
50. Shen, Z.; Wagoner, R.H.; Clark, W.A.T. Dislocation and grain boundary interactions in metals. *Acta Metall.* **1988**, *36*, 3231–3542. [CrossRef]
51. Malyar, N.V.; Grabowski, B.; Dehm, G.; Kirchlechner, C. Dislocation slip transmission through a coherent Σ3{111} copper twin boundary: Strain rate sensitivity, activation volume and strength distribution function. *Acta Mater.* **2018**, *161*, 412–419. [CrossRef]
52. Merkle, K.L. High resolution electron microscopy of grain boundaries. *Interface Sci.* **1995**, *2*, 311–345. [CrossRef]
53. Du, H.; Jia, C.L.; Houben, L.; Metlenko, V.; De Souza, R.A.; Waser, R.; Mayer, J. Atomic structure and chemistry of dislocation cores at low-angle tilt grain boundary in SrTiO3 bicrystals. *Acta Mater.* **2015**, *89*, 344–351. [CrossRef]
54. Nohara, Y.; Tochigi, E.; Shibata, N.; Yamamoto, T.; Ikuhara, Y. Dislocation structures and strain fields in [111] low-angle tilt grain boundaries in zirconia bicrystals. *J. Electron Microsc. (Tokyo)* **2010**, *59*, 117–121. [CrossRef] [PubMed]
55. Ikuhara, Y.; Nishimura, H.; Nakamura, A.; Matsunaga, K.; Yamamoto, T.; Peter, K.; Lagerlöf, D. Dislocation structures of low-angle and near-Σ3 Grain boundaries in alumina bicrystals. *J. Am. Ceram. Soc.* **2003**, *86*, 595–602. [CrossRef]
56. Oktyabrsky, S. Dislocation structure of low-angle grain boundaries in YBa2Cu3O7-δ/MgO films. *J. Mater. Res.* **1999**, *14*, 2764–2772. [CrossRef]
57. Yazyev, O.V.; Louie, S.G. Topological defects in graphene: Dislocations and grain boundaries. *Phys. Rev. B Condens. Matter Mater. Phys.* **2010**, *81*. [CrossRef]
58. Smallman, R.E.; Ngan, A.H.W. Surfaces, grain boundaries and interfaces. In *Modern Physical Metallurgy*; Elsevier: Amsterdam, The Netherlands, 2014; pp. 415–442.
59. Hasson, G.; Boos, J.Y.; Herbeuval, I.; Biscondi, M.; Goux, C. Theoretical and experimental determinations of grain boundary structures and energies: Correlation with various experimental results. *Surf. Sci.* **1972**, *31*, 115–137. [CrossRef]
60. Wolf, D.; Phillpot, S. Role of the densest lattice planes in the stability of crystalline interfaces: A computer simulation study. *Mater. Sci. Eng. A* **1989**, *107*, 3–14. [CrossRef]
61. Balluffi, R.W. Grain boundary diffusion mechanisms in metals. *Metall. Trans. A, Phys. Metall. Mater. Sci.* **1982**, *13A*, 2069–2095. [CrossRef]
62. Gleiter, H. The structure and properties of high-angle grain boundaries in metals. *Phys. Status Solid B* **1971**, *45*, 9–38. [CrossRef]
63. Read, W.T.; Shockley, W. Dislocation models of crystal grain boundaries. *Phys. Rev.* **1950**, *78*, 275–289. [CrossRef]
64. Gleiter, H. Recent developments in the understanding of the structure and properties of grain boundaries in metals. *Phys. Status Solidi* **1971**, *45*, 9–38. [CrossRef]
65. Balluffi, R.W.; Brokman, A.; King, A.H. CSL/DSC lattice model for general crystalcrystal boundaries and their line defects. *Acta Metall.* **1982**, *30*, 1453–1470. [CrossRef]
66. Grimmer, H.; Bollmann, W.; Warrington, D.H. Coincidence-site lattices and complete pattern-shift in cubic crystals. *Acta Crystallogr. Sect. A* **1974**, *30*, 197–207. [CrossRef]
67. Merkle, K.L.; Csencsits, R.; Rynes, K.L.; Withrow, J.R.; Stadelmann, P.A. The effect of three-fold astigmatism on measurements of grain boundary volume expansion by high-resolution transmission electron microscopy. *J. Microsc.* **1998**, *190*, 204–213. [CrossRef]
68. Buckett, M.I.; Merkle, K.L. Determination of grain boundary volume expansion by HREM. *Ultramicroscopy* **1994**, *56*, 71–78. [CrossRef]
69. Pénisson, J.M.; Vystavel, T. Measurement of the volume expansion of a grain boundary by the phase method. *Ultramicroscopy* **2002**, *90*, 163–170. [CrossRef]
70. Wolf, D. Correlation between structure, energy, and ideal cleavage fracture for symmetrical grain boundaries in fcc metals. *J. Mater. Res.* **1990**, *5*, 1708–1730. [CrossRef]
71. Hansen, N.; Mehl, R.F. New discoveries in deformed metals. *Metall. Mater. Trans. A* **2001**, *32*, 2917–2935. [CrossRef]
72. Kuhlmann-Wilsdorf, D.; Hansen, N. Geometrically necessary, incidental and subgrain boundaries. *Scr. Metall. Mater.* **1991**, *25*, 1557–1562. [CrossRef]
73. Hughes, D.A.; Hansen, N. Microstructure and strength of nickel at large strains. *Acta Mater.* **2000**, *48*, 2985–3004. [CrossRef]

74. Hughes, D.A.; Hansen, N.; Bammann, D.J. Geometrically necessary boundaries, incidental dislocation boundaries and geometrically necessary dislocations. *Scr. Mater.* **2003**, *48*, 147–153. [CrossRef]
75. Huang, X.; Winther, G. Dislocation structures. Part I. Grain orientation dependence. *Philos. Mag.* **2007**, *87*, 5189–5214. [CrossRef]
76. Winther, G.; Huang, X. Dislocation structures. Part II. Slip system dependence. *Philos. Mag.* **2007**, *87*, 5215–5235. [CrossRef]
77. Hong, C.; Huang, X.; Winther, G. Experimental characterization of dislocations in deformation induced planar boundaries of rolled aluminium. *Proc. Risø Int. Symp. Mater. Sci.* **2012**, *33*, 239–248.
78. Hong, C.; Huang, X.; Winther, G. Dislocation content of geometrically necessary boundaries aligned with slip planes in rolled aluminium. *Philos. Mag.* **2013**, *93*, 3118–3141. [CrossRef]
79. Liu, Q.; Hansen, N. Geometrically necessary boundaries and incidental dislocation boundaries formed during cold deformation. *Scr. Metall. Mater.* **1995**, *32*, 1289–1295. [CrossRef]
80. Hughes, D.A.; Hansen, N. Microstructure parameters: Quantitative and theoretical analysis. *ASM Int.* **2004**, *9*, 192–206.
81. Smallman, R.E.; Ngan, A.H.W. Plastic deformation and dislocation behaviour. In *Modern Physical Metallurgy*; Elsevier: Amsterdam, The Netherlands, 2014; pp. 357–414.
82. Cynn, H.; Yoo, C.S.; Baer, B.; Iota-Herbei, V.; McMahan, A.K.; Nicol, M.; Carlson, S. Martensitic fcc-to-hcp transformation observed in xenon at high pressure. *Phys. Rev. Lett.* **2001**, *86*, 4552–4555. [CrossRef] [PubMed]
83. Errandonea, D.; Schwager, B.; Boehler, R.; Ross, M. Phase behavior of krypton and xenon to 50 GPa. *Phys. Rev. B—Condens. Matter Mater. Phys.* **2002**, *65*, 1–6. [CrossRef]
84. Howe, J.M. Comparison of the atomic structure, composition, kinetics and mechanisms of interfacial motion in martensitic, bainitic, massive and precipitation face-centered cubic-hexagonal close-packed phase transformations. *Mater. Sci. Eng. A* **2006**, *438–440*, 35–42. [CrossRef]
85. Howe, J.M.; Mahon, G.J. High-resolution transmission electron microscopy of precipitate plate growth by diffusional and displacive transformation mechanisms. *Ultramicroscopy* **1989**, *30*, 132–142. [CrossRef]
86. Howe, J.M.; Aaronson, H.I.; Gronsky, R. Atomic mechanisms of precipitate plate growth in the AlAg system-II. High-resolution transmission electron microscopy. *Acta Metall.* **1985**, *33*, 649–658. [CrossRef]
87. Li, P.; Howe, J.M.; Reynolds, W.T. Atomic structure of a {1 1 1} incoherent interface in Zr-N alloy. *Acta Mater.* **2004**, *52*, 239–248. [CrossRef]
88. Hirth, J.P.; Pond, R.C. Steps, dislocations and disconnections as interface defects relating to structure and phase transformations. *Acta Mater.* **1996**, *44*, 4749–4763. [CrossRef]
89. Howe, J.M.; Pond, R.C.; Hirth, J.P. The role of disconnections in phase transformations. *Prog. Mater. Sci.* **2009**, *54*, 792–838. [CrossRef]
90. Das, S.; Howe, J.M.; Perepezko, J.H. A high-resolution transmission electron microscopy study of interfaces between the γ, B2, and α2 phases in a Ti-Al-Mo alloy. *Metall. Mater. Trans. A Phys. Metall. Mater. Sci.* **1996**, *27*, 1623–1634. [CrossRef]
91. Wang, L.; Yang, Y.; Eisenlohr, P.; Bieler, T.R.; Crimp, M.A.; Mason, D.E. Twin nucleation by slip transfer across grain boundaries in commercial purity titanium. *Metall. Mater. Trans. A Phys. Metall. Mater. Sci.* **2010**, *41*, 421–430. [CrossRef]
92. Smith, D.A. Interaction of dislocation with grain boundaries. *J. Phusique* **1982**, *43*, 225–237. [CrossRef]
93. Kacher, J.; Robertson, I.M. Quasi-four-dimensional analysis of dislocation interactions with grain boundaries in 304 stainless steel. *Acta Mater.* **2012**, *60*, 6657–6672. [CrossRef]
94. Lee, T.C.; Robertson, I.M.; Birnbaum, H.K. An in situ transmission electron microscope deformation study of the slip transfer mechanisms in metals. *Metall. Trans. A* **1990**, *21*, 2437–2447. [CrossRef]
95. Bieler, T.R.; Wang, L.; Beaudoin, A.J.; Kenesei, P.; Lienert, U. In situ characterization of twin nucleation in pure Ti using 3D-XRD. *Metall. Mater. Trans. A Phys. Metall. Mater. Sci.* **2014**, *45*, 109–122. [CrossRef]
96. Bachurin, D.V.; Weygand, D.; Gumbsch, P. Dislocation-grain boundary interaction in <1 1 1> textured thin metal films. *Acta Mater.* **2010**, *58*, 5232–5241. [CrossRef]
97. Clark, W.A.T.; Wagoner, R.H.; Shen, Z.Y.; Lee, T.C.; Robertson, I.M.; Birnbaum, H.K. On the criteria for slip transmission across interfaces in polycrystals. *Scr. Metall. Mater.* **1992**, *26*, 203–206. [CrossRef]
98. Livingston, J.D.; Chalmers, B. Multiple slip in bicrystal deformation. *Acta Metall.* **1957**, *5*, 322–327. [CrossRef]
99. Hansen, N. Hall-petch relation and boundary strengthening. *Scr. Mater.* **2004**, *51*, 801–806. [CrossRef]
100. Zhou, X.; Feng, Z.; Zhu, L.; Xu, J.; Miyagi, L.; Dong, H.; Sheng, H.; Wang, Y.; Li, Q.; Ma, Y.; et al. High-pressure strengthening in ultrafine-grained metals. *Nature* **2020**, *579*, 67–72. [CrossRef] [PubMed]
101. Chassagne, M.; Legros, M.; Rodney, D. Atomic-scale simulation of screw dislocation/coherent twin boundary interaction in Al, Au, Cu and Ni. *Acta Mater.* **2011**, *59*, 1456–1463. [CrossRef]
102. Zhu, T.; Li, J.; Samanta, A.; Kim, H.G.; Suresh, S. Interfacial plasticity governs strain rate sensitivity and ductility in nanostructured metals. *Proc. Natl. Acad. Sci. USA* **2007**, *104*, 3031–3036. [CrossRef] [PubMed]
103. Jin, Z.H.; Gumbsch, P.; Ma, E.; Albe, K.; Lu, K.; Hahn, H.; Gleiter, H. The interaction mechanism of screw dislocations with coherent twin boundaries in different face-centred cubic metals. *Scr. Mater.* **2006**, *54*, 1163–1168. [CrossRef]
104. Jin, Z.H.; Gumbsch, P.; Albe, K.; Ma, E.; Lu, K.; Gleiter, H.; Hahn, H. Interactions between non-screw lattice dislocations and coherent twin boundaries in face-centered cubic metals. *Acta Mater.* **2008**, *56*, 1126–1135. [CrossRef]
105. Zhu, T.; Gao, H. Plastic deformation mechanism in nanotwinned metals: An insight from molecular dynamics and mechanistic modeling. *Scr. Mater.* **2012**, *66*, 843–848. [CrossRef]

106. Li, N.; Wang, J.; Misra, A.; Zhang, X.; Huang, J.Y.; Hirth, J.P. Twinning dislocation multiplication at a coherent twin boundary. *Acta Mater.* **2011**, *59*, 5989–5996. [CrossRef]

107. Jeon, J.B.; Dehm, G. Formation of dislocation networks in a coherent Cu Σ3(1 1 1) twin boundary. *Scr. Mater.* **2015**, *102*, 71–74. [CrossRef]

108. Kacher, J.; Kirchlechner, C.; Michler, J.; Polatidis, E.; Schwaiger, R.; Van Swygenhoven, H.; Taheri, M.; Legros, M. Impact of in situ nanomechanics on physical metallurgy. *MRS Bull.* **2019**, *44*, 465–470. [CrossRef]

109. Wu, Z.X.; Zhang, Y.W.; Srolovitz, D.J. Dislocation-twin interaction mechanisms for ultrahigh strength and ductility in nanotwinned metals. *Acta Mater.* **2009**, *57*, 4508–4518. [CrossRef]

110. Lee, J.H.; Holland, T.B.; Mukherjee, A.K.; Zhang, X.; Wang, H. Direct observation of Lomer-Cottrell Locks during strain hardening in nanocrystalline nickel by in situ TEM. *Sci. Rep.* **2013**, *3*, 1–6. [CrossRef]

111. Idrissi, H.; Amin-Ahmadi, B.; Wang, B.; Schryvers, D. Review on TEM analysis of growth twins in nanocrystalline palladium thin films: Toward better understanding of twin-related mechanisms in high stacking fault energy metals. *Phys. Status Solidi Basic Res.* **2014**, *251*, 1111–1124. [CrossRef]

112. Malyar, N.V.; Micha, J.S.; Dehm, G.; Kirchlechner, C. Dislocation-twin boundary interaction in small scale Cu bi-crystals loaded in different crystallographic directions. *Acta Mater.* **2017**, *129*, 91–97. [CrossRef]

113. Lu, L.; Chen, X.; Huang, X.; Lu, K. Revealing the maximum strength in nanotwinned copper. *Science* **2009**, *323*, 607–610. [CrossRef] [PubMed]

114. Christiansen, E.; Marioara, C.D.; Holmedal, B.; Hopperstad, O.S.; Holmestad, R. Nano-scale characterisation of sheared β″ precipitates in a deformed Al-Mg-Si alloy. *Sci. Rep.* **2019**, *9*, 1–11. [CrossRef]

115. Louchet, F.; Pelissier, J.; Caillard, D.; Peyrade, J.P.; Levade, C.; Vanderschaeve, G. In situ TEM study of dislocation mobility in semiconducting materials. *Microsc. Microanal. Microstruct.* **1993**, *4*, 199–210. [CrossRef]

116. Wolfenstine, J.; Allen, J.L.; Sakamoto, J.; Siegel, D.J.; Choe, H. Mechanical behavior of Li-ion-conducting crystalline oxide-based solid electrolytes: A brief review. *Ionics (Kiel)* **2018**, *24*, 1271–1276. [CrossRef]

117. Bhattacharyya, J.J.; Wang, F.; Stanford, N.; Agnew, S.R. Slip mode dependency of dislocation shearing and looping of precipitates in Mg alloy WE43. *Acta Mater.* **2018**, *146*, 55–62. [CrossRef]

118. Suri, S.; Viswanathan, G.B.; Neeraj, T.; Hou, D.H.; Mills, M.J. Room temperature deformation and mechanisms of slip transmission in oriented single-colony crystals of an α/β titanium alloy. *Acta Mater.* **1999**, *47*, 1019–1034. [CrossRef]

119. Zhang, X.; Godfrey, A.; Huang, X.; Hansen, N.; Liu, Q. Microstructure and strengthening mechanisms in cold-drawn pearlitic steel wire. *Acta Mater.* **2011**, *59*, 3422–3430. [CrossRef]

120. Zhang, X.; Godfrey, A.; Hansen, N.; Huang, X. Hierarchical structures in cold-drawn pearlitic steel wire. *Acta Mater.* **2013**, *61*, 4898–4909. [CrossRef]

121. Zhang, X.; Hansen, N.; Godfrey, A.; Huang, X. Dislocation-based plasticity and strengthening mechanisms in sub-20 nm lamellar structures in pearlitic steel wire. *Acta Mater.* **2016**, *114*, 176–183. [CrossRef]

122. Liang, L.W.; Wang, Y.J.; Chen, Y.; Wang, H.Y.; Dai, L.H. Dislocation nucleation and evolution at the ferrite-cementite interface under cyclic loadings. *Acta Mater.* **2020**, *186*, 267–277. [CrossRef]

123. Zhao, Y.; Tan, Y.; Ji, X.; Xiang, Z.; He, Y.; Xiang, S. In situ study of cementite deformation and its fracture mechanism in pearlitic steels. *Mater. Sci. Eng. A* **2018**, *731*, 93–101. [CrossRef]

124. Zhang, X.; Godfrey, A.; Huang, X.; Hansen, N.; Liu, W.; Liu, Q. Characterization of the microstructure in drawn pearlitic steel wires. In Proceedings of the 30th Risø International Symposium on Materials Science, Roskilde, Denmark, 7–11 September 2009; pp. 409–416.

125. Zhang, X.; Hansen, N.; Godfrey, A.; Huang, X. Structure and strength of sub-100 nm lamellar structures in cold-drawn pearlitic steel wire. *Mater. Sci. Technol. (UK)* **2018**, *34*, 794–808. [CrossRef]

126. Zhang, X. Exploring high strength metallic materials: A lesson from pearlitic steel wire. *IOP Conf. Ser. Mater. Sci. Eng.* **2019**, *580*, 012058. [CrossRef]

127. Zhang, X.D.; Godfrey, A.; Liu, W.; Liu, Q. Study on dislocation slips in ferrite and deformation of cementite in cold drawn pearlitic steel wires from medium to high strain. *Mater. Sci. Technol.* **2011**, *27*, 562–567. [CrossRef]

128. Gershteyn, G.; Nürnberger, F.; Cianciosi, F.; Shevchenko, N.; Schaper, M.; Bach, F.W. A study of structure evolution in pearlitic steel wire at increasing plastic deformation. *Steel Res. Int.* **2011**, *82*, 1368–1374. [CrossRef]

129. Jia, N.; Shen, Y.F.; Liang, J.W.; Feng, X.W.; Wang, H.B.; Misra, R.D.K. Nanoscale spheroidized cementite induced ultrahigh strength-ductility combination in innovatively processed ultrafine-grained low alloy medium-carbon steel. *Sci. Rep.* **2017**, *7*, 1–9. [CrossRef]

130. Linz, M.; Rodríguez Ripoll, M.; Pauly, C.; Bernardi, J.; Steiger-Thirsfeld, A.; Franek, F.; Mücklich, F.; Gachot, C. Heterogeneous strain distribution and saturation of geometrically necessary dislocations in a ferritic-pearlitic steel during lubricated sliding. *Adv. Eng. Mater.* **2018**, *20*, 1–11. [CrossRef]

131. Shimokawa, T.; Oguro, T.; Tanaka, M.; Higashida, K.; Ohashi, T. A multiscale approach for the deformation mechanism in pearlite microstructure: Atomistic study of the role of the heterointerface on ductility. *Mater. Sci. Eng. A* **2014**, *598*, 68–76. [CrossRef]

132. Karkina, L.E.; Karkin, I.N.; Kabanova, I.G.; Kuznetsov, A.R. Crystallographic analysis of slip transfer mechanisms across the ferrite/cementite interface in carbon steels with fine lamellar structure. *J. Appl. Crystallogr.* **2015**, *48*, 97–106. [CrossRef]

133. Kar'kina, L.E.; Kabanova, I.G.; Kar'kin, I.N. Strain transfer across the ferrite/cementite interface in carbon steels with coarse lamellar pearlite. *Phys. Met. Metallogr.* **2018**, *119*, 1114–1119. [CrossRef]

134. Shimokawa, T.; Niiyama, T.; Okabe, M.; Sawakoshi, J. Interfacial-dislocation-controlled deformation and fracture in nanolayered composites: Toward higher ductility of drawn pearlite. *Acta Mater.* **2019**, *164*, 602–617. [CrossRef]

135. Zhang, X.; Godfrey, A.; Winther, G.; Hansen, N.; Huang, X. Plastic deformation of submicron-sized crystals studied by in-situ Kikuchi diffraction and dislocation imaging. *Mater. Charact.* **2012**, *70*, 21–27. [CrossRef]

136. Nikas, D.; Zhang, X.; Ahlström, J. Evaluation of local strength via microstructural quantification in a pearlitic rail steel deformed by simultaneous compression and torsion. *Mater. Sci. Eng. A* **2018**, *737*, 341–347. [CrossRef]

137. Feng, H.; Fang, F.; Zhou, X.; Zhang, X.; Xie, Z.; Jiang, J. Heavily cold drawn iron wires: Role of nano-lamellae in enhancing the tensile strength. *Mater. Sci. Eng. A* **2020**, *796*, 140017. [CrossRef]

138. Chen, S.L.; Hu, J.; Zhang, X.D.; Dong, H.; Cao, W.Q. High Ductility and Toughness of a Micro-duplex Medium-Mn Steel in Large Temperature Range from −196 °C to 200 °C. *J. Iron Steel Res. Int.* **2015**, *22*, 1126–1130. [CrossRef]

139. Wang, B.; Liu, B.; Zhang, X.; Gu, J. Enhancing heavy load wear resistance of AISI 4140 steel through the formation of a severely deformed compound-free nitrided surface layer. *Surf. Coatings Technol.* **2018**, *356*, 89–95. [CrossRef]

140. Zhang, X.; Liu, W.; Godfrey, A.; Liu, Q. The effect of long-time austenization on the wear resistance and thermal fatigue properties of a high-speed steel roll. *Metall. Mater. Trans. A Phys. Metall. Mater. Sci.* **2009**, *40*, 2171–2177. [CrossRef]

141. Dhar, S.; Ahlström, J.; Zhang, X.; Danielsen, H.K.; Juul Jensen, D. Multi-axial Fatigue of Head-Hardened Pearlitic and Austenitic Manganese Railway Steels: A Comparative Study. *Metall. Mater. Trans. A Phys. Metall. Mater. Sci.* **2020**, *51*, 5639–5652. [CrossRef]

142. Cao, Z.; Shi, Z.; Liang, B.; Zhang, X.; Cao, W.; Weng, Y. Melting route effects on the rotatory bending fatigue and rolling contact fatigue properties of high carbon bearing steel SAE52100. *Int. J. Fatigue* **2020**, *140*, 105854. [CrossRef]

143. Cao, Z.; Liu, T.; Yu, F.; Cao, W.; Zhang, X.; Weng, Y. Carburization induced extra-long rolling contact fatigue life of high carbon bearing steel. *Int. J. Fatigue* **2020**, *131*, 105351. [CrossRef]

144. Mallipeddi, D.; Norell, M.; Naidu, V.M.S.; Zhang, X.; Näslund, M.; Nyborg, L. Micropitting and microstructural evolution during gear testing -from initial cycles to failure. *Tribol. Int.* **2021**, *156*, 106820. [CrossRef]

145. Yang, R.; Wu, G.; Zhang, X.; Fu, W.; Hansen, N.; Huang, X. Gradient microstructure in a gear steel produced by pressurized gas nitriding. *Materials* **2019**, *12*, 3797. [CrossRef] [PubMed]

146. Navickaitė, K.; Mocerino, A.; Cattani, L.; Bozzoli, F.; Bahl, C.; Liltrop, K.; Zhang, X.; Engelbrecht, K. Enhanced heat transfer in tubes based on vascular heat exchangers in fish: Experimental investigation. *Int. J. Heat Mass Transf.* **2019**, *137*, 192–203. [CrossRef]

147. Fan, G.H.; Wang, Q.W.; Du, Y.; Geng, L.; Hu, W.; Zhang, X.; Huang, Y.D. Producing laminated NiAl with bimodal distribution of grain size by solid-liquid reaction treatment. *Mater. Sci. Eng. A* **2014**, *590*, 318–322. [CrossRef]

148. Wang, S.H.; Liu, Z.Y.; Wang, G.D.; Liu, J.L.; Liang, G.F.; Li, Q.L. Effects of twin-dislocation and twin-twin interactions on the strain hardening behavior of TWIP steels. *J. Iron Steel Res. Int.* **2010**, *17*, 70–74. [CrossRef]

149. Guo, C.; Xin, R.; Wu, G.; Liu, F.; Hu, G.; Liu, Q. Observation of twin transmission process in Mg alloys by in situ EBSD. *Adv. Eng. Mater.* **2019**, *21*, 1–8. [CrossRef]

150. Culbertson, D.; Yu, Q.; Jiang, Y. In situ observation of cross-grain twin pair formation in pure magnesium. *Philos. Mag. Lett.* **2018**, *98*, 139–146. [CrossRef]

151. Shi, D.; Liu, T.; Wang, T.; Hou, D.; Zhao, S.; Hussain, S. {10-12} twins across twin boundaries traced by in situ EBSD. *J. Alloys Compd.* **2017**, *690*, 699–706. [CrossRef]

152. Zhang, Z.Z.; Sheng, H.; Wang, Z.; Gludovatz, B.; Zhang, Z.Z.; George, E.P.; Yu, Q.; Mao, S.X.; Ritchie, R.O. Dislocation mechanisms and 3D twin architectures generate exceptional strength-ductility-toughness combination in CrCoNi medium-entropy alloy. *Nat. Commun.* **2017**, *8*, 1–8. [CrossRef]

153. Yu, Q.; Wang, J.; Jiang, Y.; McCabe, R.J.; Tomé, C.N. Co-zone (1012) twin interaction in magnesium single crystal. *Mater. Res. Lett.* **2014**, *2*, 82–88. [CrossRef]

154. Yu, Q.; Wang, J.; Jiang, Y.; McCabe, R.J.; Li, N.; Tomé, C.N. Twin-twin interactions in magnesium. *Acta Mater.* **2014**, *77*, 28–42. [CrossRef]

155. Mokdad, F.; Chen, D.L.; Li, D.Y. Single and double twin nucleation, growth, and interaction in an extruded magnesium alloy. *Mater. Des.* **2017**, *119*, 376–396. [CrossRef]

156. Hall, E.O. The deformation and ageing of mild steel: II Characteristics of the Lüders deformation. *Proc. Phys. Soc. Sect. B* **1951**, *64*, 742–747. [CrossRef]

157. Li, J.C.M. Fetch relation and grain boundary source. *Trans. TMS-AIME* **1963**, *227*, 239–242.

158. Petch, N.J. The cleavage strength of polycrystals. *J. Iron Steel Inst.* **1953**, *174*, 25–28.

159. Nabarro, F.R.N.; Basinski, Z.S.; Holt, D.B. The plasticity of pure single crystals. *Adv. Phys.* **1964**, *13*, 193–323. [CrossRef]

160. Williams, R.O. Origin of strengthening on precipitation: Ordered particles. *Acta Metall.* **1957**, *5*, 241–244. [CrossRef]

161. Kang, F.; Li, Z.; Wang, J.T.; Cheng, P.; Wu, H.Y. The activation of c + a non-basal slip in Magnesium alloys. *J. Mater. Sci.* **2012**, *47*, 7854–7859. [CrossRef]

162. Luo, X.; Feng, Z.; Yu, T.; Luo, J.; Huang, T.; Wu, G.; Hansen, N.; Huang, X. Transitions in mechanical behavior and in deformation mechanisms enhance the strength and ductility of Mg-3Gd. *Acta Mater.* **2020**, *183*, 398–407. [CrossRef]

163. Zhang, X.; Hansen, N.; Gao, Y.; Huang, X. Hall-Petch and dislocation strengthening in graded nanostructured steel. *Acta Mater.* **2012**, *60*, 5933–5943. [CrossRef]

164. Zhang, X.; Nielsen, C.V.; Hansen, N.; Silva, C.M.A.; Martins, P.A.F. Local stress and strain in heterogeneously deformed aluminum: A comparison analysis by microhardness, electron microscopy and finite element modelling. *Int. J. Plast.* **2019**, *115*, 93–110. [CrossRef]

165. Visser, W.; Ghonem, H. Dynamic flow stress of shock loaded low carbon steel. *Mater. Sci. Eng. A* **2019**, *753*, 317–330. [CrossRef]

166. Conrad, H. Thermally activated deformation of metals. *Jom* **1964**, *16*, 582–588. [CrossRef]

167. Tang, X.; Wang, B.; Huo, Y.; Ma, W.; Zhou, J.; Ji, H.; Fu, X. Unified modeling of flow behavior and microstructure evolution in hot forming of a Ni-based superalloy. *Mater. Sci. Eng. A* **2016**, *662*, 54–64. [CrossRef]

168. Appel, F.; Lorenz, U.; Oehring, M.; Sparka, U.; Wagner, R. Thermally activated deformation mechanisms in micro-alloyed two-phase titanium amminide alloys. *Mater. Sci. Eng. A* **1997**, *233*, 1–14. [CrossRef]

169. Schulze, V.; Vohringer, O. Plastic Deformation: Constitutive Distribut. In *Encyclopedia of Materials: Science and Technology*; Pergamon: Oxford, UK, 2001; pp. 7050–7064.

170. Molodov, K.D.; Al-Samman, T.; Molodov, D.A. Profuse slip transmission across twin boundaries in magnesium. *Acta Mater.* **2017**, *124*, 397–409. [CrossRef]

171. Slama, M.B.H.; Maloufi, N.; Guyon, J.; Bahi, S.; Weiss, L.; Guitton, A. In situ macroscopic tensile testing in SEM and electron channeling contrast imaging: Pencil glide evidenced in a bulk β -Ti21S polycrystal. *Materials* **2019**, *12*, 2479. [CrossRef]

172. De Hosson, J.T.M.; Soer, W.A.; Minor, A.M.; Shan, Z.; Stach, E.A.; Syed Asif, S.A.; Warren, O.L. In situ TEM nanoindentation and dislocation-grain boundary interactions: A tribute to David Brandon. *J. Mater. Sci.* **2006**, *41*, 7704–7719. [CrossRef]

173. Lohmiller, J.; Baumbusch, R.; Kraft, O.; Gruber, P.A. Differentiation of deformation modes in nanocrystalline Pd films inferred from peak asymmetry evolution using in situ X-ray diffraction. *Phys. Rev. Lett.* **2013**, *110*, 1–5. [CrossRef]

174. Lohmiller, J.; Baumbusch, R.; Kerber, M.B.; Castrup, A.; Hahn, H.; Schafler, E.; Zehetbauer, M.; Kraft, O.; Gruber, P.A. Following the deformation behavior of nanocrystalline Pd films on polyimide substrates using in situ synchrotron XRD. *Mech. Mater.* **2013**, *67*, 65–73. [CrossRef]

175. Zhang, H.; Jérusalem, A.; Salvati, E.; Papadaki, C.; Fong, K.S.; Song, X.; Korsunsky, A.M. Multi-scale mechanisms of twinning-detwinning in magnesium alloy AZ31B simulated by crystal plasticity modeling and validated via in situ synchrotron XRD and in situ SEM-EBSD. *Int. J. Plast.* **2019**, *119*, 43–56. [CrossRef]

176. Pfeifer, M.A.; Williams, G.J.; Vartanyants, I.A.; Harder, R.; Robinson, I.K. Three-dimensional mapping of a deformation field inside a nanocrystal. *Nature* **2006**, *442*, 63–66. [CrossRef]

177. Simons, H.; King, A.; Ludwig, W.; Detlefs, C.; Pantleon, W.; Schmidt, S.; Stöhr, F.; Snigireva, I.; Snigirev, A.; Poulsen, H.F. Dark-field X-ray microscopy for multiscale structural characterization. *Nat. Commun.* **2015**, *6*, 1–6. [CrossRef]

178. Polcarová, M.; Gemperlová, J.; Brádler, J.; Jacques, A.; George, A.; Priester, L. In-situ observation of plastic deformation of Fe-Si bicrystals by white-beam synchrotron radiation topography. *Philos. Mag. A Phys. Condens. Matter Struct. Defects Mech. Prop.* **1998**, *78*, 105–130.

179. Yang, C.; Kang, L.M.; Li, X.X.; Zhang, W.W.; Zhang, D.T.; Fu, Z.Q.; Li, Y.Y.; Zhang, L.C.; Lavernia, E.J. Bimodal titanium alloys with ultrafine lamellar eutectic structure fabricated by semi-solid sintering. *Acta Mater.* **2017**, *132*, 491–502. [CrossRef]

180. Jacques, A.; George, A.; Polcarova, M.; Bradler, J. Dislocation transmission through a Σ = 3 grain boundary in Fe 6 at.% Si: In situ experiments in compression specimen. *Nucl. Instrum. Methods Phys. Res. Sect. B Beam Interact. Mater. Atoms.* **2003**, *200*, 261–266. [CrossRef]

181. Li, L. In situ analysis of multi-twin morphology and growth using synchrotron polychromatic X-ray microdiffraction. *Trans. Nonferr. Met. Soc. China* **2015**, *25*, 2156–2164. [CrossRef]

article

The Effect of Adhesive Layer Thickness on Joint Static Strength

Marek Rośkowicz [1], Jan Godzimirski [1], Andrzej Komorek [2],* and Michał Jasztal [1]

[1] Faculty of Mechatronics, Armament and Aerospace, Military University of Technology, 00-908 Warsaw, Poland; marek.roskowicz@wat.edu.pl (M.R.); jan.godzimirski@wat.edu.pl (J.G.); michal.jasztal@wat.edu.pl (M.J.)
[2] Faculty of Aviation, Polish Air Force University, 08-521 Dęblin, Poland
* Correspondence: a.komorek@law.mil.pl

Abstract: One of the most relevant geometrical factors defining an adhesive joint is the thickness of the adhesive layer. The influence of the adhesive layer thickness on the joint strength has not been precisely understood so far. This article presents simplified analytical formulas for adhesive joint strength and adhesive joint coefficient for different joint loading, assuming, inter alia: linear-elastic strain of adhesive layer, elastic strain of adherends and only one kind of stress in adhesive. On the basis of the presented adhesive joint coefficient, the butt joint was selected for the tests of the influence of adhesive thickness on the adhesive failure stress. The tests showed clearly that with an increase in the thickness of the tested adhesive layers (up to about 0.17 mm), the value of their failure stress decreased quasi linearly. Furthermore, some adhesive joints (inter alia subjected to shearing) may display the optimum value of the thickness of the adhesive layer in terms of the strength of the joint. Thus, the aim of this work was to explain the phenomenon of optimal adhesive layer thickness in some types of adhesive joints. The verifying test was conducted with use of single simple lap joints. Finally, with the use of the FE method, the authors were able to obtain stresses in the adhesive layers of lap joints for loads that destroyed that joints in the experiment, and the FEM-calculated failure stresses for lap joints were compared with the adhesive failure stresses determined experimentally using the butt specimens. Numerical calculations were conducted with the use of the continuum mechanics approach (stress-based), and the non-linear behavior of the adhesive and plastic strain of the adherends was taken into account.

Keywords: adhesive joint; adhesive bond strength; adhesive layer thickness

Citation: Rośkowicz, M.; Godzimirski, J.; Komorek, A.; Jasztal, M. The Effect of Adhesive Layer Thickness on Joint Static Strength. *Materials* **2021**, *14*, 1499. https://doi.org/10.3390/ma14061499

Academic Editors: Christian Motz and Tomasz Sadowski

Received: 11 February 2021
Accepted: 15 March 2021
Published: 18 March 2021

Publisher's Note: MDPI stays neutral with regard to jurisdictional claims in published maps and institutional affiliations.

1. Introduction

Adhesive bonds are frequently used in numerous industrial sectors [1,2]. Adhesive bonding is a particularly attractive assembly technique for application in which weight gain is at a premium, e.g., air transport [3]. In the aeronautical industry, adhesive bonds are most frequently applied in the manufacturing of sandwich constructions, fiber metal laminates and fatigue-resistant hybrid adhesive-riveted joints. In pure adhesive joints, a wide zone can be used to transmit the load, and no holes are used; thus, a reduction in weight can be obtained. However, the reliability of adhesively bonded joints needs to be properly addressed during their design. The strength of adhesive joints depends upon many constructional and technological factors and service conditions. The geometry of the adhesive joint, the sort of joint materials used (both adhesive and adherends), and the load manner are ranked as constructional factors. The geometry of the structural adhesive joints is a very important aspect in the design of the joint. One of the most relevant geometrical factors is the thickness of the adhesive layer.

Numerous investigations have been carried out in order to determine the influence of adhesive thickness on the strength of adhesive joints [4–14]. However, the effect of the adhesive thickness on the bond strength, even in single lap joints, is still not precisely understood.

As early as sixty years ago, Alner [4] and Winter [5] indicated that the dependence of adhesive joint strength on adhesive thickness was different from the results shown by analytical formulas. According to the analytical formulas, an increase in adhesive thickness should lead to an increase in the strength of adhesive joints subjected to shearing. In fact, usually, the strength of such joints decreases with increasing adhesive thickness if it is more than the so-called optimum thickness. For most adhesives, the optimal adhesive thickness is between 0.05 and 0.15 mm [1].

Grant et al. [6] verified that lap joints under tension are very sensitive to adhesive thickness. They explained that as the bond line thickness increases, there is an increase in the bending stress, since the bending moment increases. Consequently, the strength of the joint is reduced. Additionally, da Silva et al. [7] verified that lap shear strength decreases as the adhesive thickness increases from 0.5 to 2 mm.

Naito et al. [8] researched the effect of adhesive thickness on the tensile and shear strength of polyimide adhesive. Their experimental test results indicated that the tensile strength of butt joints decreases with increasing adhesive thickness, and the shear strength of single lap joints is almost constant, regardless of adhesive thickness (0.1–1 mm), although a lot of scatter was observed in the tensile and shear strength of the tested adhesive.

Xu and Wei [9] researched the influence of adhesive thickness on the cohesive properties and overall strength of metallic adhesive bonding structures, employing the cohesive zone model to equivalently simulate adhesive layers with various thicknesses. The obtained results showed that both the cohesive parameters and the overall strength of metallic adhesive were highly dependent on the adhesive thickness, and the variations in overall strength resulting from the various thickness exhibited discrepancies due to the toughness and strain hardening capacity of the adhesive.

Bezemer et al. [10] tested the impact strength of the adhesive. They proved that there is an optimum adhesive thickness at different test speeds. The optimum thickness depends both on the adhesive used and on the test speed. Only for tough adhesives did increased adhesive thickness give a better impact behavior.

Arends et al. [11] took up the attempt to determine optimum adhesive thickness for structural adhesive joints. They performed experimental tests and carried out analyses concerning the influence of adhesive thickness on the tensile lap-shear strength of single overlap joints. For adhesive thicknesses between 0.4 and 0.8 mm, it was found that the failure mode was essentially cohesive, and the value of the strength increased as the adhesive thickness was reduced. For adhesive thicknesses less than 0.4 mm, it was noted that the failure mode was essentially adhesive, and the shear strength showed higher values, but with typical deviation. Because the cohesive failure mode guaranteed the greater reliability of the adhesive joint under the foreseeable working conditions, the optimal adhesive thickness for the tested adhesive should be maintained between 0.4 and 0.5 mm.

Marzi et al. [12] researched the influence of adhesive thickness on the fracture energy of the crash-optimized high-strength adhesive. All of the methods used by them gave virtually the same fracture energy for small layer thicknesses. Here, the fracture energy increased with increasing layer thickness. For larger thicknesses, the results obtained from different methods deviated.

Praksah et al. [13] investigated the influence of adhesive thickness on the high-velocity impact performance of composite targets. The research proved that parameters such as target deformation, energy transformation, depth of penetration and profile deflection of the back plate were dependent to a high degree on the adhesive layer.

Cognard et al. [14] experimentally tested and numerically analyzed the influence of relatively thick adhesive thicknesses (until 1.3 mm) on the strength and edge effects of TAST specimens. They showed that an increase in the adhesive thickness increases the influence of edge effect and increases the risk of crack initiation near the free edges of adhesive. This conclusion seems to be applicable to joints loaded in a similar fashion to the TAST specimens.

In addition to the experimental approach, prediction of the strength of adhesively bonded joints can be performed with use of three main numerical approaches: the continuum mechanics approach (stress-based), the fracture mechanics approach, and the damage mechanics approach [15].

In the continuum mechanics approach, the adhesive and adherends are modeled using continuum elements, assuming that the adhesive is perfectly bonded to the adherends. The assumption of a perfect bond means that the finite element analysis takes no account of the adhesion properties of the interface. This type of modeling approach makes it possible to take into account the influence of the thickness of the adhesive layer on adhesively bonded joint strength [16].

The fracture mechanics approach uses an energy parameter (toughness) as the failure criterion. Prediction of the joint strength is based on linear elastic fracture mechanics, which relies on the existence of a crack and linear elasticity [17,18]. However, well-manufactured joints may not have macroscopic cracks. Furthermore, large-scale plasticity in the adherends limits the use of the linear elastic fracture mechanics approach in numerical calculations.

The progressive damage approach can be implemented using either local or continuum approaches. In the continuum approach, the damage is modeled over a finite region. The local approach, in which the damage is confined to a zero thickness of the surface, is often referred to as the cohesive zone approach (CZM). The CZM approach is used to simulate the macroscopic damage along predefined crack path by specification of a traction–separation response between initially coincident nodes on either side of the predefined crack path [19]. In the FEM strength calculations of adhesively bonded joints with the use of "cohesive" elements (CZM), the problem becomes to assign appropriate values to the parameters of such elements. The analyses of the authors of the article [20] show that this requires experimental tests and numerical calculations for a specific adhesive, the method of preparing the surface for adhesive and specific thickness of the joint. "Cohesive" elements (CZM) have no thickness and make it possible to determine only normal stresses perpendicular to the bonded surfaces and shear stresses (normal stresses occurring in the adhesive plane are ignored). For this reason, they have been found to be unsuitable for the numerical analysis of the influence of the adhesive layer thickness on static strength.

The purpose of the numerical calculations in this paper is to obtain stresses in the adhesive layers of lap joints for loads that destroy that joints in the experiment, and compare the FEM-calculated failure stresses of the lap joints with the adhesive failure stresses determined experimentally. Therefore, the authors decided to apply the continuum mechanics approach.

The general topic of the present paper is the analysis of the influence of the thickness of the adhesive layer on the strength of differently loaded adhesive joints. However, the main aim of the work is to explain the phenomenon of optimal adhesive layer thickness in some types of adhesive joints. The article presents simplified analytical formulas for adhesive joint strength for different joint loading and elastic properties of adhesive and adherends, calculated on the basis of Refs. [21–26]. The experimental tests were conducted with the use of the butt joints and single simple lap joints. Finally, FE method calculations were carried out within the range of plastic deformations of both adhesive and adherend in order to verify the adhesive failure stress dependence on adhesive thickness.

2. Analytical Calculations of the Strength of Adhesive Joint

In the analytical calculations, the assumption may be made that the strength of the adhesive joint is given by the formula:

$$P = \tau_n \cdot A \cdot \psi \tag{1}$$

where:

τ_n—breaking stresses of the adhesive layer dependent on the sort of adhesive, the sort of adherend material, the adhesive bonding technology (among others surface preparation), and the adhesive thickness;

A—area of the adhesive layer;

ψ—joint coefficient, dependent on the joint geometry, load manner, mechanical properties of adhesive and mechanical properties of the adherends.

The joint coefficient may be determined analytically. Even for numerous simplifications, including

- two-dimensional models,
- elastic strain of the adherends,
- linear-elastic strain of the adhesive layer,
- constant stress along thickness of the adhesive layer, and
- only one stress component in the adhesive,

the joint coefficient for lap joints subjected to shearing is given by a quite complicated formula resulting from Volkersen's theory:

$$\psi = \sqrt{\frac{W}{\Delta}} \frac{\sinh\sqrt{\Delta W}}{\cosh\sqrt{\Delta W} + W - 1} \tag{2}$$

where:
$$\Delta = \frac{G_k l^2}{\delta_k E_2 \delta_2}, \quad W = \frac{E_1\delta_1 + E_2\delta_2}{E_1\delta_1}$$
where:
E—Youngs' modulus of the adherends,
G_k—shear modulus of the adhesive (adhesive layer),
l—length of the adhesive layer,
δ—thickness of the adherends,
δ_k—thickness of the adhesive layer.

Considering the geometry of most of the lap joints used in engineering, it is possible to simplify Equation (1) to:

$$\Psi = \sqrt{\frac{W}{A}} = \sqrt{\frac{E_1\delta_1 + E_2\delta_2}{E_1\delta_1}} \times \sqrt{\frac{\delta_k E_2 \delta_2}{G_k l^2}} \tag{3}$$

Assuming that both adherends have equal longitudinal rigidity, it is possible to make additional simplifications. In Table 1, simplified formulas of joint coefficients (calculated on the basis of papers [21–26]) are presented for different models of adhesive joint loading.

In Table 1, only failure shear stresses are presented. It follows from this that the maximal principal positive stresses hypothesis is useful to evaluate the effort in the adhesive layer with acceptable accuracy, and for this hypothesis, the failure shear stresses are equal to the failure normal stresses ($\sigma_n = \tau_n$) [27]. This results from the secular equation:

$$\sigma^3 - \sigma^2(\sigma_x + \sigma_y + \sigma_z) + \sigma\left(\sigma_x\sigma_y + \sigma_y\sigma_z + \sigma_z\sigma_x - \tau_{xy}^2 - \tau_{yz}^2 - \tau_{zx}^2\right)$$
$$- \left(\sigma_x\sigma_y\sigma_z - \sigma_x\tau_{yz}^2 - \sigma_y\tau_{zx}^2 - \sigma_z\tau_{xy}^2 + 2\tau_{xy}\tau_{yz}\tau_{zx}\right) = 0 \tag{4}$$

where σ—principal stress, σ_x, σ_y, σ_z—normal stress, τ_{xy}, τ_{xz}, τ_{yz}—shear stress
If $\sigma_x = 0$, $\sigma_y = 0$, $\sigma_z = 0$, $\tau_{yz} = 0$ and $\tau_{zx} = 0$, then

$$\sigma_I = \tau_{xy} \tag{5}$$

where P—force, M_g—bending moment, M_s—twisting moment, d—diameter, e—eccentricity, E—Youngs' modulus of the adherends, E_k—Young' modulus of the adhesive, G_k—shear modulus of the adhesive (adhesive layer), l—length of the adhesive layer, δ—thickness of the adherends, δ_k—thickness of the adhesive layer, A—area of the adhesive layer.

On the basis of the formulas shown in Table 1, it can be seen that only the strength of joints subjected to pure shearing (No. 4), to tension (No. 11) and to cleaving (No. 12,13) are not dependent upon the adhesive layer thickness. The strength of the other joints (the most important for engineering subjected to shearing) is dependent upon the adhesive thickness, and this dependence can be described by the following formula:

$$P, M_s, M_g = C_i \sqrt{\delta_k} \tag{6}$$

where C_i—constant value for specific type of joint.

Table 1. Simplified formulas of adhesive joint coefficient for different joint loading valid for elastic adhesive (calculated on the basis of papers [15–20]).

No.	Model of Joint	Strength of Joint	Joint Coefficient	References
1		$P = \tau_n A \Psi_1$	$\psi_1 = \sqrt{\dfrac{2E\delta\delta_k}{l^2 G_k}}$	[21]
2		$P = \tau_n A \Psi_2$	$\psi_2 = \sqrt{\dfrac{2G\delta\delta_k}{l^2 G_k}}$	[22]
3		$M_s = \tau_n A \Psi_3$	$\psi_3 = \dfrac{d}{l}\sqrt{\dfrac{G\delta\delta_k}{G_k}}$	[23]
4		$M_s = \tau_n A \Psi_4$	$\psi_4 = \dfrac{d}{2}$	*
5		$M_g = \sigma_n A \Psi_5$	$\psi_5 = \sqrt{\dfrac{E\delta^3 \delta_k}{24 l^2 E_k}}$	* [24]
6		$P = \sigma_n A \Psi_6$	$\psi_6 = \sqrt[4]{\dfrac{E\delta^3 \delta_k}{96 l^4 E_k}}$	* [26]
7		$M_g = \sigma_n A \Psi_7$	$\psi_7 = \sqrt{\dfrac{E\delta^3 \delta_k}{12 l^2 E_k}}$	[24,25]
8		$P = \sigma_n A \Psi_8$	$\psi_8 = \sqrt[4]{\dfrac{E\delta^3 \delta_k}{48 l^4 E_k}}$	[26]
9		$M_g = \sigma_n A \Psi_9$	$\psi_9 = \sqrt{\dfrac{E\delta^3 \delta_k}{6 l^2 E_k}}$	* [24]
10		$P = \sigma_n A \Psi_{10}$	$\psi_{10} = \sqrt[4]{\dfrac{E\delta^3 \delta_k}{6 l^4 E_k}}$	* [26]
11		$P = \sigma_n A \Psi_{11}$	$\psi_{11} = 1$	*
12		$P = \sigma_n A \Psi_{12}$	$\psi_{12} = \dfrac{d}{d+8e}$	*
13		$P = \sigma_n A \Psi_{13}$	$\psi_{13} = \dfrac{l}{l+6e}$	*
14		$P = \tau_n A \Psi_{14}$	$\psi_{14} = \sqrt{\dfrac{E\delta\delta_k}{2 l^2 G_k}}$	* [21]

*—own study, * [A]—own study based on publication [A].

According to Equation (6), the strength of the majority of adhesive joints should increase along with an increase in adhesive thickness, as shown in Figure 1.

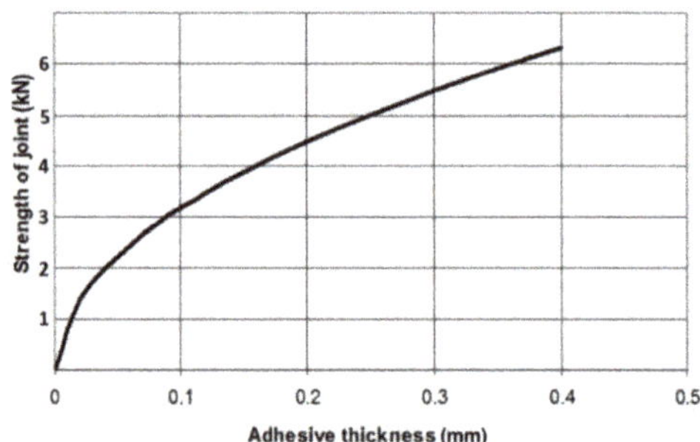

Figure 1. Theoretical dependence of lap adhesive joint strength upon adhesive layer thickness when failure stress of the adhesive layer is not dependent on adhesive layer thickness (vertical axis non-dimensional).

The experimental tests do not verify this dependence on account of the fact that the failure stress of the adhesive is dependent upon the adhesive thickness. Adams and Peppiatt [28] attribute the decrease in joint strength with adhesive thickness to the fact that thicker bond lines contain more defects, such as voids and microcracks. This theory seems to be very probable, although there are many other theories.

3. Experimental Tests

To determine the influence of adhesive thickness upon the failure stress of the adhesive layer, an experiment was carried out with the butt specimens, as shown in Figure 2. The simple butt joint was selected for the tests, because its joint coefficient is not dependent upon adhesive thickness (see Table 1).

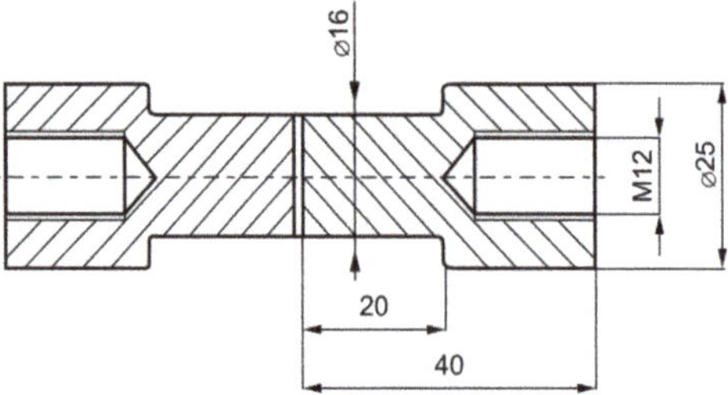

Figure 2. The specimen used in the test for estimating the dependence of adhesive failure stress upon adhesive thickness.

The FEM calculations conducted showed that in the adhesive layer of such specimens, the distribution of stresses is approximately uniform (Figure 3).

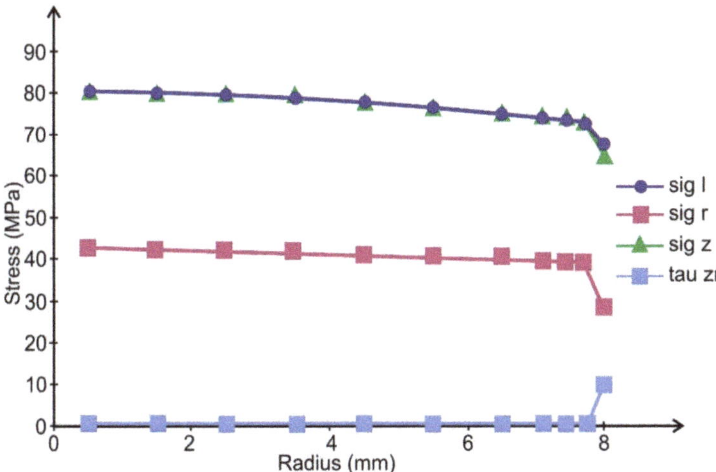

Figure 3. Stresses distribution in the adhesive layer of the butt specimen (sig I—maximal principal stress, sig r—radial stress equal circumferential stress, sig z—axial stress, tau zr—shear stress, the adhesive was modeled by three layers of elements).

The bonded surfaces of all specimens were prepared by pickling in aqueous solution of sulfuric acid and chromic acid at 60 °C. In the strength tests, six specimens were prepared for every measuring point. The results were elaborated statistically: Student-Fisher's method was applied to calculate the confidence interval for the level $\alpha = 0.95$. To obtain the required adhesive thickness, two thin wires of suitable thickness were placed in uncured adhesive layers. The thinnest adhesive thicknesses (about 0.02 mm) were obtained without distance wires. The adhesive layers were cured at a pressure of about 0.8 MPa. The adhesive thicknesses were estimated by measuring similarly adhesively bonded single simple lap specimens which thickness were measured by micrometer. The thickest bond line was 0.3 mm.

The tests were conducted for Epidian 57 adhesive (CIECH Resins, Nowa Sarzyna, Poland) cured with Z-1 hardener (CIECH Resins, Nowa Sarzyna, Poland) for 1 h at 60 °C and Araldite AW126H (Huntsman Advanced Materials, Duxford, UK) cured with HY994 hardener (Huntsman Advanced Materials, Duxford, UK) for 1 h at 80 °C. The failure stresses were calculated using the following formula:

$$\tau_n = \sigma_n = \frac{P}{A} \tag{7}$$

where *P*—average breaking force, *A*—area of adhesive layer.

The results of the tests are presented in Figure 4.

The tests showed explicitly that value of adhesive failure stress are dependent upon the adhesive thickness. Along with the increase in the adhesive thicknesses to about 0.17 mm, the value of failure stress decreases in a quasi-linear fashion for the tested adhesive. A similar dependence of failure shear stress upon adhesive thickness was obtained for heading joints with two sleeves subjected to shearing [29] and normal stress for tensile butt joints [8].

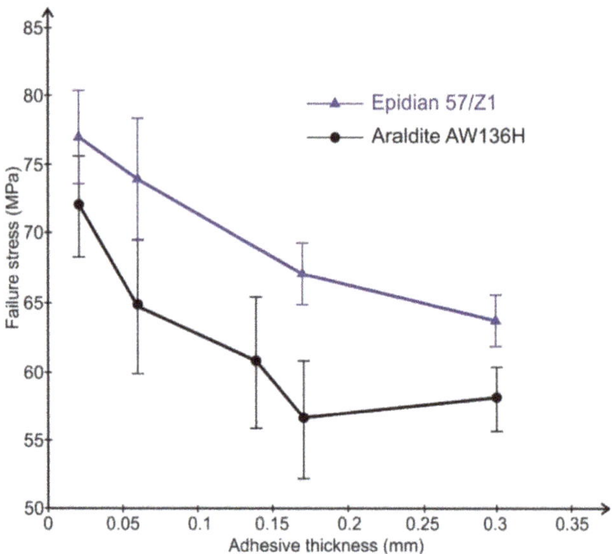

Figure 4. Dependence of adhesive failure stress upon adhesive thickness determined for two adhesives.

The linear dependence of adhesive failure stresses upon adhesive thickness can be described by the following formula:

$$\sigma_n = \tau_n = B - D \cdot \delta_k \tag{8}$$

where B and D—constants for each adhesive dependent on surface preparation manner for adhesive bending.

The insertion of Equation (8), e.g., into the formula for loading manner 1, shown in Table 1, makes it possible to obtain the dependence of the strength of the joint subjected to shearing on the adhesive layer thickness:

$$P = (B - D \cdot \delta_k) A \sqrt{\frac{2E\delta\delta_k}{l^2 G_K}} \tag{9}$$

Function $P = P(\delta_k)$ has an extreme (maximum) for:

$$\delta_k = \frac{B}{3D} \tag{10}$$

and this formula describes the optimum thickness value of the adhesive layer for load manner 1, shown in Table 1. The same formula describes the optimum thickness value of the adhesive layer for manners of adhesive bond loading 2, 3, 5, 7, 9 and 14, shown in Table 1.

For adhesive bonds loaded according to manner 6 (shown in Table 1), the strength as a function of adhesive thickness is described by the following formula:

$$P = (B - D \cdot \delta_k) A \sqrt[4]{\frac{E\delta^3\delta_k}{96l^4 E_k}} \tag{11}$$

Equation (11) has a maximum for:

$$\delta_k = \frac{3B}{7D} \tag{12}$$

The same formula describes the optimum thickness value of the adhesive layer for loading manners 8 and 10, shown in Table 1.

Due to the above, it seems that some adhesive joints may have an optimum adhesive thickness. The optimum thickness value of the adhesive layer shouldn't be dependent upon the thickness of the adherends, the Young's modulus of the bonded materials, or the Young's modulus of the adhesive, but it should be dependent upon loading manner.

On the basis of Figure 3, the linear dependences of failure stress on adhesive thickness ($\tau_n = \tau(\delta_k)$) were calculated as follows for the Epidian 57 adhesive:

$$\tau_n = 76.9 - 160.1\ \delta_k \cdot \text{MPa} \tag{13}$$

and for Araldite AV 136H:

$$\tau_n = 72.2 - 240.8\ \delta_k \cdot \text{MPa} \tag{14}$$

By inserting Dependencies (13) and (14) into Equation (10), we can calculate the values of the optimal thickness of the adhesive layer for the two tested adhesives. Therefore, the optimal adhesive thickness for joints subjected to shearing bonded with Epidian 57 adhesive should be equal, at about:

$$\delta_{kopt} = \frac{B}{3D} = \frac{76.9}{3 \cdot 160.1} \approx 0.16\ mm \tag{15}$$

and, in turn, for joints subjected to shearing bonded with Araldite AV 136 H adhesive:

$$\delta_{kopt} = \frac{72.2}{3 \cdot 240.8} \approx 0.1\ mm \tag{16}$$

A verifying test was performed. A single simple lap joint with a 25 mm overlap, 25 mm width, and 2-mm-thick adherends made of aluminum alloy 2024T4 (Henan Mingtai aluminum industry Co. LTD, Zhengzhou, Henan, China) was tested to determine the influence of adhesive thickness upon joint strength. Six samples were used for each setting. The bonded surface of the specimens was prepared for bonding by pickling. Two thin wires were placed in uncured adhesive layers to obtain the definite adhesive thickness. The specimens were bonded with the same adhesives as the butt joints. The adhesive thickness was estimated by micrometer measuring of each specimen. The spread of average adhesive thickness was not greater then ±0.01 mm for each batch. The results of the strength tests are shown in Figures 5 and 6.

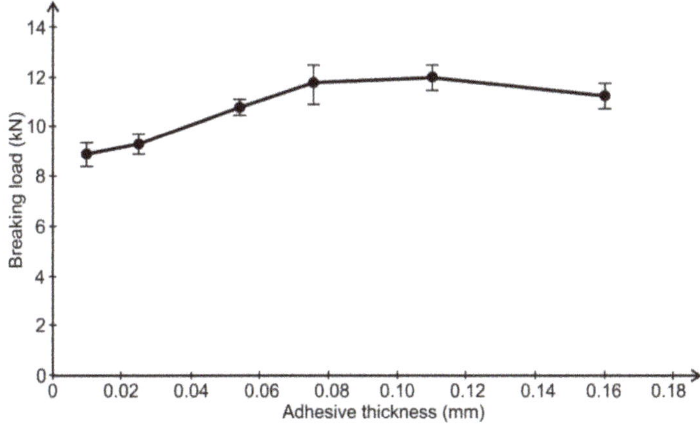

Figure 5. Dependence of lap joint strength on adhesive thickness (Epidian 57/Z1 adhesive).

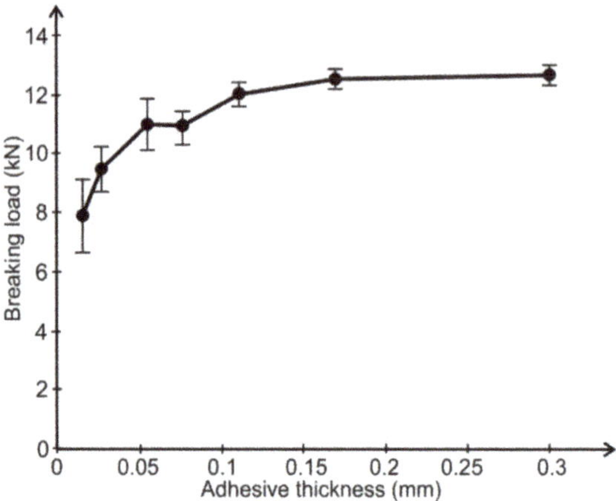

Figure 6. Dependence of lap joint strength on adhesive thickness (Araldite AW136H adhesive).

The results of the experimental tests were different from the analytical calculations—the optimal adhesive thickness for joints bonded with Epidian 57 were smaller than the calculated ones, and the joints bonded with Araldite AV 136H did not show the optimal adhesive thickness. The cause of this inaccuracy may be the simplifications made, especially the assumptions with respect to the elastic strains of adhesive and adherends.

4. Numerical Calculations

FE method calculations were conducted to estimate the adhesive failure stresses in the adhesive layer of the experimentally tested single lap joints. As stated in the introduction, the authors decided to apply the continuum mechanics approach, and the adhesive joint was modeled with eight-node quad elements. The failure stress was calculated for joints bonded with Epidian 57 and AW 136H adhesives for the actual adhesive thickness and breaking loads on the basis of the experimental test. The MSC Nastran 2018 (MSC Software Corporation, Irvine, CA, USA) for Windows program was used for numerical calculation. In the numerical calculation, constant stress along the width of the joint was assumed. Therefore, the 2D plane strain models were used throughout this study. The actual thickness of specimen sheets, overlap lengths, thickness of the adhesive layer and the spacing of the testing machine handles were used in the numerical calculations. The adhesive layers were modeled by one layer of finite elements (eight-node quad elements) [30]. The possibility of such modeling is a result of the numerical investigations of lap joints subjected to shearing. The performed calculations led to the following conclusions:

- the edges of the adhesive layer are a so-called singular point in which numerically calculated stresses approach infinity when the element net is thickened,
- small (equal in dimension to the thickness of the adhesive layer) "flashes" presenting at the edges of the adhesive layers of the lap joint cause considerable reduction of the stress concentrations in adhesive layers, as well as the uniformity of stresses along the thickness of the adhesive layer,
- the uniformity of stresses along the thickness of the adhesive layer makes it possible to model the adhesive layer with a single element layer.

The length of the adhesive layer elements was equal to 0.5 mm. The adherends were modeled using four layers of elements. The tested models of the specimens were loaded with average forces according to the experimental tests in the respective groups The calculations were performed for the non-linear characteristics of the adhesive and the adherends. The non-linear properties of the adhesive and the adherends were described for calculations using discrete functions made on the basis of their stress–strain curves The stress–strain curves of adhesive layers (Figure 7) were adopted on the basis of the compression curves determined using the cast cylindrical specimens (diameter Φ = 12.5 mm, height l = 25 mm). The stress–strain curve of alloy 2024T4 based on the set of tensile curves is presented in Figure 8. The Poisson's ratio of the alloy 2024T4 was assumed to be ν = 0.3 and the Poisson's ratio of the adhesive was ν = 0.35. The model of the single lap specimen under load is illustrated on Figure 9.

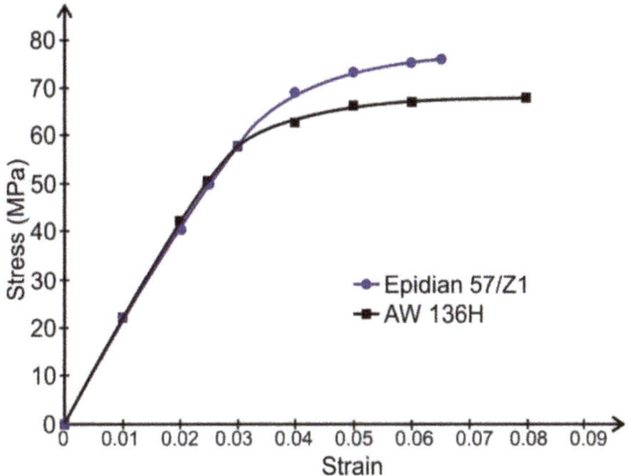

Figure 7. The stress–strain curves for Epidian 57/Z1 and AW 136H adhesive (compression curves without the considering of specimen cross-section area change).

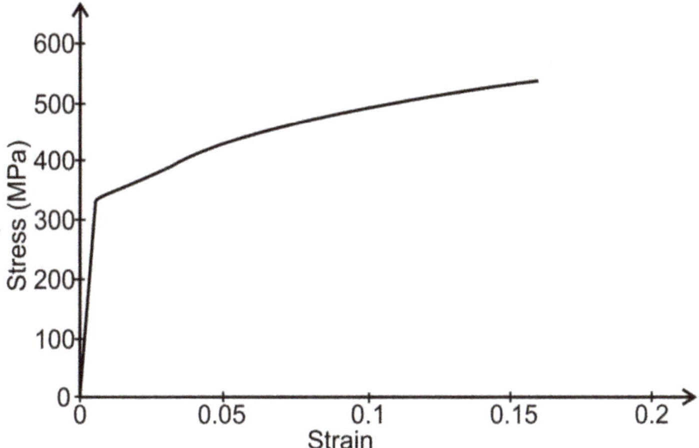

Figure 8. The stress–strain curve for 2024T4 alloy (tensile curve with the considering of specimen cross-section area change).

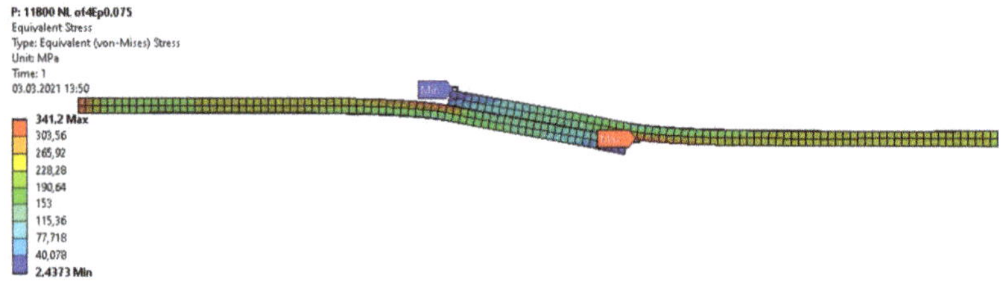

Figure 9. Deformation of a single lap specimen under shear load with a force of 11,800 N (scale 3.2:1, plastic deformation of jointed elements) and Mises stress distribution.

From the results of the conducted calculations, in all calculated cases, the yield point of adherends (Re $\approx$ 330 MPa) exceeded what caused plastic deformation. Calculation on the basis of the failure stress in the maximal principal stresses hypothesis showed results about 10% higher than the results from the tests of the tensile butt specimens (Figures 10 and 11).

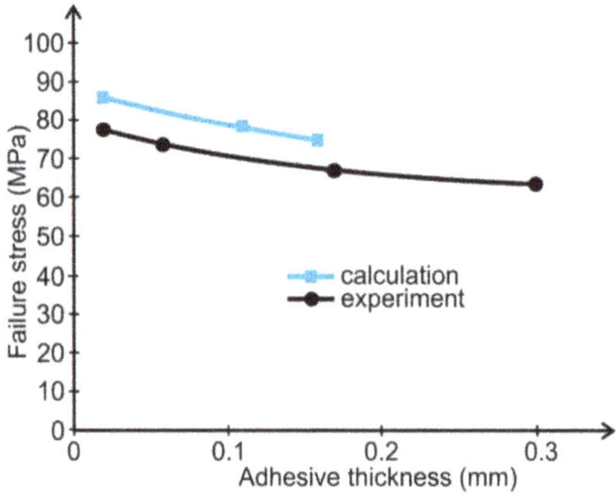

Figure 10. Comparison of Epidian 57/Z1 adhesive failure stresses determined experimentally using butt specimens with FEM calculated failure stresses for lap joints.

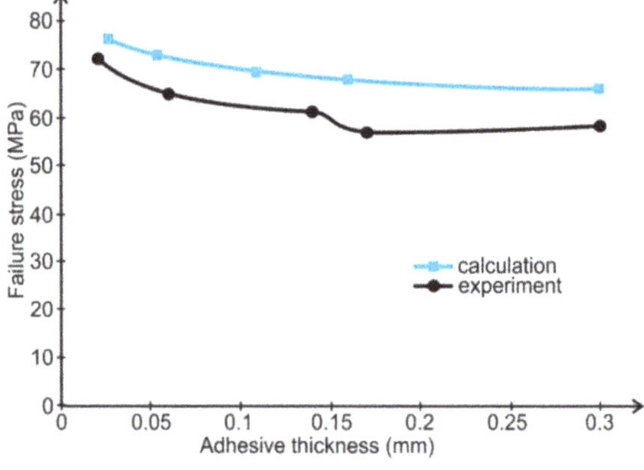

Figure 11. Comparison of AW 136H adhesive failure stresses determined experimentally using butt specimens with FEM calculated failure stresses for lap joints.

The cause of the difference in failure stress between the calculated and experimentally determined results may result from the greater area of the adhesive layer loaded with maximal stresses in the tensile butt joint than in the lap joint subjected to shearing. In the tensile butt joint, the entire adhesive is loaded by failure stress; in lap joint, only the edge of the adhesive layer is loaded. This is in accordance with defect theory relating to adhesive layers.

5. Conclusions

- The value of adhesive failure stress is dependent upon the adhesive thickness. The failure stress decreases with increasing adhesive layer thickness, at first in a quasi-linear manner, and then less intensively. This fact may be the result of the greater number of defects in the thicker adhesive layer.
- The dependence of adhesive failure stress upon adhesive thickness causes the strength of most adhesive joints (shear loaded and peel loaded) to be dependent on the thickness of the adhesive layer in different ways to those in the results shown by analytical formulas. Lap joints subjected to shearing and joints subjected to peeling may have the optimal adhesive thickness.
- The same adhesive joints, such as butt joints subjected to tensioning and joints subjected to cleaving, have the highest strength for very thin adhesive layers. For such joints, the facetious saying is "the adhesive bond is superlative if there is no adhesive at all".
- The more accurate analysis of adhesive thickness influence on adhesively bonded joint strength demands the taking into account of the non-linear behavior of adhesive and the plastic strain ability of the adherends. This is possible using FE methods. The numerical calculations verify that the failure stress of thin adhesive layers is greater than in thicker ones.

Author Contributions: Conceptualization, M.R. and J.G.; methodology, J.G.; software, J.G.; validation, M.R., A.K. and J.G.; formal analysis, M.J.; investigation M.R.; resources, M.J.; data curation, A.K.; writing—original draft preparation, M.R. and J.G.; writing—review and editing, M.R., M.J. and A.K.; visualization, A.K.; supervision, J.G.; project administration, M.R. and M.J.; funding acquisition, A.K. All authors have read and agreed to the published version of the manuscript.

Funding: This research received no external funding.

Institutional Review Board Statement: Not applicable.

Informed Consent Statement: Not applicable.

Data Availability Statement: Data are contained within the article.

Conflicts of Interest: The authors declare no conflict of interest.

References

1. Cagle, C.V. *Adhesive and Adhesive Bonding*; Wydawnictwo Naukowo Techniczne: Warsaw, Poland, 1977.
2. da Silva, L.F.M.; Öchsner, A.; Adams, R.D. (Eds.) *Handbook of Adhesion Technology*; Springer: Heidelberg, Germany, 2011.
3. Higgins, A. Adhesive bonding of aircraft structures. *Int. J. Adhes. Adhes.* **2000**, *20*, 367–376. [CrossRef]
4. Alner, D.J. *Aspect of Adhesion*; University of London Press: London, UK, 1965.
5. Winter, H. *Fertigungstechnik von Luft- und Raumfahrzeugen*; Springer: Berlin/Heidelberg, Germany, 1967.
6. Grant, L.D.R.; Adams, R.D.; da Silva, L.F.M. Experimental and numerical analysis of single-lap joints for the automotive industry. *J. Adhes. Adhes.* **2009**, *29*, 405–413. [CrossRef]
7. da Silva, L.F.M.; Carbas, R.J.C.; Critchlow, G.W.; Figueredo, M.A.V.; Brown, K. Effect of material, geometry, surface treatment an environment on the sear strength of single lap joints. *J. Adhes. Adhes.* **2009**, *29*, 621–632. [CrossRef]
8. Naito, K.; Onta, M.; Kogo, Y. The effect of adhesive thickness on tensile and shear strength of polyimide adhesive. *J. Adhes. Adhes.* **2012**, *36*, 77–85. [CrossRef]
9. Xy, W.; Wei, Y. Influence of adhesive thickness on local interface fracture and overall strength of metallic bonding structures. *J. Adhes. Adhes.* **2013**, *40*, 158–167. [CrossRef]
10. Bezemer, A.A.; Guyt, C.B.; Vlot, A. New impact specimen for adhesives: Optimization of high-speed-loaded adhesive joints. *J. Adhes. Adhes.* **1998**, *18*, 255–260. [CrossRef]

11. Arenas, J.M.; Narbon, J.J.; Alia, C. Optimum adhesive thickness in structural adhesives joints using statistical techniques on Weibull distribution. *J. Adhes. Adhes.* **2010**, *30*, 160–165. [CrossRef]

12. Marzi, S.; Biel, A.; Stigh, U. On experimental methods to investigate the effect of layer thickness on the fracture behavior of adhesively bonded joints. *J. Adhes. Adhes.* **2011**, *31*, 840–850. [CrossRef]

13. Prakash, A.; Rajasankar, J.; Anandavalli, N.; Verma, M.; Iyer, N.R. Influence of adhesive thickness on high velocity performance of ceramic/metal composite targets. *J. Adhes. Adhes.* **2013**, *41*, 186–197. [CrossRef]

14. Cognard, J.Y.; Creaćhcadec, R.; Sohier, D.; Leguillon, D. Influence of adhesive thickness on the behaviour of bonded assemblies under shear loading using a modified TAST fixture. *J. Adhes. Adhes.* **2010**, *30*, 257–266. [CrossRef]

15. Filho, S.T.A.; Blaga, L.-A. (Eds.) *Joining of Polymer-Metal Hybrid Structures: Principles and Applications*; John Wiley & Sons, Inc.: Hoboken, NJ, USA, 2018.

16. Gleich, D.M.; Van Tooren MJ, L.; Beukers, A. Analysis and evaluation of bondline thickness effects on failure load in adhesively bonded structures. *J. Adhes. Sci. Technol.* **2001**, *15*, 1091–1101. [CrossRef]

17. Hutchinson, J.W.; Suo, Z. Mixed-mode cracking in layered materials. *Adv. Appl. Mech.* **1992**, *29*, 64–187.

18. Funari, M.F.; Lonetti, P. Initiation and evolution of debonding phenomena in layered structures. *Theor. Appl. Fract. Mech.* **2017**, *92*, 133–145. [CrossRef]

19. de Moura, M.F.S.F.; Gonçalves, J.P.M.; Chousal, J.A.G.; Campilho, R.D.S.G. Cohesive and continuum mixed-mode damage models applied to the simulation of the mechanical behaviour of bonded joints. *Int. J. Adhes. Adhes.* **2008**, *28*, 419–426. [CrossRef]

20. Godzimirski, J.; Pietras, A. The modeling of an adhesive layer for FEM calculation. *Technol. I Autom. Montażu* **2013**, *4*, 40–44.

21. Volkersen, O. Die Nietkraftverteilung in zugbeanspruchten Nietverbindungen mit konstanten Laschenquerschnitten. *Luftfahrtforchung.* **1938**, *15*, 41–47.

22. Misztal, F. Distribution shear stresses in adhesive layers of adhesive joints. *Arch. Mach. Constr. III* **1956**, *1*, 35–63.

23. Grudziński, K.; Lorkiewicz, J. Torsional strength of cylindrical adhesive joints. *Mech. Rev.* **1970**, *29*, 235–239.

24. Lamb, G.E. On the theory of peeling. *J. Polym. Sci.* **1961**, *51*, 573–576. [CrossRef]

25. Gardon, J.L. Peel adhesion. A theoretical analysis. *J. Appl. Polym. Sci.* **1963**, *7*, 643–666. [CrossRef]

26. Jouwersma, G. On the theory of right-angle peeling. *J. Polym. Sci.* **1960**, *45*, 253–255. [CrossRef]

27. Jakubowicz, A.; Orłoś, Z. *Mechanics of Materials*; Wydawnictwo Naukowo Techniczne: Warsaw, Poland, 1978.

28. Adams, R.D.; Peppiatt, N.A. Stress analysis of adhesively bonded lap joints. *J. Strain Anal. Eng. Des.* **1974**, *9*, 185–196. [CrossRef]

29. Weiss, P. Adhesion and Cohesion. In Proceedings of the Symposium on Adhesion and Cohesion, General Motors Research Laboratories, Warren, MI, USA, 1 January 1961.

30. Godzimirski, J.; Tkaczuk, S. Numerical calculations of adhesives joint subjected to shearing. *J. Theor. Appl. Mech.* **2007**, *45*, 311–324.

Article

Bending Response of 3D-Printed Titanium Alloy Sandwich Panels with Corrugated Channel Cores

Zhenyu Zhao [1,2,3], Jianwei Ren [1,2], Shaofeng Du [3,*], Xin Wang [2,4], Zihan Wei [2,4], Qiancheng Zhang [2,4], Yilai Zhou [1,2], Zhikun Yang [3] and Tian Jian Lu [1,2,*]

1 State Key Laboratory of Mechanics and Control of Mechanical Structures, Nanjing University of Aeronautics and Astronautics, Nanjing 210016, China; zhenyu_zhao@nuaa.edu.cn (Z.Z.); jeawren0621@nuaa.edu.cn (J.R.); yilaizhou@nuaa.edu.cn (Y.Z.)
2 MIIT Key Laboratory of Multi-Functional Lightweight Materials and Structures, Nanjing University of Aeronautics and Astronautics, Nanjing 210016, China; wxtj_9449@stu.xjtu.edu.cn (X.W.); wzh123@stu.xjtu.edu.cn (Z.W.); zqc111999@xjtu.edu.cn (Q.Z.)
3 State Key Laboratory of Smart Manufacturing for Special Vehicles and Transmission System, Baotou 014030, China; yzkwh@163.com
4 State Key Laboratory for Strength and Vibration of Mechanical Structures, Xi'an Jiaotong University, Xi'an 710049, China
* Correspondence: dusf.yx@163.com (S.D.); tjlu@nuaa.edu.cn (T.J.L.)

Abstract: Ultralight sandwich constructions with corrugated channel cores (i.e., periodic fluid-through wavy passages) are envisioned to possess multifunctional attributes: simultaneous load-carrying and heat dissipation via active cooling. Titanium alloy (Ti-6Al-4V) corrugated-channel-cored sandwich panels (3CSPs) with thin face sheets and core webs were fabricated via the technique of selective laser melting (SLM) for enhanced shear resistance relative to other fabrication processes such as vacuum brazing. Four-point bending responses of as-fabricated 3CSP specimens, including bending resistance and initial collapse modes, were experimentally measured. The bending characteristics of the 3CSP structure were further explored using a combined approach of analytical modeling and numerical simulation based on the method of finite elements (FE). Both the analytical and numerical predictions were validated against experimental measurements. Collapse mechanism maps of the 3CSP structure were subsequently constructed using the analytical model, with four collapse modes considered (face-sheet yielding, face-sheet buckling, core yielding, and core buckling), which were used to evaluate how its structural geometry affects its collapse initiation mode.

Keywords: sandwich panels with corrugated channel core; 3D-printed sandwich; bending response; mechanism maps; geometrical optimization

Citation: Zhao, Z.; Ren, J.; Du, S.; Wang, X.; Wei, Z.; Zhang, Q.; Zhou, Y.; Yang, Z.; Lu, T.J. Bending Response of 3D-Printed Titanium Alloy Sandwich Panels with Corrugated Channel Cores. *Materials* **2021**, *14*, 556. https://doi.org/10.3390/ma14030556

Academic Editors: Tomasz Sadowski and Holm Altenbach
Received: 30 November 2020
Accepted: 21 January 2021
Published: 24 January 2021

Publisher's Note: MDPI stays neutral with regard to jurisdictional claims in published maps and institutional affiliations.

1. Introduction

Increasing demand for integration and miniaturization of load-sustaining mechanical structures calls for novel lightweight materials. Various methodologies have been developed to construct high-performance lightweight structures, e.g., selection of lightweight base materials [1], structural/topological optimization based on machine learning [2,3], bio-inspired design such as hierarchical [4] and graded [5] strategies and their combination, and so on. The essence of the above method is to remove redundant mass by optimizing the structure topology and size. Nowadays, however, a load-bearing mechanical structure is often required to possess additional attributes, e.g., energy absorption, sound attenuation, heat/mass transfer, energy storage, and so on [6–9]. Design and construction of lightweight multifunctional structures have thus become a focal point in a wide range of applications.

High-porosity cellular metals with either open or closed cells have emerged as attractive multifunctional materials [10]. Particularly, for applications requiring simultaneous load bearing and active cooling, sandwich constructions with three-dimensional (3D) lattice trusses or two-dimensional (2D) prismatic cores have been extensively exploited [11]. For

63

instance, an all-metallic corrugated sandwich panel is a typical multifunctional structure: in addition to heat dissipation [1,12], it can effectively absorb sound if micro-perforations are introduced to both its face sheets and core webs [13,14]; it can also provide effective projectile penetration resistance if ceramic or concrete prisms are inserted to the interstices of its corrugated core [15–17]. Nonetheless, in terms of specific stiffness and specific strength, a corrugated sandwich construction is inferior to other competing structures such as honeycomb-cored sandwiches, especially when subjected to bending and shearing [18,19]. More recently, to maintain the flow-through topology of the corrugated core and improve its load-carrying capability, a novel corrugated channel core (i.e., periodic fluid-through wavy passages) was envisioned for multifunctional sandwich constructions [20]. It was demonstrated, both theoretically and experimentally, that all metallic, corrugated-channel-cored sandwich panels (3CSPs) exhibit not only significantly enhanced convective heat transfer rate, but also superior mechanical performance relative to sandwich panels with parallel plate channels widely used for active cooling. Further, when subjected to the loading of out-of-plane compression, the proposed corrugated channel core exhibits superiority, particularly in the low-density regime, in comparison with competing sandwich core topologies (e.g., square honeycombs, hollow pyramidal trusses, and octet trusses) [20]. Upon inserting PMI foam blocks into the interstices of the corrugated channel core made of a purpose-made fiber-reinforced composite, it was also shown theoretically and experimentally that the resulting hybrid-cored sandwich panel (with composite face sheets) not only exhibits superior specific stiffness/strength but also is endowed with a new attribute, i.e., microwave absorption/transmission [21].

With ever increasing demand for lightweight multifunctional structures, we envision that the novel all-metallic 3CSP construction proposed in our previous study holds great potential for a wide variety of engineering applications, targeting in particular simultaneous load-bearing and heat dissipation via active cooling at ultra-lightweight. To this end, processing methods suitable for fabricating high-quality 3CSP structures need to be exploited, especially when high performance metals such as titanium alloys are used as the parent material. Further, in addition to systematically studying the heat transfer performance of an as-fabricated 3CSP specimen, its mechanical performance also needs to be investigated comprehensively. Furthermore, analytical and numerical models need to be established to explore in detail the physical mechanisms that underlie the experimentally observed mechanical responses.

The objectives of this study were therefore three-fold. Firstly, the method of selective laser melting (SLM) was employed to fabricate Ti-6Al-4V 3CSP sandwich panels for the first time. Previously, these panels were fabricated using a four-step procedure: (a) folding of corrugated sheet; (b) fabrication of corrugated web; (c) assembling of sandwich panel; and (d) vacuum brazing [20,22]. Under bending and/or shear loading, 3CSP specimens thus fabricated with relatively thin face sheets and core web plates will not reach the pre-designated strength, because it is difficult to weld the thin face sheets with the thin core webs, thus yielding inferior shear strength. The SLM technique adopted in this study can squarely address this deficiency, because the pressing demand for high performance 3CSPs as ultra-lightweight multifunctional structures in fields such as aerospace might compromise the relatively high cost of SLM. Secondly, in addition to the out-of-plane uniform compression considered in [20], the performances of as-fabricated Ti-6Al-4V 3CSP sandwich panels subjected to four-point bending were systematically characterized using a combined experimental, analytical, and numerical approach. Thirdly, the analytical model, validated against experimental measurements, was employed to construct collapse mechanism maps, thus providing an effective strategy for designing 3CSP structures with optimal bending responses. Eventually, upon performing the foregoing tasks, we demonstrated that four competing collapse initiation modes governed the failure processes of the Ti-6Al-4V 3CSP structure as its geometry was varied, and the collapse mechanism maps constructed using the developed analytical model provides an effective strategy for designing 3CSP structures with optimal mechanical responses.

The results of this study demonstrate that the proposed 3CSP structure indeed holds great potential in multifunctional applications demanding simultaneous load-bearing and heat dissipation at ultra-lightweight.

2. Methodologies

2.1. Fabrication of 3CSP Specimens

2.1.1. Topology of 3CSP Specimen

As depicted schematically in Figure 1, the all-metallic sandwich panel investigated in the current study is composed of two face sheets of equal thicknesses (t_f) and a corrugated channel core with a thickness t_c and a height h. Let L, B, and H represent the length, width, and height of the sandwich panel, respectively. The idealized core profile is also shown in Figure 1, which presents a triangular corrugation core with a wavelength l and an inclination angle θ. The equivalent neutral surfaces of the webs are parallel to each other, with spacing d. The face sheets and the core are made of the same metallic material (i.e., Ti-6Al-4V). For the present four-point bending test, as detailed in Section 2.2, the distance between the two indenters (L_p) is fixed at 34 mm, while the span between the two bearings (L_b) is fixed at 119 mm. The diameter of the indenter/bearing is 10 mm. The characteristic length of the specimen is thus given by (Figure 1): $\chi = (L_b - L_p)/2$. The weight per unit area W denotes the weight of a structure per unit surface area [18]. For the present 3CSP, it is given by $W = \frac{M_f + M_c}{a} = 2t_f\rho_s + \frac{t_c h}{d\cos\theta}\rho_s$ where $M_f = 2t_f LB\rho_s$ and $M_c = \frac{Lt_c hB}{d\cos\theta}\rho_s$ are the weight of the face sheets and the core, respectively; $a = LB$ is the surface area of the 3CSP; and ρ_s is the density of its parent material. Additionally, to facilitate the quantification of structural mass, a dimensionless metric $\psi = W/\rho_s\chi$ is defined in this study. Subject to the limits of fabrication precision and specimen dimensions, the specimens are envisioned to vary systemically in geometry, as listed in Table 1. It should be mentioned that, in Figure 1, the face sheets and the core webs are not welded together as done in our previous study [20]. Rather, the two formed an integral part as 3D printing is used in the current study to construct the 3CSP structure from scratch. Indeed, no interaction layer is present between the core web and the face sheet in the as-fabricated Ti-6Al-4V 3CSP specimen.

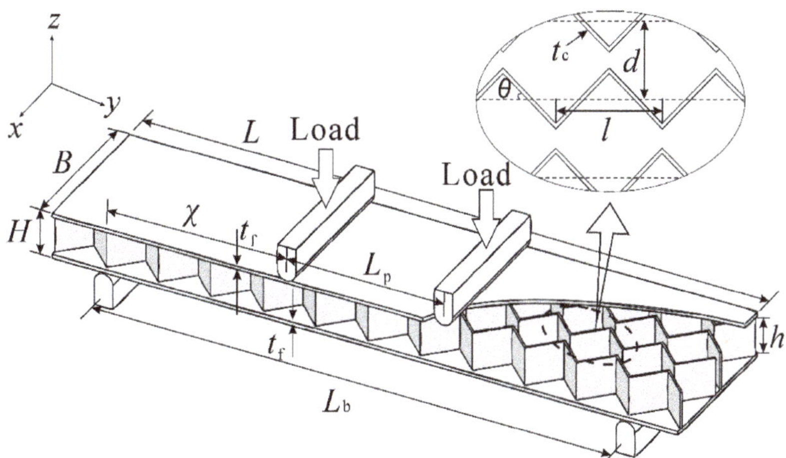

Figure 1. Schematic of sandwich panel with corrugated channel core subjected to four-point bending.

Table 1. Geometric details of sandwich panels with corrugated channel cores.

Specimen No.	Length L (mm)	Width W (mm)	Core Height h (mm)	Web Spacing d (mm)	Thickness of Face Sheet t_f (mm)	Thickness of Core Sheet t_c (mm)	Angle θ (°)	Wave Length l (mm)
I	136	24	12	12	1.0	0.4	45	17
II	136	24	12	12	1.0	0.4	45	17
III	136	24	12	12	1.0	0.6	45	17
IV	136	24	12	12	0.8	0.6	45	17
V	136	24	12	12	0.8	0.4	45	17
VI	136	24	12	12	0.8	0.6	45	17
VII	136	24	12	12	0.2	0.8	45	17

2.1.2. Fabrication Methodology

The method of selective laser melting (SLM) enables manufacturing metal structures via additive printing layer by layer [23,24]. Because it completely melts a powder material, the SLM methodology allows for a density of approximately 100% and assures series-identical properties for fabricated specimens [25]. Due to the relatively complex geometry of the proposed 3CSP (Figure 1), all specimens in this study (Table 1) were fabricated via SLM. Figure 2 presents a typical Ti-6Al-4V sandwich panel thus fabricated using EOS M290 (EOS Ltd. Germany). The device is equipped with an IPG fiber laser having a power of 400 W and an equivalent speckle diameter of 100 μm. The IPG fiber laser outputs a laser in continuous wave (CW) mode. A checkerboard scanning strategy was adopted during the printing process, and the laser heat source was moved from the top to the bottom while also moving from left to right. The corresponding processing parameters are presented in Table 2, while geometries of as-fabricated specimens are listed Table 3, together with measured deviations from those envisioned (Table 1). Subsequently, thermal treatment of the fabricated specimens was carried out to minimize stress localization (Figure 3), which is composed of three phases: a linear heating period from 0 to 800 °C, with a heating velocity of 400 °C per hour; subsequent maintaining period at a constant temperature of 800 °C for 4 h; and a furnace cooling treatment of 240 °C per hour, from furnace temperature (800 °C) to 80 °C. Lastly, air cooling proceeded until the specimen temperature reaches ambient temperature. Otherwise, a furnace vacuum condition with a base pressure less than 0.02 Pa was used during the entire thermal treatment.

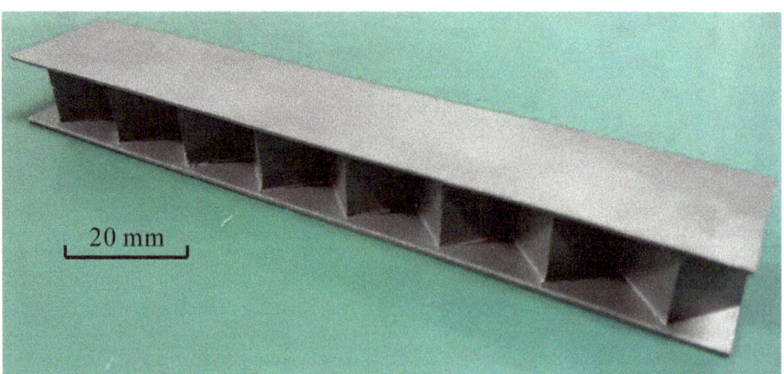

Figure 2. Ti-6Al-4V alloy 3CSP fabricated with the SLM methodology.

Table 2. SLM processing parameters for Ti-6Al-4V alloy 3CSP specimens.

Term	Layer Thickness	Laser Powder	Scan Speed	Hatch Distance	Particle Size
Parameter	40 μm	230 W	970 mm/s	100 μm	15–53 μm

Table 3. Measured dimensions of test specimens fabricated with the SLM.

Specimen No.	Thickness of Face Sheet (mm)			Thickness of Core Sheet (mm)		
	Design Size	Measured Size	Error	Design Size	Measured Size	Error
I	1.0	0.992	−0.80%	0.4	0.445	11.25%
II	1.0	1.063	6.30%	0.4	0.430	7.50%
III	1.0	1.042	4.20%	0.6	0.645	7.50%
IV	0.8	0.853	6.62%	0.6	0.551	−8.17%
V	0.8	0.831	3.87%	0.4	0.407	1.75%
VI	0.8	0.813	1.62%	0.6	0.603	0.50%
VII	0.2	0.225	12.50%	0.8	0.764	−4.50%

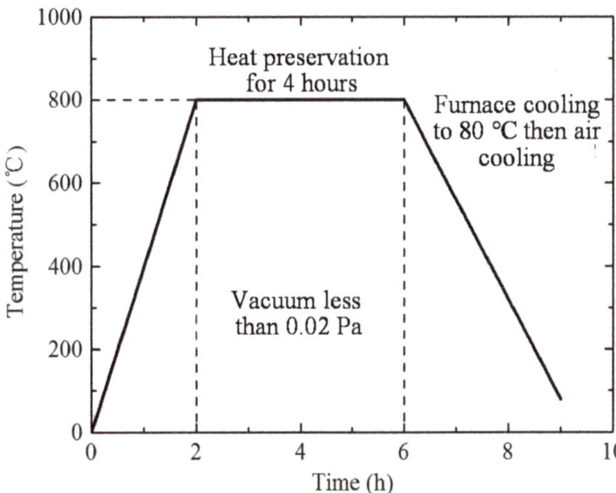

Figure 3. Thermal treatment procedure for Ti-6Al-4V alloy 3CSP specimen.

Figure 4 displays the scanning electron microscope (SEM; Hitachi TM-4000, Hitachi High-Tech Corporation, Tokyo, Japan) images for selected connection region between the face sheet and the corrugated channel core. Built upon these SEM observations, relative size offsets between the as-fabricated specimens and the envisioned specimens were measured, as summarized in Table 3. The largest discrepancy (12.5%) occurs on the face sheet of Specimen VII. In general, as shown in Figure 4, the connection between the face sheet and core exhibits excellent continuity, thus avoiding secondary connection typically needed during conventional processing. Note that all the test specimens listed in Table 1 were fabricated using the method of additive manufacturing. Due to limitations of the additive manufacturing facility used in the current study, relatively large size errors were found in specimens having either thin face sheets (e.g., Specimen VII) or thin core web sheets (e.g., Specimen I). For instance, from the SEM image displayed in Figure 4, partially non-molten particles (0.02–0.03 mm in size) are present on the surface of core web sheet, causing large roughness of the surface. This is the main reason behind the large relative size offsets between the as-fabricated specimens and the envisioned specimens shown in Table 3.

Figure 4. As-fabricated Ti-6Al-4V alloy 3CSP specimen: (**a**) SEM observation with low magnification view; and (**b**) detail of joint region in (**a**) marked by dashed line A.

Contrasted with specimens fabricated using the traditional methods such as by vacuum brazing, surface property has become a focal point in the evaluation of processing quality for specimens fabricated with additive manufacturing [24,26]. Therefore, surface roughness on the face sheets of as-fabricated specimens with/without polishing treatment was characterized using a confocal microscope (OLYMPUS 3D measuring laser microscope OLS4000; Olympus Corporation, Shinjuku, Japan). As illustrated in Figure 5, the roughness (R_a) of the outer surface that was polished is 0.40 μm, while that of the in-situ inner surface without polishing is 15.13 μm. Further analysis of the unpolished surfaces revealed non-molten powders deposited on these surfaces, thus causing roughened surfaces of face sheets and core web plates, as shown by the SEM image of Figure 4b. The surface roughness results in the relative size offsets between the as-fabricated and envisioned specimens shown in Table 3. These size offsets are one of the reasons for the relatively large deviations of the present analytical/numerical predictions from experimental measurements. Previous studies showed that reasonable surface treatment such as surface polishing contributes to the removal of non-molten parent powders [24] and the reduction of surface roughness [27], leading to improved constitutive mechanical behavior at microscopic level. Likewise, the influence of non-molten residuum on the bending performance of a 3D printed specimen is obvious [28].

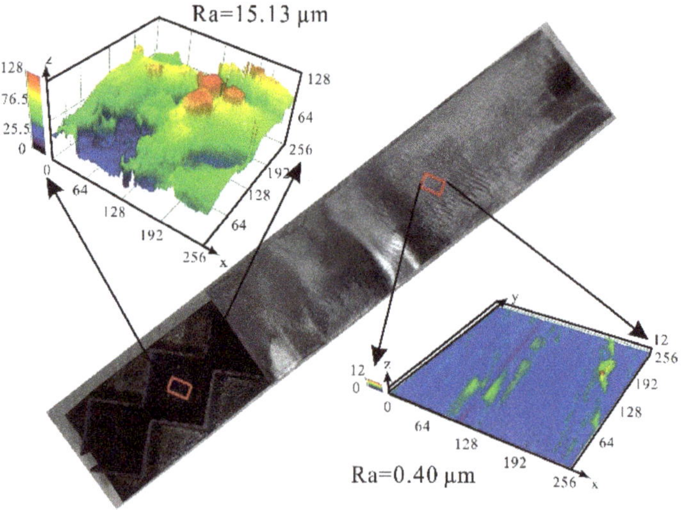

Figure 5. Typical surface characteristics of as-fabricated Ti-6Al-4V alloy 3CSP specimen.

2.1.3. Material Properties of Parent Material

To characterize the mechanical properties of the parent material, dog-bone shaped tensile specimens of Ti-6Al-4V were manufactured with SLM 3D printing. To ensure identical mechanical properties, the dog-bone tensile specimens and the 3CSP test specimens for 4-point bending were fabricated simultaneously. Tensile tests were performed using a standard servo-hydraulic test machine (MTS-858 Mini bionix; MTS Corporation, Eden Prairie, MN, USA) at ambient temperature, with a nominal strain rate set as 1×10^{-3} s^{-1} in reference to ISO 6892-1:2009 [29]. The load and the strain were measured by the loading gauge with an accuracy of 1 N and an extensometer with an accuracy of 10^{-3} mm, respectively.

Quasi-static uniaxial true strain versus true stress curve measured using the dog-bone shaped Ti-6Al-4V specimen is presented in Figure 6. It is shown that the parent material obtained with SLM 3D printing can be expressed approximately as an elastic-plastic material, with a Young's modulus of E_s =111 GPa, a yield strength of σ_{ys} =1072 MPa, and a linearly hardening modulus of E_{ts} =1.09 GPa. These measured material properties of Ti-6Al-4V were subsequently applied to analytical predictions and finite element (FE) simulations, as detailed in the following sections.

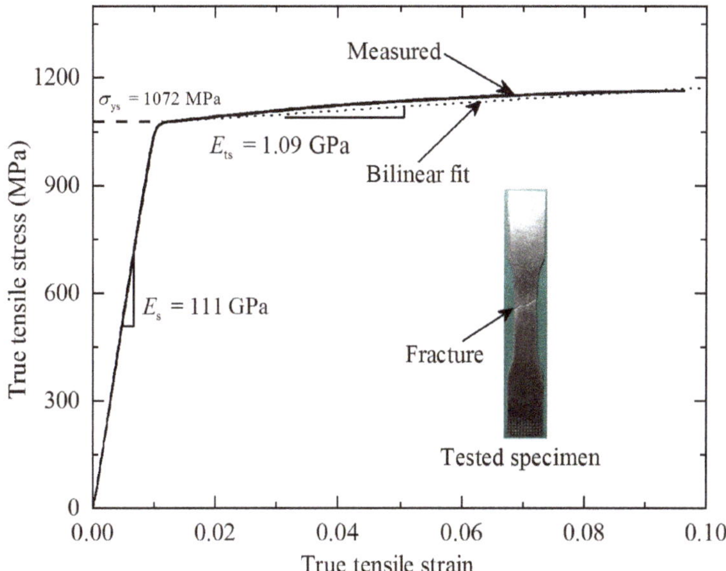

Figure 6. Uniaxial true stress versus true strain curve of 3D-printed Ti-6Al-4V alloy.

2.2. Experimental Setup for Four-Point Bending

Both four-point bending and three-point bending tests are widely used experimental methods for assessing the bending response of a structure. Compared with three-point bending test, four-point bending test produces a pure bending region between the two indenters, thus reducing stress concentration in the region in close contact with the indenter. In the current study, the four-point bending experiment setup is illustrated in Figure 7. The sandwich specimen was placed between the indenters and bearings, both having a diameter of 10 mm. Four-point bending test was performed with an MTS-880 experimental loading system (MTS Corporation, Eden Prairie, MN, USA). With reference to testing codes ASTM C393-11 and D7249, the loading rate was fixed at 0.5 mm/min to mimic quasi-static loading. Analog signals for the load and displacement data were extracted by the measure sensors of the MTS system. To capture the deformation and collapse evolution of each specimen, a high-resolution photography system was utilized to record the entire experiment process.

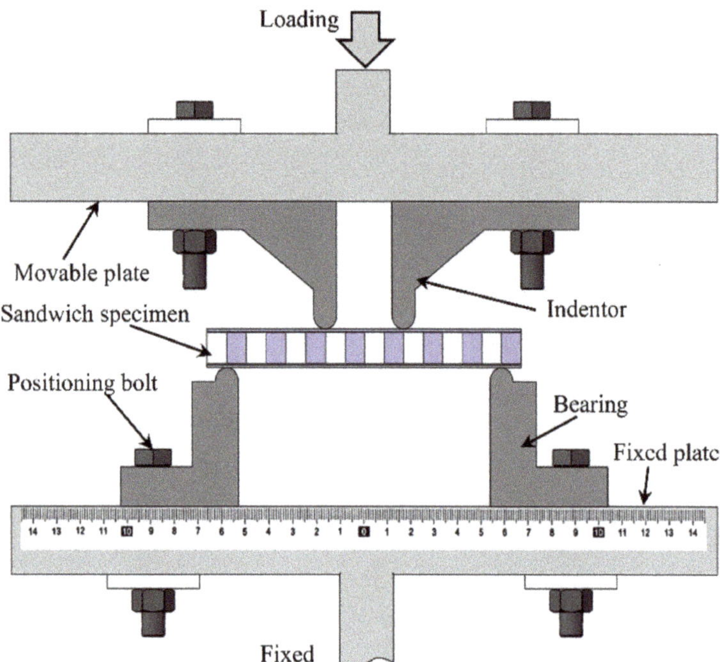

Figure 7. Set-up for four-point bending test.

2.3. Finite Element Simulation

Finite element (FE) simulations were performed with ABAQUS v6.10 (Dassault Systèmes, Vélizy-Villacoublay, France). Geometries of the FE model were identical to those of the test specimen. Ti-6Al-4V was modeled as an isotropic hardening material, as characterized in Section 2.3, with a Young's modulus of E_s = 111 GPa, a Poisson ratio of v_s = 0.34, and a yield stress of σ_{ys} = 1072 MPa (Figure 6). Both the face sheets and core webs of the sandwich specimen were meshed using four-node shell elements (S4R), with their dimensions set not according to the designated values but to the measured sizes listed in Table 3. As a result, relative size offsets and deviations were accounted for in the FE simulations. Each 3CSP specimen was made and integrated with 3D printing, requiring no secondary connection as in the construction of sandwich structures using traditional methods. From the SEM image shown in Figure 4a, it can be seen that, with additive manufacturing, no interaction layer is present between the core web and the face sheet. Therefore, in the present FE simulations, the face sheet and the core web were merged into one component using the Boolean operation: that is, it was assumed that the core web is perfectly bonded to the face sheet. It should be mentioned that, at present, there is not yet a report on the Poisson ratio of 3D printed titanium alloy. In the current study, as an approximation, we simply used the Poisson ratio (0.34) measured for traditional rolled titanium alloy sheets [20]. In a future study, we plan to evaluate systematically how 3D printing and the associated processing parameters affect the Poisson ratio of titanium alloys.

The indenters and bearings were modeled as 3D analytical rigid shells, as they are much stiffer than the sandwich specimen. A fixed displacement boundary condition was used for each bearing. An automatic face-to-face interaction algorithm based on the penalty function was used to simulate the contacts between indenter/bearings and face sheets, with the friction coefficient in the tangential direction fixed at 0.1. The selection of a meshing size of ~1 mm was determined after performing a systematic study of mesh convergence, as further refinement is found to have negligible effect on numerically simulated four-point bending responses. For Specimens II and VIII, Figure 8 plots the numerically calculated initial failure load P, normalized by $P_{ref.}$ calculated with the mesh size fixed at 1 mm, as a function of mesh size. Similar results were obtained for other specimens and hence not presented here for brevity. It is shown that the value of $P/P_{ref.}$ converges when the mesh size drops to 3 mm for Specimen II and 1 mm for Specimen VIII. Meanwhile, the stable time increment ΔT is also plotted in Figure 8, where $\Delta T_{ref.}$ is the stable time increment estimated with 1 mm mesh size. The larger is the $\Delta T_{ref.}/\Delta T$, the lower is the computational efficiency. Thus, with the numerical convergency and computational efficiency considered simultaneously, the optimal mesh size in this study was selected as 1 mm.

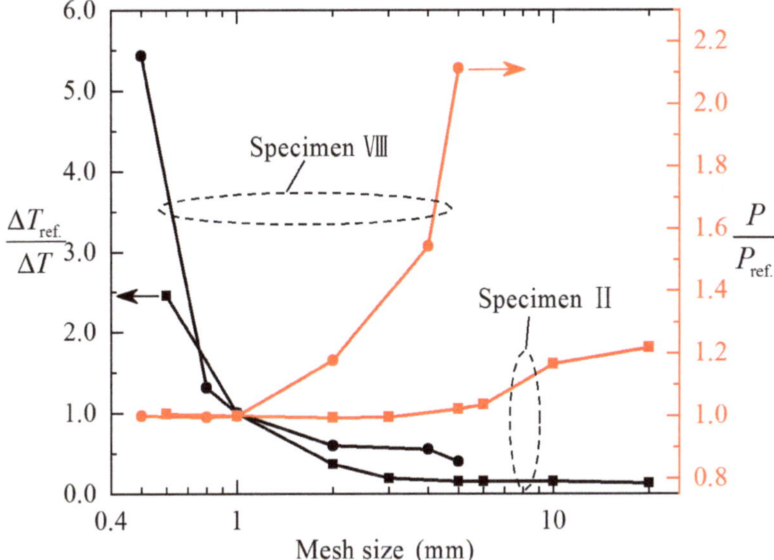

Figure 8. Mesh insensitivity on initial failure load and computational efficiency for Specimens II and VIII.

To simulate quasi-static bending with initial imperfections, the present FE simulations were conducted with linear perturbation and explicit dynamics solvers. A buckling eigenvalue analysis was first performed to select the initial geometry imperfection pattern for subsequent quasi-static four-point bending simulation. Upon performing a velocity independence analysis, the velocity of the indenters was set to be 0.1 m/s.

2.4. Analytical Modeling

This section describes how the flexural stiffness and initial collapse load of the present 3CSP sandwich structure can be predicted using analytical models. Following the classical work of Allen [30], it was assumed that the face sheets are mainly subjected to bending moments, while the cores are mainly subjected to transverse shear. To make the prediction theoretically valid, none of the specimens in this study was allowed to exceed the geometric threshold for normalized core height [18].

2.4.1. Bending Stiffness

The total deflection δ of a sandwich structure under four-point bending can be expressed as a linear superposition of deflection δ_M caused by the bending moment and deflection δ_V caused by the shear force, namely:

$$\delta = \delta_M + \delta_V \tag{1}$$

The deflection caused by the bending moment is calculated as [31]:

$$\frac{\delta_M}{P} = \frac{(L_b - \chi)^2 \chi^2}{6 L_b D_{eq}} + \frac{\chi}{12 L_b D_{eq}}\left[\frac{L_b}{\chi}(2\chi - L_b)^3 + \left(L_b^2 - 2\chi^2\right)\chi\right] \tag{2}$$

where P is the total force, half of which is transferred to the loaded specimen by each indenter. Let D_{eq} denote the equivalent flexural rigidity of the sandwich.

In practice, the corrugated channel core is in general not loaded in the transverse direction (y-direction in Figure 1) [32]. Therefore, contribution of the core to the flexural rigidity may be ignored. The equivalent flexural rigidity D_{eq} can thence be expressed as:

$$D_{eq} = E_s \frac{B t_f^3}{6} + E_s \frac{B t_f h^2}{2} \tag{3}$$

where E_s is the Young's modulus of the face sheet material.

In a simply supported beam subjected to a total force P, the shear force in the region between the left indenter and the left bearing is $P/2$. Thus, the deflection caused by the shear force is determined as:

$$\frac{\delta_V}{P} = \frac{\chi}{2 G_{eq} B h} \tag{4}$$

where represents the equivalent shear modulus of the core. Its value can be calculated as:

$$G_{eq} = \bar{\rho} G \cos^2 \theta \tag{5}$$

where $\bar{\rho} = \frac{t_c}{d \cos \theta}$ is the relative density of the corrugated channel core and G denotes the shear modulus of the parent material of the core.

2.4.2. Initial Failure Loads

Under four-point bending, a 3CSP specimen may exhibit four major collapse modes: (i) face yielding (FY); (ii) face buckling (FB); (iii) core yielding (CY); and (iv) core buckling (CB). The pure bending section of the specimen is subjected to the largest bending moment, given by:

$$M = \frac{P\chi}{2B} \tag{6}$$

where M represents the bending moment per width. As mentioned above, it was assumed that the bending moment is carried by the face sheets. Therefore, the maximum normal stress on the loaded face sheets is:

$$\sigma_f = \frac{P\chi}{2 B t_f (h + t_f)} \tag{7}$$

At the same time, the maximum shear force per width in the loaded core is determined as:

$$V = \frac{P}{2B} \tag{8}$$

such that the maximum shear stress in the loaded core is:

$$\tau_c = \frac{Pd}{2 B t_c h} \tag{9}$$

Correspondingly, the initial failure criteria of each collapse mode are summarized as shown below. The critical stress of face yielding is:

$$\sigma_f = \sigma_{ys} \tag{10}$$

where σ_{ys} is the tensile yield strength of the parent material. It follows that the face yielding load is:

$$P_{FY} = \frac{2Bt_f(h + t_f)}{\chi}\sigma_{ys} \tag{11}$$

At the onset of core yielding, the critical load is:

$$\tau_c = \tau_{ys} \tag{12}$$

where τ_{ys} is the shear yielding strength of the parent material, which was assumed to depend upon the tensile yield strength as $\tau_{ys} = \sigma_{ys}/\sqrt{3}$. Correspondingly, the core yielding load is determined as:

$$P_{CY} = \frac{2Bt_c h}{d}\tau_{ys} \tag{13}$$

The critical stress of face buckling is given by:

$$\sigma_{fb} = \frac{k_{fb}\pi^2 E_s}{12(1 - \nu_s^2)}\left(\frac{t_f}{d}\right)^2 \tag{14}$$

where E_s and ν_s denote the Young's modulus and Poisson ratio of the parent material, respectively. k_{fb} is the compression buckling coefficient, approximately equal to 6.97. Face buckling will not take place if $\sigma_f \leq \sigma_{fb}$. Hence, the critical load of face buckling is given by:

$$P_{FB} = \frac{2Bt_f(h + t_f)}{\chi}\sigma_{fb} \tag{15}$$

The shear buckling stress induced by core buckling is:

$$\tau_{cb} = \frac{k_{cb}\pi^2 E_s}{12(1 - \nu_s^2)}\left(\frac{t_s}{h}\right)^2 \tag{16}$$

where k_{cb} is the shear buckling coefficient, which depends on the geometry parameter h/s. For a rectangular core panel with an aspect ratio of $h/s =1$, the value of k_{cb} is assumed to be 14.71. Core buckling occurs when $\tau_c = \tau_{cb}$. Therefore, the critical load of core buckling is calculated as:

$$P_{CB} = \frac{2Bt_c h}{d}\tau_{cb} \tag{17}$$

3. Results and discussion

In this section, quasi-static four-point bending results obtained from analytical predictions, FE simulations, and experimental measurements are summarized and compared. Let the non-dimensional bending stiffness be represented by $S = \overline{P}/\overline{\delta}$, where $\overline{P} = P/(2E_s B\chi)$ is the non-dimensional initial failure load and $\overline{\delta} = \delta/\chi$ is the corresponding non-dimensional indenter displacement.

3.1. Observations of Structural Failure

3.1.1. Core Yielding

Figure 9 displays the experimentally measured, analytically predicted, and numerically simulated load versus deflection curves for 3CSP Specimen II (Table 1): note that the initial failure of this specimen is core yielding, as indicated by the analytical predictions. Corresponding to the circled numbers on the curves, photos of the deformed specimen are compared in Figure 10 with those numerically simulated. It should be pointed out that, during the present experiments, the method of synchronous triggering is used to ensure that data collection by the testing machine and image acquisition by the photography system have identical zero time. Thus, for any point on the load versus displacement curve, the exact photo image corresponding to this point can be extracted, thus enabling the comparisons shown in Figure 10 and Figure 12. The loading procedure is marked as Points 1–5, while the unloading as Points 6 and 7. As Figure 9 illustrates, during loading, the bending response of Specimen II is composed of three major phases: (i) the elastically loading phase (1–3); (ii) the plastically stable phase (3–4); and (iii) the plastic buckling phase (4–5). The results in Figures 9 and 10 show that the key features obtained with FE simulations agree with corresponding experimental observations. In the elastically loading phase, the initial failure load $\overline{P}$ increases linearly with indenter displacement $\overline{\delta}$. The specimen exhibits global bending, with no obvious out of plane deformation in the face sheets and core webs, as shown in Sequence 2 of Figure 10. With further loading, a plastic stable phase emerges, represented by 3 and 4 in Figure 9, wherein a gradual reduction in stiffness occurs, and the load reaches a peak at Point 4. From the corresponding Image 3 shown in Figure 10, the premature collapse initiation of core yielding can be seen in the core shear region between the indenter and the bearing. As the loading continues, shear deformation in the core becomes more visible in Image 4 of Figure 10. Then, the plastic buckling response begins and the load decreases with increasing indenter displacement. Subsequently, an unloading procedure is performed. The corresponding response is shown on the experimental and simulation curves as 5–7. During unloading, as elastic strain energy is released, springback of the specimen occurs, the radius of curvature becomes larger, and the deformation decreases. When the load $\overline{P}$ is dropped to 0 N, permanent plastic deformation characterized with core yielding dominates the specimen.

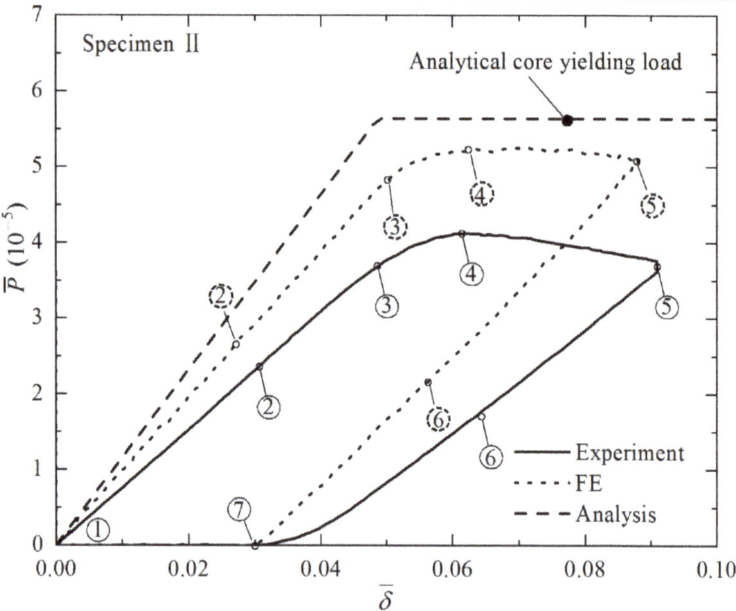

Figure 9. Non-dimensional load versus non-dimensional indenter displacement curve of Specimen II under four-point bending. Numbers circled by solid and dashed lines represent typical phases of structural responses observed in experiment and FE simulation, respectively. Corresponding deformed specimen configurations are displayed in Figure 10.

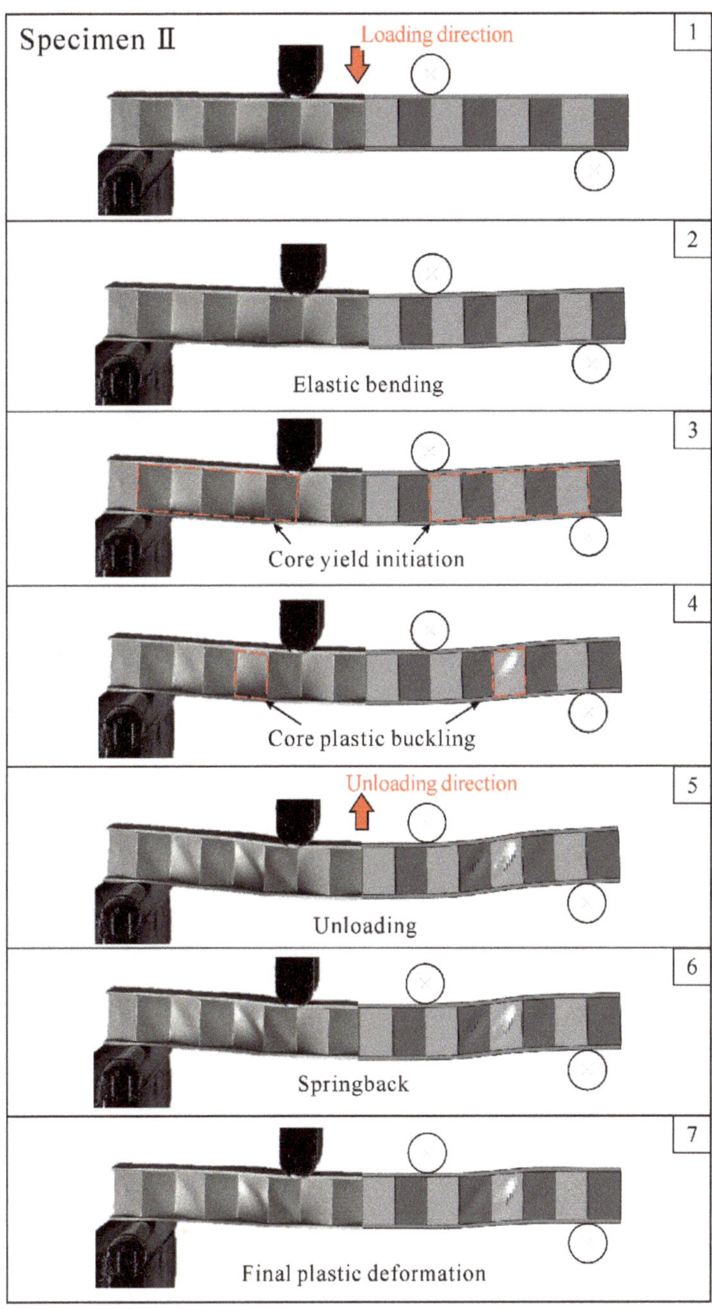

Figure 10. Comparison between numerically simulated deformation and failure patterns of Specimen II with those observed during experiment. Circled numbers correspond to those marked on load versus deflection curves of Figure 9.

3.1.2. Face Yielding

Similar to the case of core yielding, load versus displacement curves and corresponding deformed configurations are presented in Figures 11 and 12 for the case of face yielding (Specimen IV in Table 1). Again, the bending response can be characterized into three phases. After a linearly progressive elastic phase, plastic deformation dominates with the initiation of face yielding, as indicated on Images 3 and 4 of Figure 12. The plastic deformation is accompanied by stretching of the bottom face sheet and contraction of the top face sheet in the pure bending region. Unlike Specimen II, the core of Specimen IV experiences minimal plastic deformation, while its face sheets undergo dramatically plastic deformation. The top face sheet buckles, developing a number of waves, as shown in Image 5 of Figure 12. For Specimens VI, the FE simulation results agree well with those of the experiments, but the analytical predictions somewhat overestimate due to idealized assumptions made in the modeling.

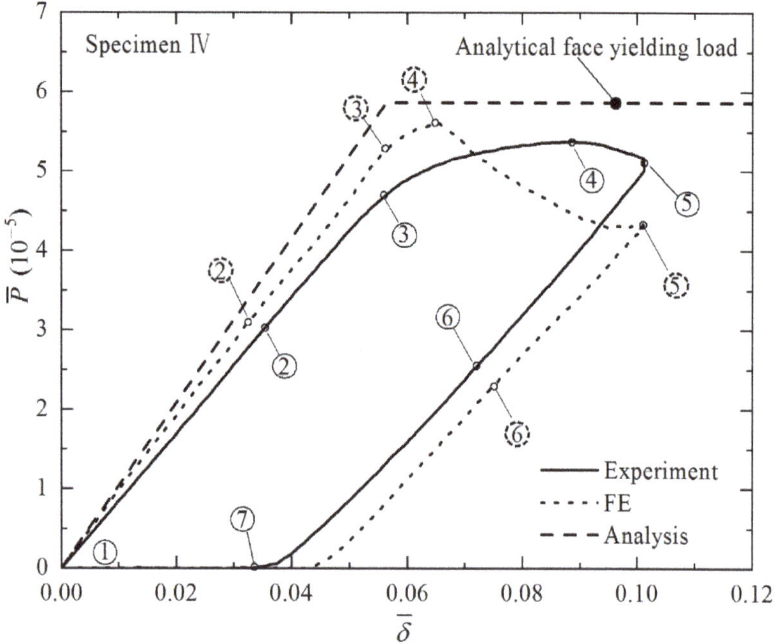

Figure 11. Non-dimensional load versus non-dimensional indenter displacement curve of Specimen IV under four-point bending. Numbers circled by solid and dashed lines represent typical phases of structural responses observed in experiment and FE simulation, respectively. Corresponding deformed specimen configurations are displayed in Figure 12.

3.1.3. Face Buckling

In addition to the collapse modes discussed above for Specimens II and IV, face buckling is captured in Specimen VII, both experimentally and numerically. Its final deformation patterns and load versus deformation curves obtained via experiment and FE simulation are displayed in Figures 13 and 14. Good agreement is obtained between numerical and experimental results for bending stiffness and initial failure load. The elastic buckling of a plate is well known to exhibit a stable post buckling response [33]. Therefore, the peak loads obtained from the present experiment and simulation are much larger than that predicted analytically, as shown in Figure 14.

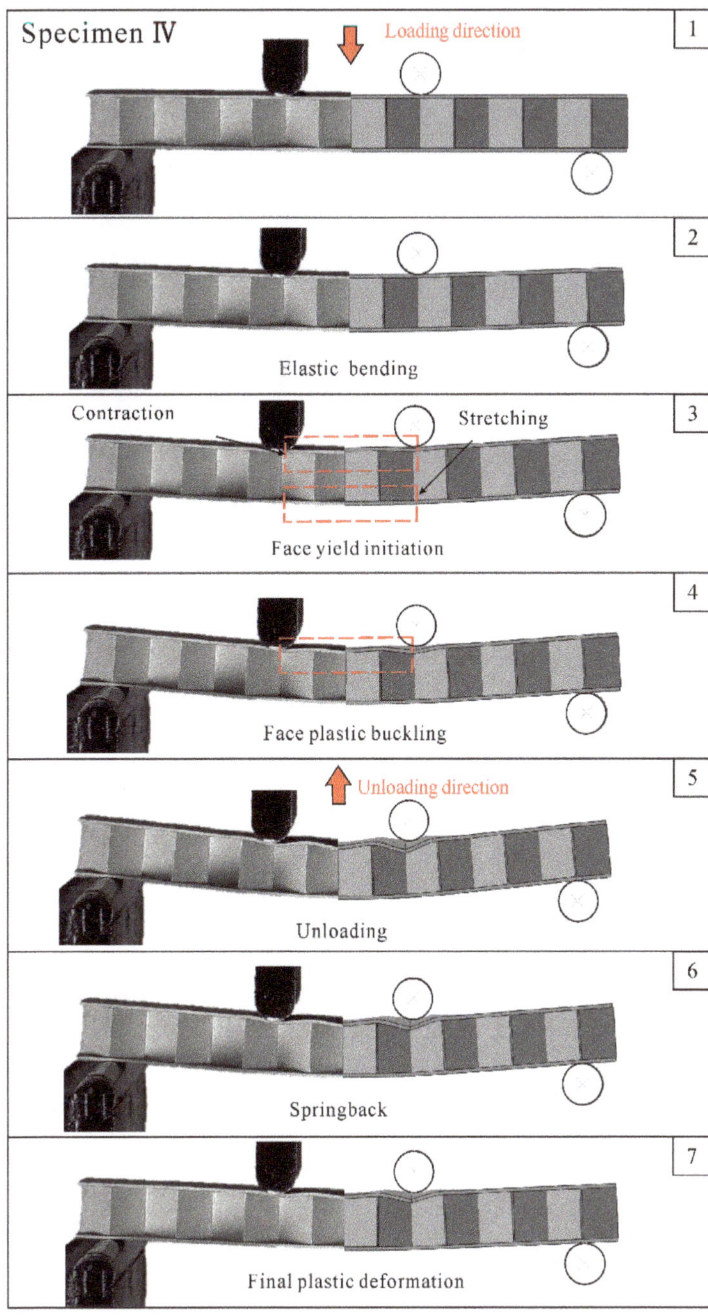

Figure 12. Comparison between numerically simulated deformation and failure patterns of Specimen IV with those observed during experiment. Circled numbers correspond to those marked on load versus deflection curves of Figure 11.

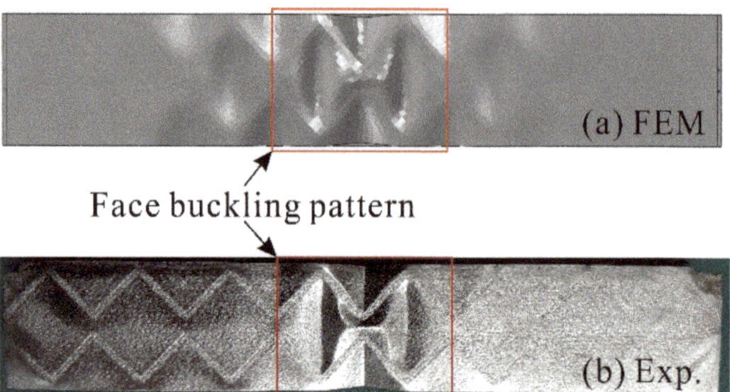

Figure 13. Final deformation pattern of Specimen VII: comparison between (**a**) FE simulation and (**b**) experiment.

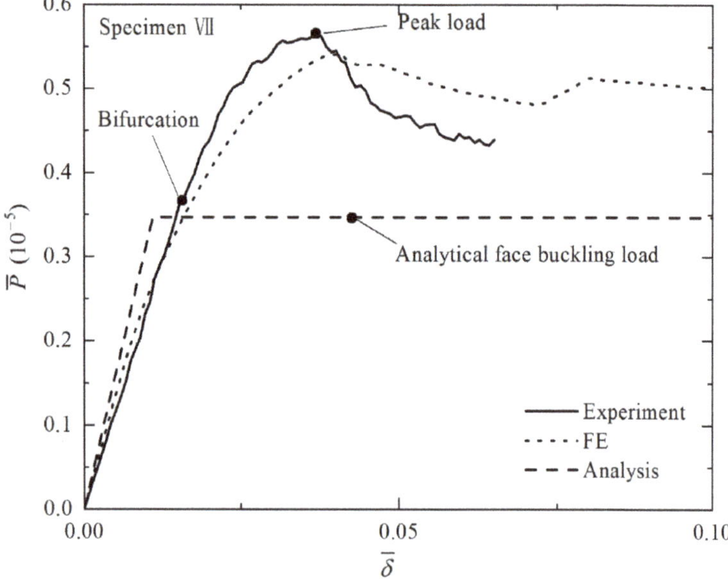

Figure 14. Non-dimensional load versus non-dimensional indenter displacement curves of Specimen VII.

3.2. Comparison among Experimental, Analytical and FE Results

For each 3CSP specimen listed in Table 1, Table 4 compares the experimental, analytical, and numerical (FE) results obtained for its non-dimensional bending stiffness, non-dimensional initial failure load, and collapse modes. It is shown that, when contrasted with experimental measurements, both the analytical and FE models predict accurately the failure modes but overestimate the bending stiffness and initial failure load. For the analytical model, the prediction error ranges 18–41%, whereas for the FE model, the error ranges 5–26%. Further, it is found that the load-carrying capacity of a 3CSP structure depends strongly on its structural geometry. To explore the underlying mechanisms, three typical experiments with different failure modes are analyzed in detail next.

Table 4. Summary of four-point bending responses from analytical prediction, FE simulation, and experimental measurement.

Specimen No.	Bending Stiffness S (10^{-5})			Initial Failure Load $\overline{P}$ (10^{-5})			Collapse Mode		
	Analysis	Simulation	Experiment	Analysis	Simulation	Experiment	Analysis	Simulation	Experiment
I	111.11	99.06	76.91	5.84	5.34	3.72	CY	CY	CY
II	116.00	97.41	77.57	5.64	4.87	3.76	CY	CY	CY
III	129.93	103.12	76.42	7.27	5.46	3.58	FY	FY	FY
IV	104.51	94.55	85.79	5.86	5.26	4.54	FY	FY	FY
V	93.51	80.82	76.19	5.34	4.29	3.74	CY	CY	CY
VI	102.53	90.55	82.82	5.57	5.11	4.65	FY	FY	FY
VII	31.79	22.28	24.02	0.35	0.32	0.36	FB	FB	FB

3.3. Discussion

3.3.1. Effect of Geometric Sizes on Bending Stiffness

The results in Table 4 reveal that relatively large discrepancies exist between experimentally measured and numerically calculated bending stiffnesses and initial failure loads: for instance, for Specimen III, the discrepancy of bending stiffness is 26%. This is mainly attributed to geometric imperfections induced during 3D printing. According to the specific morphology of the corrugated channel core, the method of SLM involves three main steps.

1. Printing: To ensure the face sheet and the core are formed in one step, the specimen needs to be placed at an inclined angle, with a support used between the supporting plate and the suspended surface to ensure the forming accuracy.
2. Heat treatment: To eliminate residual stresses induced during 3D printing, the as-printed specimen (together with the supporting plate) is put into a vacuum furnace, with the furnace temperature controlled in accordance with that shown in Figure 3.
3. Post processing: Upon removing the supporting plate via wire cutting, electrical grinding tool and sandpaper are used to manually polish the surfaces of the specimen.

Taking again Specimen III as an example, because the processing accuracy of the present wire cutting and surface polishing methods is not high, its face sheets are not even, as shown clearly in Figure 15. In the present study, the analytical model assumes that both the top and bottom face sheets of a 3CSP specimen have uniform thickness. Therefore, in accordance with the analytical model, the face sheet thickness listed in Table 3 for each specimen is the average of measurements at selected positions along the face sheet. This causes the discrepancy between model prediction and experimental measurement. Further, in the FE simulations, shell elements having uniform thickness are used to model the face sheet, which also leads to discrepancy between numerical simulation and experimental measurement. In future studies, 3D solid elements in lieu of shell elements will be employed to refine the FE model so as to quantify the influence of non-even face sheets on the mechanical behavior of a 3CSP structure.

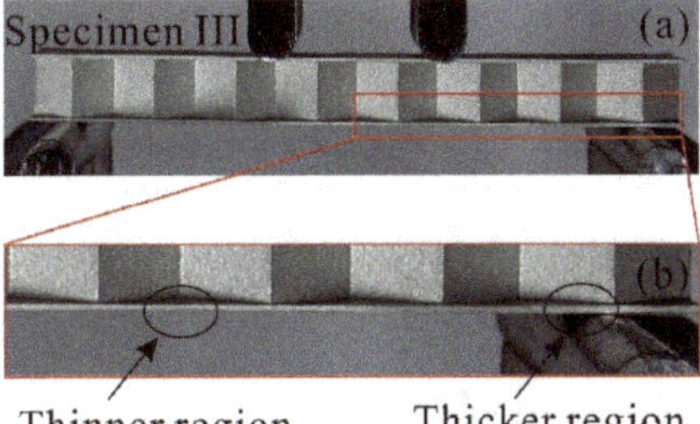

Figure 15. Geometric imperfections of Specimen III: (a) lateral view of Specimen III before experiment; and (b) the partial enlarged view of Specimen III.

When developing the present analytical model, the bending deformation of a 3CSP specimen is taken as the superposition of deformation caused by the bending moment and that induced by the shear force. The former is dictated by the bending stiffness of the face sheets, while the latter is dominated by the shear stiffness of the core. The equivalent shear stiffness of the core, as shown in Equation (5), is derived via homogenization, which requires that a sufficiently large number of cells are present within the characteristic length of the 3CSP structure [34]. In the current study, due to 3D printing accuracy and constraints on printed cell size, the ratio of characteristic length to unit cell length is $\chi/l = 2.5$ for each specimen. This causes the discrepancy between analytical model and FE simulation. Figure 16 plots the analytically predicted bending stiffness as a function of χ/l for specimens having identical total mass ($W = 13.25$ kg/m^2). For comparison, results calculated with the FE model are also presented. It is shown that the two curves converge only when the value of χ/l is sufficiently large, e.g., at $\chi/l = 5.5$. Therefore, to ensure the prediction accuracy of analytical modeling for 3CSP structures, special focus needs to be placed upon the magnitude of χ/l in future studies.

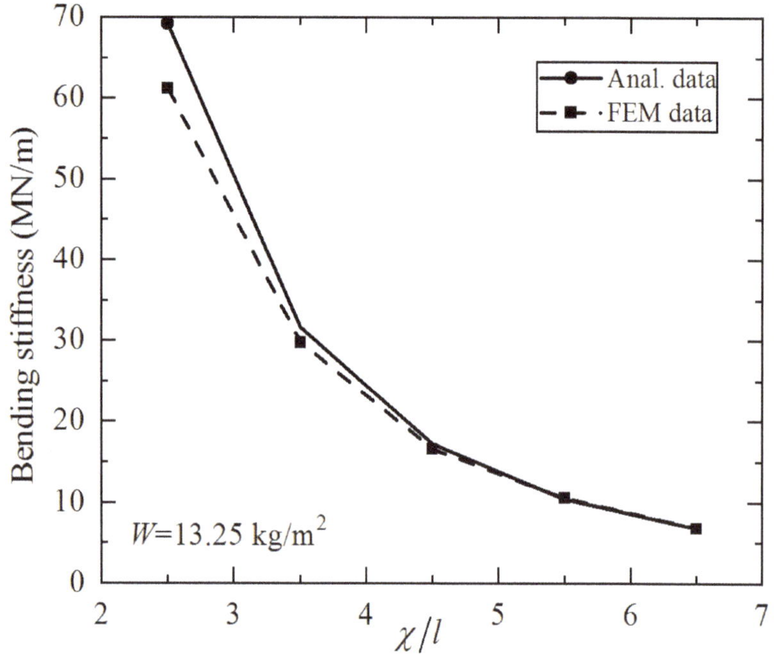

Figure 16. Effect of wave numbers of 3CSP sandwich panels with $W = 13.25$ kg/m^2 on bending stiffness.

3.3.2. Effect of Poisson Ratio

As mentioned above, in the absence of Poisson ratio for 3D printed Ti-6Al-4V, a Poisson ratio of 0.34 measured using traditional rolled titanium alloy sheets is used in the present study. In this section, how the bending response of a 3CSP structure is dependent upon the value of Poisson ratio is quantified using the FE model. As the value of Poisson ratio is varied in the range of 0.10–0.45, Figure 17 presents the errors of numerically calculated bending stiffness and peak load relative to the reference case of Poisson ratio equaling to 0.34. It is shown that, for the bending stiffness, the largest error of 5.54% occurs when the Poisson ratio is 0.1. Similarly, for the peak load, the largest error of 1.7% also occurs when the Poisson ratio is set to 0.1. Overall, the influence of Poisson ratio on the numerically simulated mechanical performance of 3CSP structures is small.

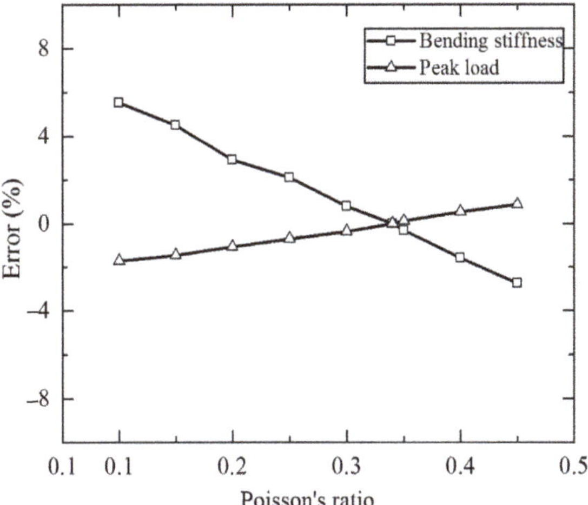

Figure 17. Effect of Poisson ratio on numerically calculated bending stiffness and peak load of 3CSP structure.

4. Collapse Mechanism Maps

Collapse mechanism maps [35,36] are used next to probe the effect of structural geometry on collapse initiation mode. To this end, dimensionless quantifications of structural response are conducted for the cases described in the previous section, with normalized geometrical parameters introduced as:

$$\bar{t}_f = \frac{t_f}{\chi}, \ \bar{t}_c = \frac{t_c}{\chi}, \ \bar{h} = \frac{h}{\chi}, n = \frac{d}{h} \tag{18}$$

Upon substituting these normalized parameters into Equations (11), (13), (15), and (17), the critical loads corresponding to the four collapse modes can be rewritten as:

$$\overline{P}_{FY} = \varepsilon_y \bar{t}_f (\bar{h} + \bar{t}_f) \tag{19}$$

$$\overline{P}_{FB} = \frac{k_{f\,b}\pi^2}{12(1 - v_s^2)} \left(\frac{1}{n\bar{h}}\right)^2 \bar{t}_f^3 (\bar{h} + \bar{t}_f), \tag{20}$$

$$\overline{P}_{CY} = \varepsilon_y \frac{\bar{t}_c}{n\sqrt{3}}, \tag{21}$$

$$\overline{P}_{CB} = \frac{k_{cb}\pi^2}{12(1 - v_s^2)} \frac{\bar{t}_c^3}{n\bar{h}^2}, \tag{22}$$

where $\varepsilon_y = \sigma_y / E_s$ is the yielding strain.

To facilitate visualization of optimization results based on failure maps, Figure 18 plots a prototypical collapse map for the scenario of $n = 1$ and $\psi = 0.0526$, where "FB" "FY", "CB", and "CY" represent abbreviations of the four collapse modes. "F" and "C" represent "Face sheet" and "Core", while "Y" and "B" represent "Yielding" and "Buckling", respectively. Non-dimensional initial failure load contour plots for the 3CSP structure with different geometries are delineated via the dashed lines. With these contour plots, the optimal geometries can be obtained by searching for those associated with higher $\overline{P}$ values. Optimal geometries thus obtained are marked with and without considering geometry limitation, represented by red and blue solid circles, respectively. For a 3CSP specimen with $\psi = 0.0526$, excellent bending resistance can be achieved by a geometry lying at the intersection point (red solid circle) of face buckling, face yielding, and core yielding, if geometrical limitation is ignored. The optimal geometry lies on the confluence (blue solid circle) of face yielding, core yielding, and geometric limitation (if geometric limit is considered). Further, given the geometric parameters of Specimen V in Table 1, its bending resistance can be effectively enhanced by decreasing the normalized thickness of face sheet and increasing the normalized height of core (black solid circle in Figure 18).

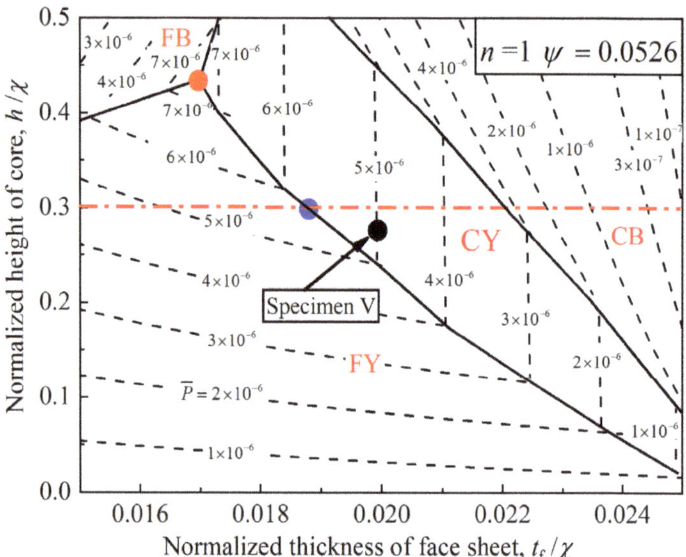

Figure 18. Prototypical collapse map for 3CPS structure and the corresponding optimal design strategy for the scenario of $n = 1$ and $\psi = 0.0526$. Geometric threshold of core height is represented by dash-dotted line.

Figure 19 presents the collapse maps for three 3CSPs having different values of ψ. For the given range of normalized geometries, the initial collapse modes for sandwich structures with different geometries (characterized with different values of n and ψ) vary. Solid circles marked in Figure 19 represent the geometries of experimental specimens described separately in the previous section, namely Specimens V, VI, and VII. Regardless of the geometry of a 3CSP specimen, the four competing collapse modes can all occur. For the scenario of $n = 1$ and $\psi = 0.0363$, the dominating initial collapse modes are collapses produced from the face sheets, which can be validated via the four-point bending test of Specimen VII. Within the $\bar{t}_f$ range of 0.001–0.015 and $\bar{h}$ range of 0.0–0.3, the collapse mechanism mode is presented by face sheet bucking and yielding when $\bar{t}_f$ is less than 0.013. However, when $\bar{t}_f$ exceeds 0.013, the collapse mechanism mode changes from face sheet yielding to core yielding and core buckling. Between these two modes of core collapse, core buckling increasingly dominates as the core height is increased.

The domination of collapse mode(s) also varies with varying structural mass. As the value of ψ is increased, the region representing core collapse grows, as shown in Figure 19a–c. For the scenario of $n = 1$ and $\psi = 0.0526$ (Figure 19b), the component of the collapse mechanism map assembles with the four competing collapse modes: i.e., face sheet buckling, face sheet yielding, core buckling, and core yielding, consistent with the scenario of $n = 1$ and $\psi = 0.0583$ (Figure 19c). However, Figure 19b,c shows that the core buckling and core yielding collapse regions gradually expand relative to those of Figure 19a. Additionally, the solid circles marked in each collapse mechanism map of Figure 19 reveal that the 3CSP specimens tested in this study do not have optimal geometries.

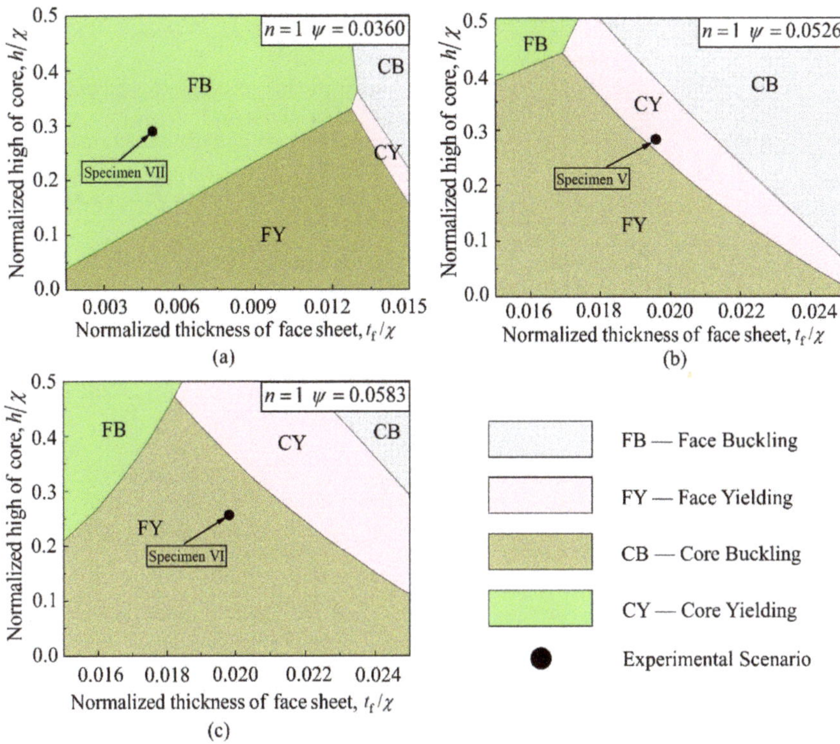

Figure 19. Prototypical collapse maps for each 3CSP structure with: (**a**) $\psi = 0.0363$; (**b**) $\psi = 0.0526$; and (**c**) $\psi = 0.0583$. In all cases, the alternative key parameter was fixed at $n = 1$.

It must be pointed out that the present analytical predictions deviate, in some cases quite significantly, from experimental results. As previously discussed, these large discrepancies are likely caused by the idealized assumptions (e.g., sufficiently large number of unit cells to ensure the prediction accuracy of homogenization) made to simplify the analytical modeling, as well as by the relatively low surface quality of 3CSP specimens fabricated using the 3D printing facility at hand. As a preliminary study, while the idealized analytical model was employed to carry out the current structural optimization, the results need be applied in caution, especially when the core height is large where the intersection points may shift by as much as 20–30%. More comprehensive and accurate collapse mechanism maps will be constructed in future studies with significantly improved analytical models.

5. Conclusions

Novel ultra-lightweight corrugated-channel-cored sandwich structures are envisioned and fabricated from Ti-6Al-4V alloy using the selective laser melting (SLM) methodology. Their performances (e.g., bending resistance and initial collapse modes) when subjected to four-point bending are subsequently experimentally measured. A combined approach of analytical modeling and numerical simulation based on the method of finite elements (FE) is employed to further explore in detail physical mechanisms underlying the bending performance. The main conclusions are summarized as follows.

1. Four competing collapse initiation modes, i.e., face-sheet yielding, face-sheet buckling, core yielding, and core buckling, govern the failure processes of a 3CSP structure as its geometry is varied.
2. Both the analytical and FE models predict accurately the failure modes but overestimate the bending stiffness and initial failure load.
3. Collapse mechanism maps constructed using the developed analytical model provide an effective strategy for designing 3CSP structures with optimal bending responses.
4. The collapse mechanism maps can be employed to quantify the influence of 3CSP structural topology on collapse initiation modes.

The proposed 3CSP structures hold great potential in a wide variety of multifunctional applications targeting simultaneous load-bearing, and heat dissipation via active cooling at ultra-lightweight. Nonetheless, several issues of the present work need to be addressed in future studies, including the relatively large deviations of analytical and FE predictions from experimental measurements, the determination of Poisson ratio for 3D printed titanium alloys, the improvement in processing quality with 3D printing, and the characterization of heat transfer performance of 3CSP structures.

Author Contributions: Conceptualization, Z.Z. and T.J.L.; methodology, Z.Z., Q.Z. and T.J.L.; software, Z.Z. and Y.Z.; investigation, Z.Z. and J.R.; resources, S.D., Z.Y.; data curation, J.R., X.W. and Y.Z.; writing—original draft preparation, Z.Z., J.R. and Z.W.; writing—review and editing, Z.Z., J.R. and T.J.L.; supervision, T.J.L.; and funding acquisition, Z.Z. and T.J.L. All authors have read and agreed to the published version of the manuscript.

Funding: This work was supported by the National Natural Science Foundation of China (11972185, 12002156, and 11902148), China Postdoctoral Science Foundation (2020M671473), State Key Laboratory of Smart Manufacturing for Special Vehicles and Transmission System (GZ2019KF015), Open Fund of State Key Laboratory of Mechanics and Control of Mechanical Structures (MCMS-E0219K02 and MCMS-I-0219 K01), and the Priority Academic Program Development of Jiangsu Higher Education Institutions (PAPD).

Institutional Review Board Statement: Not applicable.

Informed Consent Statement: Not applicable.

Data Availability Statement: The data presented in this study are available on request from the corresponding author.

Acknowledgments: The authors would like to thank Shenzhen Huayang Laser Co. Ltd. for supporting the 3D-printed specimens applied in this study.

Conflicts of Interest: The authors declare no conflict of interest.

References

1. Valdevit, L.; Vermaak, N.; Zok, F.W.; Evans, A.G. A Materials Selection Protocol for Lightweight Actively Cooled Panels. *J. Appl. Mech.* **2008**, *75*, 061022. [CrossRef]
2. Liu, F.; Jiang, X.; Wang, X.; Wang, L. Machine learning-based design and optimization of curved beams for multistable structures and metamaterials. *Extreme Mech. Lett.* **2020**, *41*, 101002. [CrossRef]
3. Abueidda, D.W.; Koric, S.; Sobh, N. Topology optimization of 2D structures with nonlinearities using deep learning. *Comput. Struct.* **2020**, *237*, 106283. [CrossRef]
4. Lv, W.; Li, D.; Dong, L. Study on mechanical properties of a hierarchical octet-truss structure. *Compos. Struct.* **2020**, *249*, 112640. [CrossRef]
5. Peng, C.; Tran, P. Bioinspired functionally graded gyroid sandwich panel subjected to impulsive loadings. *Compos. Part B Eng.* **2020**, *188*, 107773. [CrossRef]
6. Ashby, M.F.; Evans, A.G.; Fleck, N.A.; Gibson, L.J.; Hutchinson, J.W.; Wadley, H.N.G. *Metal Foams A Design Guide*; Butterworth-Heinemann: Burlington, MA, USA, 2000.
7. Ashby, M.F.; Cebon, D. Materials selection in mechanical design. *Le J. De Phys. IV* **1993**, *3*, C7-1. [CrossRef]
8. Nguyen, N.T.; Siegmund, T.; Tsutsui, W.; Liao, H.; Chen, W. Bi-objective optimal design of a damage-tolerant multifunctional battery system. *Mater. Des.* **2016**, *105*, 51–65. [CrossRef]
9. Attar, P.; Galos, J.; Best, A.; Mouritz, A. Compression properties of multifunctional composite structures with embedded lithium-ion polymer batteries. *Compos. Struct.* **2020**, *237*, 111937. [CrossRef]

10. Wadley, H.N.G. Multifunctional periodic cellular metals. *Philos. Trans. R. Soc. A Math. Phys. Eng. Sci.* **2005**, *364*, 31–68. [CrossRef]
11. Lu, T.J.; Valdevit, L.; Evans, A. Active cooling by metallic sandwich structures with periodic cores. *Prog. Mater. Sci.* **2005**, *50*, 789–815. [CrossRef]
12. Valdevit, L.; Pantano, A.; Stone, H.; Evans, A. Optimal active cooling performance of metallic sandwich panels with prismatic cores. *Int. J. Heat Mass Transf.* **2006**, *49*, 3819–3830. [CrossRef]
13. Meng, H.; Galland, M.A.; Ichchou, M.; Bareille, O.; Xin, F.X.; Lu, T.J. Small perforations in corrugated sandwich panel sig-nificantly enhance low frequency sound absorption and transmission loss. *Compos. Struct.* **2017**, *182*, 1–11. [CrossRef]
14. Tang, Y.; Ren, S.; Meng, H.; Xin, F.; Huang, L.; Chen, T.; Zhang, C.; Lu, T. Hybrid acoustic metamaterial as super absorber for broadband low-frequency sound. *Sci. Rep.* **2017**, *7*, 1–11. [CrossRef]
15. Wadley, H.; Dharmasena, K.; O'Masta, M.; Wetzel, J.; O'Masta, M.R. Impact response of aluminum corrugated core sandwich panels. *Int. J. Impact Eng.* **2013**, *62*, 114–128. [CrossRef]
16. Wadley, H.; O'Masta, M.; Dharmasena, K.; Compton, B.; Gamble, E.; Zok, F. Effect of core topology on projectile penetration in hybrid aluminum/alumina sandwich structures. *Int. J. Impact Eng.* **2013**, *62*, 99–113. [CrossRef]
17. Ni, C.; Hou, R.; Xia, H.; Zhang, Q.; Wang, W.; Cheng, Z.; Lu, T. Perforation resistance of corrugated metallic sandwich plates filled with reactive powder concrete: Experiment and simulation. *Compos. Struct.* **2015**, *127*, 426–435. [CrossRef]
18. Rathbun, H.; Zok, F.W.; Evans, A. Strength optimization of metallic sandwich panels subject to bending. *Int. J. Solids Struct.* **2005**, *42*, 6643–6661. [CrossRef]
19. Cotea, F.; Deshpande, V.S.; Fleck, N.A.; Evans, A.G. The compressive and shear responses of corrugated and diamond lattice materials. *Int. J. Solids Struct.* **2006**, *43*, 6220–6242. [CrossRef]
20. Zhao, Z.-Y.; Han, B.; Wang, X.; Zhang, Q.; Lu, T.J. Out-of-plane compression of Ti-6Al-4V sandwich panels with corrugated channel cores. *Mater. Des.* **2018**, *137*, 463–472. [CrossRef]
21. Jiang, W.; Ma, H.; Yan, L.; Wang, J.; Han, Y.; Zheng, L.; Qu, S. A microwave absorption/transmission integrated sandwich structure based on composite corrugation channel: Design, fabrication and experiment. *Compos. Struct.* **2019**, *229*, 111425. [CrossRef]
22. Wang, X.; Zhao, Z.-Y.; Li, L.; Zhang, Z.-J.; Zhang, Q.-C.; Han, B.; Lu, T.J. Free vibration behavior of Ti-6Al-4V sandwich beams with corrugated channel cores: Experiments and simulations. *Thin-Walled Struct.* **2019**, *135*, 329–340. [CrossRef]
23. Simonelli, M.; Tse, Y.Y.; Tuck, C. Effect of the build orientation on the mechanical properties and fracture modes of SLM Ti–6Al–4V. *Mater. Sci. Eng. A* **2014**, *616*, 1–11. [CrossRef]
24. Tancogne-Dejean, T.; Spierings, A.B.; Mohr, D. Additively-manufactured metallic micro-lattice materials for high specific energy absorption under static and dynamic loading. *Acta Mater.* **2016**, *116*, 14–28. [CrossRef]
25. Bremen, S.; Meiners, W.; Diatlov, A. Selective Laser Melting: A manufacturing technology for the future? *Laser Tech. J.* **2012**, *2*, 33–38. [CrossRef]
26. Fousova, M.; Vojtech, D.; Kubasek, J.; Jablonska, E.; Fojt, J. Promising charcteristics of gradient porosity TI-6Al-4V alloy pre-pared by SLM process. *J. Mech. Behav. Bioed.* **2017**, *69*, 368–376. [CrossRef] [PubMed]
27. Gautam, R.; Idapalapati, S.; Feih, S. Printing and characterisation of Kagome lattice structures by fused deposition modelling. *Mater. Des.* **2018**, *137*, 266–275. [CrossRef]
28. Vayssette, B.; Saintier, N.; Brugger, C.; El May, M. Surface roughness effect of SLM and EBM Ti-6Al-4V on multiaxial high cycle fatigue. *Theor. Appl. Fract. Mech.* **2020**, *108*, 102581. [CrossRef]
29. *ISO 6892-1:2009 Metallic materials—Tensile Testing. In Part 1: Method of Test at Room Temperature*; ISO: Geneva, Switzerland, 2009.
30. Allen, H.G. *Analysis and Design of Structural Sandwich Panels*; Pergamon Press: Oxford, London, UK, 1969.
31. Roylance, D. Beam Displacement. Department of Materials Science and Engineering, Massachusetts Institute of Technology, 2000. Available online: https://ocw.mit.edu/courses/materials-science-and-engineering/3-11-mechanics-of-materials-fall-1999/modules/MIT3_11F99_bdisp.pdf (accessed on 26 September 2018).
32. Guo, Y.L.; Tong, J.Z.; Jiang, Z.X. *Design fundamentals and application of corrugated-web steel structures*; China Science Publishing & Media Ltd.: Beijing, China, 2015.
33. Wilbert, A.; Jang, W.-Y.; Kyriakides, S.; Floccari, J. Buckling and progressive crushing of laterally loaded honeycomb. *Int. J. Solids Struct.* **2011**, *48*, 803–816. [CrossRef]
34. Hohe, J.; Becker, W. Effective stress-strain relations for two-dimensional cellular sandwich cores: Homogenization, material models, and properties. *Appl. Mech. Rev.* **2002**, *55*, 61–87. [CrossRef]
35. Yan, L.; Han, B.; Yu, B.; Chen, C.; Zhang, Q.-C.; Lu, T. Three-point bending of sandwich beams with aluminum foam-filled corrugated cores. *Mater. Des.* **2014**, *60*, 510–519. [CrossRef]
36. Wei, X.Y.; Wu, Q.Q.; Gao, Y.; Xiong, J. Bending characteristics of all-composite hexagon honeycomb sandwich beams: Experimental tests and a three-dimensional failure mechanism map. *Mech. Mater.* **2020**, *148*, 103401. [CrossRef]

materials

Article

The Use of Neural Networks in the Analysis of Dual Adhesive Single Lap Joints Subjected to Uniaxial Tensile Test

Jakub Gajewski [1],* , Przemysław Golewski [2] and Tomasz Sadowski [2]

1   Department of Machine Design and Mechatronics, Faculty of Mechanical Engineering, Lublin University of Technology, Nadbystrzycka 36, 20-618 Lublin, Poland
2   Department of Solid Mechanics, Faculty of Civil Engineering and Architecture, Lublin University of Technology, Nadbystrzycka 38, 20-618 Lublin, Poland; p.golewski@pollub.pl (P.G.); t.sadowski@pollub.pl (T.S.)
*   Correspondence: j.gajewski@pollub.pl

Abstract: Adhesive bonding are becoming increasingly important in civil and mechanical engineering, in the field of mobile applications such as aircraft or automotive. Adhesive joints offer many advantages such as low weight, uniform stress distribution, vibration damping properties or the possibility of joining different materials. The paper presents the results of numerical modeling and the use of neural networks in the analysis of dual adhesive single-lap joints subjected to a uniaxial tensile test. The dual adhesive joint was created through the use of adhesives with various parameters in terms of stiffness and strength. In the axis of the overlap, there was a point bonded joint characterized by greater stiffness and strength, and on the outside, there was a bonded joint limited by the edges of the overlap and characterized by lower stiffness and strength. It is an innovative solution for joining technology and the influence of such parameters as the thickness of one of the adherends, the radius of the point bonded joint and the material parameters of both adhesive layers were analyzed. The joint is characterized by a two-stage degradation process, i.e., after the damage of the rigid adhesive, the flexible adhesive ensures the integrity of the entire joint. For numerical modeling, the Finite Element Method (FEM) and cohesive elements was used, which served as input data to an Artificial Neural Network (ANN). The applied approach allowed the impact of individual parameters on the maximum force, initiation energy, and fracture energy to be studied.

Keywords: dual adhesive; single lap joints; numerical modeling; artificial neural networks

Citation: Gajewski, J.; Golewski, P.; Sadowski, T. The Use of Neural Networks in the Analysis of Dual Adhesive Single Lap Joints Subjected to Uniaxial Tensile Test. *Materials* **2021**, *14*, 419. https://doi.org/10.3390/ma14020419

Academic Editor: Pasquale Fernando Fulvio
Received: 13 November 2020
Accepted: 9 January 2021
Published: 15 January 2021

Publisher's Note: MDPI stays neutral with regard to jurisdictional claims in published maps and institutional affiliations.

1. Introduction

Adhesive joints are used in assembly technology in almost every field of engineering; it is increasingly used as an alternative method to welding, riveting and other conventional fasteners. In order to increase their strength and thus reduce the joining surface and/or the amount of adhesive, many techniques are used, such as: appropriate preparation of the substrate surface [1,2], chamfering the edges [3], selecting the shape of the overlap [4] or the use of hybrid joints [5,6].

One of the methods is also the use of connections of the "dual adhesive" or "mixed adhesive" type [7,8]. This method involves the use of two layers of adhesive with different properties in one joint. Layers of less stiff adhesive are placed at the ends of the overlap, and the stiff adhesive is used inside the overlap. The use of a flexible layer at the ends of the overlap allows for a significant reduction of stresses, thus increasing the strength of the entire joint. Intensive research into this type of connection is currently underway.

In [9], optimization of single lap joints of steel sheets was achieved through the selection of an appropriate set of adhesives. Four adhesives of different strengths and breaking strains were used. The best results were achieved by combining a very stiff with a very flexible and tough adhesive. An analytical solution was proposed for calculating the allowable breaking force of an adhesive connection.

A significant problem in dual adhesive joints is the phenomenon of mixing of both adhesives and the way they separate. Therefore, the authors in [10] carried out numerical simulations; two independent methodologies were proposed for selecting the intermediate material between the adhesive bands in mixed adhesive joints, attending to the singularity impact. Another approach is to allow the phenomenon of mixing. In the work [11], a Computational Fluid Dynamics (CFD) simulation concerning the flow of adhesive before curing and structural calculations after curing was performed. It has also been proposed to use a special nozzle for the simultaneous application of both adhesive layers.

In order to further optimize a mixed-adhesive connection and the separation of both adhesive layers, notches in the joined sheets can be used, an approach that was presented in [12]. Five types of connections were analyzed: a control without notches and test pieces with two, three, four and five notches. The aim was to find the optimal solution. The tangential stresses were found to be reduced by 34.5% and normal stresses by 26.4%. The best results were obtained for the model with five notches and a layer of epoxy and polyurethane adhesive. However, the disadvantage of this solution is the additional CNC machining operations required.

Depending on the choice of materials and adhesives, the advantages of dual adhesive joints may only become apparent when they are tested in a wider temperature range. The work [13] presents the results for single adhesive and dual adhesive joints using AV 138 (Huntsman Advanced Materials (Switzerland) GmbH Klybeckstrasse 200 CH - 4057 Basel, Switzerland) and SikaFast 5211 NT (Sika Deutschland GmbH Stuttgarter Str. 139, 72574 Bad Urach, Germany). The tests were carried out in the temperature range from $-30\,^\circ$C to $80\,^\circ$C for both quasi-static and dynamic loads. In the temperature range from approx. $-8\,^\circ$C to $58\,^\circ$C, higher breaking force was achieved for the dual adhesive joints; however, the absorbed energy was lower than for the single adhesive joints with the SikaFast 5211 NT flexible layer. Varying temperatures can occur in aviation and space applications. The authors in [14] glued ceramics to metal, which can be used in the installation of thermal barriers. The benefits of using a double adhesive layer were minimal. However, if we consider a temperature range from $-65\,^\circ$C to $100\,^\circ$C, this type of layer allows the joint to work at the level of 50–60% strength in relation to the joint operating at an ambient temperature.

Due to the fact that in dual adhesive connections, both adhesives are mixed at contact, voids or weak bonds may occur [15]. Therefore, it is also important to conduct tests with the use of X-rays to analyze the internal structure [16].

Current research has been limited to analyzing a small number of commercial adhesives. However, in order to fully analyze the influence of material properties and geometric features, it is necessary to use neural networks [17–19]. Neural networks are a component machine learning, which can boost the efficacy of monitoring tools. Multi-layered neural network can be easily understood by an designers and engineers. These machine learning models can be directly deployed due to their increased universality and transparency compared to other methods used in exploratory data analysis and for making predictive models. Artificial neural networks are efficient computing and approximation models. The advantage of using neural networks models is also the ability to work with incomplete data. An analysis of the literature indicates that the research problem is topical. The use of artificial intelligence (AI) methods to analyze the strength of joints, including adhesive ones, is the subject of much research [20–22].

Numerical simulations were carried out in the Abaqus program; the results were used as input data to the neural network. The research was carried out on the novel concepts of joints with a rigid point adhesive joint surrounded by an elastic joint. The research results allowed the optimal range of parameters to be determined.

2. Dual Adhesive Model Description

Two types of adhesives are used in the single lap model: 1—an adhesive with lower stiffness and lower strength, and 2—an adhesive with higher stiffness and greater strength. This type of connection, called "dual adhesive" or "mixed adhesive," is characterized by a two-step operation, which is described later in this article. In order to be able to properly select the strength of the joint and to influence its behavior after exceeding the load capacity, the proportions of the share of the adhesive surface 1 and 2 in the overlap should be selected appropriately. In the models considered, this was done by changing the radius "r" of the point adhesive joint. The second variable parameter was the thickness "g" of one of the adherends. The other dimensions were constant and are shown in Figure 1.

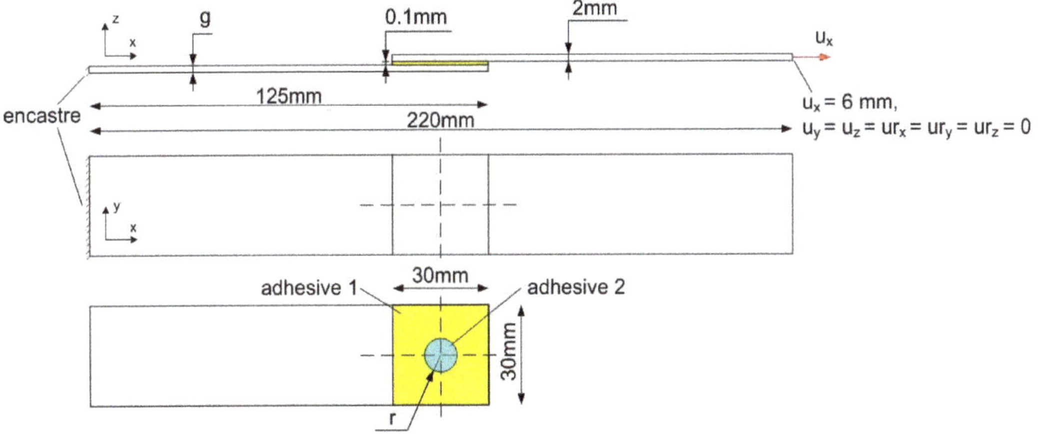

Figure 1. Single lap joint geometry.

In addition to changes in the geometry, modifications were also made to the material properties of both adhesive joints. Data from [23] for such adhesives as Araldite 2015 and Araldite AV138 were used as input data for variable parameters of adhesive joints. The use of two liquid layers of adhesive is a technologically difficult issue, and currently the authors use only rectangular-shaped joints. To prevent the adhesives from mixing, different methods are used—e.g., by using silicone gaskets [24] or by adding fibers [25]. The technology of making this type of connection is not yet fully resolved. Therefore, the current work focuses only on the numerical model. However, the authors are working on the use of double-sided adhesive tape and liquid epoxy as a point joint. Figure 2 shows a practical embodiment of this type of connection. Initially, double-sided tape is applied to one of the joined parts and a hole is made with a die (Figure 2A). In the next step, the liquid epoxy is applied until the hole is filled and the protective film is removed (Figure 2B). Finally, the connection is made by adding a second adherend (Figure 2C). The use of double-sided tape as a layer of lower rigidity has the advantage that the joint formed by the epoxy point joint has the same geometry in each case, since no mixing occurs.

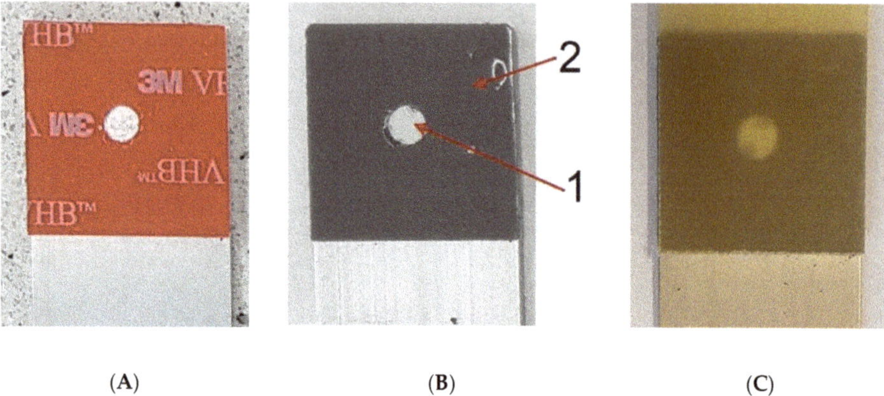

(A)   (B)   (C)

Figure 2. Example of dual adhesive sample preparation. (**A**) double side adhesive tape and liquid epoxy application; (**B**)removing of protective film and (**C**) adding a second adherend.

3. Numerical Modeling

Performing static, dynamic, or cyclic testing is always time-consuming considering samples and laboratory test stand preparation as well as equipment maintenance during the test. Therefore, with a large number of samples, it is necessary to use numerical methods [26].

In this work, 100 numerical simulations were carried out in the Abaqus 6.16 program (Dassault Systemes SIMULIA), which then served as input data to the neural network. Numerical modeling was performed in the Abaqus Explicit program. The joint geometry is shown in Figure 1. In the analyzed model, the following geometric variables were assumed:

- the thickness of one of the adherends "g" (2; 4; 6; 8; 10; 12; 14; 16; 18; 20) (mm),
- the radius of the point adhesive joint "r" (1; 2.25; 3.5; 4.75; 6; 7.25; 8.5; 9.75; 11; 12.25; 13.5; 14.75) (mm).

The analysis of the influence of thickness is due to the fact that in a lap joint there is an eccentricity (distance between the lines of action of the load), which causes the joint to bend and additionally loads the adhesive joint. When the thickness of one or both of the adherends increases, the bending effect is reduced as shown in the Figure 3 for model 6 and model 57 for a similar load level. To better show the bending effect, the deformations of both models were scaled five times.

(**A**)

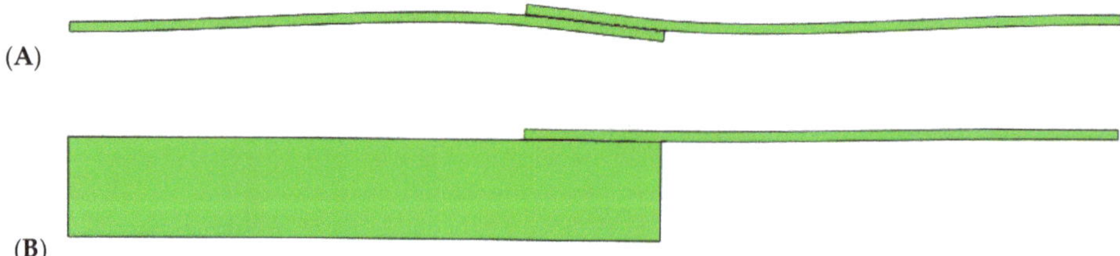

(**B**)

Figure 3. Deformation of single lap joints: (**A**) model 6 and (**B**) model 57.

In addition to geometric changes, the properties of both adhesive joints were modified for Young's modulus E, Kirchoff G modulus, shear strength and tensile strength. The scope of these changes is presented in Table 1. Parameter changes for 100 models are included in Appendix A.

Table 1. Ranges of changes in the parameters of dual adhesive joints.

	Modulus E (MPa)	Modulus G (MPa)	Shear Strength (MPa)	Tensile Strength (MPa)
Adhesive 1	185–436.22	56–132.05	1.7–4.01	2.2–5.19
Adhesive 2	716.73–1850	216.96–700	6.97–18	8.52–22

The quadratic nominal stress criterion (QUADS) was used to describe the initiation of damage of the adhesive joint material. This criterion considers concurring quadratic ratios between nominal stress and allowable stress acting in different directions:

$$\left(\frac{\sigma_n}{\sigma_n^{\max}}\right)^2 + \left(\frac{\sigma_t}{\sigma_t^{\max}}\right)^2 + \left(\frac{\sigma_s}{\sigma_s^{\max}}\right)^2 = 1 \tag{1}$$

where:

- σ_n is the normal stress applied to the surface of the adhesive layer;
- σ_t and σ_s are the shear stress components along the adhesive layer;
- σ_{nmax}, σ_{tmax} and σ_{smax} are the critical values of the normal and shear stress components corresponding to appropriate damage mode initiation.

Damage is assumed to initiate when the maximum nominal stress ratio reaches a value of one.

The elastic/plastic properties for the adherend material were adopted as for aluminum 2017 from TABAL LTD, Poland: Young's modulus E = 72.5 GPa, Poisson ratio ν = 0.33, yield stress σ_y = 250 MPa, peak stress σ_u = 400 MPa and deformation at break A = 10%.

Before making the FEM mesh, a literature analysis was performed. J.J.M. Machado et al. [25] investigated the mesh refinement in the mixed adhesive single lap joints. The calculations were made for 4 mesh densities from 0.2 to 4 mm and from 1 to 10 elements along the thickness. The result clear show reduced mesh dependency of the results with mesh configuration, even with very coarse meshes. This demonstrates that cohesive element modelling is suitable for modelling of large and complex structures, as the need for refinement is minimal.

In the current model, the adherends was built on the basis of C3D8R (eight-node brick element with reduced integration) elements, the global size of the element was 1 mm and 4 elements along the thickness of the sheets. COH3D8 elements (eight-node three-dimensional cohesive element) were used for both adhesive layers, the global element size for the outer layer was 1 mm, and for the point joint 0.5 mm (Figure 4).

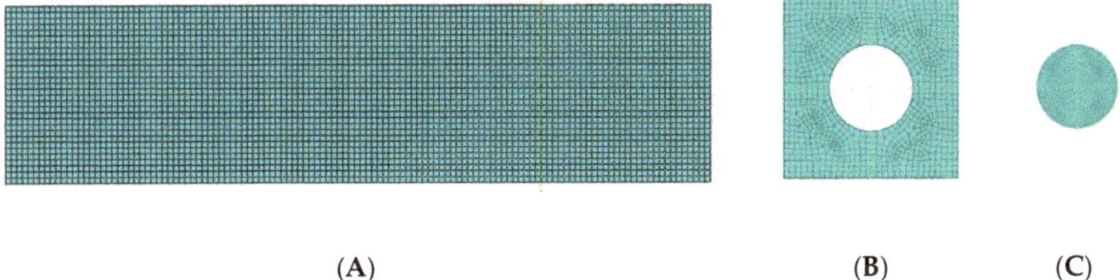

(A)   (B)   (C)

Figure 4. (A–C) Mesh of finite elements.

For all parts of the assembly, "tie" constraints were used, thereby removing all degrees of freedom between the surfaces in contact.

In order to be able to read the reaction values and assign boundary conditions, the RP-1 and RP-2 reference points were created on the front surfaces of both laps (Figure 5).

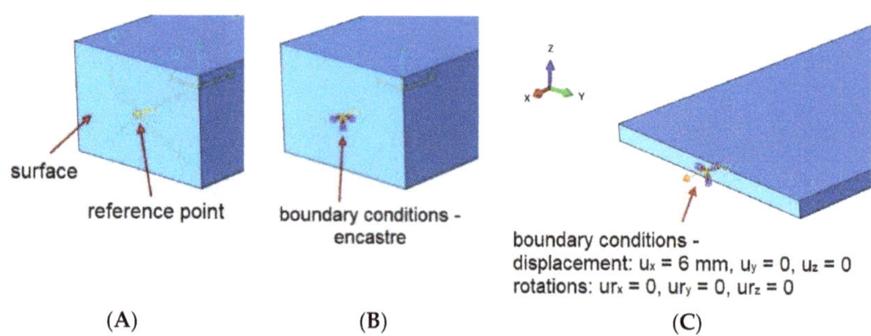

Figure 5. Boundary conditions. (**A**) Reference point connected to the surface; (**B**) Encastre boundary conditions and (**C**) Displacement boundary conditions.

The reference points were connected to the appropriate surfaces with a "coupling" type constraint, which allowed all the degrees of freedom to be transmitted (Figure 5A). One end of the sample was encastre (Figure 5B), while the other end could only move along the "x" axis up to 6 mm (Figure 5C). The simulations were carried out in Abaqus Explicit for a time of 1 s; therefore, it was also necessary to use mass scaling to shorten the calculation time. The target time increment was set to 2×10^{-6}. After the simulation, the values of elastic energy and kinetic energy were controlled (Figure 6).

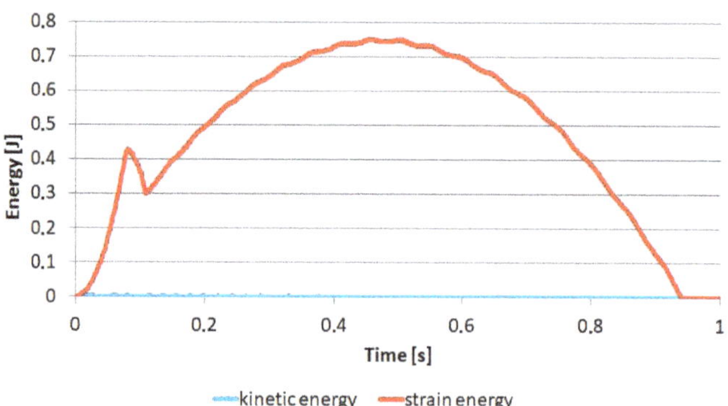

Figure 6. The change of energy over time.

An example force-displacement diagram is shown in Figure 7. For each model, the total energy required to fracture the sample and the damage initiation energy were calculated. The work done or the energy consumed for a given displacement is computed as:

$$W_{(s)} = \int_0^s F_{(s)} ds \tag{2}$$

The maximum force value (F_m) and corresponding displacement (s_m) were assumed as the limit point for calculating the initiation energy. Figure 7 additionally shows points from 1 to 6, which correspond to the stress fields in the adhesive layer in Figure 8.

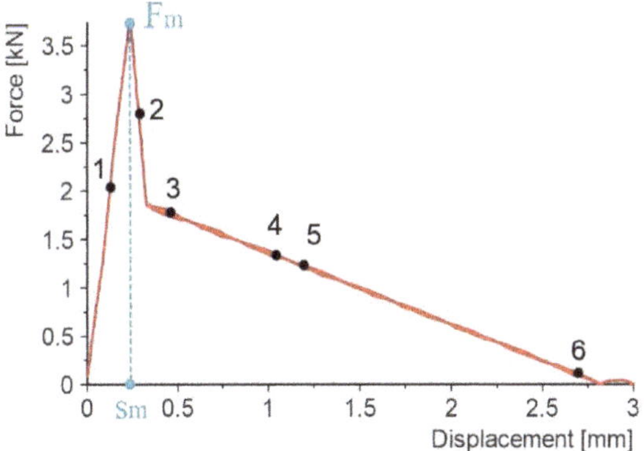

Figure 7. Force displacement graph for model 6.

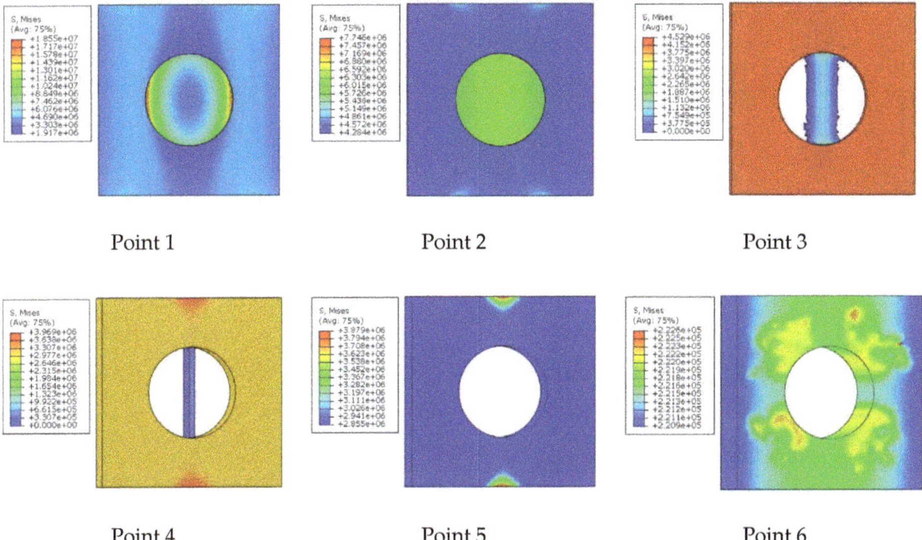

Figure 8. Degradation of adhesive layers for model 6 in [Pa] (1e6 = 1×10^6 Pa).

In point 1, the load is transferred mainly through the point adhesive, which is responsible for the stiffness of the entire joint. A rigid and brittle connection fails with a slight displacement—in this case the maximum value was reached for a value of 0.25 mm. After reaching the maximum force, the stiffness of the point adhesive material gradually degrades. From point 3 onwards, the load is mainly carried by the outer joint. A small fragment in the axis of the overlap is still visible in point 4. In point 6, significant deformation of the outer joint is visible, which is damage for a displacement of 2.8 mm.

Until the external joint is damaged, the entire joint is an integral whole. As soon as the connection is relieved with a significant displacement of, e.g., 2 mm, both parts of the plate will still be connected. This type of connection can be used wherever it is required not only for a large force to be transmitted by the connection but also for a large displacement to be damaged, and thus a large fracture energy. Therefore, further work is planned for dynamic loads.

The computational time for all models was presented in Figure 9.

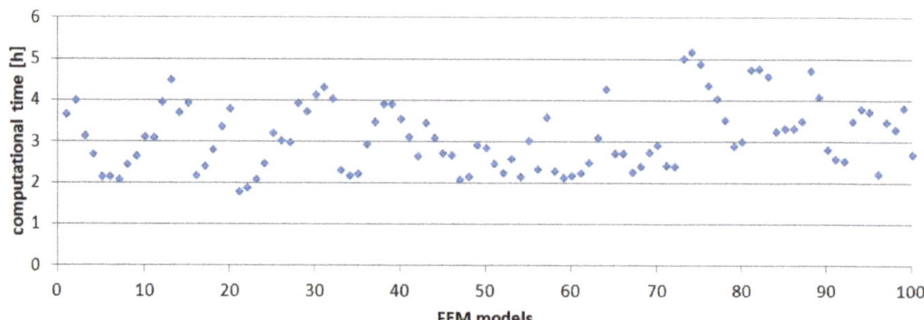

Figure 9. Computational time of FEM models.

Maximum and minimum values were, respectively, 5 h 11 min and 1 h 48 min; the average value for all models was 3 h 25 min. The differences in the computational time resulted mainly from the number of finite elements in the models due to the variable thickness of one of the adherends. The calculations were performed on a workstation equipped with two Intel(R) Xeon(R) 2.3 GHz processors (64 logical cores) and 256 GB of RAM.

4. Application of Neural Networks

The artificial neural network model was created based on analogies to biological counterparts. Neural networks are currently widely used in technical issues, among others [27–29]. They are a good solution for forecasting and regression problems. Neural networks are signal processing mathematical models. The main advantages of the neural network are that it works in conditions of incomplete information, it works automatically (does not require knowledge of the algorithm to solve the task), it can generalize (works in areas outside the input information).

Since each FEM simulation of the model take a long computational time, particularly, considering the computational costs [30], in order to be able to quickly simulate the adhesive joint parameters, the ANN modelling was used. In this research, the neural network is a model for predicting maximum connection strength, initiation energy and fracture energy. In numerical research a network with 10 input variables was applied. These are the parameters of adhesives and geometric dimensions of the connection (Table 2).

Table 2. Neural network input parameters.

Adhesive 1	Modulus E_1 (MPa)	Modulus G_1 (MPa)	Shear strength (k1) (MPa)	Tensile strength (k1) (MPa)
Adhesive 2	Modulus E_2 (MPa)	Modulus G_2 (MPa)	Shear strength (k2) (MPa)	Tensile strength (k2) (MPa)
Geometrical parameters	Radius r (mm)	Thickness g (mm)		

In the hidden network layer, 12 neurons were modelled. Output parameters are: (i) maximum force, (ii) initiation energy and (iii) fracture energy.

The graphs in Figures 10–12 show the dependence of the network output parameters on data from the numerical experiment.

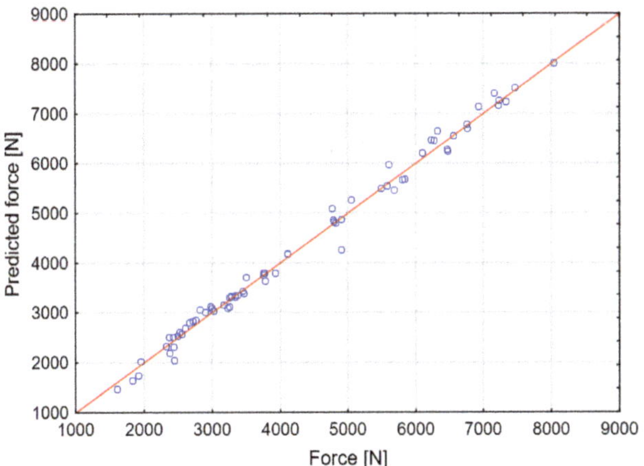

Figure 10. Actual versus predicted values: maximum force.

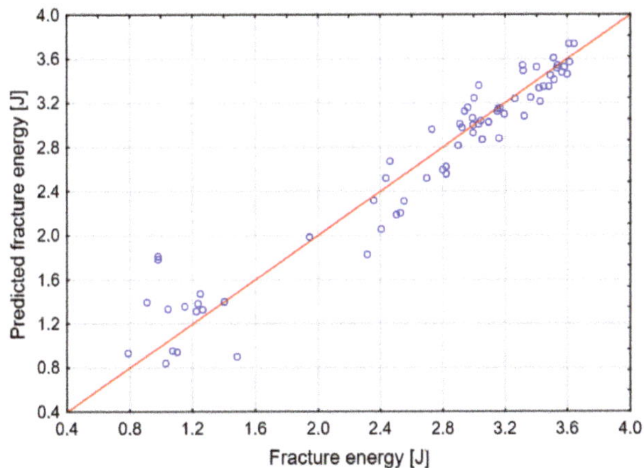

Figure 11. Actual versus predicted values: fracture energy.

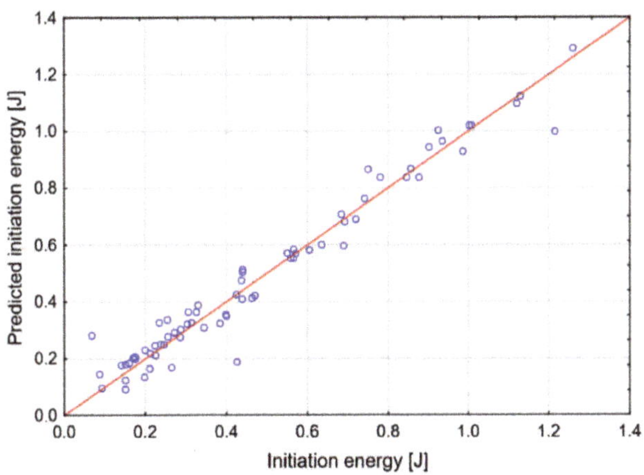

Figure 12. Actual versus predicted values: initiation energy.

The neural network is trained in such a way that its parameters are changed using the selected learning algorithm. The best known example of such an algorithm is the backpropagation algorithm. This algorithm, based on the collected data, modifies the weights and thresholds of the network in such a way as to ensure that the error made by the network (in this case, the prediction error) for all the data included in the training set is minimal. The data set included 100 cases from the numerical experiment. The data were classified as follows: 80% training, 10% test, 10% validation. The parameters and outputs of the network are presented in Table 3. It was found that a multilayer perceptron (MLP) provides the best quality, which is why this network model was qualified for calculation. The sigmoid network is the most frequently used artificial neural network. This network model has a multilayered placement of neurons; neurons calculate the sum of the inputs. These values are arguments of the function that calculates the output of the neurons. The predictive effectiveness of the neural network is based on data from the FEM experiment. It is calculated using the sum of squares error function. The predictive effectiveness of the MLP network for all output parameters was approx. 98%.

Table 3. Artificial neural networks operation parameters.

Network	Quality (Training)	Quality (Testing)	Quality (Validation)	Training Algorithm	Error Function	Activation (Hidden)	Activation (Output)
MLP 10–12–3	0980	0971	0981	Broyden-Fletcher–Goldfarb-Shanno	Sum ofsquares	Tanh	Linear

5. Discussion of the Results

The use of neural networks enables effective forecasting of force, initiation and fracture energy values. Determining the key parameters of the connection will allow its parameters to be optimized. Research has shown that the most important parameter for the strength of the joint is the size of the hole in which the more rigid adhesive is used. It has been shown that in the tested joint, in order to increase the maximum joint strength, it is advisable to use an internal adhesive with higher Young's modulus values. Kirchoff G modulus values are important in relation to fracture energy.

Several conclusions emerge from the artificial neural network (ANN) sensitivity analysis. A sensitivity analysis makes it possible to distinguish important variables from those that are not relevant. The input variables in the analyzed case are not completely independent. The sensitivity analysis shows the loss we incur when we reject a particular variable. To carry out the analysis, the data is presented to the network repeatedly. In each test, all values of one variable are converted into missing data and the total error is calculated, similarly to the standard learning of a network. The key factor with respect to maximum force, fracture energy and initiation energy is the joint radius r. The value of radius r affects the results of numerical analyses in about 70%.

Figures 13–15 show diagrams of the dependence of ANN output parameters as a function of the diameter (radius r) and the thickness g of the connection.

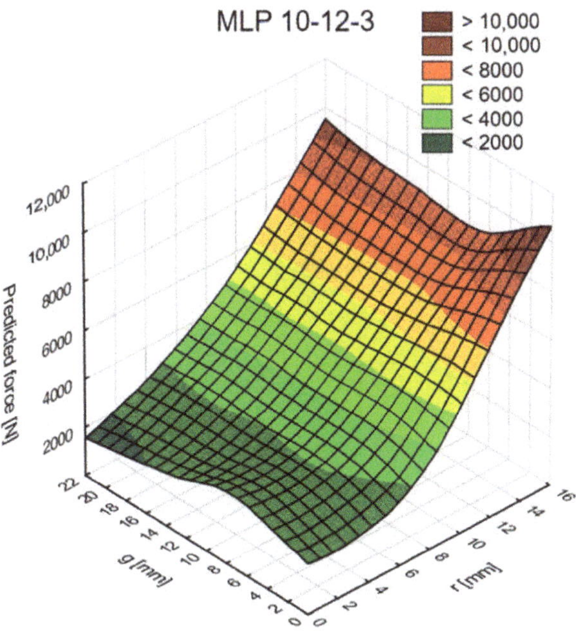

Figure 13. The graph of force values as a function of radius (r) and thickness (g).

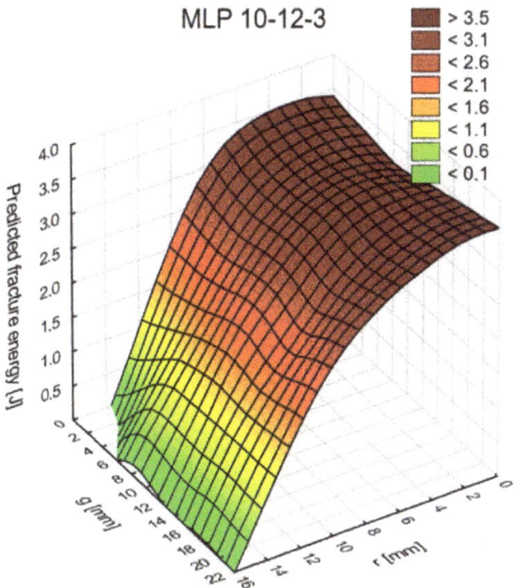

Figure 14. The graph of fracture energy values as a function of radius (r) and thickness (g).

MLP 10-12-3

Figure 15. The graph of initiation energy values as a function of radius (r) and thickness (g).

Separate tests were conducted to determine the most important input parameters of the network, in relation to individual output variables. The research assumed an average connection diameter value. The Young's modulus (Figure 16) and the stress values of the adhesives used are the most important for the maximum joint force.

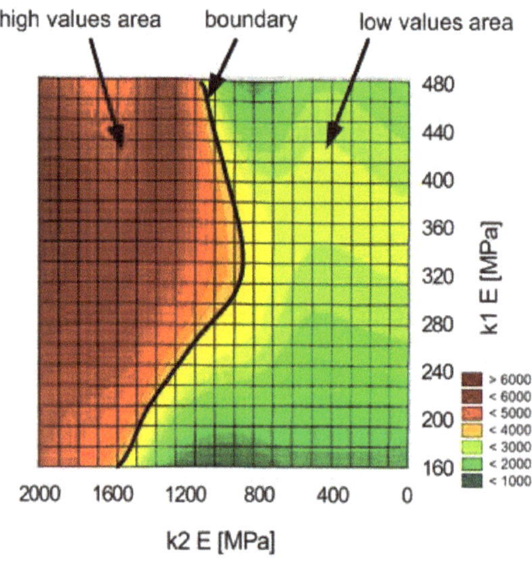

Figure 16. Diagram of force values [N] in function of Young's moduli of both adhesives.

Figure 16 shows that the maximum force increases significantly for adhesive 2 Young's modulus values above 1200 MPa.

For the forecasted energy values, the most important input variables are both adhesives' rigidity moduli (Figure 17).

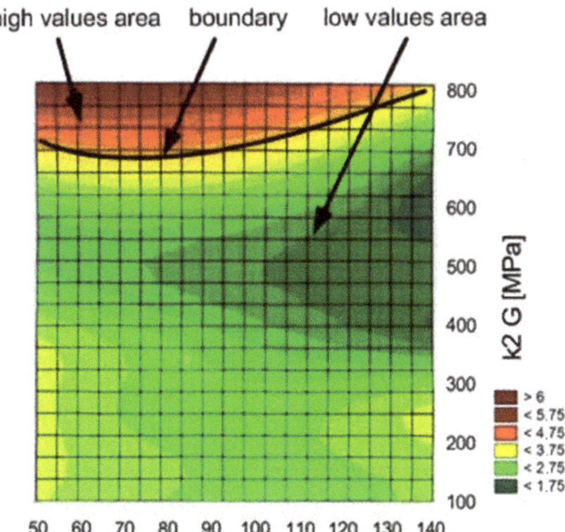

Figure 17. Diagram of fracture energy [J] in function of rigidity modulus of both adhesives.

6. Conclusions

The strength of an adhesive joint is crucial for design and optimization of its mechanical response. Optimal selection of geometric parameters of the joint and properties of the adhesives improves safety and reduces costs. The application of ANN allows the assessment of the significance of individual connection parameters and their modification; the use of neural networks in joint design allows the correctness of numerical calculations to be verified. The Finite Element Method calculation time for different connection parameters is significantly reduced.

The tests conducted provide valuable information for future numerical and laboratory tests. An important conclusion concerns the thickness "g" of one of the adherends, the increase in which has little effect on changes in both the maximum force and the fracture energy. Moreover, testing the joints in the range of the value of the radius "r" from 2 to 6 mm is not advisable as the changes in the value of the maximum force and fracture energy are insignificant. Only in the "r" range from 8 to 16 mm does the sample response become more sensitive. The research also proves that dual adhesive joints are sensitive to the appropriate selection of the material parameters of both adhesives. In the work, the graphs define the limits for both the E and G module; when exceeded, the maximum force and fracture energy suddenly increase.

The presented work does not exhaust all the possibilities of using FEM and neural networks in relation to adhesive joints. Important parameters are also defects, excess adhesive or variable thickness of the adhesive. These types of parameters have a practical relationship with the strength of real structures subjected to the bonding technique. Knowledge about the size of the defect obtained, for example, from a tomograph, in conjunction with the knowledge resulting from the use of FEM and neural networks, can be used to assess the condition of the structure and prevent failures.

Author Contributions: Conceptualization, J.G. and P.G.; methodology, J.G. and P.G.; software, J.G. and P.G.; validation, J.G., P.G. and T.S.; formal analysis, J.G. and T.S.; investigation, J.G. and P.G.; resources, J.G. and T.S.; data curation, J.G. and P.G.; writing—original draft preparation, J.G. and P.G.; writing—review & editing, J.G. and T.S.; visualization, J.G. and P.G.; supervision, J.G. and T.S.; project administration, J.G. and T.S.; funding acquisition, J.G. All authors have read and agreed to the published version of the manuscript.

Funding: This research received no external funding.

Institutional Review Board Statement: Not applicable.

Informed Consent Statement: Not applicable.

Data Availability Statement: The data presented in this study are available on request from the corresponding author.

Acknowledgments: This work was financially supported by Ministry of Science and Higher Education (Poland) within the statutory research number FN-82/IS/2020, FN 20/ILT/2020, FN 9/IM/2020. The project/research was financed in the framework of the project Lublin University of Technology-Regional Excellence Initiative, funded by the Polish Ministry of Science and Higher Education (contract no. 030/RID/2018/19).

Conflicts of Interest: The authors declare no conflict of interest.

Appendix A

Table A1. Parameter values for each model.

No.	"r" (mm)	"g" (mm)	Adhesive 1				Adhesive 2			
			E (MPa)	G (MPa)	σ (MPa)	τ (MPa)	E (MPa)	G (MPa)	σ (MPa)	τ (MPa)
1	1	20	185	56	2.2	1.7	1213.79	367.42	14.43	11.81
2	2.25	16	203.50	61.60	2.42	1.87	983.17	297.61	11.69	9.57
3	3.5	12	223.85	67.76	2.66	2.06	1850.00	560.00	22.00	18.00
4	4.75	8	246.24	74.54	2.93	2.26	1092.41	330.67	12.99	10.63
5	6	4	270.86	81.99	3.22	2.49	884.85	267.85	10.52	8.61
6	7.25	2	297.94	90.19	3.54	2.74	1213.79	367.42	14.43	11.81
7	8.5	4	327.74	99.21	3.90	3.01	1850.00	560.00	22.00	18.00
8	9.75	6	360.51	109.13	4.29	3.31	1665.00	504.00	19.80	16.20
9	11	8	396.56	120.04	4.72	3.64	1498.50	453.60	17.82	14.58
10	12.25	10	436.22	132.05	5.19	4.01	1348.65	408.24	16.04	13.12
11	1	12	223.85	67.76	2.66	2.06	1213.79	367.42	14.43	11.81
12	2.25	14	297.94	90.19	3.54	2.74	1092.41	330.67	12.99	10.63
13	3.5	16	396.56	120.04	4.72	3.64	983.17	297.61	11.69	9.57
14	4.75	18	360.51	109.13	4.29	3.31	884.85	267.85	10.52	8.61
15	6	20	327.74	99.21	3.90	3.01	796.36	241.06	9.47	7.75
16	7.25	2	185.00	56.00	2.20	1.70	716.73	216.96	8.52	6.97
17	8.5	6	203.50	61.60	2.42	1.87	1092.41	330.67	12.99	10.63
18	9.75	10	223.85	67.76	2.66	2.06	884.85	267.85	10.52	8.61
19	11	14	246.24	74.54	2.93	2.26	1665.00	504.00	19.80	16.20
20	12.25	18	270.86	81.99	3.22	2.49	716.73	216.96	8.52	6.97
21	1	2	297.94	90.19	3.54	2.74	645.06	195.26	7.67	6.28
22	2.25	4	327.74	99.21	3.90	3.01	580.55	175.73	6.90	5.65
23	3.5	6	360.51	109.13	4.29	3.31	522.49	158.16	6.21	5.08
24	4.75	8	396.56	120.04	4.72	3.64	470.25	142.34	5.59	4.58
25	6	10	436.22	132.05	5.19	4.01	423.22	128.11	5.03	4.12
26	7.25	12	327.74	99.21	3.90	3.01	380.90	115.30	4.53	3.71
27	8.5	14	360.51	109.13	4.29	3.31	342.81	103.77	4.08	3.34
28	9.75	16	185.00	56.00	2.20	1.70	1850.00	560.00	22.00	18.00
29	11	18	203.50	61.60	2.42	1.87	1665.00	504.00	19.80	16.20
30	12.25	20	223.85	67.76	2.66	2.06	1498.50	453.60	17.82	14.58
31	1	20	246.24	74.54	2.93	2.26	1092.41	330.67	12.99	10.63
32	2.25	12	270.86	81.99	3.22	2.49	1092.41	330.67	12.99	10.63
33	3.5	6	297.94	90.19	3.54	2.74	1850.00	560.00	22.00	18.00
34	4.75	2	327.74	99.21	3.90	3.01	716.73	216.96	8.52	6.97
35	6	4	360.51	109.13	4.29	3.31	983.17	297.61	11.69	9.57

Table A1. *Cont.*

			Adhesive 1				Adhesive 2			
No.	"r" (mm)	"g" (mm)	E (MPa)	G (MPa)	σ (MPa)	τ (MPa)	E (MPa)	G (MPa)	σ (MPa)	τ (MPa)
36	7.25	8	396.56	120.04	4.72	3.64	796.36	241.06	9.47	7.75
37	8.5	12	436.22	132.05	5.19	4.01	1498.50	453.60	17.82	14.58
38	9.75	20	297.94	90.19	3.54	2.74	1348.65	408.24	16.04	13.12
39	11	18	327.74	99.21	3.90	3.01	884.85	267.85	10.52	8.61
40	12.25	16	360.51	109.13	4.29	3.31	1213.79	367.42	14.43	11.81
41	1	10	396.56	120.04	4.72	3.64	983.17	297.61	11.69	9.57
42	2.25	6	396.56	120.04	4.72	3.64	1498.50	453.60	17.82	14.58
43	3.5	18	185.00	56.00	2.20	1.70	1348.65	408.24	16.04	13.12
44	4.75	14	203.50	61.60	2.42	1.87	884.85	267.85	10.52	8.61
45	6	10	223.85	67.76	2.66	2.06	1665.00	504.00	19.80	16.20
46	7.25	2	246.24	74.54	2.93	2.26	716.73	216.96	8.52	6.97
47	8.5	4	270.86	81.99	3.22	2.49	1092.41	330.67	12.99	10.63
48	9.75	4	297.94	90.19	3.54	2.74	1213.79	367.42	14.43	11.81
49	11	12	327.74	99.21	3.90	3.01	983.17	297.61	11.69	9.57
50	12.25	10	360.51	109.13	4.29	3.31	1665.00	504.00	19.80	16.20
51	1	8	396.56	120.04	4.72	3.64	1498.50	453.60	17.82	14.58
52	2.25	2	396.56	120.04	4.72	3.64	1348.65	408.24	16.04	13.12
53	3.5	8	203.50	61.60	2.42	1.87	796.36	241.06	9.47	7.75
54	4.75	4	223.85	67.76	2.66	2.06	1850.00	560.00	22.00	18.00
55	6	12	246.24	74.54	2.93	2.26	1213.79	367.42	14.43	11.81
56	7.25	2	223.85	67.76	2.66	2.06	1665.00	504.00	19.80	16.20
57	8.5	20	246.24	74.54	2.93	2.26	983.17	297.61	11.69	9.57
58	9.75	6	270.86	81.99	3.22	2.49	1850.00	560.00	22.00	18.00
59	11	2	327.74	99.21	3.90	3.01	1498.50	453.60	17.82	14.58
60	12.25	4	360.51	109.13	4.29	3.31	1348.65	408.24	16.04	13.12
61	1	4	294	81.99	3.51	2.71	1204.00	367.42	14.33	11.72
62	2.25	8	294	90.19	3.51	2.71	1204.00	297.61	14.33	11.72
63	3.5	14	294	56	3.51	2.71	1204.00	560.00	14.33	11.72
64	4.75	20	294	61.60	3.51	2.71	1204.00	330.67	14.33	11.72
65	6	10	294	67.76	3.51	2.71	1204.00	267.85	14.33	11.72
66	7.25	2	294	74.54	3.51	2.71	1204.00	367.42	14.33	11.72
67	8.5	4	294	81.99	3.51	2.71	1204.00	560.00	14.33	11.72
68	9.75	6	294	90.19	3.51	2.71	1204.00	504.00	14.33	11.72
69	11	8	294	99.21	3.51	2.71	1204.00	453.60	14.33	11.72
70	12.25	10	294	109.13	3.51	2.71	1204.00	408.24	14.33	11.72
71	13.5	12	294	120.04	3.51	2.71	1204.00	580.00	14.33	11.72
72	14.75	14	294	132.05	3.51	2.71	1204.00	600.00	14.33	11.72
73	1	16	294	90.19	3.51	2.71	1204.00	620.00	14.33	11.72
74	2.25	18	294	81.99	3.51	2.71	1204.00	640.00	14.33	11.72
75	3.5	20	294	56	3.51	2.71	1204.00	660.00	14.33	11.72
76	4.75	16	294	120.04	3.51	2.71	1204.00	680.00	14.33	11.72
77	6	12	294	56	3.51	2.71	1204.00	367.42	14.33	11.72
78	7.25	8	294	61.60	3.51	2.71	1204.00	297.61	14.33	11.72
79	8.5	4	294	67.76	3.51	2.71	1204.00	560.00	14.33	11.72
80	9.75	2	294	74.54	3.51	2.71	1204.00	330.67	14.33	11.72
81	11	20	294	81.99	3.51	2.71	1204.00	267.85	14.33	11.72
82	12.25	16	294	90.19	3.51	2.71	1204.00	367.42	14.33	11.72
83	13.5	18	294	99.21	3.51	2.71	1204.00	560.00	14.33	11.72
84	14.75	4	294	109.13	3.51	2.71	1204.00	504.00	14.33	11.72
85	1	8	294	120.04	3.51	2.71	1204.00	453.60	14.33	11.72
86	2.25	12	294	132.05	3.51	2.71	1204.00	408.24	14.33	11.72

Table A1. *Cont.*

			Adhesive 1				Adhesive 2			
No.	**"r"** **(mm)**	**"g"** **(mm)**	**E** **(MPa)**	**G** **(MPa)**	**σ** **(MPa)**	**τ** **(MPa)**	**E** **(MPa)**	**G** **(MPa)**	**σ** **(MPa)**	**τ** **(MPa)**
87	3.5	10	294	90.19	3.51	2.71	1204.00	580.00	14.33	11.72
88	4.75	20	294	81.99	3.51	2.71	1204.00	620.00	14.33	11.72
89	6	12	294	56	3.51	2.71	1204.00	660.00	14.33	11.72
90	7.25	2	294	120.04	3.51	2.71	1204.00	700.00	14.33	11.72
91	8.5	18	294	56	3.51	2.71	1204.00	367.42	14.33	11.72
92	9.75	6	294	61.60	3.51	2.71	1204.00	297.61	14.33	11.72
93	11	12	294	67.76	3.51	2.71	1204.00	560.00	14.33	11.72
94	12.25	20	294	74.54	3.51	2.71	1204.00	330.67	14.33	11.72
95	13.5	14	294	81.99	3.51	2.71	1204.00	267.85	14.33	11.72
96	14.75	4	294	90.19	3.51	2.71	1204.00	367.42	14.33	11.72
97	11	18	294	99.21	3.51	2.71	1204.00	560.00	14.33	11.72
98	12.25	10	294	109.13	3.51	2.71	1204.00	504.00	14.33	11.72
99	13.5	14	294	120.04	3.51	2.71	1204.00	453.60	14.33	11.72
100	14.75	8	294	132.05	3.51	2.71	1204.00	408.24	14.33	11.72

References

1. Moroni, F.; Musiari, F.; Favi, C. Effect of the surface morphology over the fatigue performance of metallic single lap-shear joints. *Int. J. Adhes. Adhes.* **2020**, *97*, 102484. [CrossRef]
2. Hirulkar, N.; Jaiswal, P.; Alessandro, P.; Reis, P. Influence of Mechanical surface treatment on the strength of mixed adhesive joint. *Mater. Today Proc.* **2018**, *5*, 18776–18788. [CrossRef]
3. Golewski, P.; Sadowski, T. Investigation of the effect of chamfer size on the behaviour of hybrid joints made by adhesive bonding and riveting. *Int. J. Adhes. Adhes.* **2017**, *77*, 174–182. [CrossRef]
4. Golewski, P.; Sadowski, T. The Influence of Single Lap Geometry in Adhesive and Hybrid Joints on Their Load Carrying Capacity. *Materials* **2019**, *12*, 1884. [CrossRef] [PubMed]
5. Sadowski, T.; Golewski, P. Numerical Study of the Prestressed Connectors and Their Distribution on the Strength of a Single Lap, a Double Lap and Hybrid Joints Subjected to Uniaxial Tensile Test. *Arch. Met. Mater.* **2013**, *58*, 579–585. [CrossRef]
6. Sadowski, T.; Golewski, P. Effect of Tolerance in the Fitting of Rivets in the Holes of Double Lap Joints Subjected to Uniaxial Tension. *Key Eng. Mater.* **2014**, *607*, 49–54. [CrossRef]
7. Bouchikhi, A.; Megueni, A.; Gouasmi, S.; Boukoulda, F. Effect of mixed adhesive joints and tapered plate on stresses in retrofitted beams bonded with a fiber-reinforced polymer plate. *Mater. Des.* **2013**, *50*, 893–904. [CrossRef]
8. Machado, J.; Gamarra, P.-R.; Marques, E.; Da Silva, L.F. Numerical study of the behaviour of composite mixed adhesive joints under impact strength for the automotive industry. *Compos. Struct.* **2018**, *185*, 373–380. [CrossRef]
9. Da Silva, L.F.; Lopes, M.J.C. Joint strength optimization by the mixed-adhesive technique. *Int. J. Adhes. Adhes.* **2009**, *29*, 509–514. [CrossRef]
10. Breto, R.; Chiminelli, A.; Lizaranzu, M.; Rodríguez, R. Study of the singular term in mixed adhesive joints. *Int. J. Adhes. Adhes.* **2017**, *76*, 11–16. [CrossRef]
11. Chiminelli, A.; Breto, R.; Izquierdo, S.; Bergamasco, L.; Duvivier, E.; Lizaranzu, M. Analysis of mixed adhesive joints considering the compaction process. *Int. J. Adhes. Adhes.* **2017**, *76*, 3–10. [CrossRef]
12. Kanani, A.Y.; Hou, X.; Ye, J. The influence of notching and mixed-adhesives at the bonding area on the strength and stress distribution of dissimilar single-lap joints. *Compos. Struct.* **2020**, *241*, 112136. [CrossRef]
13. Machado, J.; Marques, E.A.D.S.; Da Silva, L.F. Influence of low and high temperature on mixed adhesive joints under quasi-static and impact conditions. *Compos. Struct.* **2018**, *194*, 68–79. [CrossRef]
14. Marques, E.A.D.S.; Da Silva, L.; Flaviani, M. Testing and simulation of mixed adhesive joints for aerospace applications. *Compos. Part B Eng.* **2015**, *74*, 123–130. [CrossRef]
15. Jairaja, R.; Naik, G.N. Numerical studies on weak bond effects in single and dual adhesive bonded single lap joint between CFRP and aluminium. *Mater. Today Proc.* **2020**, *21*, 1064–1068. [CrossRef]
16. Jairaja, R.; Naik, G.N. Single and dual adhesive bond strength analysis of single lap joint between dissimilar adherends. *Int. J Adhes. Adhes.* **2019**, *92*, 142–153. [CrossRef]
17. Gajewski, J.; Golewski, P.; Sadowski, T. Geometry optimization of a thin-walled element for an air structure using hybrid system integrating artificial neural network and finite element method. *Compos. Struct.* **2017**, *159*, 589–599. [CrossRef]
18. Rogala, M.; Gajewski, J.; Ferdynus, M. The Effect of Geometrical Non-Linearity on the Crashworthiness of Thin-Walled Conical Energy-Absorbers. *Materials* **2020**, *13*, 4857. [CrossRef]
19. Szklarek, K.; Gajewski, J. Optimisation of the Thin-Walled Composite Structures in Terms of Critical Buckling Force. *Materials* **2020**, *13*, 3881. [CrossRef]

20. Tosun, E.; Çalik, A. Failure load prediction of single lap adhesive joints using artificial neural networks. *Alex. Eng. J.* **2016**, *55*, 1341–1346. [CrossRef]
21. Almeida, S.A.; Guner, S. A hybrid methodology using finite elements and neural networks for the analysis of adhesive anchors exposed to hurricanes and adverse environments. *Eng. Struct.* **2020**, *212*, 110505. [CrossRef]
22. Zgoul, M.H. Use of artificial neural networks for modelling rate dependent behaviour of adhesive materials. *Int. J. Adhes. Adhes.* **2012**, *36*, 1–7. [CrossRef]
23. Zaeri, A.R.; Googarchin, H.S. Experimental investigation on environmental degradation of automotive mixed-adhesive joints. *Int. J. Adhes. Adhes.* **2019**, *89*, 19–29. [CrossRef]
24. Da Silva, L.; Adams, R. Adhesive joints at high and low temperatures using similar and dissimilar adherends and dual adhesives. *Int. J. Adhes. Adhes.* **2007**, *27*, 216–226. [CrossRef]
25. Machado, J.; Marques, E.; Silva, M.; Da Silva, L.F. Numerical study of impact behaviour of mixed adhesive single lap joints for the automotive industry. *Int. J. Adhes. Adhes.* **2018**, *84*, 92–100. [CrossRef]
26. Zhu, S.-P.; Liu, Q.; Peng, W.; Zhang, X.-C. Computational-experimental approaches for fatigue reliability assessment of turbine bladed disks. *Int. J. Mech. Sci.* **2018**, *142–143*, 502–517. [CrossRef]
27. Pavlenko, I.; Sága, M.; Kuric, I.; Kotliar, A.; Basova, Y.; Trojanowska, J.; Ivanov, V. Parameter Identification of Cutting Forces in Crankshaft Grinding Using Artificial Neural Networks. *Materials* **2020**, *13*, 5357. [CrossRef]
28. Merayo, D.; Rodríguez-Prieto, A.; Camacho, A.M. Prediction of Mechanical Properties by Artificial Neural Networks to Characterize the Plastic Behavior of Aluminum Alloys. *Materials* **2020**, *13*, 5227. [CrossRef]
29. Machrowska, A.; Szabelski, J.; Karpiński, R.; Krakowski, P.; Jonak, J.; Jonak, K. Use of Deep Learning Networks and Statistical Modeling to Predict Changes in Mechanical Parameters of Contaminated Bone Cements. *Materials* **2020**, *13*, 5419. [CrossRef]
30. Zhu, S.; Foletti, S.; Beretta, S. Probabilistic framework for multiaxial LCF assessment under material variability. *Int. J. Fatigue* **2017**, *103*, 371–385. [CrossRef]

materials

MDPI

Article

Direct Evaluation of Mixed Mode I+II Cohesive Laws of Wood by Coupling MMB Test with DIC

Jorge Oliveira [1,2], José Xavier [3,*], Fábio Pereira [1], José Morais [1] and Marcelo de Moura [4]

1   CITAB, School of Science and Technology, University of Trás-os-Montes e Alto Douro, 5001-801 Vila Real, Portugal; jqomarcelo@demad.estv.ipv.pt (J.O.); famp@utad.pt (F.P.); jmorais@utad.pt (J.M.)
2   Escola Superior de Tecnologia do Instituto Superior Politécnico de Viseu, Campus Politécnico de Repeses, 3504-510 Viseu, Portugal
3   UNIDEMI, Department of Mechanical and Industrial Engineering, NOVA School of Science and Technology, NOVA University Lisbon, 2829-516 Caparica, Portugal
4   Faculdade de Engenharia da Universidade do Porto, Departamento de Engenharia Mecânica e Gestão Industrial, Rua Roberto Frias, 4200-465 Porto, Portugal; mfmoura@fe.up.pt
*   Correspondence: jmc.xavier@fct.unl.pt; Tel.: +351-212-948 567

Abstract: Governing cohesive laws in mixed mode I+II loading of *Pinus pinaster* Ait. are directly identified by coupling the mixed mode bending test with full-field displacements measured at the crack tip by Digital Image Correlation (DIC). A sequence of mixed mode ratios is studied. The proposed data reduction relies on: (i) the compliance-based beam method for evaluating strain energy release rate; (ii) the local measurement of displacements to compute the crack tip opening displacement; and (iii) an uncoupled approach for the reconstruction of the cohesive laws and its mode I and mode II components. Quantitative parameters are extracted from the set of cohesive laws components in function of the global phase angle. Linear functions were adjusted to reflect the observed trends and the pure modes (I and II) fracture parameters were estimated by extrapolation. Results show that the obtained assessments agree with previous experimental measurements addressing pure modes (I and II) loadings on this wood species, which reveals the appropriateness of the proposed methodology to evaluate the cohesive law under mixed mode loading and its components.

Keywords: wood; cohesive law; digital image correlation; fracture mechanics; mixed mode I+II loading

Citation: Oliveira, J.; Xavier, J.; Pereira, F.; Morais, J.; de Moura, M. Direct Evaluation of Mixed Mode I+II Cohesive Laws of Wood by Coupling MMB Test with DIC. *Materials* **2021**, *14*, 374. https://doi.org/10.3390/ma14020374

Received: 7 December 2020
Accepted: 10 January 2021
Published: 14 January 2021

Publisher's Note: MDPI stays neutral with regard to jurisdictional claims in published maps and institutional affiliations.

1. Introduction

Structural applications of wood and wood products have been increasing recently, owing to economic and ecological reasons. As a result, the development of adequate failure criteria models becomes an imperative task. In this context, fracture mechanics-based approaches reveal to be most suitable, taking into consideration the anisotropic and heterogeneous nature of wood. Most of the works addressing this topic have been dedicated to fracture under pure mode loading, I or II [1–6]. In real structural applications, however, mixed mode conditions arise from external loading and/or because of the anisotropy and heterogeneity of the material. In this context, the mixed mode fracture characterization of wood becomes a relevant research topic, which has been addressed by several authors proposing different test methods. Jernkvist [7] employed the Double Cantilever Beam (DCB) with asymmetrical arms and the Single Edge Notched (SEN) tensile tests to investigate mixed mode I+II in Picea abies. The asymmetrical Wedge Splitting test was proposed by Tschegg et al. [8] to fracture the characterization of spruce under mixed mode I+II. In this test, the mode ratio can be altered using asymmetrical wedges with different angles. The referred tests only provide a very limited range of mode ratio values. In addition, fracture characterization was performed at initiation and results may suffer from some uncertainty on the definition of critical load.

A valuable alternative is the employment of the mixed mode bending (MMB) test, which consists of a combination of the DCB and End Notched-Flexure (ENF) tests through a specific conceived setup. The MMB was firstly proposed to characterize the fracture behavior in mixed mode I+II of polymeric composite materials [9,10]. The original setup was then redesigned to mitigate erroneous toughness measurements introduced by geometric non-linearity during the test [11], and by the weight of some components [12]. Eventually, the MMB test turned out a standard for unidirectional composite materials [13]. The major advantage of the MMB test, over counterpart mixed mode tests, is allowing a spectrum of different mixed mode ratios by simply varying the length of the loading lever. The MMB test has already been applied to wood and wood bonded joints fracture characterization under mixed mode I+II loading, allowing the definition of their fracture envelopes [14,15].

Several full-field deformation techniques have been proposed and developed in recent decades for experimental solid mechanics [16]. A milestone was achieved by allowing contact free full-field kinematic measurements over a spectrum of length scales. The availability of this type of information has been shifting boundaries in several research domains. Different engineering problems have already been tackled such as experimental evidence of local gradients due to material heterogeneities [17,18], cracking characterization [19,20], damage and fracture model evaluation [21,22], high strain rate characterization of advanced composite material [23,24], validation of phenomenological numerical models [25] and multi-parameter identification from single test configurations [26,27]. Among the universe of full-field optical techniques, digital image correlation (DIC) was selected in this work [28]. DIC can be conveniently coupled with conventional apparatus such as testing machines, and typically only requires a careful preparation of a random pattern (with suitable speckled size and distribution) over the surface of interest. Moreover, the spatial resolution of the technique is flexibly adjusted by selecting the optical system (camera and lens), allowing to image a spectrum of different length scales of interest. Taking advantage of non-contact and full-field data, DIC can be suitable for fracture mechanics studies [29]. It can be used to access the local displacements field near the crack tip and to evaluate the crack tip opening displacement during the fracture tests [30]. This local information can therefore be used to determine the so-called cohesive law relating the tractions and relative displacements occurring at the crack tip. The cohesive law is representative of material fracture behavior, and is currently used in finite element analysis of materials fracture. The utilization of DIC to measure local displacement at the crack tip provides a direct approach, with the advantage of not imposing a priori the shape of the softening law.

The direct evaluation of the cohesive laws describing the mixed mode I+II fracture behavior of *Pinus pinaster* Ait. was experimentally assessed by coupling the MMB test to DIC measurements. Wooden specimens oriented in the RL propagation system were analyzed. The controlled parameters in the configuration of the MMB setup were set to address a range of mixed mode ratios. The compliance-based beam method (CBBM) was applied to independently determine the total strain energy release rate during the fracture tests, by only processing specimen dimensions, load and applied displacement. DIC measurements were post-processed to inspect the crack tip opening displacements at the initial crack location. Combining this information, the direct evaluation of the cohesive laws was assessed by a numerical approximation and differentiation approach. The evolution of the cohesive laws and its mode I and II components as function of the global phase angle was obtained, as well as the relations representing the evolution of the relevant cohesive parameters with the mode mixity.

2. Materials and Methods

2.1. Mixed Mode Bending (MMB) Test

The MMB test was employed to study mixed mode I+II fracture loading. The geometry and nominal dimensions of the MMB specimen are shown in Figure 1a. The setup and free-body diagrams associated with the MMB test are presented in Figure 1b.

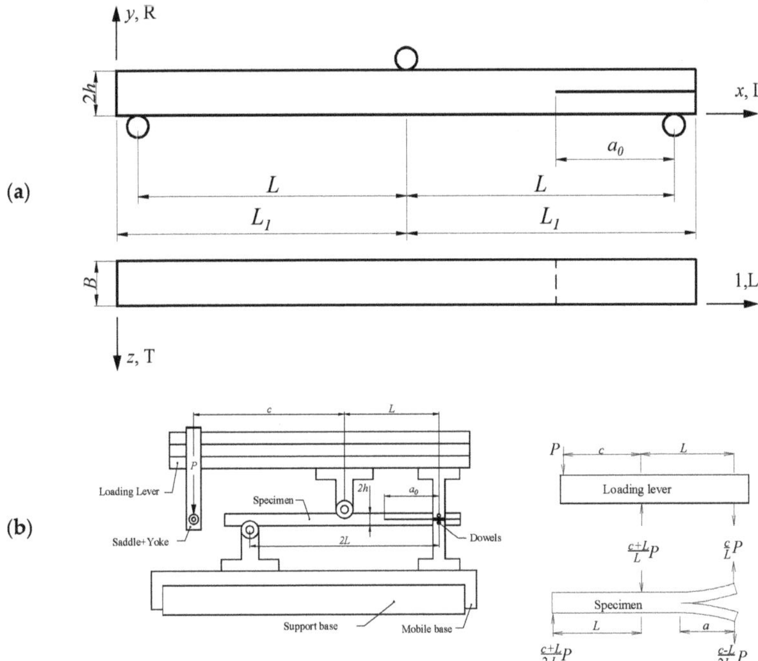

Figure 1. Mixed mode bending (MMB) test: (**a**) specimen geometry ($2h$ = 20 mm, L = 230 mm, L_1 = 250 mm, a_0 = 162 mm and B = 20 mm); (**b**) set-up and free-body diagrams (P—applied load).

From the loading applied to the specimen, the energy of deformation due to bending can expressed as

$$U = \int_0^a \left(\frac{c-L}{2L}\right)^2 \frac{(Px)^2}{2E_L I} dx + \int_0^a \left(\frac{c}{L}\right)^2 \frac{(Px)^2}{2E_L I} dx$$
$$+ \int_a^L \left(\frac{c+L}{2L}\right)^2 \frac{[P(x-a)]^2}{16 E_L I} dx + \int_L^{2L} \left(-\frac{c+L}{2L}\right)^2 \frac{[P(x-L)]^2}{16 E_L I} dx \tag{1}$$

where P is the applied load, E_L is the longitudinal modulus, I is the second moment of area of each arm given by $I = Bh^3/12$, and c and L are geometric dimensions as defined in Figure 1. Algebraic manipulation yields the following expression:

$$U = \frac{P^2}{16 E_L Bh^3 L^2} \left[a^3 \left(39c^2 - 18cL + 7L^2 \right) + 2L^3 (c+L)^2 \right] \tag{2}$$

Applying the Castigliano Theorem, it is possible to obtain the displacement of the left extremity of the loading lever (Figure 1b), i.e., $\delta = dU/dP$. The ratio between this displacement (δ) and applied load (P) is a measure of the specimen compliance (C):

$$C = \frac{1}{8 E_L Bh^3 L^2} \left[a^3 \left(39c^2 - 18cL + 7L^2 \right) + 2L^3 (c+L)^2 \right] \tag{3}$$

In order to avoid the difficult and inaccurate crack length monitoring during its propagation, an equivalent crack length-based procedure can be employed. With this aim,

Equation (3) can be used to estimate the actual crack length (a_e) as function of the current specimen compliance:

$$a_e = \left[\frac{8E_L B h^3 L^2 C - 2L^3 (c+L)^2}{(39c^2 - 18cL + 7L^2)} \right]^{\frac{1}{3}} \tag{4}$$

The evolution of the total strain energy release rate (G_T) as function of the equivalent crack length can be obtained combining the Irwin-Kies relation

$$G_T = \frac{P^2}{2B} \frac{dC}{da} \tag{5}$$

where B stands for the width of the beam (Figure 1a), with Equation (3), which leads to:

$$G_T = \frac{3P^2 a_e^2}{16E_L B^2 h^3 L^2} \left(39c^2 - 18cL + 7L^2 \right) \tag{6}$$

This equation provides the evolution of the total strain energy release rate under mixed mode I+II loading as a function of a_e for the MMB test, only measuring the applied load and displacement of the left extremity of the loading lever in the course of the test.

2.2. Evaluation of Cohesive Law

In the identification of a mixed mode cohesive law, it is typically assumed that total the strain energy release rate is a function of both components of the crack tip opening displacement thorough a given potential function Φ [31]:

$$G_T = \Phi(v, u) \tag{7}$$

Assuming an uncoupled approach, mode I (σ) and mode II (τ) stress components of the mixed mode cohesive law can be determined from the partial derivatives of the potential function as:

$$\sigma(v, u) = \frac{\partial \Phi}{\partial v} \qquad \text{and} \quad \tau(v, u) = \frac{\partial \Phi}{\partial u} \tag{8}$$

The phase angle (ϕ) is given by the ratio between the crack tip opening displacements (CTOD) components: $\tan \varphi = v/u$. This term can be normalized, taking a global phase angle (θ) defined as

$$\tan \theta = \frac{u_c}{v_c} \tan \varphi = \frac{u_c}{v_c} \cdot \frac{v}{u} \tag{9}$$

where v_c and u_c are the (average) critical or ultimate values of the CTOD in mode I and mode II, respectively. These reference values were determined independently from DCB [5] and ENF [6] tests carried out on the same wood species. The measurement of the crack tip opening displacement was achieved by post-processing displacements over a pair of subsets, selected regarding a coordinate system located at the initial crack tip with a spatial resolution of about 0.5 mm [5,6]. The assumption under the uncoupled modelling is supported by the scenario of a monotonically increment of the opening displacement at the crack tip [31]. Therefore, the components of the cohesive laws under mixed mode loading can be determined independently by using Equation (8) for a spectrum of phase angles.

The identification method proposed by Högberg [32] for the cohesive laws in mixed-mode I+II was used in this work. One assumption of this approach is the linearity of local deformation path during the mixed fracture test (i.e., constant mode ratio during the test). Hence, the evolution of the strain energy release rate will be only dependent on the magnitude of the total displacement (Δ), simple defined as:

$$\Delta^2 = v^2 + u^2 \tag{10}$$

A second assumption on the theory is the existence of a potential function $G^*(\lambda)$ for each global phase angle θ (Equation (9)), from which the mixed mode cohesive law can be directly computed:

$$S_\theta(\lambda) = \frac{dG^*(\lambda)}{d\lambda} \tag{11}$$

In this equation G^* represents the normalized total strain energy release rate defined as:

$$G^*(\lambda) = \frac{G_T(\lambda)}{2G_c} \tag{12}$$

The 2 factor is introduced by the linear assumption on the cohesive law and λ is the normalized crack tip displacement, which was assumed in the present work as:

$$\lambda = \frac{\Delta}{\Delta_c} \tag{13}$$

The explicit form of Equation (13), where the normalization is performed considering the specimen displacement measurements (Δ), instead of using reference critical values for mode I (v_c) and mode II (u_c) as initially proposed by Högberg [32], is justified by the dispersion typically found in biological materials such as wood. Finally, the individual mode I and mode II components of the mixed mode cohesive law can be decomposed according to the following relationships [32]

$$\sigma(v) = \sigma_u \left(\frac{v}{v_c}\right)\frac{S}{\lambda} \qquad \text{and} \qquad \tau(u) = \tau_u \left(\frac{u}{u_c}\right)\frac{S}{\lambda} \tag{14}$$

where σ_u and τ_u are the cohesive strength values in mode I and mode II, respectively.

$$G_I = \int_0^v \sigma(v,u)dv \qquad \text{and} \qquad G_{II} = \int_0^u \tau(v,u)du \tag{15}$$

3. Experimental Work

3.1. Material and Specimens

Logs from a single *Pinus pinaster* Ait. tree were selected for this work. In the first stage, boards were conventionally kiln-dried, and then left drying for about four weeks in the hygrothermal conditions of the laboratory environment (temperature between 20 and 25 °C and relative humidity between 60 and 65%). An average wood density of 643 kg/m^3 was determined for an equilibrium moisture content of 12.3%.

Wooden specimens (Figure 2a) for the MMB set-up (Figure 2b) were manufactured with nominal dimensions of $2h$ = 20 mm, L = 230 mm, L_1 = 250 mm, a_0 = 162 mm and B = 20 mm, as shown in Figure 1a. Specimens were oriented to respect the RL propagation system. The initial crack on the specimens was introduced in two steps. Firstly, a notch of 1 mm thickness was sawn. Secondly, a final sharp crack with an extension of about 2–5 mm was introduced by manual impact using a thin blade.

(a)

Figure 2. *Cont.*

Figure 2. (**a**) MMB specimens oriented in the RL propagation system; (**b**) MMB set-up coupled with digital image correlation (DIC) (speckle pattern and its histogram over a region of interest of 29 × 20 mm^2).

3.2. Full-Field Deformation Measurements by DIC

The MMB set-up was coupled with digital image correlation. A suitable speckled pattern was initially painted over the wooden specimens (Figure 2b). A regular and thin layer of matt white was firstly painted over the natural surface of wood. A textured, random pattern was then created using an airbrush (IWATA, model CM-B, Anesta Iwata Iberica SL, Barcelona, Spain) with a 0.18-mm nozzle. The typical size of the speckles was imaged over at least three pixels to avoid image aliasing. The ARAMIS v6.0.2-6 DIC-2D system was used for image grabbing and image processing. A 8-bit Baumer charge coupled device (CCD) camera coupled with a TC2336 bi-telecentric lens was selected for the optical system. A region of interest of 29.3 × 22.1 mm^2 was imaged, circumscribing the initial crack tip. The image focus was achieved by adjusting the working distance between specimen and lens to 103 mm. This image system defines a fixed conversion factor of 0.018 mm/pixel. Images were recorded during the test at a frequency of 1 Hz. The analogical load signal was synchronized with images during the tests. The DIC parameters were selected to enhance spatial resolution, since crack tip displacement measurements were required. A subset facet of 15 × 15 pixels^2 and subset step of 13 × 13 pixels^2 were selected. This set of parameters defined a displacement spatial resolution of 0.270 mm. The resolution or accuracy of the DIC measurements was estimated as the standard deviation of the Gaussian noise signal, which is typically measured on pair of images recorded on an in-plane rigid-body translation of the speckled pattern [33]. A sub-pixel displacement resolution in the order of 10^{-2} pixel was obtained.

3.3. MMB Fracture Tests

The MMB set-up used in this work is described in [14] (see Figure 2b). This setup has the advantage of allowing the variation of mixed mode I+II ratio by adjusting the distance c in the setup (Figure 1b). In this study, several mixed mode ratios were selected. The MMB fracture tests were carried out in a universal testing machine (model 1125, Instron, Barcelona, Spain) at a controlled displacement rate of 0.5 mm/min. The applied load was measured by means of a 100 kN load cell. The signals of the load were simultaneously recorded by a HBM SPIDER 8 with a frequency of 10 Hz.

4. Results and Discussion

From the raw load-displacement (P–δ) curves, the resistance-curves (R–curves) were determined by means of the CBBM approach described in Section 2.1. As discussed in Section 2.2, the methodology adopted to evaluate the cohesive laws assumes uncoupled behavior between mode I and mode II.

For the purpose of the direct evaluation of the cohesive laws, both P–δ and P–$CTOD$ curves were obtained from raw data by coupling the MMB test with DIC measurements. The CTOD was naturally decomposed into mode I (v) and mode II (u) components to define P–v and P–u curves, respectively. For this evaluation, a pair of points were just selected above and below the initial crack tip location to evaluate the relative displacement during test [5,6]. Figure 3 shows the average evolution of CTOD as a function of δ, for each interval of global phase angles analyzed in this work. As expected, both u and v components of the CTOD are strictly increasing functions of δ. These examples are representative of the expected relative amplitudes of u and v regarding the mixed mode ratios of the MMB specimens. It was noticed that the amplitude of v compared to u increases when the phase angle increases revealing a predominance of mode I loading.

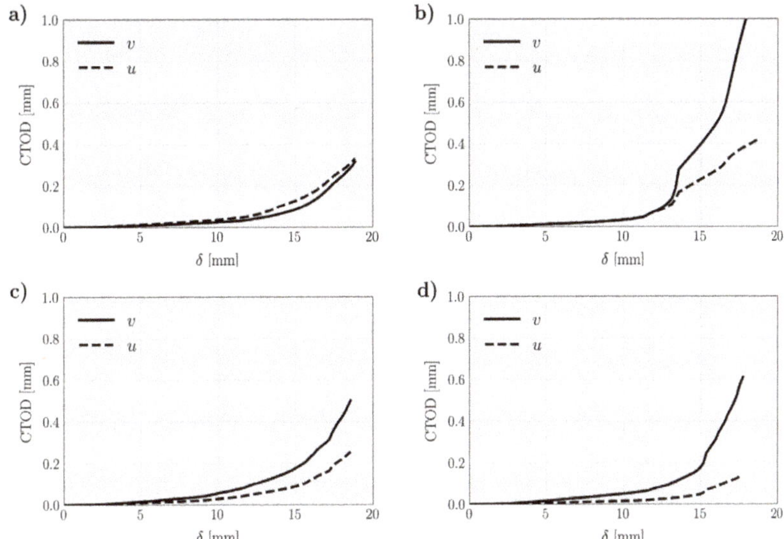

Figure 3. Average crack tip opening displacement (CTOD) as a function of the applied displacement (v: mode I component of the CTOD; u: mode II component of CTOD), for the global phase angles (**a**) $45° < \theta < 60°$; (**b**) $60° < \theta < 70°$; (**c**) $70° < \theta < 80°$; and (**d**) $80° < \theta < 90°$.

The local deformation path at the crack tip was expressed by the u–v relationship, as summarized in Figure 4 for the given phase angles ranges. As it can be seen, the monotonic variation was typically linear, which reinforces the statement of an almost constant mode-mixity during loading. Consequently, the global phase angle (Equation (9)) for each mixed mode ratio was determined by linear regression over the data points until the maximum load, as shown in Figure 4.

Figure 5 shows a typical example of the $G^*(\lambda)$ function (Equation (12)). To compute the derivative of this function with regard to λ (Equation (11)), three types of continuous functions were firstly fitted to the raw data using the least-square regression method. The purpose of this fitting is both filtering intrinsic experimental noise and providing a robust mathematical framework for the differentiation. Since the $G^*(\lambda)$ curve has a typical S-shape (sigmoid curve), theoretically converging to a plateau which represents the critical strain energy release rate, the logistic function was initially selected

$$G^* = \frac{A_1 - A_2}{1 + (\lambda/\lambda_0)^p} + A_2 \tag{16}$$

where A_1, A_2, p and λ_0 are constants determined by the least-square regression method. Secondly, a cubic smoothing spline estimate $\hat{G}^*(\lambda)$ of the function $G^*(\lambda)$ was selected, as defined to be the minimizer of the objective function [34]

$$p \sum_i w_i \left[G_i^* - \hat{G}^*(\lambda_i)\right]^2 + (1 - p) \int \left(\frac{d^2 \hat{G}^*}{d\lambda^2}\right)^2 d\lambda \tag{17}$$

for specified smoothing parameter p $(0 \leq p \leq 1)$ and weights (w_i). For small values of p, the fitted curve tends to be too rigid, and for values close to unity (with a significant number of digits), the regression curve loses the filtering capacity. In the analysis, a smoothing parameter $p = 0.989$ was chosen, along with weights $w_i = 1$ for all data points.

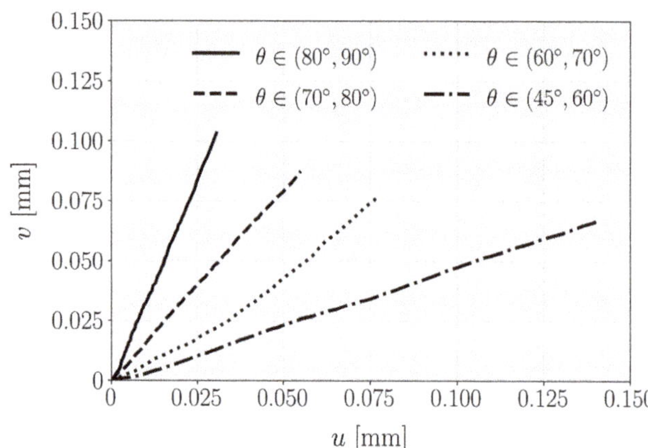

Figure 4. Pattern of the local deformation for each mixed mode ratio of the MMB tests.

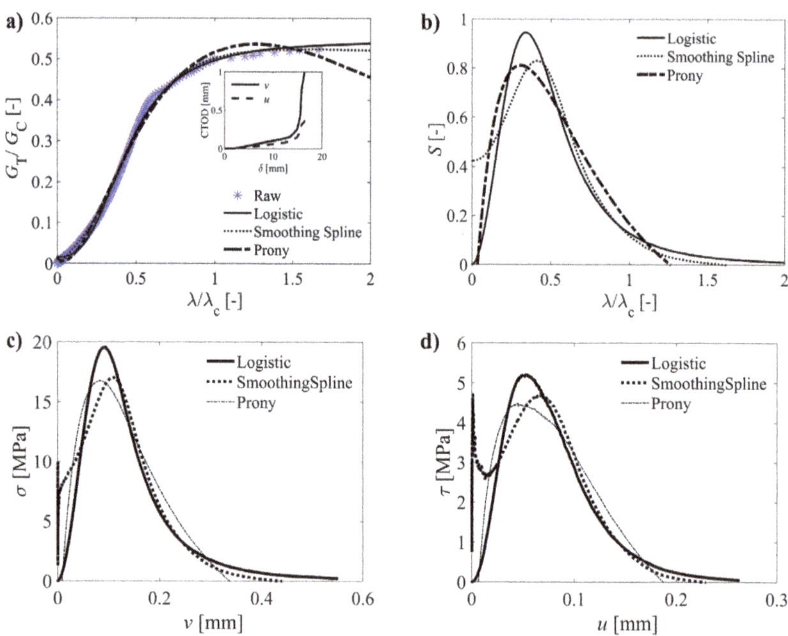

Figure 5. For a typical specimen: (**a**) potential function; (**b**) normalized cohesive law; (**c**) mode I; and (**d**) mode II components of the reconstructed cohesive laws.

Lastly, the Prony-series was also considered, expressed by the following relationship [34]

$$\frac{G^*}{G^*_{max}} = \sum_{i=1}^{n} B_i exp^{\left(-\frac{n\lambda}{i\lambda_{max}}\right)} \tag{18}$$

where the parameters B_i in the series are determined by least-square regression. In Equation (18), G^*_{max} represents the maximum value of G^*, and λ_{max} is the corresponding value of λ. In a sensitivity study, the terms of n were taken between 7 and 17. A convergence was achieved by considering $n = 10$.

An application example of the typical curve fitting analysis by least-square regression and reconstructed curves (by differentiation) are shown in Figure 5. The resulting cohesive laws components for mode I and mode II are shown, respectively, in Figure 5c,d. It can be observed that the Prony-series tend to underestimate the initial and final parts of the raw data. On the other hand, the flexibility of the smoothing spline to follow the raw data comes at the cost of a reduced filtering effect. Finally, the logistic function guarantees a certain degree of smoothing, and moreover, forces the regularization of the typical $G^*(\lambda)$ S-shape curve. Furthermore, the differentiation can be computed directly from the fitted constants defining the function (Equation (15)). Consequently, the logistic curve was considered for the systematic analysis of the data.

The mode I and mode II components of the reconstructed cohesive laws were determined according to Equation (14). Figure 6 shows an overview of mean curves for each phase angle intervals considered. Overall, it can be stated that the mode I component of the cohesive law increases with the increase of the global phase angle (Figure 6a), in contrast to what happens with the mode II component (Figure 6b). This behavior agrees with the anticipated trend, which reinforces the appropriateness of the proposed methodology.

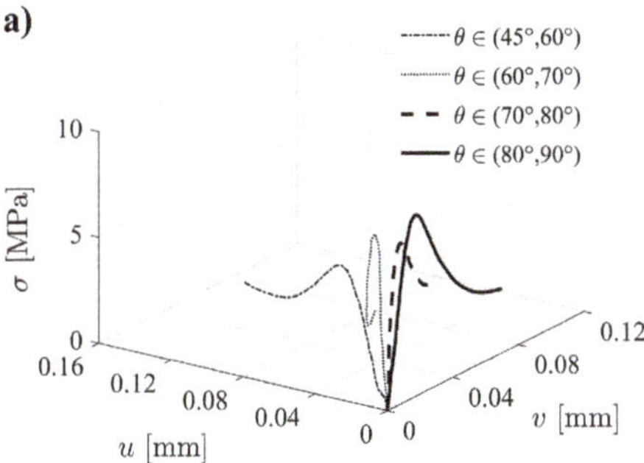

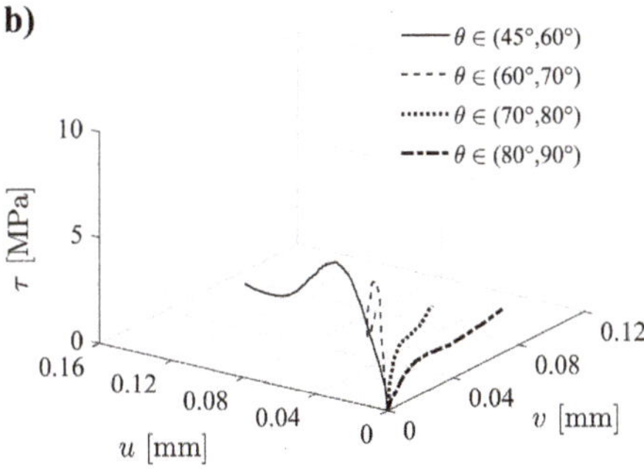

Figure 6. Mean cohesive laws distribution over the range of phase angles: (**a**) $\sigma(v,u)$; (**b**) $\tau(v,u)$.

In more detail, Figure 7 plots the evolution of the several cohesive parameters a function of the global phase angle. The rising trend of the normal peak stresses (Figure 7a and the decreasing tendency observed for the shear ones were approximated by linea functions. Extrapolating for $\theta = 90°$ in the linear fitted equation ($\sigma = f(\theta)$) Figure 7a and for $\theta = 0°$ in the relation of Figure 7b provides an estimation of the local strengths under pure mode I and pure mode II, respectively. The obtained values point to $\sigma_u = 8.96$ MP. and $\tau_u = 16.56$ MPa. These values compare well with the ones reported in [4] for the same wood species ($\sigma_u = 7.93$ MPa and $\tau_u = 16.0$ MPa), revealing that the followed procedure effectively captures these material parameters.

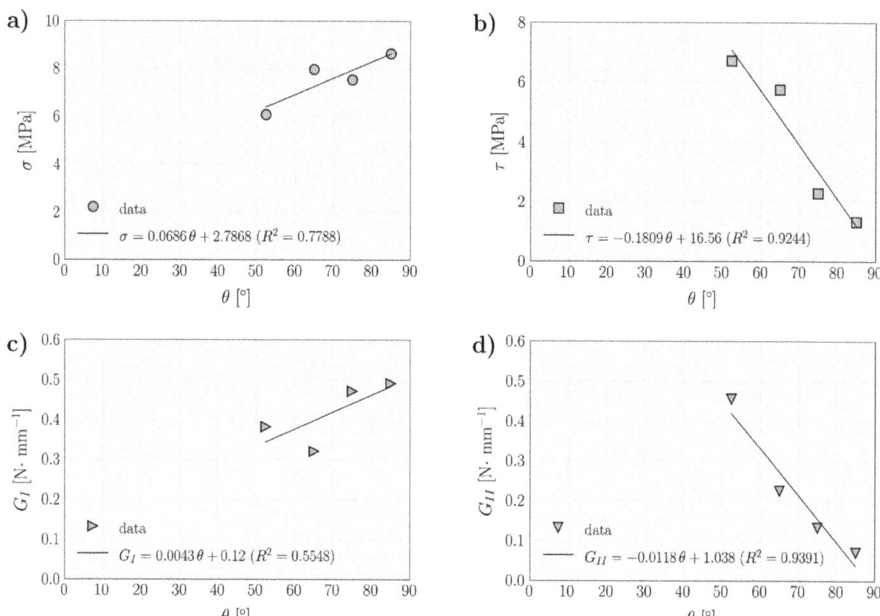

Figure 7. Evolution of cohesive law parameters as a function of global phase angle (**a**) ultimate normal stress; (**b**) ultimate shear stress; (**c**) mode I strain energy component; and (**d**) mode II strain energy component.

The same strategy was followed for the evolution of the strain energy release rate components (G_I and G_II) in function of θ. Considering the linear relationships obtained in Figure 7c,d, the critical fracture energies point to $G_{IC} = 0.5$ N/mm and $G_{IIC} = 1.0$ N/mm respectively. In the mode I case, the obtained value is somewhat higher than expected since a recent characterization of this wood species under mode I loading [35] pointed to values in the range of 0.3–0.4 N/mm. In contrast, the G_{IIC} value is in close agreement with recent pure mode II fracture characterization, which points to $G_{IIC} = 0.97$ N/mm [36].

5. Conclusions

This work addresses the experimental identification of cohesive laws in mixed mode I+II of *Pinus pinaster* Ait. The approach combines the MMB test with DIC measurements. The MMB setup was configured to target different mixed mode ratios. The obtained results showed that the ratio of the local mixed mode ($tan\phi = v/u$), determined by opening displacements in mode I (v) and mode II (u), is almost constant throughout the test. This observation allowed the definition of the global phase angle for each mixed mode ratio assuming linear regressions. It was found that mode I components of the cohesive laws under mixed mode reveal an increase, with the global phase angle in opposition to what happens with the mode II component. The evolution of the peak stresses and strain energy release rate components as functions of the global phase angle was analyzed in more detail. It was observed that normal peak stress and mode I strain energy release rate component increase with the global phase angle, contrasting to shear peak stress and mode II strain energy release rate component that decrease alongside it. These relations were fitted by linear functions that were subsequently used to estimate the pure mode I and II parameters by extrapolation. The values obtained for the local strengths (σ_u and τ_u) and critical fracture energies (G_{IC} and G_{IIC}) are globally in agreement with previous experimental determinations for this wood species.

These results validate the proposed procedure as a valuable tool to assess the cohesive laws and its components, and serve as a support for development of cohesive zone models appropriate to deal with the mixed mode fracture behavior of wood and other materials.

Author Contributions: Conceptualization, J.X., F.P., J.M. and M.d.M.; methodology, J.O., J.X., F.P., J.M. and M.d.M.; formal analysis, J.O., J.X., F.P., J.M. and M.d.M.; investigation, J.X., F.P., J.M. and M.d.M.; resources, J.M. and M.d.M.; writing—original draft preparation, J.O., J.X. and F.P.; writing—review and editing, J.X., F.P., J.M. and M.d.M.; supervision, J.M. and M.d.M. All authors have read and agreed to the published version of the manuscript.

Funding: This research was funded by European Union Funds (FEDER/COMPETE—Operational Competitiveness Programme) and by Portuguese Foundation for Science and Technology (FCT) grant number FCOMP-01-0124-287 FEDER-022692, UIDB/04033/2020 and UIDB/00667/2020 (UNIDEMI).

FCT Fundação para a Ciência e a Tecnologia

MINISTÉRIO DA CIÊNCIA, TECNOLOGIA E ENSINO SUPERIOR

Data Availability Statement: Data sharing not applicable.

Conflicts of Interest: The authors declare no conflict of interest.

References

1. Reiterer, A.; Sinn, G.; Stanzl-Tschegg, S.E. Fracture characteristics of different wood species under mode I loading perpendicular to the grain. *Mater. Sci. Eng. A* **2002**, *332*, 29–36. [CrossRef]
2. Yoshihara, H.; Ohta, M. Measurement of mode II fracture toughness of wood by the end-notched flexure test. *J. Wood Sci.* **2000**, *46*, 273–278. [CrossRef]
3. De Moura, M.F.S.F.; Morais, J.J.L.; Dourado, N. A new data reduction scheme for mode I wood fracture characterization using the DCB test. *Eng. Fract. Mech.* **2008**, *75*, 3852–3865. [CrossRef]
4. De Moura, M.F.S.F.; Silva, M.A.L.; de Morais, A.B.; Morais, J.J.L. Equivalent crack based mode II fracture characterization of wood. *Eng. Fract. Mech.* **2006**, *73*, 978–993. [CrossRef]
5. Xavier, J.; Oliveira, J.; Monteiro, P.; Morais, J.J.L.; de Moura, M.F.S.F. Direct evaluation of cohesive law in mode I of Pinus pinaster by digital image correlation. *Exp. Mech.* **2014**, *54*, 829–840. [CrossRef]
6. Xavier, J.; Oliveira, M.; Morais, J.; de Moura, M.F.S.F. Determining mode II cohesive law of Pinus pinaster by combining the end-notched flexure test with digital image correlation. *Constr. Build. Mater.* **2014**, *71*, 109–115. [CrossRef]
7. Jernkvist, L.O. Fracture of wood under mixed-mode loading II. Experimental investigation of Pices abies. *Eng. Fract. Mech.* **2001**, *68*, 565–576. [CrossRef]
8. Tschegg, E.K.; Reiterer, A.; Pleschbergers, T.; Stanzl-Tschegg, E. Mixed mode fracture energy of spruce wood. *J. Mater. Sci.* **2001**, *36*, 3531–3537. [CrossRef]
9. Reeder, J.R.; Crews, J.H., Jr. Mixed-mode bending method for delamination testing. *AIAA J.* **1990**, *28*, 1270–1276. [CrossRef]
10. Crews, J.H.; Reeder, J.R. *A Mixed-Mode Bending Apparatus for Delamination Testing*; NASA Technical Memorandum 100662; National Aeronautics and Space Administration, Langley Research Center: Hampton, VA, USA, 1988.
11. Reeder, J.; Crews, J. Redesign of the Mixed-Mode Bending Delamination Test to Reduce Nonlinear Effects. *ASTM J. Compos. Technol. Res.* **1992**, *14*, 12–19.
12. Chen, J.H.; Sernow, R.; Schulz, E.; Hinrichsen, G.A. modification of the mixed-mode bending test apparatus. *Compos. Part A Appl. Sci.* **1999**, *30*, 871–877. [CrossRef]
13. ASTM D6671/D6671M-19. *Standard Test Method for Mixed Mode I-Mode II Interlaminar Fracture Toughness of Unidirectional Fiber Reinforced Polymer Matrix Composites*; ASTM International: West Conshohocken, PA, USA, 2019.
14. De Moura, M.F.S.F.; Oliveira, J.M.Q.; Morais, J.J.L.; Xavier, J. Mixed-mode I/II wood fracture characterization using the mixed-mode bending test. *Eng. Fract. Mech.* **2010**, *77*, 144–152. [CrossRef]

15. De Moura, M.F.S.F.; Oliveira, J.M.Q.; Morais, J.J.L.; Dourado, N. Mixed-mode (I+II) fracture characterization of wood bonded joints. *Constr. Build. Mater.* **2011**, *25*, 1956–1962. [CrossRef]
16. Grédiac, M.; Hild, F. (Eds.) *Full-Field Measurements and Identification in Solid Mechanics*; John Wiley and Sons: Hoboken, NJ, USA, 2012.
17. Pereira, J.; Xavier, J.; Morais, J.; Lousada, J. Assessing wood quality by spatial variation of elastic properties within the stem: Case study of P. pinaster in the transverse plane. *Can. J. For. Res.* **2014**, *44*, 1–11. [CrossRef]
18. Pereira, J.L.; Xavier, J.; Ghiassi, B.; Lousada, J.; Morais, J. On the identification of earlywood and latewood radial elastic modulus of P. pinaster by digital image correlation: A parametric analysis. *J. Strain Anal. Eng.* **2018**, *53*, 566–574. [CrossRef]
19. Catalanotti, G.; Camanho, P.P.; Xavier, J.; Dávila, C.G.; Marques, A.T. Measurement of resistance curves in the longitudinal failure of composites using digital image correlation. *Compos. Sci. Technol.* **2010**, *70*, 1986–1993. [CrossRef]
20. Cappello, R.; Pitarresi, G.; Xavier, J.; Catalanotti, G. Experimental determination of mode I fracture parameters in orthotropic materials by means of Digital Image Correlation. *Theor. Appl. Fract. Mec.* **2020**, *108*, 102663. [CrossRef]
21. Dias, G.F.; de Moura, M.F.S.F.; Chousal, J.A.G.; Xavier, J. Cohesive laws of composite bonded joints under mode I loading. *Compos. Struct.* **2013**, *106*, 646–652. [CrossRef]
22. Fernandes, R.M.R.P.; Chousal, J.A.G.; de Moura, M.F.S.F.; Xavier, J. Determination of cohesive laws of composite bonded joints under mode II loading. *Compos. Part B Eng.* **2013**, *52*, 269–274. [CrossRef]
23. Koerber, H.; Xavier, J.; Camanho, P.P. High strain rate characterisation of unidirectional carbon-epoxy IM7-8552 in transverse compression and in-plane shear using digital image correlation. *Mech. Mater.* **2010**, *42*, 1004–1019. [CrossRef]
24. Catalanotti, G.; Kuhn, P.; Xavier, J.; Koerber, H. High strain rate characterisation of intralaminar fracture toughness of GFRPs for longitudinal tension and compression failure. *Compos. Struct.* **2020**, *240*, 112068. [CrossRef]
25. Xavier, J.; Pereira, J.C.R.; de Jesus, A.M.P. Characterisation of steel components under monotonic loading by means of image-based methods. *Opt. Lasers Eng.* **2014**, *53*, 142–151. [CrossRef]
26. Xavier, J.; Avril, S.; Pierron, F.; Morais, J. Novel experimental approach for longitudinal-radial stiffness characterisation of clear wood by a single test. *Holzforschung* **2007**, *61*, 573–581. [CrossRef]
27. Xavier, J.; Pierron, F. Characterisation of orthotropic bending stiffness components of P. pinaster by the virtual fields method. *J. Strain Anal. Eng.* **2018**, *53*, 556–565. [CrossRef]
28. Sutton, M.; Orteu, J.-J.; Schreier, H. *Image Correlation for Shape, Motion and Deformation Measurements: Basic Concepts, Theory and Applications*; Springer: Berlin/Heidelberg, Germany, 2009.
29. Réthoré, J.; Hild, F.; Roux, S. Extended digital image correlation with crack shape optimization. *Int. J. Numer. Meth. Eng.* **2008**, *732*, 248–272. [CrossRef]
30. Sousa, A.M.R.; Xavier, J.; Morais, J.J.L.; Filipe, V.M.J.; Vaz, M. Processing discontinuous displacement fields by a spatio-temporal derivative technique. *Opt. Lasers Eng.* **2011**, *49*, 1402–1412. [CrossRef]
31. Sørensen, B.F.; Kirkegaard, P. Determination of Mixed Mode Cohesive Laws. *Eng. Fract. Mech.* **2006**, *73*, 2642–2661. [CrossRef]
32. Högberg, J. Mixed mode cohesive law. *Int. J. Fract.* **2006**, *141*, 549–559. [CrossRef]
33. Xavier, J.; de Jesus, A.M.P.; Morais, J.J.L.; Pinto, J.M.T. Stereovision measurements on evaluating the modulus of elasticity of wood by compression tests parallel to the grain. *Constr. Build. Mater.* **2012**, *26*, 207–215. [CrossRef]
34. Hastie, T.J.; Tibshirani, R.J. *Generalized Additive Models*; Chapman and Hall: London, UK, 1990.
35. De Moura, M.F.S.F.; Dourado, N. Mode I fracture characterization of wood using the TDCB test. *Theor. Appl. Fract. Mec.* **2018**, *94*, 40–45. [CrossRef]
36. De Moura, M.F.S.F.; Silva, M.A.L.; Morais, J.J.L.; Dourado, N. Mode II fracture characterization of wood using the Four-Point End-Notched Flexure (4ENF) test. *Theor. Appl. Fract. Mec.* **2018**, *98*, 23–29. [CrossRef]

materials

MDPI

Article

Material Origins of the Accelerated Operational Wear of RD-33 Engine Blades

Adam Kozakiewicz [1], Stanisław Jóźwiak [2], Przemysław Jóźwiak [3] and Stanisław Kachel [1,*]

[1] Faculty of Mechatronics, Armament and Aerospace, Institute of Aviation Technology, Military University of Technology, 00-908 Warszawa, Poland; adam.kozakiewicz@wat.edu.pl
[2] Faculty of Advanced Technology and Chemistry, Institute of Materials Science and Engineering, Military University of Technology, 00-908 Warszawa, Poland; stanislaw.jozwiak@wat.edu.pl
[3] Air Force Institute of Technology, 01-494 Warszawa, Poland; przemyslaw.jozwiak@itwl.pl
* Correspondence: stanislaw.kachel@wat.edu.pl; Tel.: +48-261-839-170

Abstract: The structural and strength analysis of the materials used to construct an important engine element such as the turbine is of great significance, at both the design stage and during tests and training relating to emergency situations. This paper presents the results of a study on the chemical composition, morphology, and phased structure of the metallic construction material used to produce the blades of the high- and low-pressure turbines of the RD-33 jet engine, which is the propulsion unit of the MiG-29 aircraft. On the basis of an analysis of the chemical composition and phased structure, the data obtained from tests of the blade material allowed the grade of the alloy used to construct the tested elements of the jet engine turbine to be determined. The structural stability of the material was found to be lower in comparison with the engine operating conditions, which was shown by a clear decrease in the resistance properties of the blade material. The results obtained may be used as a basis for analyzing the life span of an object or a selection of material replacements, which may enable the production of the analyzed engine element.

Keywords: turbine jet engine; material tests; ember-resistant alloys

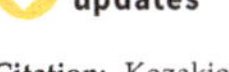

check for **updates**

Citation: Kozakiewicz, A.; Jóźwiak, S.; Jóźwiak, P.; Kachel, S. Material Origins of the Accelerated Operational Wear of RD-33 Engine Blades. *Materials* **2021**, *14*, 336. https://doi.org/10.3390/ma14020336

Received: 2 November 2020
Accepted: 8 January 2021
Published: 11 January 2021

Publisher's Note: MDPI stays neutral with regard to jurisdictional claims in published maps and institutional affiliations.

1. Introduction

Aviation turbine engines must meet very high requirements relating to reliability, strength, minimum weight, serviceability, the acceptable period of use in service, noise, ecology, and cost-effectiveness [1]. In turbine jet engines, one of the essential components that must meet these criteria is the turbine assembly. A turbine assembly is a system that is heavily loaded mechanically and thermally. This is caused by tension and bending of the blades as a result of centrifugal forces, bending, and twisting resulting from the exhaust gas mass flow. It is also caused by high temperature. This results in a number of complex stress states, especially in the blade palisades of rotor wreaths, including the occurrence of variable stress. Therefore, in order to increase the operational properties, two changes in the blade design are made: changes in the design that improve the cooling efficiency and changes in the material of the blade matrix and protective coatings that protect from overheating (Figure 1). One particular question that arises when it comes to the employment of aircraft jet engines is what should be done to prolong the engine's service life, specifically the lifespan of the engine's turbine. The problem becomes more critical when the engines in question are used and replaced on the basis of hours in service. Complicating this problem are factors concerning the engine component quality and fatigue. A good example of this would be the Klimov RD-33 engine, used by the Polish Air Force. If we analyze the operating costs of two RD-33 engines that power the MiG-29 in comparison with those of the F-100-PW-229 engine used to power the F-16C, the costs of repairs are much higher in the case of the RD-33 (about 18 million dollars in total costs [2]), considering an operational lifespan of approximately 3200 h for both units.

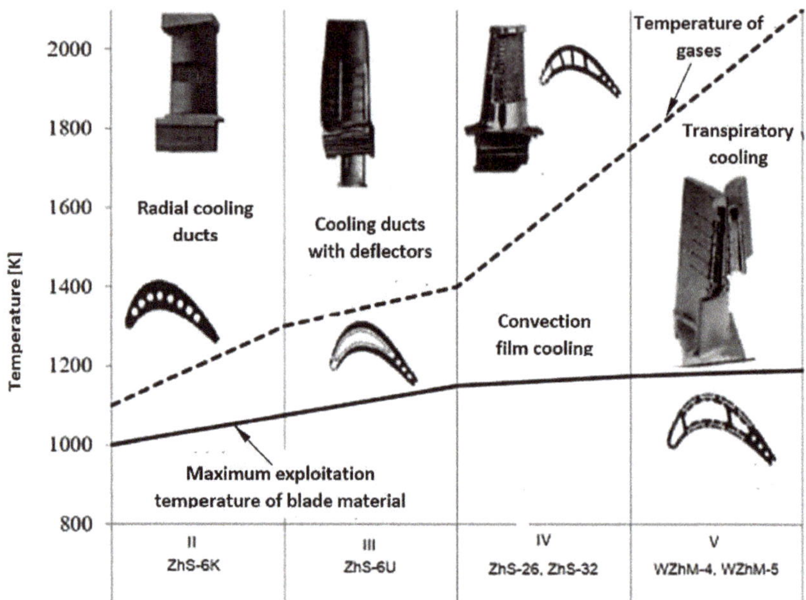

Figure 1. Development of the cooling technology and change in materials used to construct Russian turbine jet engines [3].

These costs are further increased in the event of premature engine failure, resulting in removal, as occurred in the case we analyzed. For instance, approximately 13% of engines are prematurely decommissioned due to damage to the high-pressure turbine [4]. Studies have found that the most common causes of turbine element damage are mechanical and high-temperature damage [5]. The main mechanics behind mechanical damage are the propagation of cracks, specifically fatigue-related ones, in addition to abrasive wear on the surface of turbine blades [6,7]. However, based on S. Szczeciński's papers [8], the problem of abrasive wear applies mostly to turbine engines installed in helicopters, which often use unimproved "dusty" airfields. This paper focuses on turbofan engines fitted to attack aircraft, which generally operate at improved airfields, where the problem of dust and foreign object damage is notably less of a concern. Moreover, S. Szczeciński claims that low-bypass turbofan 2-spool jet engines—the analyzed RD-33 being an example of such an engine due to the placement of the turbine within the profile of the core of the engine—are less susceptible to abrasive wear of the turbine's elements [8]. The development of cracks on the leading edge or the trailing edge of the turbine blade is usually caused by changes in the structure of the blade material both in diffusion zones of the coating of the core and in the overheated core (HAZ), which lead to the creation of hard and fragile phases in grain boundaries, resulting in a lowered fatigue strength due to an easier generation of cracks [9]. The cause of changes in the material structure is usually damage to thermal barrier coatings (TBC), which occurs due to high-temperature gaseous corrosion [10,11] Based on the analysis in [12], the most common cause of first-degree damage of the turbines in RD-33 engines is high-temperature gaseous corrosion (totaling 45% of all kinds of damage), while the development of cracks is the cause in only 35% of cases. In light of the abovementioned facts, this paper focuses on structural changes caused by turbine blade overheating, which might result in various damage mechanisms, including accelerated fatigue of the material. Accelerated wear of engine components generates higher costs of maintenance, and many attempts at elucidating the causes of damage and finding preventive measures have been made, although it has been impossible to acquire precise data from the manufacturer. Additionally, the turbine blades of the RD-33 engine are made of various types of high-temperature superalloys, with a variety of thermal performance characteristics. It was therefore necessary to determine the properties of the materials used in the turbine blades, after the engine's premature removal from service, in order to accurately analyze the mechanisms underlying their damage. This analysis would imply the benefit of modifications to the thermal barrier coatings, effectively expanding the service time of the engine.

2. Analysis of the State of the Issue

The basic parameter characterizing the ember-resistance of nickel-based super alloys, intended for the construction of the turbine blades of jet engines, is their creep resistance. The increase of this parameter, depending on the phased structure of the alloy, and in particular, the share of γ/γ' phases, reinforcement carbides, Laves phases, etc., is obtained by modifying the chemical composition and morphology of the crystal structure of the blade matrix material. The same also applies in the case of Russian superalloys. An increase in the creep resistance of the materials used to produce turbine blades was obtained by using Re and Ta as additives, which, together with directional crystallization (ZhS-26) or monocrystalline crystallization (ZhS-32), allowed, unlike ZhS-6U alloy, for a significant increase of this parameter (Table 1).

Table 1. Creep resistance of super ZhS alloys [12].

Alloy	σ_{100}^{900} (MPa)	σ_{100}^{1000} (MPa)	σ_{100}^{1100} (MPa)
ZhS-6U	350	170	65
ZhS-26	400	200	85
ZhS-32	480	250	120

An increase of the creep resistance of ZhS-32 alloy, according to the authors in [13], allowed the temperature of the gases on the blades in the first stage of the turbine to be increased from 1263 °C (for the ZhS-6U-VI alloy [14]) up to 1400 °C. However, due to the acceptable period of use in service of the RD-33 engine, the prolonged temperature of the blade material, exceeding 1100 °C, is not recommended.

The preservation of these exploitation regimes allowed the acceptable period of use in service of the RD-33 engine blades made of ZhS-32 superalloy to be extended from 300 h (ZhS-6) up to 1000 h of exploitation. Nevertheless, one must keep in mind that prolonged exposure to hot exhaust gases can lead to structural changes within the blade material, especially if the protective coating is damaged, thus leading to its destruction. A detailed analysis of the RD-33 engine damage showed that over 42% of failures were caused by damage to the blades, and damage to blades in the first stage of the turbine accounted for over 60% [13]. The main causes of damage are gaseous corrosion at the leading edge, cracks at the leading and trailing edges, and thermo-mechanical damage (Figure 2).

The scope of application of the alloys used to construct the blades of the RD-33 engine and, in particular, the creep resistance, depending on the temperature and duration of its long-term impact on the material, are best represented by the Larson–Miller parametric coordinates, as defined by Formula (1) (Figure 3):

$$P = (T + 273) \times (20 + \log t) * 10^{-3} \tag{1}$$

where:

T—temperature [°C]
t—exploitation time [h]

The higher permissible exploitation stress of the ZhS-32 alloy, compared to ZhS-6U, should be explained by its higher volume content of phase γ' in the material structure, which is 67% (compared to 57%), and its higher thermal stability, caused by the Re and

Ta additives, which increase the temperature of the solidus from 1240 °C (ZhS-6U) up to 1310 °C (ZhS-32) [15]. It is obvious that the long-drawn process of exploitation, linked with the high-temperature influence of exhaust gases, causes the grain coarsening and decrease of the phase γ' content in the blade material structure (Figure 4).

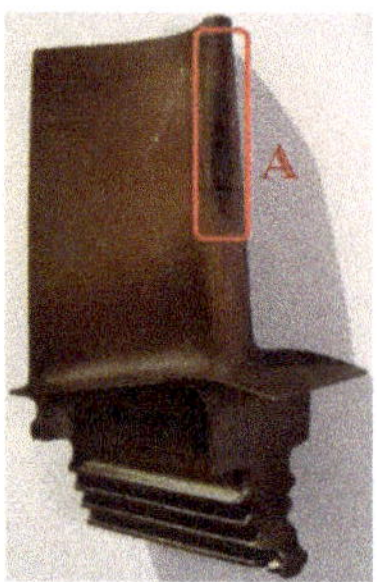

Figure 2. Blade from the rotor wreath of a high-pressure turbine, with a marked zone (A) of gas corrosion damage and crack damage of the leading edge of the blade.

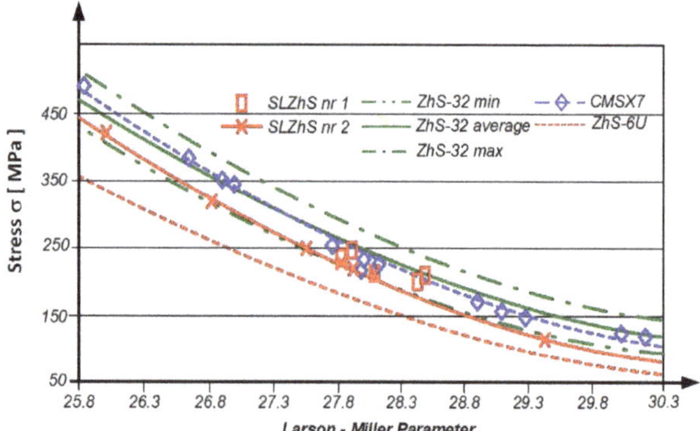

Figure 3. Influence of temperature and exploitation time on the exploitation stress of ZhS alloys used in the construction of RD-33 engine turbine blades, as defined by the Larson–Miller parameter P [3].

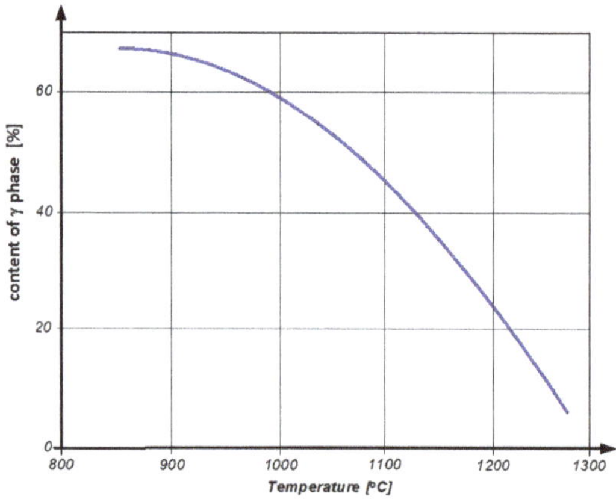

Figure 4. Influence of exploitation temperature on the share of phase γ' in the structure of the ZhS-32 alloy [14].

These phenomena result in a decrease in strength parameters in nickel-based superalloys, manifested by a significant decrease in the fatigue strength of the blade material (Figure 5).

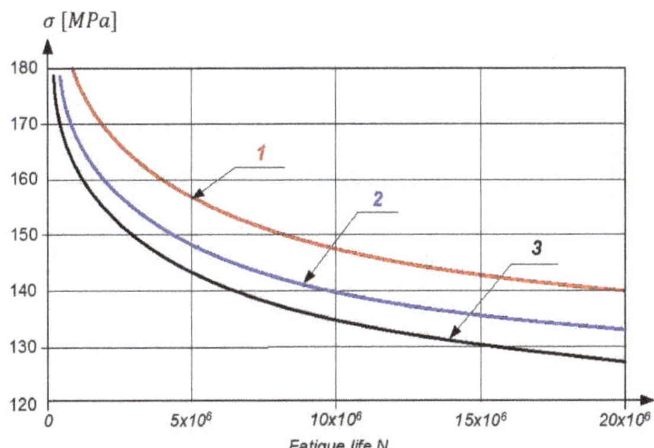

Figure 5. The fatigue curves of turbine blades made of EI-437B, where the working time t = 0 h (1), t = 200 h (2), and t = 400 h (3) [15].

In connection with the abovementioned conditionalities/exploitation and strength rationale, for the purpose of research on the construction of the aircraft engine assemblies in the area of structural analysis, a thorough analysis of material degradation shall be carried out. This must take into account tests of the grain structure and the chemical composition that determines the phased structure, which directly affects the operational properties, including the mechanical properties that determine the performance of such a complex structure as the modern turbine system. Incorrect use of a given construction material, or incorrect exploitation or other factors causing the limits (resulting from strength calculations) of the applied materials to be exceeded, may lead to a reduction in strength. This, in turn, may lead to damage to the turbine assembly and engine.

The above assumptions form the basis for the research undertaken to determine the causes of premature damage to the blades of the turbines of the RD-33 engine of the MiG-29 airplane, resulting in the necessity to dismount the engines and perform engine overhauls.

3. Sample Preparation and Test Methodology

The mass loads and, in particular, the heat loads of the turbine blades caused by the exhaust gas stream from the combustion chamber (1560 K) require cooling of this engine component. In the RD-33 motor, under analysis, the blades of both the LPT and HPT are cooled by means of internal channels (Figure 6), and the blades of the HPT, due to the higher gas temperature, also have channels for cooling the trailing edge.

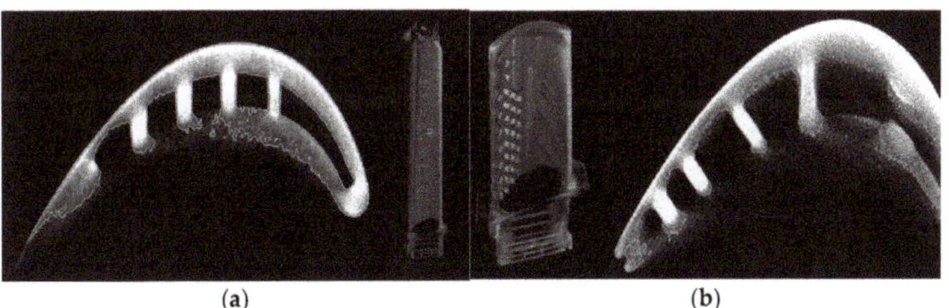

(a)   (b)

Figure 6. The shape of the LPT blade (**a**) and the HPT blade (**b**), with visible cooling channels, obtained by the computer-based micro-tomography CT.

This blade design poses serious problems relating to the selection of a place to take the strength test specimens in order to properly perform a static tensile test. Therefore, taking into account the nature of the mechanical loads carried by the material of the component tested, it was assumed that the specimens for the static tensile tests will be cut along the blade feather of the LPT, i.e., in the direction of the centrifugal forces and, in the case of the HPT, from the lock material, i.e., the component subjected to the highest mass loads (Figure 7).

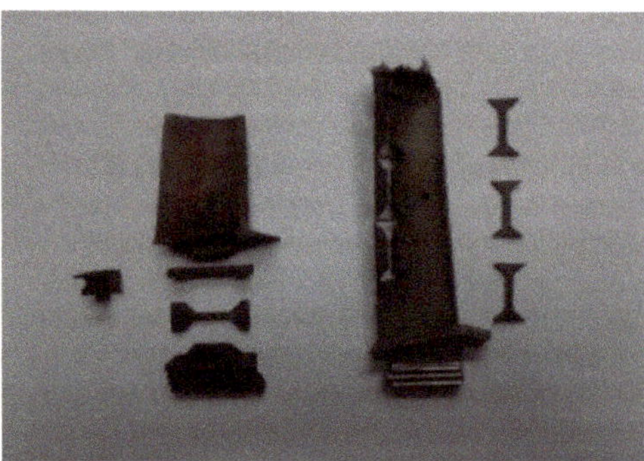

Figure 7. View of specimens cut out (using the electro-discharge method) of a material of the LPT blade and view of specimens for static tensile testing cut out of the material of the HPT blade and LPT blade.

Nevertheless, the three-dimensional shape, with variable cross-sections and curvatures, made it impossible to clamp and thus to make the strength specimens with classical machines for cavity machining (edging method machining). Therefore, in order to preserve the assumed shape, dimensions, and geometric quality of the surface of the specimens, as well as to minimize the impact of the preparation process on the structural changes of the blade material, electro-discharge machining was used as the cutting method. All of the samples were made using the electro-discharge wire rod numerically controlled machine, and the assumed geometric dimensions have been respected.

In order to make microscopic observations to reveal the grain structure of the blade material, metallographic samples were prepared (Figure 8). They were prepared, first, by mounting them in a conductive resin and then grinding them with abrasive papers of varying granulations in the range of 100–1000 and, second, by a final polishing with a diamond slurry of polishing powder using the Struers polishing machine.

Figure 8. Metallographic specimen of the material of the LPT blade.

Microscopic observations at magnifications of 100–10,000 times were made by using the Quanta 3 D FEG (SEM/FIB) high-resolution scanning microscope, which, in addition to grain structure morphology analysis, also enables, with the use of EDS/WDS/EBSD attachments, complex studies of the chemical composition in micro-areas and the analysis of crystallographic orientation.

Since chemical composition microanalysis indirectly allows only for the identification of the phased structure of the material under investigation, it was necessary to perform X-ray phase analysis (XRD). Identifying the type of crystal lattice and measuring the parameters of the elementary crystal cell allows, in a precise and unambiguous way, for the identification of the phased structure of the material under investigation, on which the functional properties, including the strength properties, depend. The phase X-ray analysis was carried out with the ULTIMA IV Rigaku diffractometer using a parallel beam. The measurements were carried out using a cobalt lamp, with a radiation length CoKα of 1 = 1.78892 Å, power of 1600 W, scanning step of 0.02°, and scanning speed of 2°/min in the angular range of 2 Θ = 20–140°. The identification of the obtained reflections and phased analysis were carried out on the basis of the crystallographic database, PDF-4.

The evaluation of the mechanical properties of the material of the tested blades was based on the hardness, microhardness, and the static tensile tests. Due to the ultrafine grain structure of the tested material, in order to unequivocally confirm the correctness of the hardness measurements, the testing of this strength parameter was carried out using three measurement methods (HBW2.5/187.5, HV10—Wolpert Wilson Testor, and HRC—Rockwell PW 106 hardness meter), and then (using appropriate nomograms, enabling a comparison of the hardness values measured by different methods) the results obtained were evaluated, taking into account the statistical analysis to determine the correctness of the data obtained. In order to further assess the homogeneity of the matrix structure, Vickers microhardness tests were performed in various areas of the blades using the SHIMADZU-DORERNST M hardness tester, with a load of 25 G.

A static tensile test, considered a destructive test, was carried out with the Instron 8501 Plus universal testing machine in accordance with the PN-EN ISO 6892–1:2010 standard. Due to the non-standardized, small dimensions of the strength specimens obtained (Figure 7), in order to execute the static tensile test, it was necessary to design and manufacture special holders to give the test a proper performance (Figure 9).

Figure 9. View of the test sample in specially designed holders.

The manufactured tooling made it possible to determine the following basic strength parameters based on the obtained tensile curves:

- 0.2% offset yield strength $R_{0,2}$;
- ultimate tensile strength R_m;
- fracture strain of sample A;
- Young's modulus E.

4. The Results of the Tests

The microscopic observations carried out at magnifications of 100–10,000× (Figure 10) showed that the TWC blade material is characterized by a multiphase structure typical for nickel-based heat-resistant alloys. After the supersaturation and aging process, it consists of the primary, dendritic structure, resulting from the crystallization of the alloy (Figure 10a); the solid solution γ, formed during supersaturation; cuboid intermetallic precipitates of the γ' phase, released during the aging process (Figure 10b); and reinforcing carbides, with a "Chinese script" morphology (Figure 10c).

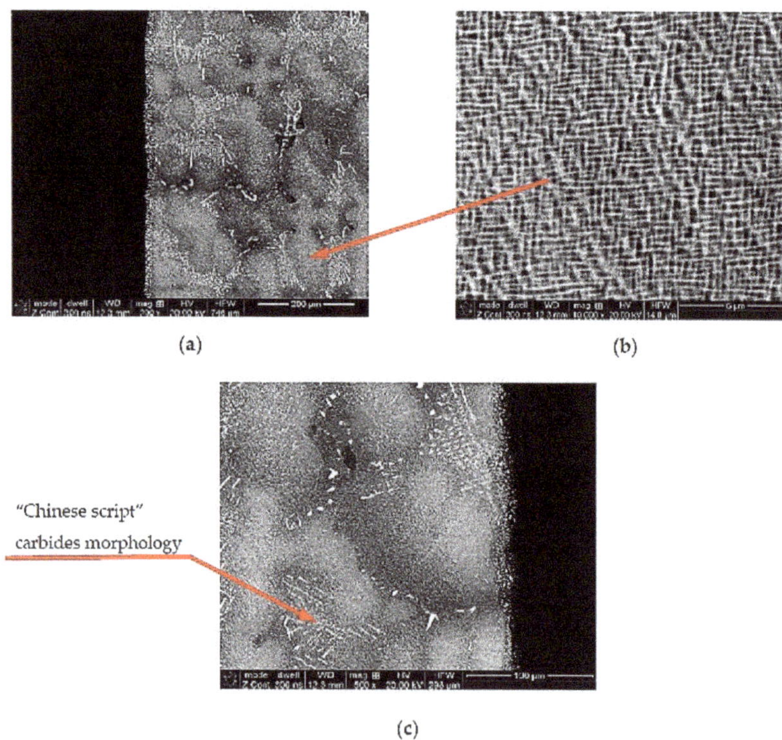

(a)

(b)

(c)

Figure 10. Multiphase structure of the TWC blade material built on the matrix of the γ phase (**a**), with cuboidal precipitates of the intermetallic phase γ' (**b**) and reinforcing carbide precipitates (**c**).

The available data show that the blades of the turbine of the RD-33 engine are currently made of two types of alloys: monocrystalline ZhS-32 or directionally crystallized Zhs-26.

Nevertheless, the microanalysis of the chemical composition of the multi-phase blade structure allowed for the identification of significant discrepancies, mainly in terms of the content of chromium and titanium, between the tested material and the ZhS-32 and ZhS-26 alloys. The chemical composition of the analyzed material of the blades suggests that they are made of an alloy with a chemical composition corresponding to the heat-resistant nickel-based superalloy, MAR-M 200 [16], which also corresponds to the chemical composition of the Russian ZhS-6U alloy [17] (Figure 11) (Table 2.).

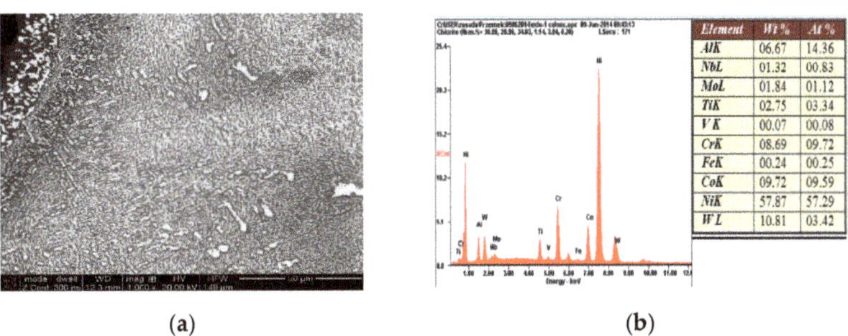

Element	Wt %	At %
Al K	06.67	14.36
Nb L	01.32	00.83
Mo L	01.84	01.12
Ti K	02.75	03.34
V K	00.07	00.08
Cr K	08.69	09.72
Fe K	00.24	00.25
Co K	09.72	09.59
Ni K	57.87	57.29
W L	10.81	03.42

(a) (b)

Figure 11. Result of the microanalysis of the chemical composition of the material of the HPT blade. The area subject to microanalysis (**a**), the spectrum of identified elements (**b**).

The maximum exploitation temperature of the MAR-M 200 alloy is (according to the CES Edu Pack database) in the range of 815–983 °C, and the temperature of the exhaust gases before the rotor wreath of a high pressure turbine (HPT) of the RD-33 engine reaches the upper level of the temperature range of the material used for the blades of the HPT. Therefore, in order to protect the material of the blade core, it was covered with a barrier coating, based on the NiAl intermetallic phase (Figure 12), created using diffusional aluminizing technology.

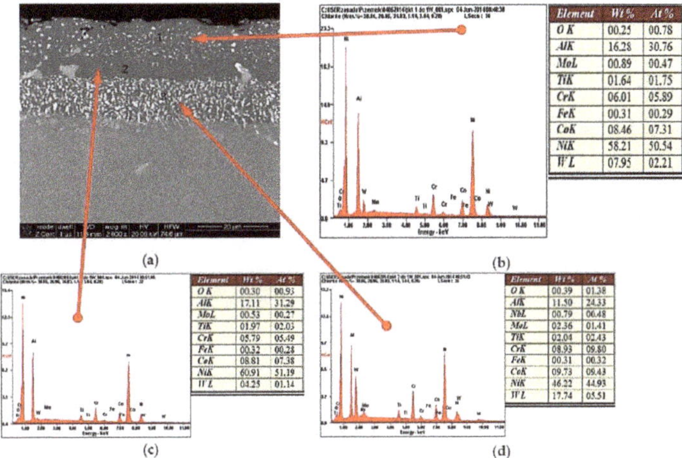

Element	Wt %	At %
O K	00.25	00.78
Al K	16.28	30.76
Mo L	00.89	00.47
Ti K	01.64	01.75
Cr K	06.01	05.89
Fe K	00.31	00.29
Co K	08.46	07.31
Ni K	58.21	50.54
W L	07.95	02.21

(a) (b)

Element	Wt %	At %
O K	00.30	00.93
Al K	17.11	31.29
Mo L	00.53	00.27
Ti K	01.97	02.05
Cr K	05.79	05.49
Fe K	00.32	00.28
Co K	08.81	07.38
Ni K	60.91	51.19
W L	04.25	01.14

Element	Wt %	At %
O K	00.39	01.38
Al K	11.50	24.33
Nb L	00.79	06.48
Mo L	02.36	01.41
Ti K	02.04	02.43
Cr K	08.93	09.80
Fe K	00.31	00.32
Co K	09.73	09.43
Ni K	46.22	44.93
W L	17.74	05.51

(c) (d)

Figure 12. Three-layer barrier nickel aluminides (NiAl), with the coating diffusively applied to the TWC blade material (**a**), and the results of the microanalysis of the chemical composition in individual areas (**b–d**).

This process has led to the constitution on the surface of the blade material of a three-layer heat-resistant coating of NiAl intermetallic phase protecting the blade material from overheating.

Table 2. Weight percentage (%) of individual alloy elements.

Elements:	Ni	Cr	Co	Fe	Al	Ti	Mo	Nb	W	Ta	Re
Tested alloy	58.49	8.68	9.86	0.77	6.19	2.70	1.75	1.26	10.69	-	-
MAR-M 200	59	9	10	<1	5	2	-	1	12.5	-	-
ZhS-6U	59.5	8.8	10.3	2	5.5	2.6	1.6	-	9.7	-	-
ZhS-32 Monocrystalline	residue	4.3–5.6	8.0–10.0	-	5.6–6.3	-	0.8–1.4	1,4–1.8	7.7–9.5	3.5–4.5	3.5–4.5
ZhS-26 directionally crystallized	residue	4.1–5.3	8.7–9.3	-	5.6–6.1	0.8–1.2	0.8–1.2	1.4–1.8	11.2–12	-	-

Nevertheless, in spite of the material of the core of the HPT blade being protected against the temperature influence of exhaust gases, the barrier coating was damaged at the leading edge, which, due to temperature increase, led to changes in the grain structure of the blade at the overheating point (Figure 13a). In addition to a small but noticeable growth of the grains of the γ matrix from 80 up to 95 μm, an anomalous selective growth and coagulation of the precipitations of the γ' phase (area 1 Figure 13b) growth and coagulation (area 2 Figure 13b) and an increase in the content of the carbides in the structure are also observed (Figure 13c), which may lead to a decrease of the creep resistance of the blade material.

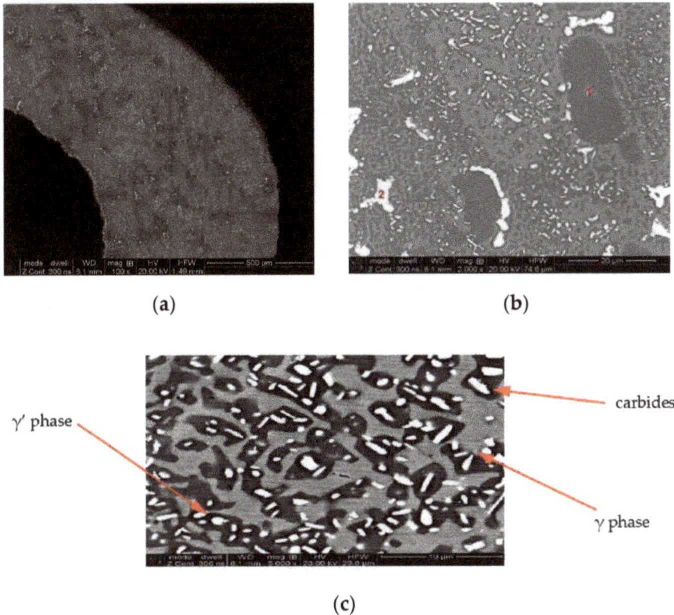

(a)

(b)

(c)

Figure 13. Damaged barrier coating at the leading edge of the HPT blade (**a**) and the effect of the anomalous growth of the γ' phase (**b**) and increase in the share of carbides in the matrix of the core material (**c**) due to overheating.

While the less thermally loaded material of the LPT blade, coated with a diffusion barrier layer made of aluminides, is also constructed of a multi-phased structure ($\gamma + \gamma' + MC$), no abnormal growth of the superstructure γ' and carbides was observed in material of the LPT blade.

A microanalysis of the chemical composition suggests that a similar phased composition to that of the material of the HPT blade is the result of a similar chemical composition, suggesting that the same alloy was used for the material of the LPT blade, as in the case of the HPT blade (Figure 14) and (Table 3).

Table 3. Comparison with the microanalysis of the material of the HPT blade.

Elements	Weight Percentage (%)								
	Ni	Cr	Co	Fe	Al	Ti	Mo	Nb	W
HPT blade	58.09	8.54	9.57	0.08	6.40	2.74	1.80	1.20	11.57
LPT blade	57.87	8.69	9.72	0.24	6.67	2.75	1.84	1.32	10.81

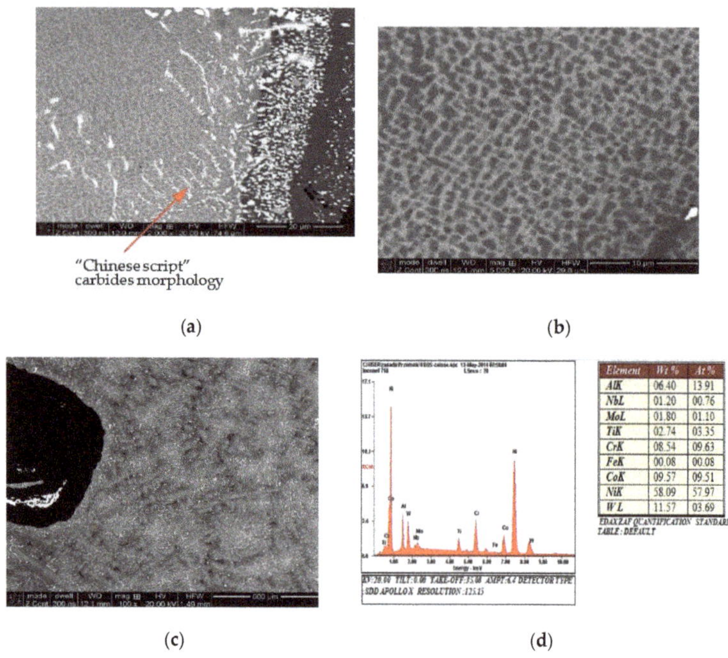

(a)

(b)

(c)

(d)

Figure 14. View of the structure of the LPT blade with the barrier coating (**a**), characteristic grain of the cuboidal phase γ′ (**b**), and uniformly distributed reinforcing precipitations of carbides (**c**). The result of the microanalysis of the chemical composition of the core material (**d**).

The conducted microscopic observations made it possible to measure, by means of microscopic image analysis, the volume content of individual phases in the structures of the analyzed blades and to assess the size of the sections of the dendrite arms of the primary γ solid solution and the average grain size of the other identified structural components' −γ′ intermetallic phase and carbides. From the collected data presented in Table 4, it can be concluded that the structure of the material of the blades is stable so long as it is protected by a barrier coating.

Table 4. Volume content and grain size of individual phases i006E of the structure of the analyzed blades.

Phases		HPT				HPT—Overheated			LPT			
		LE Int.	LE Ext.	TE Int.	TE Ext.	LE Int.	LE Ext.	TE	LE Int.	LE Ext.	TE Int.	TE Ext.
γ	Content (%)	residue				residue			residue			
	Size (μm)	84.01		80.48		95.61		93.08	20.62	43.11	35.57	37.81
γ′	Content Share (%)	43.43	49.53	50.36	46.46	51.62	43.07	43.64	41.94		38.73	43.73
	Size (μm)	1.33	3.72	0.22	0.24	1.03	2.24	0.89	1.03	1.07	1.00	1.00
carbides	Content (%)	3.62	3.15	2.37	1.45	9.04	6.32	1.04	2.4		1.56	2.75
	Size (μm)	0.83	1.09	2.42	2.55	0.70	0.75	0.74	1.91	2.52	1.63	2.65

LE—leading edge; TE—trailing edge; Int.—internal area (cooled); Ext.—external area flushed by the exhaust gas stream.

The damage to the protective layer, as already indicated, leads to a noticeable growth of the matrix grains and a three-fold increase of the carbide volume content.

The microscopic observations were confirmed by the XRD analysis. Based on the X-ray phase analysis measurements, by analyzing the positions of the obtained reflections, it was possible to unequivocally confirm the multi-phase structure of the blade material observed during the microscopic observations (Figure 15).

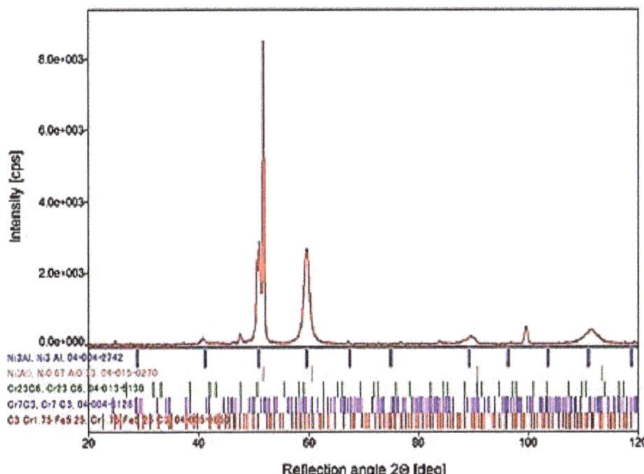

Figure 15. X-ray diffraction (XRD) pattern of the blade material with the identified phases.

The distribution of the alloying elements suggested the occurrence of the three-phased structure of the material based on the solid solution of aluminum in nickel γ and of the areas of intermetallic phase γ' on the matrix of the Ni_3Al superstructure and carbide precipitates, mainly $Cr_{23}C_6$. This was fully confirmed by the comparative analysis of the 2 Θ angular position of the obtained diffraction reflections and the data contained in the PDF-4 database. Additional evidence confirming the comparable structural-phase composition of the analyzed blades are microhardness measurements [18] carried out on the materials of the tested blades.

These tests were carried out with the semi-automatic microhardness tester, SHIMADZU-DORERNST M, and the obtained results were subjected to statistical analysis, which confirmed the structural homogeneity of the analyzed material areas, characterized by an average value of microhardness at a statistically uniform level within the range of 410–430 HV 0.025 (Figure 16).

The observed structural changes were reflected in the strength parameters of the alloy used for the construction of the tested blades. As can be seen in Figure 17, where the hardness values obtained are marked with the determined standard deviation, the intersection points (marked in green) of the hardness obtained are located on the curves, which represent the dependency between the Brinell and Rockwell hardness as a function of the Vickers hardness (curves HB = f(HV) and HRB = f(HV)), which clearly proves the correctness of the measurements carried out. It also proves the high homogeneity of the tested material, caused by the refinement of the grain structure, despite its multiphase nature.

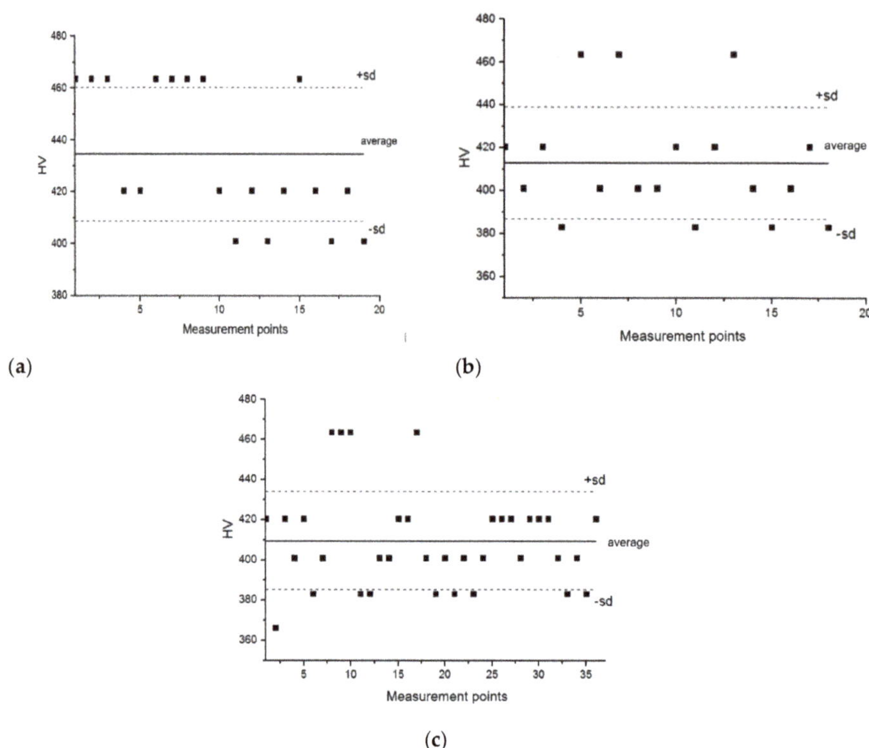

Figure 16. Statistical distribution of the microhardness of the material of the HPT blade in the overheated zone (**a**), in the non-overheated zone (**b**), and in the material of the LPT blade (**c**).

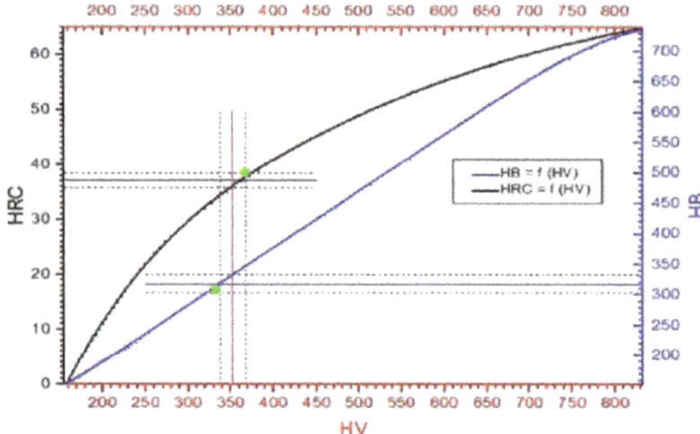

Figure 17. Comparison of the hardness measurements obtained with the Rockwell PW 106 hardness tester and the Wolpert Wilson Testor 751 universal hardness tester.

However, the average hardness value of the analyzed material, after its exploitation, decreased to the level of 350 HV10, compared to 450 HV10, which is typical in the case of a correctly structured ZhS-6U alloy. It should also be noted that the hardness (Figure 17) and microhardness (Figure 16) values measured by the same Vickers method differ significantly, reaching 70 HV hardness units in extreme cases. However, the observed phenomenon, the indentation size effect (ISE), is in accordance with the variable hardness law, which describes the influence of elastic deformation on the value of the obtained results, i.e., the smaller the load, the greater the influence of elastic deformation and thus the greater the hardness value. At this point, it should be recalled that the load for the microhardness measurements was 25 g, and for the hardness measurements, it was 10 kg. The decrease in the strength properties is confirmed by the results of the static tensile test (Figure 18).

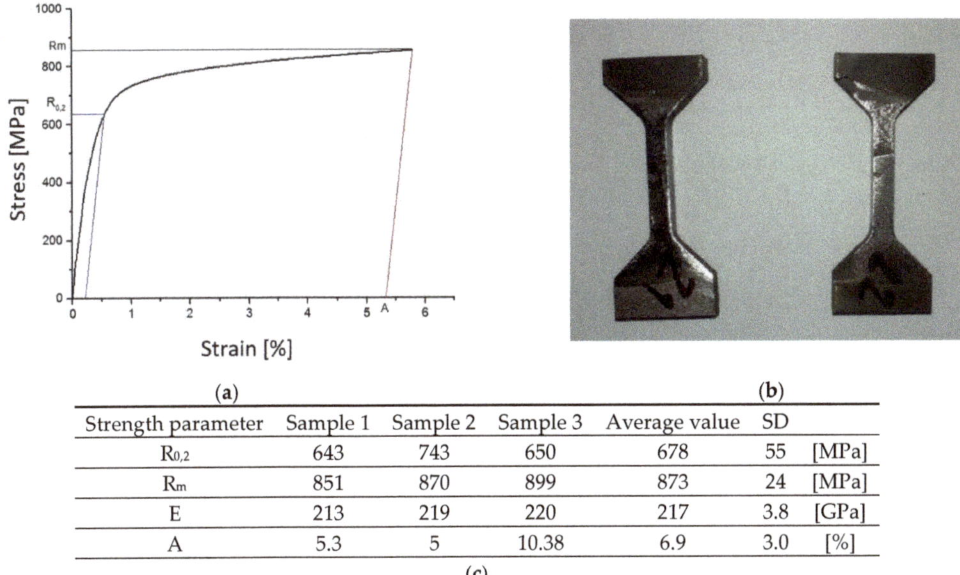

Strength parameter	Sample 1	Sample 2	Sample 3	Average value	SD	
$R_{0,2}$	643	743	650	678	55	[MPa]
R_m	851	870	899	873	24	[MPa]
E	213	219	220	217	3.8	[GPa]
A	5.3	5	10.38	6.9	3.0	[%]

(c)

Figure 18. Tensile curve of the blade material (**a**), view of specimens before and after the tensile test (**b**), and the determined strength parameters (**c**).

In accordance with the Hall–Petch relationship, the decrease in the share of the γ' phase in the material structure and the growth there caused a decrease in the yield point from the original level of 770 MPa for the ZhS-6U alloy to 678 MPa for the material of the tested blade (Table 5).

Table 5. Strength properties at room temperature of nickel-based superalloys potentially used to construct the HPT blades and LPT blades of the RD33 engine.

Alloy	E [GPa]	HV	R_{02} [MPa]	R_m [MPa]	A
Żs-32	244	466	850	880	13.0
Żs-26	253	480	790	1000	6.8
Żs-6U	240	450	770	830	3.0
MAR-M200	230	-	860	960	7.0
Material tested	217	350	678	873	6.9

5. Summary

The data presented in this study were obtained as a result of material tests of the blades of the rotors of the low- and high-pressure turbines of the RD-33 engine, taking into account the chemical composition and the phase structure corresponding to the Russian alloy, the Zhs-6U type, used in the early versions of the RD-33 engine. The deployment of blades made of the ZhS-26 and ZhS-32 alloys allowed for an increase of the temperature of exhaust gases before the turbine, thus increasing the engine performance. However, the conscious or unconscious use, under these exploitation conditions, of blades made of ZhS-6U alloy, with a noticeably lower structural stability, combined with local damage to the barrier protective coating, resulted in a distinctive reduction of the strength properties (primarily the plasticity limit) of more than 10 per cent, which could ultimately lead to engine failure and even damage to the engine.

The values of the determined strength parameters will be used in further work related to numerical analyses in the field of strength issues concerning the life and failure of engines manufactured by Russia, including the RD-33 engine. The importance and need for such works [4,12,19] may be proved by the number of RD-33 engines dismounted from the airplanes in one of the air bases due to damage to the high-pressure turbine, which is shown in the diagram in Figure 19. In addition, the analysis of structural changes caused by exploitation conditions can be helpful in the search for and selection of material substitutes used to construct the engine element under analysis, which will meet increasingly higher, mainly temperature-related design requirements.

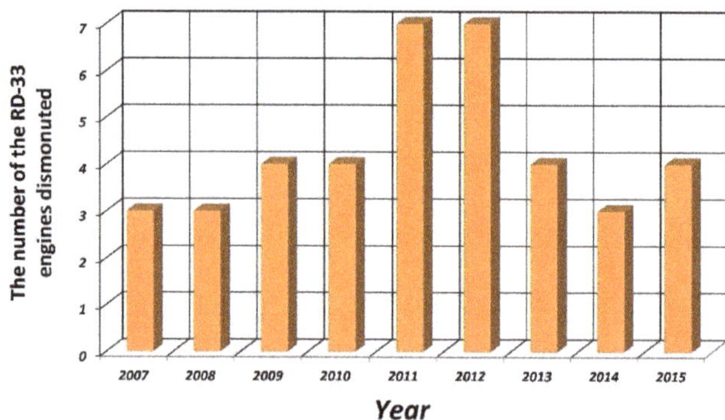

Figure 19. The number of RD-33 engines dismounted from airplanes due to damage to the high-pressure turbine in one of the air bases in Poland during the years 2007–2015 [4].

Author Contributions: A.K.–conceptualization: data analysis of life problems of turbine engines, assessment of turbine engine components in terms of research areas, and defining the research area in terms of the needs to increase the life and performance of the engine. S.J.–investigation: defining research areas, recognition of materials, identifying chemical composition, structural testing, micro-analysis of chemical composition, an analysis of multiphase structures, an alloy structure analysis, and recognition of physical properties of the jet engine material. P.J.–methodology: development of research methodology, development of the composition identification algorithm, preparation of samples, experimental research. S.K.–formal analysis: an analysis of strength and stiffness parameters of the tested material, an analysis of substitutes for jet engine materials. All authors have read and agreed to the published version of the manuscript.

Funding: This research received no external funding.

Institutional Review Board Statement: Not applicable.

Informed Consent Statement: Not applicable.

Acknowledgments: The authors would like to express their gratitude for the funding from the statutory sources of the Military University Technology (project No. UGB 767/2020/WAT) and from the research sources of the National Center for Research and Development (project No. DOB-BIO8/04/01/2016).

Conflicts of Interest: The authors declare no conflict of interest.

References

1. Koff, B.L. Gas Turbine Technology Evolution: A Designer's Perspective. *J. Propuls. Power* **2004**, *20*, 577–595. [CrossRef]
2. Kozakiewicz, A. Comparative analysis of turbojet performance of combat aircraft used in Polish Air Force. *Biuletyn WAT* **2009**, *LVIII*, 65–83.
3. Igorevich, H.I. Development of economically alloyed high-temperature nickel alloys for single-crystal casting of GTE blades. Ph.D. Thesis, Rybinsk State Aviation Technical University, Rybinsk, Russia, 2017.

4. Trelka, M.; Bartoszewicz, J.; Urbaniak, R. Selected problems of RD-33 engine reliability in operation. *Combust. Eng.* **2016**, *165*, 33–40.
5. Carter, T.J. Common failures in gas turbine blades. *Eng. Fail. Anal.* **2005**, *12*, 237–247. [CrossRef]
6. Swain, B.; Mallick, P.; Patel, S.; Roshan, R.; Mohapatra, S.; Bhuyan, S.; Priyadarshini, M.; Behera, B.; Samal, S.; Behera, A. Failure analysis and materials development of gas turbine blades. *Mater. Today* **2020**, *33*, 5143–5146. [CrossRef]
7. Laguna-Camacho, J.R.; Villagrán-Villegas, L.Y.; Martínez-García, H.; Juárez-Morales, G.; Cruz-Orduña, M.I.; Vite-Torres, M.; Ríos-Velasco, L.; Hernández-Romero, I. A study of the wear damage on gas turbine blades. *Eng. Fail. Anal.* **2016**, *61*, 88–99. [CrossRef]
8. Szczeciński, S.; Balicki, W. Possibilities for Limiting Aircraft Turbine Engine Abrasive Wear of Interior Duct Elements. *Trans. Inst. Aviat.* **2007**, *4*, 3–12.
9. Chamanfar, A.; Jahazi, M.; Bonakdar, A.; Morin, E.; Firoozrai, A. Cracking in fusion zone and heat affected zone of electron beam welded Inconel-713LC gas turbine blades. *Mater. Sci. Eng. A* **2015**, *642*, 230–240. [CrossRef]
10. Dewangan, R.; Patel, J.; Dubey, J.; Sen, P.K.; Bohidar, S.K. Gas Turbines Blades—A Critical Review of Failure on First and Second Stages. *Int. J. Mech. Eng. Rob. Res.* **2015**, *4*, 216–223.
11. Bansal, L.; Rathi, V.K.; Mudafale, K. A Review on Gas Turbine Blade Failure and Preventive Techniques. *Int. J. Eng. Res. Gen. Sci.* **2018**, *6*, 54–62.
12. Kablov, E.N.; Opekhov, H.G.; Tolopaia, B.N.; Demonis, I.M. Casting heat-resistant alloys and technology of obtaining monocrystalline turbine blades GTE. In *Light Alloy Technology*; Institute of Aviation Materials: Moscow, Russia, 2001; Volume 4.
13. Čerňan, J.; Janovec, M.; Hocko, M.; Cúttová, M. Damages of RD-33 Engine Gas Turbine and their Causes. *Transp. Res. Procedia* **2018**, *35*, 200–208. [CrossRef]
14. Jakubowski, R. Teoria Silników Lotniczych—Rozwój i Przegląd Konstrukcji. Available online: http://jakubowskirobert.sd.prz.edu.pl/file/MjMsNjcsMjcxLHRlb3JpYV9zaWxuaWtvd19sb3RuaWN6eWNoX3dwcm93YWR6ZW5pZS5wZGY= (accessed on 14 November 2019).
15. Strotin, N.N. *Monitoring and Diagnostics of Technical Condition of Gas Turbine Engines*; Science: Moscow, Russia, 2007; p. 469. ISBN 5-02-35300-0.
16. Hernas, A.; Maciejny, A. *Żarowytrzymałe Stopy Metali*; Wydawnictwo PAN: Katowice, Poland, 1989; ISBN 8304031140-9788304031142.
17. Kaibyshev, O.A. *Superplasticity of Alloys, Intermetallides and Ceramics*; Springer: Berlin/Heidelberg, Germany, 1992; ISBN 139-78-3-642-84675-5.
18. Kurzydłowski, K.J. "Rozwój Materiałów Konstrukcyjnych: Historia Stopów Aluminium" Sesja z Okazji 75-lecia Przeglądu Mechanicznego, 12 luty 2010, Warszawa. Available online: www.iinte.edu.pl/PM75/sesja2.pdf (accessed on 5 December 2014).
19. Siladic, M.; Rasuo, B. On-Condition Maintenance for Non-Modular Jet ENGINES—An Experience, ICAS 2008. In Proceedings of the 26th International Congress of The Aeronautical Sciences, Anchorage, AK, USA, 14–19 September 2008; Paper ICAS 2008-4.5.2. Available online: http://www.icas.org/ICAS_ARCHIVE/ICAS2008/PAPERS/195.PDF (accessed on 11 January 2021).

 *materials*

Article

Comparative Investigations of AlCrN Coatings Formed by Cathodic Arc Evaporation under Different Nitrogen Pressure or Arc Current

Adam Gilewicz [1], Tatyana Kuznetsova [2], Sergei Aizikovich [3,*], Vasilina Lapitskaya [2], Anastasiya Khabarava [2], Andrey Nikolaev [3] and Bogdan Warcholinski [1]

[1] Faculty of Mechanical Engineering, Koszalin University of Technology, 2, Śniadeckich, 75-453 Koszalin, Poland; adam.gilewicz@tu.koszalin.pl (A.G.); bogdan.warcholinski@tu.koszalin.pl (B.W.)

[2] Nanoprocesses and Technology Laboratory, A.V. Luikov Institute of Heat and Mass Transfer of National Academy of Science of Belarus, 15, P. Brovki str., 220072 Minsk, Belarus; kuzn06@mail.ru (T.K.); vasilinka.92@mail.ru (V.L.); av.khabarova@mail.ru (A.K.)

[3] Research and Education Center "Materials", Don State Technical University, 1, Gagarin sq., 344003 Rostov-on-Don, Russia; andreynicolaev@eurosites.ru

* Correspondence: s.aizikovich@sci.donstu.ru; Tel.: +7-863-238-15-58

Abstract: Tools and machine surfaces are subjected to various types of damage caused by many different factors. Due to this, the protecting coatings characterized by the best properties for a given treatment or environment are used. AlCrN coatings with different compositions, synthesized by different methods, are often of interest to scientists. The aim of the presented work was the deposition and investigation of two sets of coatings: (1) formed in nitrogen pressure from 0.8 Pa to 5 Pa and (2) formed at arc current from 50 A to 100 A. We study relationships between the above technological parameters and discuss their properties. Coatings formed at nitrogen pressure (p_{N2}) up to 3 Pa crystallize both in hexagonal AlN structure and the cubic CrN structure. For $p_{N2} > 3$ Pa, they crystallize in the CrN cubic structure. Crystallite size increases with nitrogen pressure. The coatings formed at different arc currents have a cubic CrN structure and the crystallite size is independent of the current. The adhesion of the coatings is very good, independent of nitrogen pressure and arc current.

Keywords: AlCrN; arc current; structure; hardness; adhesion; wear

Citation: Gilewicz, A.; Kuznetsova, T.; Aizikovich, S.; Lapitskaya, V.; Khabarava, A.; Nikolaev, A.; Warcholinski, B. Comparative Investigations of AlCrN Coatings Formed by Cathodic Arc Evaporation under Different Nitrogen Pressure or Arc Current. *Materials* **2021**, *14*, 304. https://doi.org/10.3390/ma14020304

Received: 10 December 2020
Accepted: 4 January 2021
Published: 8 January 2021

Publisher's Note: MDPI stays neutral with regard to jurisdictional claims in published maps and institutional affiliations.

1. Introduction

During the last few decades, extensive works have been carried out on the production and testing of the properties of thin coatings which can be applied in industry, medicine, electronics as protective and decorative coatings, memory technology, etc. [1,2].

Currently, it is recommended to eliminate cooling and lubricating liquids from technological processes due to economic factors and ecological requirements. Hard coated tools are used in the vast majority of metal machining operations. The most commonly used coatings are based on titanium nitride and chromium nitride [1,3]. CrN, however, is characterized by a better resistance to high temperatures compared to TiN [3]. Such coatings improve the wear resistance of cutting tools [4,5]. They have high hardness [6], a relatively low coefficient of friction [7,8], good adhesion to the substrate [9,10], chemical stability, and resistance to corrosion [11]. It was also shown that coatings may reduce the heat generated in the cutting process due to friction, and thus increase the durability of tool blades [12]. The relatively low resistance to oxidation of two-component coatings, e.g., CrN, means that they are not able to meet the expectations of the most demanding customers.

The improvement of the thin coating properties was achieved by doping with elements, such as Si, B, C, O, Al, V, and others [13–17]. The addition of Si enables to enhance the hardness and Young's modulus of the Cr–N system, but Si has a negative influence on the

135

tribological properties of the coatings—the friction coefficient increases and wear resistance decreases. This effect depends on the silicon concentration in the coating [13]. CrCN shows denser structures and significantly increased hardness and adhesion strength. It is characterized by improved corrosion resistance, the lowest corrosion current density, and the highest corrosion potential [14]. The addition of boron to CrN results in notable refinement in grains. As a result of the fine-grained structure strengthening and the solution strengthening caused by boron, CrBN coating shows improved hardness compared to CrN but relatively low adhesion [15]. Vanadium addition improve the hardness and toughness of CrN film. The coefficient of friction is also reduced for CrVN coatings [16]. Oxygen can decrease of the mean crystallite size, increase the hardness of the coating up to 28 GPa, and improve the thermal stability [17]. One of the most promising coatings seems to be AlCrN due to its very good resistance to oxidation, high hardness at elevated temperature, and good tribological properties [18]. This is because AlCrN can oxidize and form on the surface aluminum oxide Al_2O_3, blocking oxygen diffusion into the coating. Additionally, α-phase Al_2O_3 of aluminum oxide provides good tribological properties [19].

Investigations assessing the importance of the chemical composition of the coating to its structure and mechanical properties revealed the transformation of the metastable $Al_xCr_{1-x}N$ solid solution with a cubic structure (fcc) type B1 into a hexagonal structure B4, occurring at the Al/(Al + Cr) ratio from 0.6 to 0.75 (depending on the deposition method) [20]. The transformation of c-AlN $\rightarrow$ h-AlN occurs spontaneously and irreversibly. The coatings containing the AlN phase with a cubic structure (c-AlN) are characterized by better mechanical properties than the hexagonal AlN (h-AlN) phase [20,21]. Therefore, in order to obtain the coating with fcc structure and good mechanical properties, i.e., high hardness and good resistance to wear, the aluminum concentration in the coating ought to be attended to.

ALCrN coatings were formed from individual Al and Cr [22–24] cathodes and AlCr alloy cathodes with different fixed composition: $Al_{50}Cr_{50}$ [25], $Al_{60}Cr_{40}$ [26], $Al_{70}Cr_{30}$ [27,28], $Al_{80}Cr_{20}$ [29], and other [17,25,27]. They can be formed by many methods, including magnetron sputtering [19,20], high-power impulse magnetron sputtering (HIPIMS) [1,22,25], or cathodic arc evaporation [20,24,27,28,30]. The high degree of plasma ionization, high deposition rate, high density and good adhesion of the coatings to the steel substrate are the advantages of the latter method. Its disadvantage is a large number of defects (macroparticles, craters) on the coating surface, ranging in size from nanometers to single micrometers [30]. Cathodic arc evaporation systems are widely used as efficient devices in coating technology.

Coating deposition parameters such as nitrogen pressure [25,30], substrate bias voltage [24] and arc current in cathodic arc evaporation [31], as well as AlCr cathode composition [20], determine the coating properties. Although AlCrN coatings with different chemical composition are the subject of many investigations, they mainly include structure and phase composition [20,22,24,30], thermal resistance [20], and some mechanical [22,24,30] and tribological properties [22,24,27]. It has been proven that the composition of coatings [20,23] strongly influences their mechanical properties and microstructure. The technological parameter of deposition, substrate bias voltage [24,27,28], has a similar effect on the properties of the coatings. Along with the increase of Al concentration in the coating to approx. 70%, its hardness increases [20]. The coatings with a higher concentration of aluminum are characterized by lower hardness. There are few works describing the influence of nitrogen pressure on the properties of the coatings [25,30,32]. They indicate an increase in hardness and Young's modulus with increasing nitrogen pressure during coating formation [25,30], regardless of the coating formation method, as well as changes in their phase structure. With increasing pressure, the formation of the hexagonal Cr_2N phase [30] is possible as a result of a greater number of collisions of particles emitted from the cathode with gas particles. However, no phase transformation is also observed in the coatings formed at different nitrogen pressure, and only a change in texture [25]. A similar lack of changes in the dominant phase is observed in the coatings formed at different arc currents [31]. Until now, few authors have also presented systematic and detailed tests of coating adhesion to the substrate [23,30]. As adhesion is one of the basic factors determining the functional properties of coatings, it seems advisable to take up this topic, i.e., to assess the effect of nitrogen pressure during deposition on the adhesion of the coatings.

Another technological parameter that can effectively improve the surface quality of coatings formed by the cathodic arc evaporation is the arc current [31]. The arc current can reduce the macroparticle amount, but the arc current change results in the plasma density change. This can alter the properties of the coatings. This effect has not been systematically tested for AlCrN coatings, and the presented results are inconclusive [31] in relation to our preliminary research. Lin et al. found that, with increasing arc current, the grain size in the coating and its hardness decreased. According to the Hall–Petch relationship, the opposite effect should be observed.

The goal of this study was to check the influence of nitrogen pressure and arc current during deposition of AlCrN coatings from $Al_{70}Cr_{30}$ cathode on their structure, morphology, and mechanical properties including adhesion.

2. Experimental

2.1. Deposition of the Coating

To synthesize AlCrN coatings the cathodic arc evaporation method was applied. Mechanically finished substrates (Ra about 0.02 mm) were cleaned in an ultrasonic bath. After drying, they were mounted on the rotating holder in vacuum chamber. The arc sources were equipped with $Al_{70}Cr_{30}$ cathodes and were located approximately 18 cm from the substrates. The working chamber was evacuated to the pressure of 10^{-3} Pa. The substrate temperature during ion etching and coating deposition was about 350 °C. Ion etching was performed with the following parameters: argon pressure of 0.5 Pa, voltage of −600 V, time of 10 min. The first stage of forming the coating consisted of creating an adhesive layer (gradient CrN_x) improving the coating adhesion. This process was carried out using the Cr cathode and the following parameters: arc current 80 A, nitrogen pressure increase rate—about 0.1 Pa/min to reach a pressure of 1.8 Pa. In the second stage, two sets of AlCrN coatings were deposited under following parameters:

(a) Substrate bias voltage −100 V, arc current 80 A, and nitrogen pressure (p_{N2}) in the range from 0.8 Pa to 5 Pa.
(b) Substrate bias voltage −100 V, nitrogen pressure of 4 Pa, and arc current (Ic) in the range from 50 A to 100 A.

The flow of both gases used (argon and nitrogen) and gas pressure were controlled, respectively, using an MKS flow controller (MKS Instruments, Inc., Austin, TX, USA) and by a Baratron type capacity gauge (MKS Instruments, Inc., Austin, TX, USA).

Taking into account the nitrogen pressure (X) during coating formation, they were marked as AlCrN(X). This means that, for example, an AlCrN(1.8) coating was formed at p_{N2} = 1.8 Pa. The coatings formed at various arc current (Y) are marked as AlCr(Y)N. Thus, the coating designated AlCr(60)N was formed at an arc current of 60 A.

2.2. Coating Investigations

The coating thickness was investigated using spherical abrasion test. Surface morphology and microstructure were studied using scanning electron microscopy (JEOL JSM-5500LV, JEOL Ltd., Tokyo, Japan). The Hommel Tester T8000 contact profilometer (Hommelwerke GmbH, Schwenningen, Germany) was used to measure the surface roughness of the coatings. For each sample, the test was conducted five times.

The chemical composition of the coatings was determined by energy dispersive x-ray spectroscopy (EDX, Oxford Link ISIS 300, Link Analytical/Oxford Instruments, High Wycombe, UK) and wavelength-dispersive X-ray spectroscopy (WDX, ThermoScientific's Magnaray system, Thermo Fisher Scientific, Waltham, MA, USA) methods. The crystalline structure of the coatings was investigated using grazing incidence X-ray diffraction

(Empyrean PANalytical, Malvern Panalytical Ltd., Malvern, UK) with Cu-Kα radiation (0.154056 nm) and the grazing incidence geometry at 3°. For data processing, HighScore Plus with ICDD PDF 4+ Database software (The Powder Diffraction File) was applied. Diffraction images made it possible to calculate the crystallite size in the coatings using the Scherrer equation [33]. Due to the instrumental broadening of the diffraction line, the Warren and Biscoe correction method was taken into account [33].

The semi-automatic Fischerscope HM2000 hardness tester (Fischer Technology Inc., Windsor, CT, USA) equipped in WIN-HCU® software (3.0, Windsor, CT, USA) made it possible to determine the hardness (H) and Young's modulus (E). The device was equipped with a Bercovich intender three-sided pyramid, with a tip radius of 150 nm and a total included angle of 142.3°. Taking into account the coating thickness, the measurements were conducted for a fixed penetration depth of 0.3 μm, which is less than 10% of the thickness of the coating. This indentation depth avoids a change in the measured hardness due to relatively soft substrates. The coatings with high surface roughness make it impossible correct hardness measurement. Therefore, prior to the measurements, the coatings were treated using the procedure described in Ref. [12]. After polishing the coatings with fine-grained (1 μm) aluminum oxide powder to a roughness of 0.04–0.05 μm, a much smaller dispersion of the results was obtained. The average value of hardness and Young's modulus was calculated from at least 20 measurements from all tested surface.

Coating adhesion was determined using two methods:

(a) The scratch method (Revetest® by CSEM Instruments, Peseux, Switzerland) with Rockwell C type diamond indenter with a tip radius of 200 μm. At least three scratches, each of a length of 10 mm and a distance at least 3 mm from each other, were done by moving the indenter at a speed of 10 mm/min, and simultaneously increasing the load linearly from 0 to 100 N. Two characteristic damages were observed and marked: Lc_1 for first cracks of the coatings appeared and Lc_2 for total delamination of the coating,

(b) Daimler–Benz test. In this method, it was assumed that the assessment of coating adhesion is determined on the basis of the form and intensity of damage resulting from the indentation [34]. In this test the Rockwell indenter is pressed into the sample (coating) with the normal load of 1471 N. It is a comparative method (six-grade scale of adhesion quality) in which it was assumed that damage in the coating corresponding to the HF1-HF4 patterns (cracks and small chips) proves the proper adhesion of the coating. The occurrence of coating defects in the area surrounding the indentation, corresponding to the HF5-HF6 patterns, is evidence of insufficient adhesion of the coating.

The studies of the friction and wear of the coatings were conducted in the ball-on-disc system under the normal load L = 20 N and a sliding speed of about 0.2 m/s for about 29000 cycles (distance s = 2000 m) in dry friction conditions. As a counterpart the mirror finished Al_2O_3 ball with a diameter of 10 mm was applied. The measurements were conducted in moisture of about 40% at ambient temperature. The known equation $kv = Vs^{-1}L^{-1}$ was applied to calculate the wear rate (kv). In this equation V means the volume of the removed coating material. Five randomly selected cross-sectional wear profiles were applied to calculate the wear rate.

The surface investigations of the coatings were carried out using Dimension FastScan atomic-force microscope (Bruker, Santa Barbara, CA, USA) in the PeakForce Tapping QNM (Quantitative Nanoscale Mechanical Mapping) mode. The standard silicon cantilevers of NSC11 (Micromasch, Tallinn, Estonia) type with the stiffness k = 4.11 N/m and tip radius R = 10 nm were used [35].

Hysitron 750 Ubi nanoindenter (Bruker, Minneapolis, MN, USA) was applied to construct microhardness (H) and elasticity modulus (E) maps. A diamond Berkovich indenter with a tip radius of 200 nm was applied. A fused quartz calibration sample was used to calibrate the tip radius. The E and H maps with the size of 20 × 20 μm^2 were obtained from 400 indentations at a load of 2 mN on the surface. Before measurements, the coatings surface was polished to reduce the roughness.

3. Results

As already mentioned, the coating deposition time is the same for all coatings and amounts to 120 min. Nitrogen pressure definitely affects the thickness of the resulting coatings, as shown in Figure 1a. The coating thickness ranges from 2.60 μm (p_{N2} = 5.0 Pa) to 3.9 μm (p_{N2} = 3.0 Pa). Coatings synthesized at higher p_{N2} = 4 Pa and 5 Pa present decreasing thickness. In the case of another parameter of the coating technology, the arc current, a clear tendency of the coating thickness increase with increasing current can be observed. The smallest is for coatings deposited at Ic = 50 A and is about 3 μm. As the arc current increases, the thickness increases almost linearly to a value of about 4 μm (Ic = 100 A), Figure 1b.

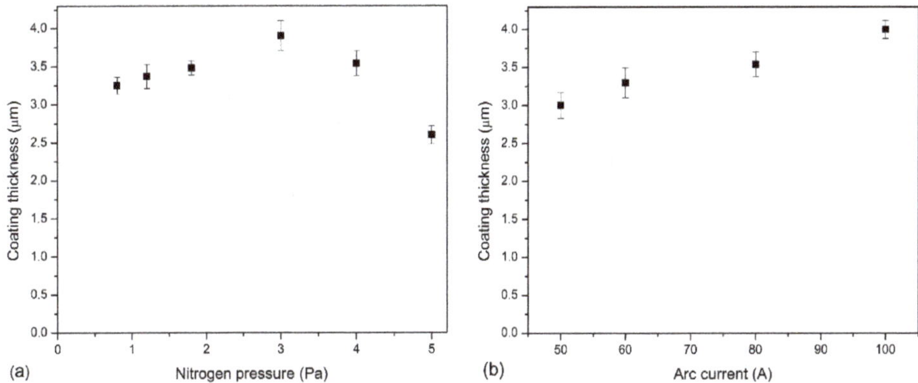

Figure 1. Thickness of AlCrN coatings formed at: (**a**) various nitrogen pressure, (**b**) various arc current.

A similar change in coating thickness was observed by Wang et al. [30]. In the Cr-Al-N coatings formed under nitrogen pressure from 1 Pa to 2 Pa, by the multi-arc ion plating method, the maximum thickness was for the coating formed at a pressure of 1.5 Pa and then its reduction. It was attributed to the resputtering effect. In the case of coatings formed at various arc currents, the observed changes in thickness are also consistent with the literature, as an increase in arc current results in an increase in the thickness by about 30% [31], which is similar to the results shown in the presented work.

3.1. Chemical and Phase Composition of AlCrN Coatings

In Figure 2, the results of the investigations on the elemental composition of the tested coatings are shown. They indicate the presence of chromium, aluminum, and nitrogen in the coatings. The tests also confirmed the presence of oxygen in all coatings in the amount of up to about 1.2% at. The nitrogen concentration in the coatings increases with the nitrogen pressure during their formation. Simultaneously, a noticeable reduction in the metallic elements chromium and aluminum is observed. The reduction in aluminum concentration is higher. This is confirmed by the calculated Al/(Al+Cr) rate in the coatings. For coatings synthesized at p_{N2} = 0.8 Pa is about 0.70, as for pure cathode and drops to about 0.666 for coatings synthesized at p_{N2} = 5 Pa, Figure 2a. This effect is probably related to lower atomic mass of aluminum. It is characterized by lower vapor density because in collisions with nitrogen it is more dispersed [36].

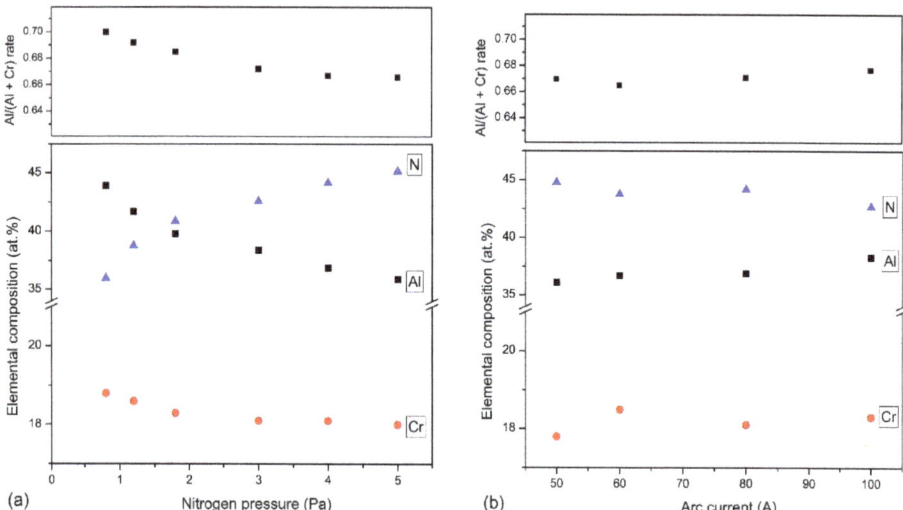

Figure 2. Elemental composition and Al/(Al + Cr) ratio of the coatings deposited at: (**a**) various nitrogen pressure, (**b**) various arc current.

In AlCrN coatings formed by HIPIMS from $Al_{50}Cr_{50}$ cathodes with an increase in the N_2/Ar ratio from 20% to 140%, an increase in nitrogen concentration in the coating from about 26% to about 51% was observed. An approximately 30% reduction in the concentration of metallic elements aluminum and chromium is also visible [25]. A similar effect is observed in the case of the coatings formed by the multi-arc ion plating from the $Al_{67}Cr_{33}$ cathode [30].

The coatings deposited at p_{N2} = 4 Pa and various arc currents are characterized by a similar chemical composition, Figure 2b. Only the coating formed at Ic = 50 A is characterized by a slightly higher, by about 2 at.%, nitrogen concentration compared to the coating formed at Ic = 100 A. In all coatings, the Al/(Al+Cr) rate is similar and amounts to 0.66–0.67.

Structure characterization clearly indicates that the AlCrN coatings are polycrystalline with structure dependent on nitrogen pressure during deposition. In Figure 3a, the grazing incidence XRD patterns of AlCrN coatings formed at nitrogen pressure ranging from 0.8 Pa to 5 Pa is shown. In the coating formed at the pressure of 0.8 Pa, the occurrence of h-AlN phase diffraction lines (ref. code 01-070-0354), for the angle 2Θ of approximately 32.9° (100), 36.0° (002), 37.7° (111), 49.6° (102), 58.8° (110) and 70.9° (112). Additionally, diffraction line from the hexagonal Cr_2N phase (ref. code 04-014-1025) for an angle of 2Θ of about 42.5° (111) is visible. It can also be assumed that the line Cr_2N (110) about 37.3° overlaps with the line (111) h-AlN. A distinct line of approximately 44.6° can be attributed to chromium (ref. code 01-077-7589). A decrease in the intensity of lines originating from the hexagonal phases of AlN and Cr_2N, and from chromium is observed in coatings desposited at higher nitrogen pressure. At the same time, the intensity of the diffraction lines, which can be attributed to the cubic phase of chromium nitride (ref.code 04-004-6868), increases. These are lines positioned at about 37.6°—(111) plane, 43.8°—(200) plane, 63.5°—(220) plane, 76.1°—(311) plane and 80.1°—(222) plane. In the coatings formed under nitrogen pressure in the range from 1.8 Pa to 4 Pa, diffraction lines from three phases, hexagonal AlN and Cr_2N and cubic CrN are observed. The lines from the hexagonal phase disappear almost completely in the coatings synthesized at p_{N2} = 4 Pa. Formation of the AlN cubic phase cannot be excluded, however, its identification is difficult. These phases are characterized by the same Fm-3m space group (225) due to the fact that the diffraction line positions in the cubic AlN and CrN phases differs only slightly.

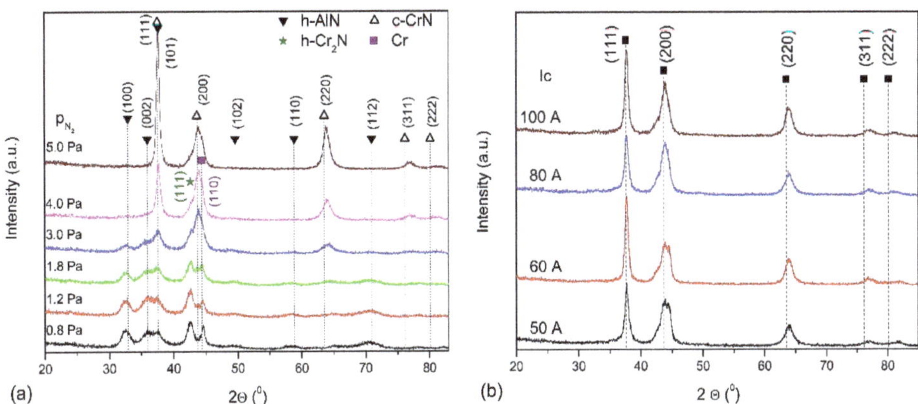

Figure 3. Diffraction patterns of AlCrN coatings formed at: (**a**) various nitrogen pressure, (**b**) various arc current.

For the lowest N_2/Ar ratio, the resulting structure is rather amorphous with hardness of around 11–13 GPa and a high wear rate of around 10^{-5} mm^3/Nm. An increase in this ratio in the process of AlCrN coating forming by HIPIMS, diffraction lines are observed only for cubic CrN phase. The intensity of the (111) CrN line with an increase in N_2/Ar decreases as well as the (220) CrN line, while the intensity of the (200) CrN line increases [25]. For coatings formed by the mult-arc ion plating method, with increasing nitrogen pressure, the intensities of lines (111) and (200) decrease, but the line (111) Cr$_2$N appears [30].

The coatings formed at different arc currents have a similar diffraction patterns, Figure 3b. The diffraction lines correspond to the major lines of the cubic CrN phase. According to the above standard, the ratio of the intensities of the most intense (200)/(111) diffraction lines is 1.22, while for coatings formed at different arc currents ranges from 0.61 to 0.88. This shows that these coatings are strongly textured.

The structure of AlCrN coatings formed at different arc currents is fcc-CrN phase. The increase in the arc current does not cause significant changes in the intensity of the observed diffraction lines, while their widening may result in grain refinement [31].

The crystallite sizes of both phases present in coatings investigated were calculated using the Scherrer equation, as shown in Figure 4. Crystallites in the coatings formed under nitrogen pressure up to 3 Pa are small, up to about 7 nm, Figure 4a. The size of the crystallites determined from the (100) h-AlN plane increases slightly from about 5 nm (p_{N2} = 0.8 Pa) to about 7 nm (p_{N2} = 1.8 Pa), and then decreases to about 3 nm (p_{N2} = 4.0 Pa). Figure 3 shows a significant reduction in the intensity of the diffraction line and its widening for this coating. The crystallites determined in planes (111) and (220) of the c-CrN phase increase from about 4–5 nm (p_{N2} = 1.8 Pa) and reach about 11 nm and 22 nm (p_{N2} = 5.0 Pa), respectively.

The coatings formed at different arc currents under nitrogen pressure of 4 Pa crystallize in the CrN cubic structure, Figure 3b. Therefore, the crystallite sizes for the diffraction lines from planes (111) and (220) were calculated, Figure 4b. Regardless of the arc current used to produce the coatings, the crystallite sizes are similar to each other and amount to about 7–8 nm and 9–12 nm, respectively. The trend of changes in the size of crystallites with the change of the arc current was not recorded.

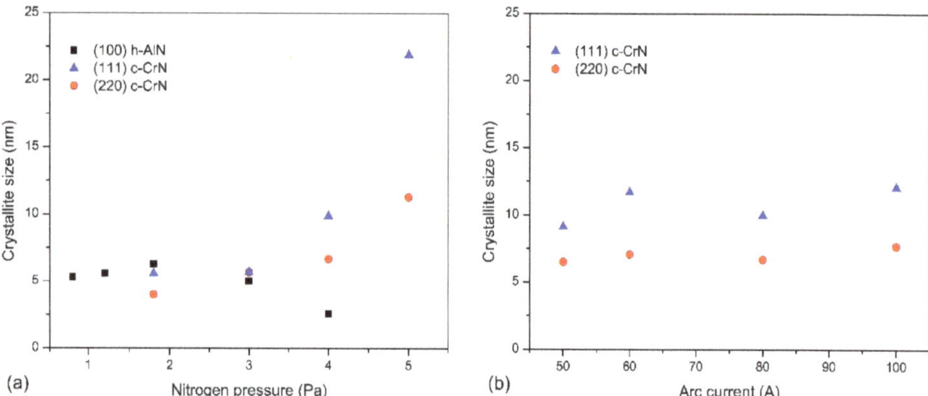

Figure 4. The crystallite size of (100) h-AlN and (111) and (220) c-CrN phases in AlCrN coating synthesized at: (**a**) different nitrogen pressure, (**b**) different arc current.

3.2. Surface Morphology

The surface morphology of the coatings formed at p_{N2} = 0.8 Pa (Figure 5a), p_{N2} = 5 Pa (Figure 5b) and Ic = 50 A (Figure 6a) and Ic =100 A (Figure 6b) is shown below. Each picture shows the area recorded at magnification: 1000× (bottom) and 3000× (top). A large number of surface defects, macroparticles, and craters were observed on the surface of all coatings. These surface defects are a disadvantage of the commonly used coating deposition method—cathodic arc evaporation. Cathodic arc discharge occurred at cathode spots. In these spots, the material of the cathode material is converted from solid to plasma extremely quickly as a result of ion bombardment and Joule heating, and macroparticles or droplets are emitted [37]. These droplets can group in the plasma and deposit on the surface of the coating [38]. Many macroparticles of different sizes were observed on the coating surface. There are also visible craters in places of removed macroparticles. The macroparticle shapes are diverse: elongated, irregular, but mostly spherical. The sizes of the particles vary greatly from fractions of a micrometer to several micrometers, with the vast majority reaching dimensions up to 1 μm. The macroparticles and their size distribution do not depend on the cathode, but on the method of coating formation, i.e., cathodic arc evaporation. AlTiN coatings have a similar size distribution of macroparticles on the coating surface [38].

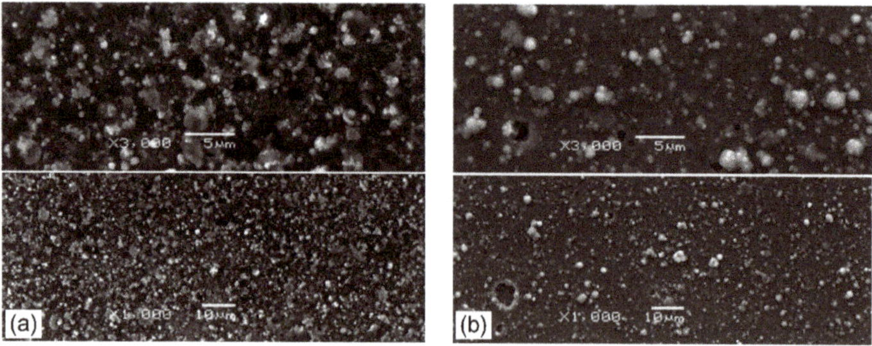

Figure 5. Morphology of of AlCrN coatings synthesized at: (**a**) p_{N2} = 0.8 Pa and (**b**) p_{N2} = 5 Pa. Each picture shows the area recorded at magnification: 1000× (**bottom**) and 3000× (**top**).

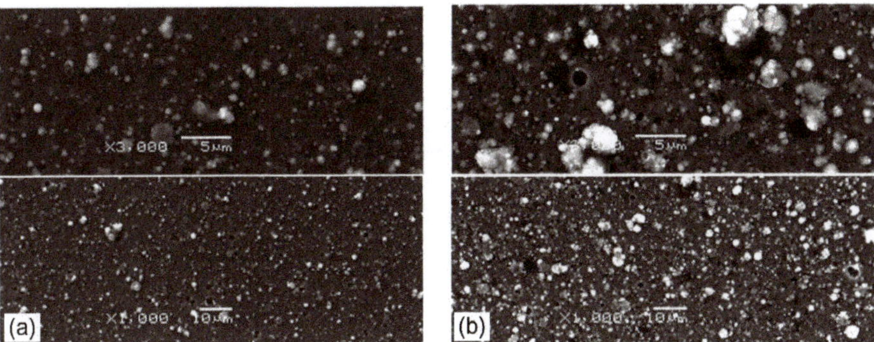

Figure 6. Morphology of of AlCrN coatings synthesized at: (**a**) Ic = 50 A and (**b**) Ic = 100 A. Each picture shows the area recorded at magnification: 1000× (**bottom**) and 3000× (**top**).

The analysis of Figures 5 and 6 shows that the coatings formed with higher nitrogen pressure have fewer surface defects. A similar effect for AlTiN coatings was observed by Cai et al. [38], while Wang et al. [30], who investigated AlCrN coatings, found the opposite effect.

For coatings synthesized at different arc currents, the opposite effect is visible. On the coating surface formed at the arc current of 50 A, fewer defects are observed, as shown in Figure 6a, than on the coating formed at higher current, shown in Figure 6b.

The roughness of the coatings depends on the development of its surface, e.g., the amount of surface defects of the coatings, i.e., macroparticles and craters. The analysis of Figure 5a,b shows a certain regularity in the coatings synthesized at higher arc current, where a higher amount of macroparticles is observed. The opposite effect, i.e., a smaller amount of macroparticles, is observed in the second set of coatings—synthesized at higher nitrogen pressure. This change in the surface quality of the coating is reflected in the roughness parameter Ra. There is a noticeable reduction in the Ra parameter, as shown in Figure 7. The initial high Ra value, about 0.24 μm, for the coating synthesized at p_{N2} = 0.8 Pa, reduces almost twice to 0.13 μm (p_{N2} = 5 Pa), as shown in Figure 7a. In the case of coatings deposited at various arc current, the opposite effect is observed, increase Ra value from about 0.12 (Ic = 50 A) to about 0.18 μm (Ic = 100 A). Such a high roughness value is comparable with the results presented by other authors for aluminum-doped chromium nitrides or titanium nitrides [38,39] deposited using the cathodic arc evaporation. CrN coatings formed by the same method at p_{N2} = 1.8 Pa are characterized by lower Ra roughness parameter of 0.08 μm [40]. This means that doping the aluminum of the CrN coating causes more than a 200% increase in surface roughness.

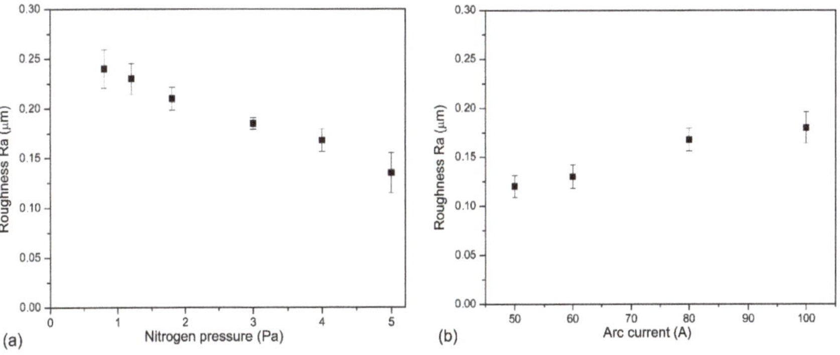

Figure 7. Roughness parameter Ra of AlCrN coatings formed at: (**a**) various pressure of nitrogen, (**b**) various arc current.

3.3. Mechanical Properties

The durability of tools or machine surfaces modified with thin coatings depends on the properties of these coatings. They determine, among other things, the reliability of the devices of load-carrying at the tribological point. The deposited coating generally promotes a favorable compressive stress. In the case of thin coatings, the most frequently used methods of measuring their hardness and modulus of elasticity are the nano-and micro-indentation methods. It should also be remembered that the penetration depth should be selected so that the substrate properties do not affect the results of the coating hardness measurement [41–43].

The indentation test allowed to determine hardness (H) and Young's modulus (E), the basic mechanical properties of the coatings. A clear dependence of these values on the nitrogen pressure in which the coatings are formed was observed, Table 1. The coating deposited at p_{N2} = 0.8 Pa has the lowest hardness, about 17.4 GPa. It increases almost linearly with the nitrogen pressure during their formation, reaching the value of 27.7 GPa at p_{N2} = 5 Pa. Young's modulus is characterized by a similar tendency and is 205 PGa the coating formed at p_{N2} = 0.8 Pa, and for the coating synthesized in p_{N2}—5 Pa it is almost 50% more—304 GPa. The wear resistance can be predicted from the coating hardness, one of the most important mechanical features. The results of many authors' tests show that the combination of the two parameters H and E enables a much better assessment of the coatings in terms of their wear resistance. The two best known are H/E [44] and H^3/E^2 [45] rates. H/E ratio (knows as elastic strain to failure) ratio and shows the elastic behavior of the coating in contact with the load used, while H^3/E^2 is known as resistance to the plastic indentation. There are also other indicators, e.g., ($H^2/2E$—modulus of resilience or $H/E)^2$—transition on mechanical contact—elastic to plastic) that have some importance for the abrasion resistance of materials [46].

Table 1. Mechanical properties of AlCrN coatings formed at various pressure of ntrogen.

Parameter	AlCrN(0.8)	AlCrN(1.2)	AlCrN(1.8)	AlCrN(3.0)	AlCrN(4.0)	AlCrN(5.0)
Hardness H (GPa)	17.4 ± 1.3	18.0 ± 2.3	18.7 ± 2.9	21.8 ± 2.1	24.6 ± 2.2	27.7 ± 2.6
Young's modulus, E (GPa)	205 ± 13	212 ± 15	218 ± 16	239 ± 14	261 ± 9	304 ± 16
H/E	0.085 ± 0.011	0.085 ± 0.017	0.086 ± 0.019	0.091 ± 0.014	0.094 ± 0.014	0.091 ± 0.013
H^3/E^2 (GPa)	0.12 ± 0.04	0.13 ± 0.07	0.15 ± 0.08	0.18 ± 0.07	0.22 ± 0.07	0.23 ± 0.09
Lc_2 (N)	77.2 ± 1.8	85.0 ± 2.3	84.5 ± 3.5	98.0 ± 3.8	91.0 ± 2.1	79.6 ± 1.8

The H/E ratio is about 0.085 for coatings formed at pressures not higher than 1.8 Pa. The coatings formed in atmosphere with nitrogen pressure from 3 Pa to 5 Pa are characterized by a higher ratio, ranging from 0.091 to 0.094. In the case of the coating resistance to plastic deformation H^3/E^2, an almost linear increase is observed from 0.12 GPa (p_{N2} = 0.8 Pa) to 0.23 GPa (p_{N2} = 5Pa).

The mechanical properties of the coatings, both hardness and Young's modulus formed at various arc currents show a similar tendency, as shown in Table 2. Above features increase with arc current in deposition process. This is the reason for the nearly identical H/E and H^3 E^2 ratios for the coatings. They are in the range from 0.092 to 0.099 and from 0.22 GPa to 0.24 GPa. Due to the uncertainty of measurement, it can be assumed that they are the same.

Table 2. Mechanical properties of f AlCrN coatings formed at various arc current.

Parameter	AlCr(50)N	AlCr(60)N	AlCr(80)N	AlCr(100)N
Hardness, H (GPa)	23.3 ± 1.7	24.9 ± 1.9	24.6 ± 2.2	25.4 ± 1.3
Young's modulus, E (GPa)	235 ± 9	252 ± 10	261 ± 9	275 ± 8
H/E	0.099 ± 0.011	0.099 ± 0.011	0.094 ± 0.014	0.092 ± 0.007
H^3/E^2 (GPa)	0.23 ± 0.07	0.24 ± 0.07	0.22 ± 0.07	0.22 ± 0.05
Lc_2 (N)	88.1 ± 3.3	89.4 ± 1.0	91.0 ± 2.1	97.2 ± 0.9

The trend of changes in hardness of coatings formed under different nitrogen pressures is the same, and the hardness values are slightly lower than those presented in Ref. [30], where it ranges from about 22 GPa to about 33 GPa. The HIPIMS coatings are characterized by a hardness ranging from about 11 GPa for low nitrogen flow rates to about 29 GPa for $N_2/Ar = 140\%$ [25]. Differences in hardness may be related to the method of forming the coatings, but also to the use of a cathode with a different chemical composition—$Al_{50}Cr_{50}$. As already mentioned, the coatings formed at different arc currents show a decrease in hardness with increasing arc current [31], which is contrary to our observations. The decrease in hardness is explained by a significant increase in the cathode temperature (arc current up to 150 A), which leads to a larger amount of the droplet phase [31].

The surface of the AlCrN coating formed at $p_{N2} = 3$ Pa, in terms of roughness, occupies a middle position among the coatings deposited at other nitrogen pressures (Figure 7), but it still contains quite a lot of macroparticles and craters (Figure 6). The surface can be described as completely covered with macroparticles with a diameter of 100 nm–3 µm, among which there are free areas of the surface no more than 1–2 µm. The probability of indentation into a macroparticle is much higher than into a free surface during NI. Therefore, to determine E and H without particles, the surfaces of the coatings were polished. After polishing, the macroparticles disappear and depressions appear on the topography in place of the largest of them, as is always the case with softer phases on the metallographic polished surface (Figure 8a). The fact that large macroparticles can spread throughout the thickness of the coating is shown in [47]. In the topography image, most of the coating is free from particles and much less of the depression from the macroparticles on the remaining surface. E and H maps show the distribution of values corresponding to this surface. Most of the surface is orange (Figure 8b,c), which in these images corresponds to high values (240–360 GPa for E and 26–40 GPa for H). There are areas of lilac color between them, corresponding to the remains of large particles, some of which left inside the coating (Figure 8b,c) (20–160 GPa for E and 2–16 GPa for H). The mean value of E for the polished surface is 241 ± 40 GPa. The mean value of H is 27.5 ± 7.4 GPa. Lower E and H correspond to particles, possibly containing metals Cr and Al (Cr was established by XRD), or traces of the metastable γ-phase Al_2O_3, which coincides in peaks with c-CrN and Cr.

The adhesion of a coating to the surface of a substrate (connected with the coating bond strength) determines the difficulty of removing the coating from the substrate. The more difficult it is to remove the coating, the better its adhesion. To improve the adhesion of the coatings, an intermediate layer (metallic adhesive layer) is often used between the substrate and the coating. One of the most common methods of testing the strength of the bond between the coating and the substrate is the scratch test. The critical load Lc_2 for coatings formed under various nitrogen pressure is from about 77 N to about 98 N (Table 1). A distinct Lc_2 maximum is observed. An increase in the critical load of Lc_2 under nitrogen pressure from about 77 N ($p_{N2} = 0.8$ Pa) to about 98 N ($p_{N2} = 3$ Pa) can be noticed, and then a decrease to about 80 N. AlCrN and CrN coatings generally present very good adhesion, the critical load Lc_2 exceeds 80 N [30]. The coatings formed at various arc current are characterized by an almost linear increase in critical load Lc_2 from about 88 N to about 97 N, Table 2.

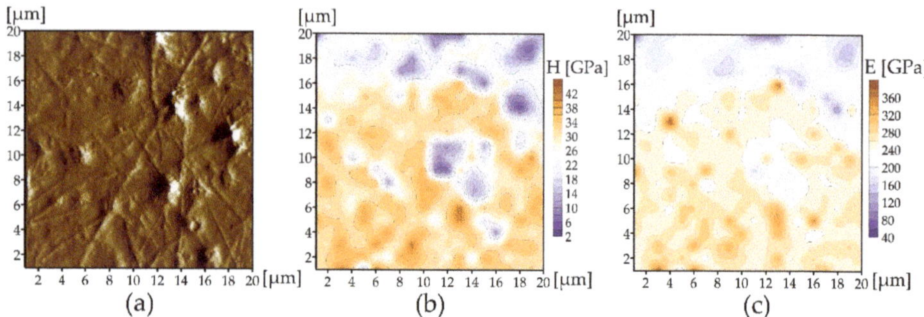

Figure 8. AlCrN coating deposited at p_{N2} = 3 Pa, Ic = 80 A: (**a**) surface morphology after polishing, (**b**) microhardness map, (**c**) Young's modulus. Coating area 20 × 20 μm^2.

The spherical abrasion test allows to determine the coating thickness. The combination of the spherical abrasion test and the scratch test proposed by Panjan et al. [48] allows the assessment of both the characteristics of the coating structure and the deformation of the coating and the substrate under proposed load. The above double test was applied to the coating characterized by the lowest critical force Lc$_2$ (Table 1) formed at p_{N2} = 0.8 Pa and Ic = 80 A, as shown in Figure 9. Figure 9a shows an image of abrasion by a steel ball in an abrasion test and a scratch in the place where the normal load of the Rockwell indenter was 70 N, which is slightly below Lc$_2$. Despite relatively low critical load Lc$_2$ this coating shows excellent adhesion both in the scratch area and in the substrate-ball crater area. No detachment was observed at both the ball abrasion—substrate and the ball abrasion—scratch interface, as shown in Figure 9b. There is noticeable deformation of the coating as well as the substrate, especially in the part including the scratch. In the scratch test, a plastic deformation of the substrate occurs as a result of a gradual increase in the normal force, and thus the scratch depth. As a result, the pressed-out substrate material and cracking of the coating occur at the scratch boundaries. The high stress on the interface between the coating and the substrate may result in the coating delamination. This effect did not occur in the coatings tested. Therefore, it can be concluded that these coatings are characterized by high adhesion, and the AlCrN coating–steel substrate systems have high mechanical stability. The coating adheres tightly to the surface, even the one pressed into the bottom of the scratch.

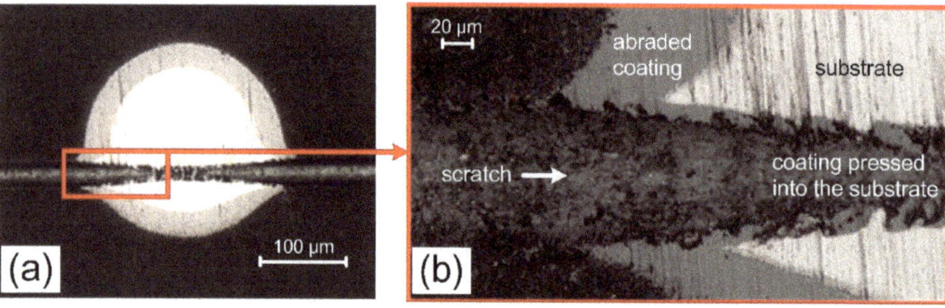

Figure 9. Optical microscope micrographs of the result of the combined test (scratch test and spherical abrasion test) for AlCrN coating deposited at p_{N2} = 0.8 Pa and Ic = 80 A (**a,b**) the enlarged image of the red rectangle in Figure 9a. Image of the coating from the top.

The standard Daimler–Benz test involves pressing a Rockwell indenter into the coating with a load of approximately 1470 N. Optical analysis of the of cracks and peeling of the coating at the edges of the impression allows these failures (corresponding to coating adhesion) to be assigned to one of the six adhesion patterns. Figure 10 shows images of indentations of the coatings formed at the lowest (Figure 10a) and highest nitrogen pressure, Figure 10b. In the coating deposited at nitrogen pressure of 0.8 Pa few short radial cracks, small areas of coating delamination and circular cracks in the peripheral region of indentation are observed, Figure 10a. Due to it, the coatings are characterized by adhesion consistent with the HF2 damage pattern. In coating formed at the nitrogen pressure of 5 Pa, the above defects are not visible.

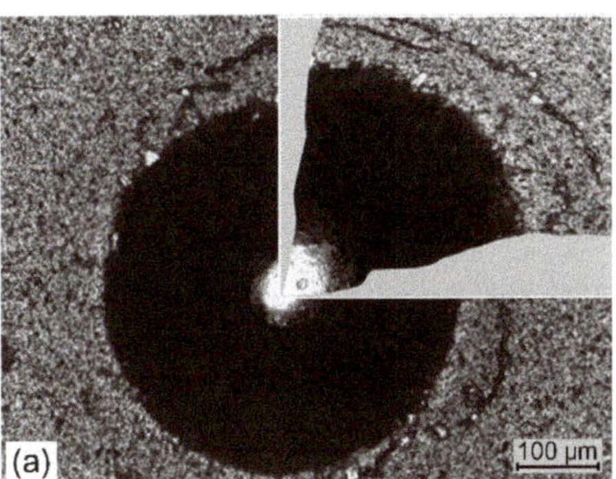

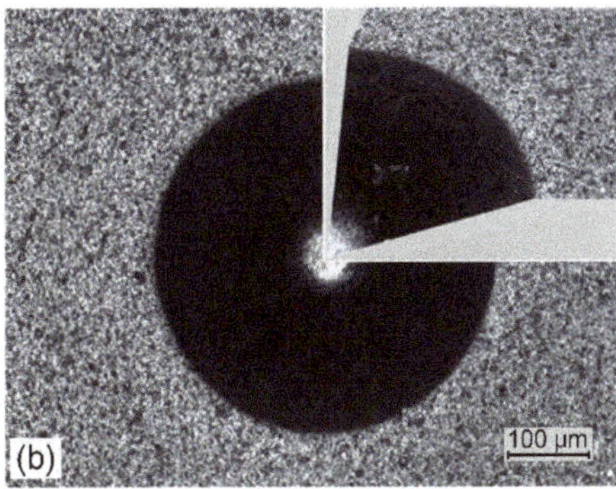

Figure 10. Daimler-Benz adhesion test results, SEM images of Rockwell indents for coatings formed at: (**a**) p_{N2} = 0.8 Pa and (**b**) p_{N2} = 5 Pa.

Only a slight pile-up of material can be observed around the indents, especially in Figure 10a. Adhesion of the coatings is consistent with the HF1 damage pattern. Likewise, the coatings formed at different arc currents have similar properties. Only the coatings deposited at the extreme arc currents of 50 A (Figure 11a) and 100 A (Figure 11b) are shown. These coatings also show an adhesion consistent with the HF1 damage pattern. In all the images it is difficult to even notice the occurrence of the most common coating defects—short radial cracks. This type of damage, or rather the lack of it, qualifies the coatings for industrial applications.

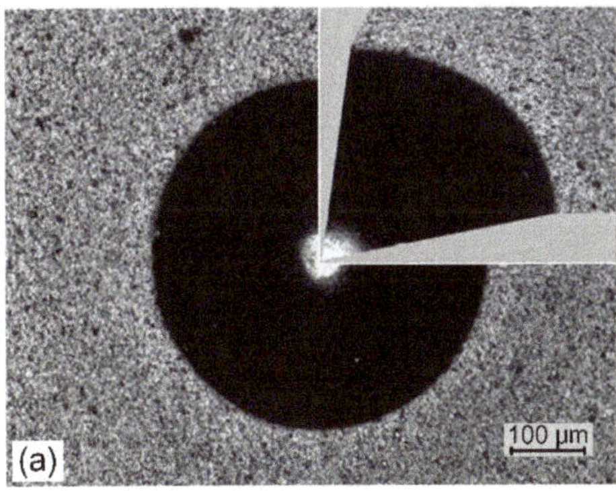

Figure 11. Daimler-Benz adhesion test results, SEM images of Rockwell indents for coatings formed at: (**a**) Ic = 50 A and (**b**) Ic = 100 A.

3.4. Friction and Wear

The tests carried out on all coatings as a function of the friction distance showed that their coefficient of friction fluctuated without any clear tendency. The changes in the friction coefficient during the test were similar for all coatings. There were two stages of the friction process: the running-in stage (distance approx. 300 m), where there were larger changes and an increase in the friction coefficient to a value of approx. 0.6–0.7, and the steady-state wear stage, where only slight fluctuations of the coefficient were observed. No sudden changes in the friction coefficient were observed, which would suggest a change in the friction environment or coating damage. The coefficient of friction of coatings formed at different nitrogen pressure, determined in the pin-on-disk system against the countersample—Al_2O_3 ball, shows a fine tendency to decrease from 0.68 ± 0.03 ($p_{N2} = 0.8$ Pa) to 0.63 ± 0 0.02 ($p_{N2} = 5$ Pa). The coatings formed at different arc currents are characterized by a similar tendency. The coefficient of friction reaches the highest value of 0.70 ± 0.03 for coatings formed at arc current of 50 A and the lowest value of about 0.64 ± 0.03 for coatings formed at arc current of 100 A. It means that, in both cases, the coefficient of friction only slightly depends on deposition parameters, i.e., nitrogen pressure and arc current.

The coatings formed at low nitrogen pressure, i.e., 0.8 Pa and 1.2 Pa show relatively high wear rates, about 6.4×10^{-7} mm³/Nm and 4.0×10^{-7} mm³/Nm, respectively, Figure 12a. One can observe that wear rate of the coatings decreases, and the coating deposited at 3 Pa shows the lowest wear rate 2.1×10^{-8} mm³/Nm. Coatings formed at higher nitrogen pressure as well as coatings deposited at various arc currents are characterized by similar values of wear rate ranging from 1.7×10^{-7} mm³/Nm to 4.7×10^{-7} mm³/Nm with a small reducing trend with arc current (Figure 12b). The wear rate of the coatings investigated is low, except for those formed under a low nitrogen pressure. The value of the wear rate, from about 2×10^{-7} mm³/Nm to about 2×10^{-8} mm³/Nm is very low and may suggest possible industrial application of the coatings. For comparison, Reiter et al. [20], examining the coatings formed with the same method, determined wear rates from 2×10^{-6} mm³/Nm (71at% of Al in AlCrN coating) to 8×10^{-6} mm³/Nm (CrN). Lower values of the wear rate, about 3×10^{-7} mm³/Nm, regardless of the applied substrate bias voltage during the formation of CrAlN coatings are presented in Ref. [24]. Antonov, examining the tribological properties of $Al_{60}Cr_{40}N$ coatings in relation to various counter-samples as well as various test parameters (normal load, sliding speed, measurement temperature), stated that for the Al_2O_3 counter-sample, the wear rate was from 3×10^{-6} mm³/Nm to 2×10^{-7} mm³/Nm depending on sliding speed.

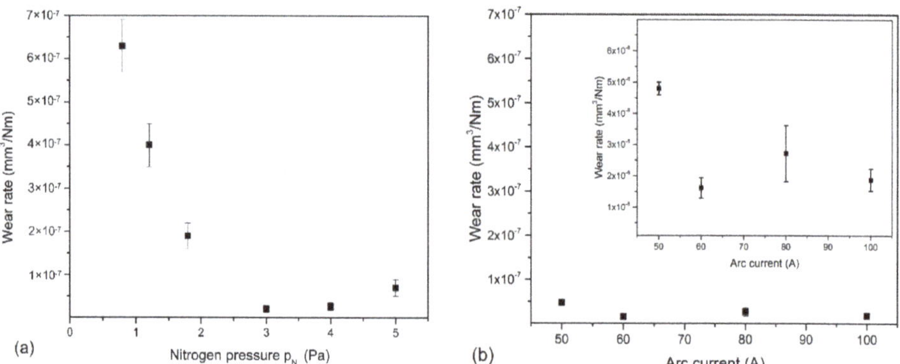

Figure 12. Wear rate of AlCrN coatings formed at: (**a**) various nitrogen pressure, (**b**) various arc current.

4. Discussion

4.1. Effect of Nitrogen Pressure

The coating thickness should be greater with increasing nitrogen pressure. Such an increase to a pressure of 3 Pa is observed, and then the thickness of the coating decreases, Figure 1a. During the movement of ions and particles from the cathode to the substrate, collisions with atoms of reactive gas, i.e., nitrogen, occur. At higher pressures, the probability of collisions is higher, which reduces the mobility and energy of particles, so the free path is shorter and the time to reach the substrate is longer [14]. This reduces the deposition rate of the coatings, i.e., coating thickness. Similar results for the coatings were presented previously for the AlCrN [24,30] and TiAlN [38] coatings. With the nitrogen pressure increase, the chemical composition of the deposited coating also changes, Figure 2a. The nitrogen concentration in the coating increases, reaching the value of about 46 at.%, and the concentration of aluminum and chromium decreases, as well as Al/(Al + Cr) rate. This means that there is loss of aluminum in the coating. The lower atomic mass of aluminum in comparison to chromium in the case of collisions with nitrogen ions results in greater scattering of aluminum ions, and thus less of its condensation on the substrate, as shown earlier [49].

In the coatings formed at p_{N2} not higher than 4 Pa, the presence of the hexagonal AlN phase was found. At low pressures, this is due to a nitrogen deficiency during coating formation. With the increase of nitrogen pressure, the intensity of the h-AlN and h-Cr$_2$N diffraction lines decreases and is small in the coatings formed at p_{N2} = 3 and 4 Pa. At the same time, an increase in the intensity of diffraction lines originating from cubic CrN is observed. The transformation of the B1 → B4 structure takes place when the AlN content in the AlCrN coatings is from 60% to 77% [20,22,23]. The tested coatings were formed from the Al$_{70}$Cr$_{30}$ cathode, and their Al/(Al + Cr) rate in the coating was only slightly different from that in the cathode. It means that, in the above conditions, the formation of h-AlN phase or mixture of h-AlN and c-AlN is authorized.

The phase composition of the coating is determined not only by the aluminum concentration. It has been found that the hexagonal phase may also arise as a result of other factors, such as vacancies or compressive stresses [50], which usually occur in PVD coatings, as well as the size of crystallites. It was found shown that the formation of the hexagonal phase is preferred in the case of coatings characterized by fine crystallites [51]. In TiAlN coating with aluminum content in the range 0.55 to 0.70, crystallites with dimensions of 3–5 nm favor the formation of the hexagonal AlN phase, while crystallites with dimensions above 8 nm favor the formation of the cubic phase [51]. Thus, it is the crystallite size (Figure 4a) that may determine the hexagonal phase formation during coating growth.

This phase adversely affects the mechanical properties of the coating. The coatings with c-AlN phase are characterized by higher hardness than those with h-AlN phase. Presumably, it is also influenced by density, and the hexagonal phase, compared to the cubic one, is about 18% lower [20].

The increase in hardness can be related to the increase in crystallinity of the c-CrN phase, visualized by the intensity of diffraction lines. The H/E ratio of coatings formed at pressures of 0.8–1.8 Pa is about 0.085, while for $p_{N2} \geq 3$ Pa it increases to 0.094. In the H/E ratio, two areas can be distinguished: elastic (H/E > 0.1) and plastic (H/E < 0.1) and [52]. Better wear resistance is predicted for higher H/E and H^3/E^2 ratios. This suggests that coatings deposited at higher nitrogen pressure should be more wear resistant. Indeed, the wear rates of coatings formed at $p_{N2} \geq 3$ Pa are significantly lower than other coatings, Figure 12.

The surface roughness of rubbing elements affects the tribological properties of the coatings. Greater roughness intensifies friction processes, accelerating the wear process. There is a clear correlation between the wear results of the coatings (Figure 12) and their roughness, Figure 7a. Another factor that determines wear resistance is hardness. The greater the hardness, the greater the wear resistance is. Although the coating is heterogeneous in terms of mechanical properties (Figure 8) and the hardness ranges from 26 GPa to 40 GPa, good wear resistance of the coatings should be associated with their hardness. This large hardness dispersion is related to surface defects (Figure 5) characteristic of the coating deposition method.

Two methods of assessing the adhesion of the coating, the scratch test enabling numerical determination of adhesion as critical load (Table 1) and the Daimler–Benz test (Figure 10), where the assessment is based on microscopic analysis of the damage to the coating around the Indentation, showed the same result—all coatings are characterized by very good properties. The Daimler–Benz test uses a high normal load of approximately 1470 N. Plastic deformation of the coating and the substrate results from the transformation of the kinetic energy of the moving indenter. This leads to pile up around the indenter which can generate high tensile stresses around the indentation, as in Figure 10a. As a result of these stresses, apart from coating delamination, circular cracks of the coating may occur [53]. This effect may be a consequence of both the high roughness of the coating and its low hardness. On the contrary, the coating formed at p_{N2} does not show similar damage.

4.2. Effect of Arc Current

Thickness of the coatings increases with arc current, Figure 1b. This is due to the greater energy of the arc spots and the plasma density, which is a result of the higher arc current used, as reported by Lan et al. [31] investigating AlCrN coatings obtained by the ion arc method. In the case of the coatings synthesized with a higher arc current, the surface is more developed. More surface defects can be observed there, Figure 6b. Higher target energy and temperature, which helps to increase the emission of the droplets from cathode material is connected with higher arc current. Figure 7b confirms this unfavorable arc current effect on the surface quality of the coating.

The coatings formed at different currents are characterized by the cubic structure of B1 of the CrN phase, Figure 4b. This is consistent with reports that the solubility of Al in solid fcc-CrN is 60–77% [20,21], so the Al atom occupies the Cr sites in the CrN lattice, forming a substitution solid solution, leading to solid solution strengthening. Despite using an arc current of 50 A to 100 A, the diffraction patterns remain similar. The mutual ratio of the intensities of the diffraction lines is similar. This suggests that an increase in the arc current does not significantly modify the crystal lattice. The analysis of the position of the (111) and (200) diffraction lines shows only a slight shift towards higher angles (Figure 4b), which suggests no change in the stress state in the coating. The full width at half maximum intensity of the line seems similar, so an increase in the arc current does not change the crystallite size, Figure 5b. This effect is the opposite to that described in Ref [31], although the determined crystallite sizes are similar. The increase in the arc current did not increase the plasma energy allowing for grain size differentiation and internal stresses in the coating.

As mentioned earlier, plasma with higher energy and density that can alter the properties of the coating is associated with a higher arc current. Under such conditions, it is possible to increase the density of the coating, also by reducing the size of the crystallites. As a result, this should result in coating hardness increase. Indeed, the hardness of the coatings was greater for higher the current value. However, the increase, although obvious, was within the range of the hardness measurement uncertainty, Table 2. This may be due to the fact that the crystallite sizes were comparable for all coatings (Figure 4b), so it would be difficult to connect a change in hardness with the Hall–Petch relationship. Probably one should take into account another effect related to the higher energy of the incident ions, the increase in the density of the coatings. Lan et al. [31] formed coatings with arc currents from 100 A to 150 A and recorded the opposite effect, reducing the hardness with increasing current during coating formation. They found that at such a high arc current, increased droplet ejection from the cathode and deterioration of the surface of the coating quality occurs. It seems that the reduction of hardness by defects on the surface and in the whole coating by macroparticles is stronger than the improvement connected with plasma energy and density. As mentioned, the coatings described in this study were formed at arc currents from 50 A to 100 A. This probably resulted in the observation of opposite changes in hardness. The amount of macroparticles on the surface increases significantly, which was also observed in Figure 3b. The macroparticles on the surface are also accompanied by similar defects in the volume of the coating as well as by voids [46]. Sometimes larger macroparticles extend through the entire thickness of the coating, deteriorating the mechanical and electrochemical properties.

Both H^3/E^2 rate and H/E rate as a predictor of the wear resistance of the coatings have similar values, of about 0.23 and about 0.1 (Table 2), regardless of the arc current during formation, respectively. The values obtained for the tested coatings indicate that they should have good wear resistance, comparable for all coatings. Indeed, the wear resistance shown in Figure 12b is high, the wear rate about (2 to 4) $\times$ 10^8 mm^3/Nm close to each other.

The adhesion of the coatings (Table 2) shows an increasing trend from about 88 N (Ic = 50 A) to about 97 N (Ic = 100 A). This means that the adhesion is very good. The increase in adhesion may be related to many factors: an increase in the coating density and its hardness as well as higher energy of incident particles. The higher energy of the particles can even result in their shallow implantation into the substrate, which creates a well-adherent adhesive layer. The implantation depth is usually 5–15 nm [53], although it can be as high as 50 nm. It depends on the energy of the ions. With higher energy, the stresses in the coating can be increased, which can lead to reduced adhesion.

The results of Daimler–Benz test confirm the very good adhesion of investigated coatings. Regardless of the arc current value during their formation, the damage image showed the highest class of HF1 adhesion, Figure 11. It is possible that one of the reasons for this was the use of a thin gradient Cr-N adhesive layer with an increasing nitrogen concentration in this layer. A similar result was reported in [54] where pure chromium was used.

The above technological works of forming coatings at various nitrogen pressures and arc currents with a wide range of their values, as well as examining their structure, mechanical and tribological properties, were to indicate a set of technological parameters enabling the formation of the coatings with the best properties. The coatings were applied to HS6-5-2 high speed steel substrates (EU EN and DIN, M2—USA), which is widely used for tools. Comparing the properties of the coatings obtained in this way with those of commercially used coated tools may contribute to their incorporation into the set of coatings used in practice.

5. Conclusions

In this work, we investigated two sets of coatings: deposited at different nitrogen pressure (constant arc current) and different arc currents (constant nitrogen pressures). The deposition method was cathodic arc evaporation. The effect of these factors on structure features, i.e., phase composition and crystallite size, as well as coating microstructure, mechanical, and tribological properties, was studied. The results indicate that with increasing nitrogen pressure during coating formation the h-AlN phase gradually disappears. Changing the arc current does not affect the type of phase formed. The coatings formed under both increasing nitrogen pressure and arc current are characterized by increasing hardness. They show very good adhesion, the critical force from the scratch test is above 77 N, and according to the Daimler–Benz test, most coatings show HF1 class. The coatings formed at higher pressure exhibit good wear resistance.

The effect of nitrogen pressure during AlCrN coating deposition from AlCr cathodes of different chemical composition is known. The further direction of technological and research works is the doping of such coatings with other metallic and non-metallic elements, creating four-and more-component coatings with excellent mechanical, tribological, and electro-chemical properties for application on tools for dry machining.

Author Contributions: Conceptualization, A.G., B.W. and T.K.; methodology, B.W.; software, A.N.; validation, A.G., S.A., V.L. and A.K.; formal analysis, B.W. and A.G.; investigation, A.G., T.K. and V.L.; resources, T.K. and V.L.; data curation, A.G. and T.K.; writing—original draft preparation, B.W.; writing—review and editing, A.N., A.G. and T.K.; visualization, A.G., V.L., A.K. and A.N.; supervision, B.W. and S.A.; project administration, A.G.; funding acquisition, S.A. All authors have read and agreed to the published version of the manuscript.

Funding: This research was funded by the National Center for Research and Development, Poland, grant number Biostrateg3/344303/14/NCBR/2018. Research was partly financed by the grants of Belarusian Republican Foundation for Fundamental Research BRFFR–SCST–Poland No. T18PLDG-002 and No. F18R-239. A.N. and S.A. were supported by the Government of Russia (grant No. 14.Z50.31.0046).

Institutional Review Board Statement: Not applicable.

Informed Consent Statement: Not applicable.

Data Availability Statement: The data presented in this study are available on request from the corresponding author.

Conflicts of Interest: The authors declare no conflict of interest.

References

1. Bobzin, K. High-performance coatings for cutting tools. *CIRP J. Manuf. Sci. Technol.* **2017**, *18*, 1–9.
2. Aizikovich, S.; Krenev, L.; Sevostianov, I.; Trubchik, I.; Evich, L. Evaluation of the elastic properties of a functionally-graded coating from the indentation measurements. *ZAMM J. Appl. Math. Mech. Z. Angew. Math. Mech.* **2011**, *91*, 493–515. [CrossRef]
3. Navinšek, B.; Panjan, P.; Cvelbar, A. Characterization of low temperature CrN and TiN (PVD) hard coatings. *Surf. Coat. Technol.* **1995**, *74–75*, 155–161. [CrossRef]
4. Ürgen, M.; Cakır, A. The effect of heating on corrosion behavior of TiN- and CrN-coated steels. *Surf. Coat. Technol.* **1997**, *96*, 236–244.
5. Fox-Rabinovich, G.S.; Kovalev, A.I.; Afanasyev, S.N. Characteristic features of wear in tools made of high-speed steels with surface engineered coatings I. Wear characteristics of surface engineered high-speed steel cutting tools. *Wear* **1996**, *201*, 38–44.
6. Hurkmans, T.; Lewis, D.B.; Paritong, H.; Brooks, J.S.; Munz, W.D. Influence of ion bombardment on structure and properties ofunbalanced magnetron grown CrN_x coatings. *Surf. Coat. Technol.* **1999**, *114*, 52–59.
7. Mo, J.L.; Zhu, M.H. Tribological characterization of chromium nitride coating deposited by filtered cathodic vacuum arc. *Appl. Surf. Sci.* **2009**, *25*, 7627–7634. [CrossRef]
8. Kudish, I.I.; Volkov, S.S.; Vasiliev, A.S.; Aizikovich, S.M. Some Criteria for Coating Effectiveness in Heavily Loaded Line Elastohydrodynamically Lubricated Contacts—Part I: Dry Contacts. *J. Tribol.* **2016**, *138*. [CrossRef]
9. van Essen, P.; Hoy, R.; Kamminga, J.D.; Ehiasarian, A.P.; Janssen, G.C.A.M. Scratch resistance and wear of CrN_x coatings. *Surf. Coat. Technol.* **2006**, *200*, 3496–3502.
10. Kudish, I.I.; Volkov, S.S.; Vasiliev, A.S.; Aizikovich, S.M. Effectiveness of coatings with constant, linearly, and exponentially varying elastic parameters in heavily loaded line elastohydrodynamically lubricated contacts. *J. Tribol.* **2017**, *139*. [CrossRef]
11. Navinsek, B.; Panjan, P.; Milosev, I. Industrial applications of CrN (PVD) coatings, deposited at high and low temperatures. *Surf. Coat. Technol.* **1997**, *97*, 182–191.
12. Rech, J.; Kusiak, A.; Battaglia, J.L. Tribological and thermal functions of cutting tool coatings. *Surf. Coat. Technol.* **2004**, *186*, 364–371. [CrossRef]
13. Benlatreche, Y.; Nouveau, C.; Aknouche, H.; Imhoff, L.; Martin, N.; Gavoille, J.; Rousselot, C.; Rauch, J.Y.; Pilloud, D. Physical and Mechanical Properties of CrAlN and CrSiN Ternary Systems for Wood Machining Applications. *Plasma Process. Polym.* **2009**, *6*, S113–S119. [CrossRef]
14. Wang, Y.; Zhang, J.; Zhou, S.; Wang, Y.; Wang, C.; Wang, Y.; Sui, Y.; Lan, J.; Xue, Q. Improvement in the tribocorrosion performance of CrCN coating bymultilayered design for marine protective application. *Appl. Surf. Sci.* **2020**, *528*, 147061.
15. Chen, W.; Hu, T.; Hong, Y.; Zhang, D.; Meng, X. Comparison of microstructures, mechanical and tribological properties of arc-deposited AlCrN, AlCrBN and CrBN coatings on Ti-6Al-4V alloy. *Surf. Coat. Technol.* **2020**, *404*, 126429. [CrossRef]
16. Uchida, M.; Nihira, N.; Mitsuo, A.; Toyoda, K.; Kubota, K.; Aizawa, T. Friction and wear properties of CrAlN and CrVN films deposited by cathodic arc ion plating method. *Surf. Coat. Technol.* **2004**, *177–178*, 627–630. [CrossRef]
17. Castaldi, L.; Kurapov, D.; Reiter, A.; Shklover, V.; Schwaller, P.; Patscheider, J. Effect of the oxygen content on the structure, morphology and oxidation resistance of Cr–O–N coatings. *Surf. Coat. Technol.* **2008**, *203*, 545–549.
18. Barshilia, H.C.; Selvakumar, N.; Deepthi, B.; Rajam, K.S. A comparative study of reactive direct current magnetron sputtered CrAlN and CrN coatings. *Surf. Coat. Technol.* **2006**, *201*, 2193–2201.
19. Vityaz', P.A.; Komarov, A.I.; Komarova, V.I.; Kuznetsova, T.A. Peculiarities of triboformation of wear-resistant layers on the surface of a MAO-coating1 modified by fullerenes. *J. Frict. Wear* **2011**, *32*, 231–241.
20. Reiter, A.E.; Derflinger, V.H.; Hanselmann, B.; Bachmann, T.; Sartory, B. Investigation of the properties of $Al_{1-x}Cr_xN$ coatings prepared by cathodic arc evaporation. *Surf. Coat. Technol.* **2005**, *200*, 2114–2122.
21. Haršáni, M.; Ghafoor, N.; Calamba, K.; Zacková, P.; Sahul, M.; Vopát, T.; Satrapinskyy, L.; Čaplovičová, M.; Čaplovič, L. Adhesive-deformation relationships and mechanical properties of nc-AlCrN/a-SiNx hard coatings deposited at different bias voltages. *Thin Solid Films* **2018**, *650*, 11–19.
22. Lin, J.; Mishra, B.; Moore, J.J.; Sproul, W.D. Microstructure, mechanical and tribological properties of Cr1−xAlxN films deposited by pulsed-closed field unbalanced magnetron sputtering (P-CFUBMS). *Surf. Coat. Technol.* **2006**, *201*, 4329–4334. [CrossRef]

23. Li, T.; Li, M.; Zhou, Y. Phase segregation and its effect on the adhesion of Cr–Al–N coatings on K38G alloy prepared by magnetron sputtering method. *Surf. Coat. Technol.* **2007**, *201*, 7692–7698. [CrossRef]
24. Romero, J.; Gómez, M.A.; Esteve, J.; Montalà, F.; Carreras, L.; Grifol, M.; Lousa, A. CrAlN coatings deposited by cathodic arc evaporation at different substrate bias. *Thin Solid Films* **2006**, *515*, 113–117. [CrossRef]
25. Tang, J.F.; Lin, C.Y.; Yang, F.C.; Chang, C.L. Influence of Nitrogen Content and Bias Voltage on Residual Stress and the Tribological and Mechanical Properties of CrAlN Films. *Coatings* **2020**, *10*, 546.
26. Antonov, M.; Afshari, H.; Baronins, J.; Adoberg, E.; Raadik, T.; Hussainova, I. The effect of temperature and sliding speed on friction and wear of Si_3N_4, Al_2O_3, and ZrO_2 balls tested against AlCrN PVD coating. *Tribol. Int.* **2018**, *118*, 500–514. [CrossRef]
27. Mo, J.L.; Zhu, M.H. Sliding tribological behavior of AlCrN coating. *Tribol. Int.* **2008**, *41*, 1161–1168.
28. Sabitzer, C.; Paulitsch, J.; Kolozsvári, S.; Rachbauer, R.; Mayrhofer, P.H. Impact of bias potential and layer arrangement on thermal stability of arc evaporated Al-Cr-N coatings. *Thin Solid Films* **2016**, *610*, 26–34.
29. Tillmann, W.; Kokalj, D.; Stangier, D.; Paulus, M.; Sternemann, C.; Tolan, M. Investigation of the influence of the vanadium content on the high temperature tribo-mechanical properties of DC magnetron sputtered AlCrVN thin films. *Surf. Coat. Technol.* **2017**, *328*, 172–181.
30. Wang, L.; Zhang, S.; Chen, Z.; Li, J.; Li, M. Influence of deposition parameters on hard Cr–Al–N coatings deposited by multi-arc ion plating. *Appl. Surf. Sci.* **2012**, *258*, 3629–3636.
31. Lan, R.; Wang, C.; Ma, Z.; Lu, G.; Wang, P.; Han, J. Effects of arc current and bias voltage on properties of AlCrN coatings by arc ion plating with large target. *Mater. Res. Express* **2019**, *6*, 116457.
32. Gong, Z.; Chen, R.; Li, J.; Cao, P.; Geng, H. Effect of N_2 Flow Rate on Structure and Corrosion Resistance of AlCrN Coatings Prepared by Multi-Arc Ion Plating. *Int. J. Electrochem. Sci.* **2020**, *15*, 1117–1127.
33. Cullity, B.D.; Weymouth, J.W. Elements of X-Ray Diffraction. *Am. J. Phys.* **1957**, *25*, 394–395. [CrossRef]
34. Vidakis, N.; Antoniadis, A.; Bilalis, N. The VDI 3198 indentation test evaluation of a reliable qualitative control for layered compounds. *J. Mater. Process. Technol.* **2003**, *143*, 481–485.
35. Lapitskaya, V.A.; Kuznetsova, T.A.; Chizhik, S.A.; Sudzilouskaya, K.A.; Kotov, D.A.; Nikitiuk, S.A.; Zaparozhchanka, Y.V. Lateral force microscopy as a method of surface control after low-temperature plasma treatment. *J. Phys. Conf. Ser.* **2018**, *443*, 012019.
36. Anders, A. A review comparing cathodic arcs and high power impulse magnetron sputtering (HiPIMS). *Surf. Coat. Technol.* **2014**, *257*, 308–325.
37. Chang, Y.Y.; Wang, D.Y.; Hung, C.Y. Structural and mechanical properties of nanolayered TiAlN/CrN coatings synthesized by a cathodic arc deposition process. *Surf. Coat. Technol.* **2005**, *200*, 1702–1708.
38. Cai, F.; Zhang, S.; Li, J.; Chen, Z.; Li, M.; Wang, L. Effect of nitrogen partial pressure on Al–Ti–N films deposited by arc ion plating. *Appl. Surf. Sci.* **2011**, *258*, 1819–1825. [CrossRef]
39. Reiter, A.E.; Mitterer, C.; de Figueiredo, M.R.; Franz, R. Abrasive and adhesive wear behavior of arc-evaporated Al1-xCrxN hard coatings. *Tribol. Lett.* **2010**, *37*, 605–611. [CrossRef]
40. Warcholinski, B.; Gilewicz, A.; Ratajski, J.; Kuklinski, Z.; Rochowicz, J. An analysis of macroparticle-related defects in the CrCN and CrN coatings in dependence on the substrate bias voltage. *Vacuum* **2012**, *86*, 1235–1239.
41. Nikolaev, A.L.; Mitrin, B.I.; Sadyrin, E.V.; Zelentsov, V.B.; Aguiar, A.R.; Aizikovich, S.M. Mechanical Properties of Microposit S1813 Thin Layers. *Adv. Struct. Mater.* **2020**, 137–146.
42. Vasiliev, A.S.; Sadyrin, E.V.; Mitrin, B.I.; Aizikovich, S.M.; Nikolaev, A.L. Nanoindentation of ZrN Coatings on Silicon and Copper Substrates. *Russ. Eng. Res.* **2018**, *38*, 735–737. [CrossRef]
43. Sadyrin, E.; Swain, M.; Mitrin, B.; Rzhepakovsky, I.; Nikolaev, A.; Irkha, V.; Aizikovich, S. Characterization of Enamel and Dentine about a White Spot Lesion: Mechanical Properties, Mineral Density, Microstructure and Molecular Composition. *Nanomaterials* **2020**, *10*, 1889. [CrossRef] [PubMed]
44. Leyland, A.; Matthews, A. On the significance of the H/E ratio in wear control: A nanocomposite coating approach to optimized tribological behaviour. *Wear* **2000**, *246*, 1–11.
45. Musil, J.; Kunc, F.; Zeman, H.; Polakova, H. Relationships between hardness, Young's modulus and elastic recovery in hard nanocomposite coatings. *Surf. Coat. Technol.* **2002**, *54*, 304–313. [CrossRef]
46. Pintaude, G. Introduction of the Ratio of the Hardness to the Reduced Elastic Modulus for Abrasion. In *Tribology—Fundamentals and Advancements*, 1st ed.; Gegner, J., Ed.; IntechOpen: London, UK, 2013.
47. Panjan, P.; Drnovšek, A.; Gselman, P.; Čekada, M.; Panjan, M. Review of Growth Defects in Thin Films Prepared by PVD Techniques. *Coatings* **2020**, *10*, 447. [CrossRef]
48. Panjan, P.; Čekada, M.; Navinšek, B. A new experimental method for studying the cracking behaviour of PVD multilayer coatings. *Surf. Coat. Technol.* **2003**, *174–175*, 55–62.
49. Birol, Y.; İsler, D. Response to thermal cycling of CAPVD (Al,Cr)N-coated hot work tool steel. *Surf. Coat. Technol.* **2010**, *205*, 275–280.
50. Jäger, N.; Meindlhumer, M.; Spor, S.; Hruby, H.; Julin, J.; Stark, A.; Nahif, F.; Keckes, J.; Mitterer, C.; Daniel, R.; et al. Microstructural evolution and thermal stability of AlCr(Si)N hard coatings revealed by in-situ high-temperature high-energy grazing incidence transmission X-ray diffraction. *Acta Mater.* **2020**, *186*, 545–554. [CrossRef]
51. Hans, M.; Music, D.; Chen, Y.-T.; Patterer, L.; Eriksson, A.O.; Kurapov, D.; Ramm, J.; Arndt, M.; Rudigier, H.; Schneider, J.M. Crystallite size-dependent metastable phase formation of TiAlN coatings. *Sci. Rep.* **2017**, *7*, 16096.

52. Pogrebnjak, A.; Beresnev, V.; Bondar, O.; Postolnyi, B.; Zaleski, K.; Coy, E.; Jurga, S.; Lisovenko, M.; Konarski, P.; Rebouta, L.; et al. Superhard CrN/MoN coatings with multilayer architecture. *Mater. Des.* **2018**, *153*, 47–59.
53. Rebholz, C.; Ziegele, H.; Leyland, A.; Matthews, A. Structure, mechanical and tribological properties of nitrogen-containing chromium coatings prepared by reactive magnetron sputtering. *Surf. Coat. Technol.* **1999**, *115*, 222–229. [CrossRef]
54. Helmersson, U.; Lattemann, M.; Bohlmark, J.; Ehiasarian, A.P.; Gudmundsson, J.T. Ionized physical vapor deposition (IPVD): A review of technology and applications. *Thin Solid Films* **2006**, *513*, 1–24. [CrossRef]

 *materials*

Article

Effects of Loaded End Distance and Moisture Content on the Behavior of Bolted Connections in Squared and Round Timber Subjected to Tension Parallel to the Grain

Antonin Lokaj [1], Pavel Dobes [1,*] and Oldrich Sucharda [2,3]

[1] Department of Structures, Faculty of Civil Engineering, VSB-Technical University of Ostrava, 70800 Ostrava-Poruba, Czech Republic; antonin.lokaj@vsb.cz
[2] Department of Building Materials and Diagnostics of Structures, Faculty of Civil Engineering, VSB-Technical University of Ostrava, 70800 Ostrava-Poruba, Czech Republic; oldrich.sucharda@vsb.cz
[3] Centre of Building Experiments, Faculty of Civil Engineering, VSB-Technical University of Ostrava, 70800 Ostrava-Poruba, Czech Republic
* Correspondence: pavel.dobes1@vsb.cz; Tel.: +42-059-732-1396

Received: 31 October 2020; Accepted: 29 November 2020; Published: 3 December 2020

Abstract: This article presents the results of static tests on bolted connections in squared and round timber with inserted steel plates. The experiment evaluates structural timber connections with different distances between the fastener and the loaded end at different moisture contents. Specimens were loaded by tension parallel to the grain and load–deformation diagrams were recorded. Fifty-six specimens with three different distances between the fastener and the loaded end, at different moisture contents, were tested. The results were statistically evaluated using regression analysis, complemented with load–deformation curves, and compared with calculations according to the valid standard for design of timber structures. A decrease in the evaluated load-carrying capacity with increasing moisture content was confirmed experimentally. A slight increase in the evaluated load-carrying capacity with increasing fastener distance from the loaded end was found.

Keywords: connection; timber; test; bolt; steel plate; moisture content; failure

1. Introduction

Currently, the use of timber as a building material is becoming increasingly popular. The general trend of using natural, renewable and easily recyclable materials in construction practice contributes to this fact. Increasingly, environmental requirements and long-term sustainability in construction are gaining prominence and application (e.g., in buildings [1], bridges and footbridges [2]). Timber structures, which are commonly used as a substitute for steel and concrete structures, considerably mitigate the impact on the environment thanks to their smaller carbon footprint [3].

One of the most important areas in the design of timber structures is connections [4–6]. Connections affect the overall composition of the load-carrying structure and dimensions of the main load-carrying members [7]. The load-carrying capacity and the stiffness of connections are crucial to the serviceability and durability of the whole structure, especially for large-span structures with many connections [1,7] and structures with heavily loaded connections [2]. The most common types of connections in timber structures use metal mechanical fasteners, which are often used in combination with steel plates slotted into cut-outs in timber members (see Figure 1) [8–16].

The stiffness of connections is closely related to the physical and mechanical properties of timber and fasteners [17,18], as well as to the design of individual details (i.e., placement of individual fasteners and the whole connection geometry) [19–21]. The influence of connection stiffness manifests in increased deformations and redistribution of internal forces, which can significantly affect the static design of the whole load-carrying structure [7].

One of the most important physical property of timber that significantly affects its strength and deformation properties [13,22], and its durability especially [23], is the moisture content. Due to atmospheric influences, timber, as a hygroscopic material, constantly exchanges moisture with the surrounding environment [24,25]. The effects of moisture content on the load-carrying capacity and stiffness of connections has been proven in long-term research by forest researchers in the US [26]. Fluctuations in moisture content also result in volume changes (shrinkage and swelling), which, due to the formation of cracks and reduction in embedment strength, negatively affects the load-carrying capacity of connections [25]. As a result, it is necessary to consider the influence of the environmental conditions (especially humidity) under which the structure will be built, as well as the type of load that will be carried. The hygroscopic properties of timber can be improved, for example, by temperature treatment [24].

Figure 1. Practical use of connections in timber structures with mechanical fasteners and steel plates.

The contemporary European standard for the design of timber structures EN 1995-1 (Eurocode 5) [27] includes calculations for timber-to-timber or steel-to-timber connections using mechanical fasteners such as bolts, dowels, nails, screws, etc. Timber-to-timber connections also include combinations of different timber materials [28]. Calculations of load-carrying capacity and connection stiffness do not take into account the edge and end distances nor spacings for fasteners; they are only given as the minimum standard requirement to prevent brittle failure [19,29,30]. Rahim et al. investigated the influence of double-shear steel-to-timber connection geometry (different spacings, end distances) on the initial stiffness [20]. It was found that the initial stiffness can be affected by many factors that are not taken into account in Eurocode 5. Awaludin and Saputro pointed to the possibility of using values smaller than those recommended for the distance between the fastener and the loaded end for a dowel-type connection in laminated veneer lumber [21].

Eurocode 5 and current design codes in many other countries are not optimal, although they are conservative [31]. The European standard only describes three modes, which do not represent total connection failure. Those failure modes are based on the so-called European yield model according to Johansen's theory from 1949 [32], which represents ductile failure modes. The main objective under this model is to design connections for which ductile failure comes before brittle failure, i.e. to design connections with the highest possible deformation capacity to prevent an unexpected collapse [33].

This approach is supported by several scientific studies [19,29,34]. Various connection reinforcement methods have also been used to prevent brittle splitting [12,14,35].

Current research indicates a disagreement between experiments (brittle failure) and design standards (ductile failure) [30,31]. Real total failure of bolted connections occurs due to tension or shear stress [36] and in some cases (multiple shear connections with several slotted-in steel plates) may even be crucial [16]. Brittle failure modes of steel-to-timber bolted connections loaded by tension parallel to the grain were described in [34]. Jorissen developed a new design approach based on fracture mechanics [29].

In this study, static tests were performed on bolted connections in squared and round timber specimens with inserted steel plates. Several connection variants with different distances between the fastener and the loaded end and with different moisture contents (square timber only) were chosen in order to verify if the real load-carrying capacity was influenced by distance. The specimens were subjected to tension parallel to the grain, until the connections failed.

2. Research Significance

Determining the load-carrying capacity of connections in timber structures and applying new knowledge to the European standard for timber structures design [27] are areas which are still under development. This research describes and compares the behavior of connections in squared and round timber, loaded in tension parallel to the grain, with different distances between the fastener and the loaded end and for different moisture contents.

Some studies from the scientific literature have investigated the design and behavior of dowel-type connections in timber structures, mainly based on experimental research. For example, important findings can be found in publications [4–6]. The relevance of this issue is also evidenced by the number of research papers written by authors mentioned in this paper.

The aim of this study is also to build on previous research activities conducted at the Faculty of Civil Engineering of the VSB Technical University of Ostrava by Klajmonova et al. A series of experiments was carried out to compare the load-carrying capacity of bolted connections with inserted steel plates, both for squared timber and round timber, but always for the same placement and number of fasteners [8–11]. Methods of effective reinforcement for bolted connections subjected to brittle failure were also designed and tested [12].

Some important new findings from our experimental testing have already been published and discussed at international conferences [37,38] and should be further extended.

Finally, the influence of moisture content on the behavior of a similar type of connection (bolt M12, glued laminated timber) loaded parallel to the grain has recently been investigated [13].

3. Design and Analysis of Timber Structure Connections in Europe

In most European countries, according to the current legislation, timber structures are designed and constructed based on Eurocode 5 [27]. Selected important knowledge about connections in timber structures is described in this chapter (for more detailed information, see the EN 1995-1 standard).

The geometric arrangement of fasteners in a connection (spacings, edge and end distances) has to meet minimum values to achieve the expected strength and stiffness. The most important value is the end distance from the loaded end (designation $a_{3,t}$ according to the standard). The minimum value depends on the fastener diameter d and is given according to Equation (1).

$$a_{3,t} = \max(7d; 80\ mm) \tag{1}$$

When determining the characteristic load-carrying capacity of connections with dowel-type fasteners made of metal, it is necessary to take into account the embedment strength of the timber, the yield strength of the fastener and less commonly the withdrawal strength of the fastener.

Connections are distinguished based on the type of materials to be connected. There are timber-to-timber, panel-to-timber and steel-to-timber connections.

For a double shear steel-to-timber connection with a steel plate of any thickness as the central member, the characteristic load-carrying capacity of one fastener for a single shear plane is determined as the minimum value according to Equation (2) for the failure modes shown in Figure 2 (failure mode *a* represents embedment of the timber, failure mode *b* represents bending of the fastener and failure mode *c* is the combination of both mentioned failure modes) or in [27]. The first member in the formula is the load-carrying capacity based on the Johansen yield theory [32] and the second member represents the so-called rope effect [39,40], which in Eurocode 5 has to be limited to 25% of the Johansen part for bolts. If the contribution from the rope effect is unknown, then it is taken as zero.

$$
F_{v,Rk} = \begin{cases}
f_{h,1,k} \cdot t_1 \cdot d & (a) \\
f_{h,1,k} \cdot t_1 \cdot d \cdot \left[\sqrt{2 + \dfrac{4 \cdot M_{y,Rk}}{f_{h,1,k} \cdot d \cdot t_1^2}} - 1 \right] + \dfrac{F_{ax,Rk}}{4} & (b) \\
2.3 \cdot \sqrt{M_{y,Rk} \cdot f_{h,1,k} \cdot d} + \dfrac{F_{ax,Rk}}{4} & (c)
\end{cases}
\tag{2}
$$

here,

$F_{v,Rk}$ is the characteristic load-carrying capacity for a single shear per fastener [N];
$f_{h,k}$ is the characteristic embedment strength in the timber member [N/mm^2];
t_1 is the smaller of the thicknesses of the timber side member [mm];
d is the diameter of the fastener [mm];
$M_{y,Rk}$ is the characteristic yield moment of the fastener [N/mm];
$F_{ax,Rk}$ is the characteristic withdrawal capacity of the fastener [N].

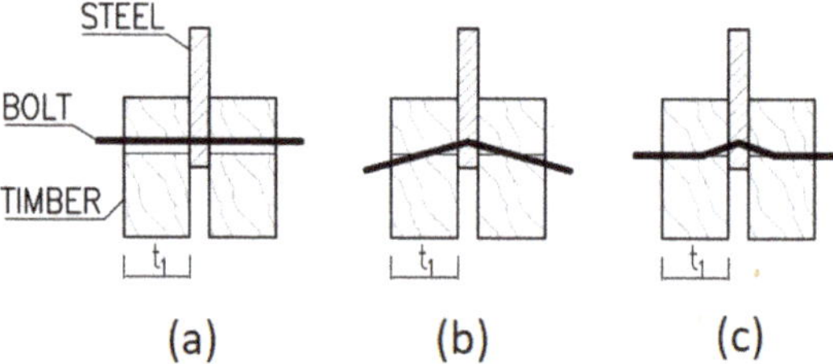

Figure 2. Failure modes for steel-to-timber connections: (**a**) Embedment of the timber; (**b**) Bending of the fastener; (**c**) Combination of both failures.

For bolts, the characteristic value of the yield moment depends on the bolt diameter d and the characteristic tensile strength $f_{u,k}$ and is determined according to Equation (3).

$$
M_{y,Rk} = 0.3 \cdot f_{u,k} \cdot d^{2.6}
\tag{3}
$$

For bolts, the characteristic embedment strength value in timber parallel to the grain depends on the bolt diameter d and the characteristic density of timber ρ_k and should be calculated according to Equation (4).

$$
f_{h,0,k} = 0.082 \cdot (1 - 0.01 \cdot d) \cdot \rho_k
\tag{4}
$$

Moisture content is given by the ratio of water weight to dry weight. When stored in air, European spruce timber dries to the so-called hygroscopic equilibrium moisture content (around 20%). The ideal moisture content for spruce timber used in exterior construction is considered to be 12% [2].

The effect of moisture on the mechanical properties of timber is significant. As the moisture content increases, the strength and stiffness of the timber decrease. The linear relationship between mechanical properties and moisture content can be considered to be in the range of 8% to 20% for moisture content [26].

When determining design values of mechanical properties according to [27], the influence of air humidity can be taken into account by assigning the structure to one of the three service classes, which affects the choice of the relevant modification factor k_{mod} and deformation factor k_{def}.

4. Methodology of Testing

4.1. Description of Materials for Test Specimens

For tension tests parallel to the grain, specimens were made of squared timber with cross-sectional dimensions of 60 × 120 mm and round timber with a diameter of 60 mm. All specimens consisted of two equal parts due to easier production of final specimens (placement of steel plates between individual parts). The specimens were made of spruce timber of the C24 strength class (determined on the basis of the standard [41]). Specimens of three different lengths were tested (see Figures 3 and 4): 400 mm (designation H1 and K1), 480 mm (designation H2 and K2) and 560 mm (designation H3 and K3). The circular holes for placing fasteners had a diameter of 20 mm.

High tensile bolts, grade 8.8 (f_y = 640 MPa, f_u = 800 MPa) with a diameter of 20 mm, were used as fasteners. The bolts were placed at three different distances from the loaded end: 140 mm (=7× bolt diameter), 180 mm (=9× bolt diameter) and 220 mm (=11× bolt diameter). The axial spacing between two bolts in one specimen was 120 mm.

The steel plates (see Figure 5) were made of structural-grade steel S235J0; their thickness was 10 mm. The dimensions of the steel plates also varied depending on the distance between the fastener and the loaded end. The protruding part of the plates needed for clamping specimens into the jaws of the testing machine was 180 mm long (see Figure 6). The plates were provided with circular holes with a diameter of 22 mm for placing bolts (a clearance of 2 mm was left).

For detailed specifications of dimensions, see Figures 3 and 5 and Table 1.

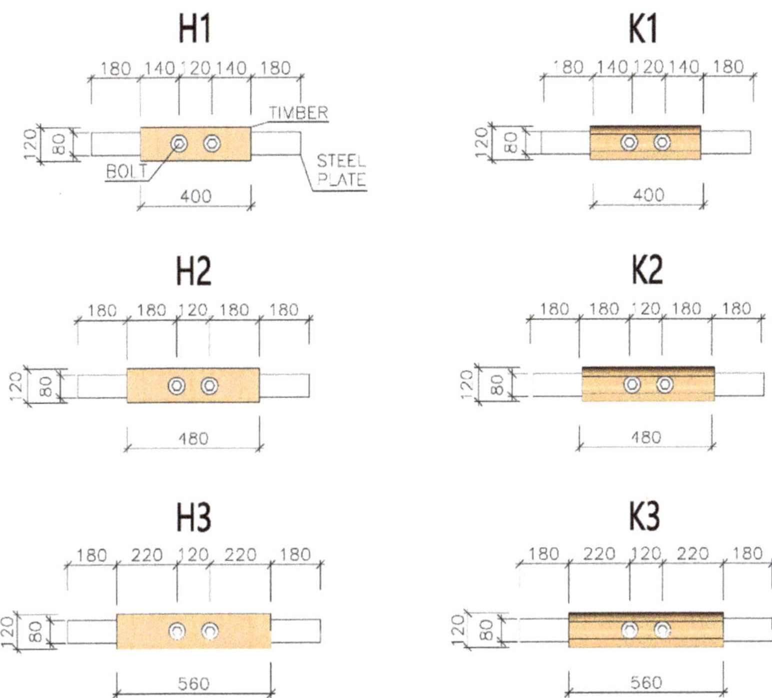

Figure 3. Types of specimens for tension tests parallel to the grain.

Figure 4. All types of squared specimens before testing.

P10 80x370 with a hole Ø22mm
for bolt M20 8.8

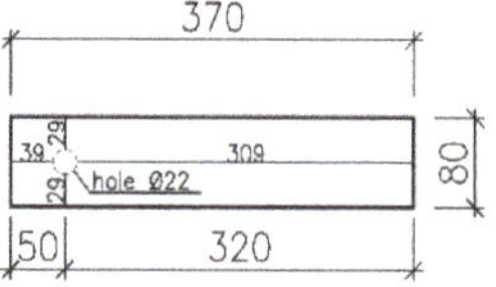

P10 80x410 with a hole Ø22mm
for bolt M20 8.8

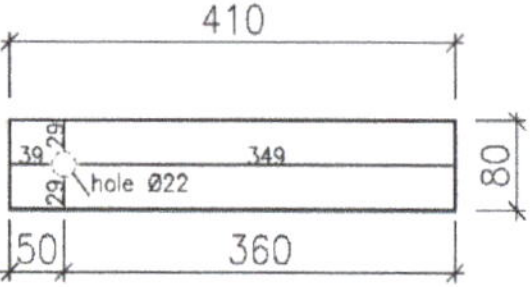

P10 80x450 with a hole Ø22mm
for bolt M20 8.8

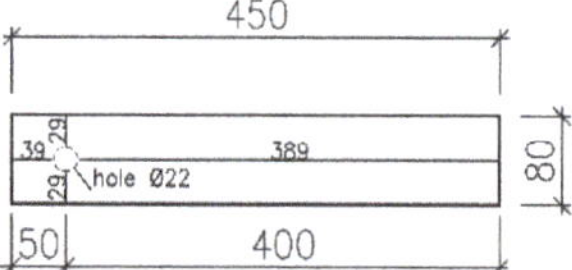

Figure 5. Dimensions of steel plates for tension tests parallel to the grain.

All timber specimens were produced and industrially manufactured at approximately equilibrium moisture content (about 12%). Some squared timber specimens were then moistened. The aim of this was not to achieve an accurate level of moisture in the tested timber, but to simulate adverse conditions that could occur during the service life of an actual supporting structure using these types of elements and connections. The specimens were exposed to changing atmospheric influences for six months (February–July) without sunlight. Some were protected against rainwater penetration inside (this corresponds to specimens with a moisture content between approximately 18% and 22%). Some were not protected (this corresponds to specimens with a moisture content of about 30% and higher).

Table 1. Dimensions of all specimen types for tension tests parallel to the grain.

Type of Specimen	H1	H2	H3	K1	K2	K3
Number of specimens [-]	15	13	13	5	5	5
Length of timber specimen [mm]	400	480	560	400	480	560
Height of timber specimen [mm]		2 × 60			2 × 60	
Width/diameter of timber specimen [mm]		120			ø120	
Loaded end distance [mm]	140	180	220	140	180	220
Length of steel plate [mm]	320	360	400	320	360	400
Thickness of steel plate [mm]				10		
Width of steel plate [mm]				80		
Diameter of fastener [mm]				20		

4.2. Description of the Testing Course

Several non-destructive tests were carried out before the main test to determine the quality of timber material, its moisture content and its density. The test specimens were weighed on a laboratory scale and their dimensions and moisture content were measured. Based on these data, the bulk density of the specimens and the density derived therefrom were determined according to [42,43]. Moisture content was measured using a capacitive material moisture meter according to [44].

The experiments were performed using the LabTest 6.1200 electromechanical testing machine (LABORTECH s.r.o., Opava, Czech Republic) with 1200 kN maximum electrohydraulic cylinder force (see Figure 6). The machine and testing procedure were controlled by computer software.

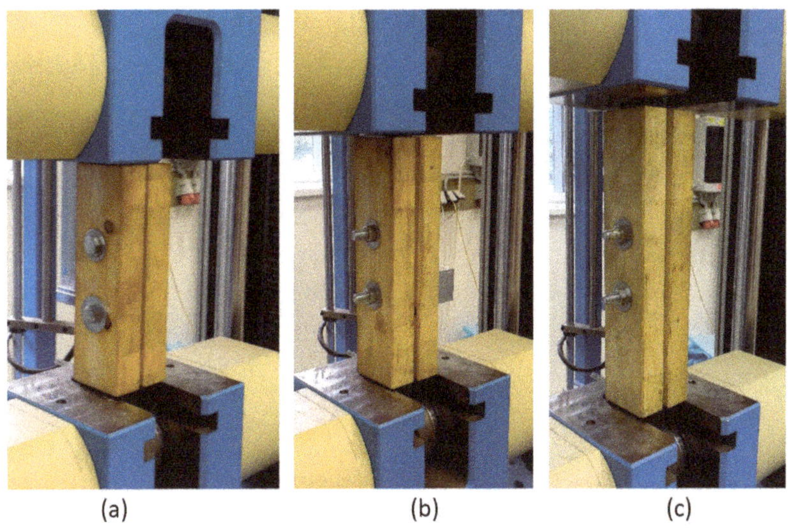

Figure 6. Specimens of squared timber clamped in the jaws of the testing machine: (**a**) H1 specimen–length 400 mm; (**b**) H2 specimen–length 480 mm; (**c**) H3 specimen–length 560 mm.

The specimens were clamped into jaws with special serrations for better adhesion, so the slip in the jaws was eliminated. The lower jaw was static, the upper jaw was movable. The tested connections were subjected to an axial tensile load with possible additional effects from imperfections for which no significant effect on the evaluated parameters was expected. Possible imperfections included only geometric inaccuracies in cutting and drilling of the timber specimens. The arrangement and loading of the test specimens were designed to correspond to the actual state of connections in real supporting structures. The aim was not to simulate ideal conditions but to verify the behavior of the connections under realistic conditions.

For this reason, the bolts were also tightened by hand, using a spanner without specific tightening force, to eliminate the rope effect [39]. The tensile load was generated by electrohydraulic cylinders with a capacity of 1200 kN. During the test, the time, tensile force and deformations of the connection in the longitudinal direction (i.e., cross-head displacement) were continuously recorded. The loading course (see Figure 7) was carried out in accordance with the standard for testing the connections in timber structures with mechanical fasteners [45]. The following loading procedure was prescribed:

1. estimate the maximum force F_{est}, for the tested connection based on experience, calculation or pre-tests;
2. load the specimen to 40% of the estimated maximum force, $0.4 \cdot F_{est}$, then hold for 30 s;
3. decrease the load to 10% of the estimated maximum force, $0.1 \cdot F_{est}$, then hold for 30 s;
4. continue loading until the specimen fails.

A constant loading speed was chosen (20 kN/min). The total testing time for one specimen was about 10 to 15 min until failure (see Figure 8).

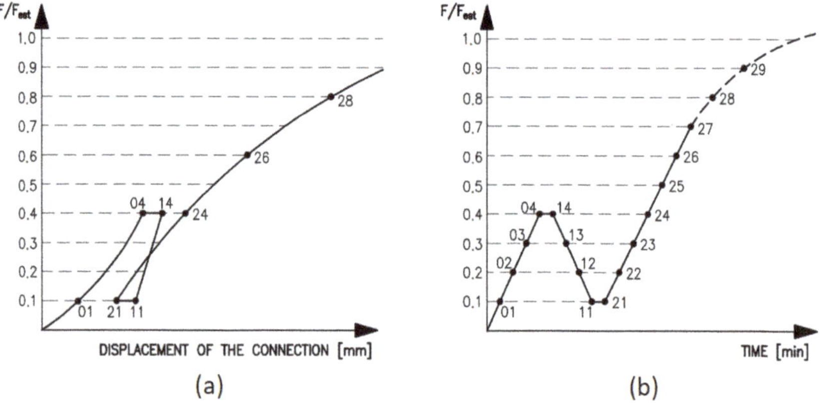

Figure 7. (**a**) Idealized load–displacement diagram and measured values; (**b**) The course of testing during time.

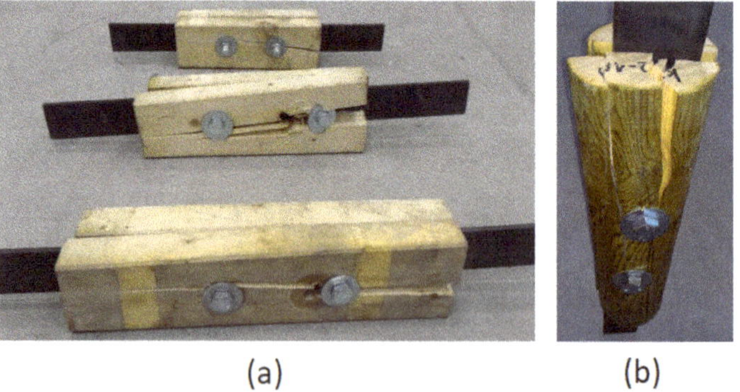

(a)   (b)

Figure 8. Specimens after failure: (**a**) Squared timber; (**b**) Round timber.

4.3. Evaluation of the Testing

The measured displacements v_{01}, v_{04}, v_{14}, v_{11}, v_{21}, v_{24}, v_{26}, v_{28} and the displacements at the maximum load were recorded. Based on the recorded data, it is possible to determine the initial displacement $v_i = v_{04}$, the modified initial displacement (Equation (5)), the permanent initial displacement (Equation (6)) derived therefrom and the initial slip modulus (Equation (7)) and the slip modulus (Equation (8)) derived therefrom [45].

$$v_{i,mod} = \frac{4}{3} \cdot (v_{04} - v_{01}) \tag{5}$$

$$v_s = v_i - v_{i,mod} \tag{6}$$

$$k_i = 0.4 \cdot \frac{F_{est}}{v_i} \tag{7}$$

$$k_s = 0.4 \cdot \frac{F_{est}}{v_{i,mod}} \tag{8}$$

5. Results

For this work, six types of bolted connections in squared timber and round timber with inserted steel plates were tested. Only squared timber specimens were tested with different levels of moisture content. In total, fifty-six specimens were tested.

Figures 9–14 show the load–deformation curves of selected H1, H2, H3, K1, K2 and K3 specimens. In the graphs the initial loading at 40% of the estimated maximum force, the subsequent unloading to 10% of the estimated maximum force and the loading to failure can be seen. It is also possible to see the actual maximum force and the corresponding deformation at failure. The legends in the graphs for squared timber also contain information about the moisture content of the individual specimens.

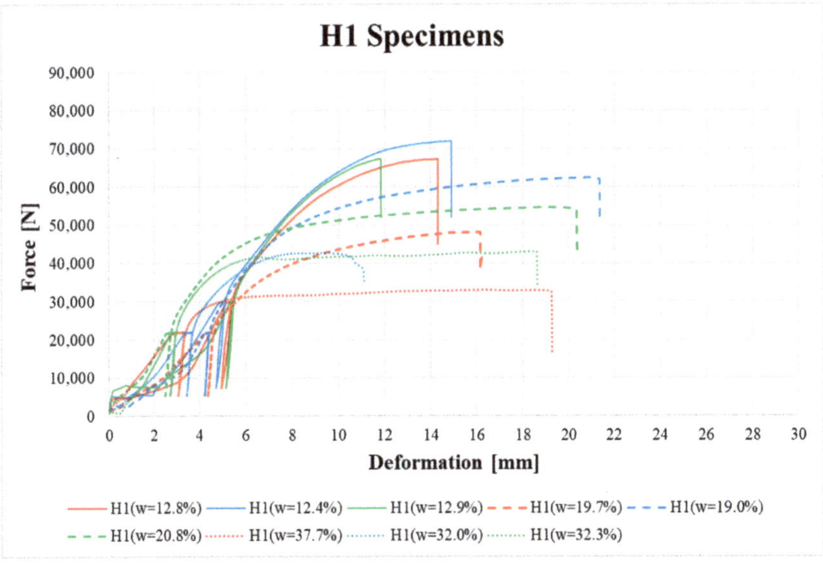

Figure 9. Load–deformation curves of H1 specimens.

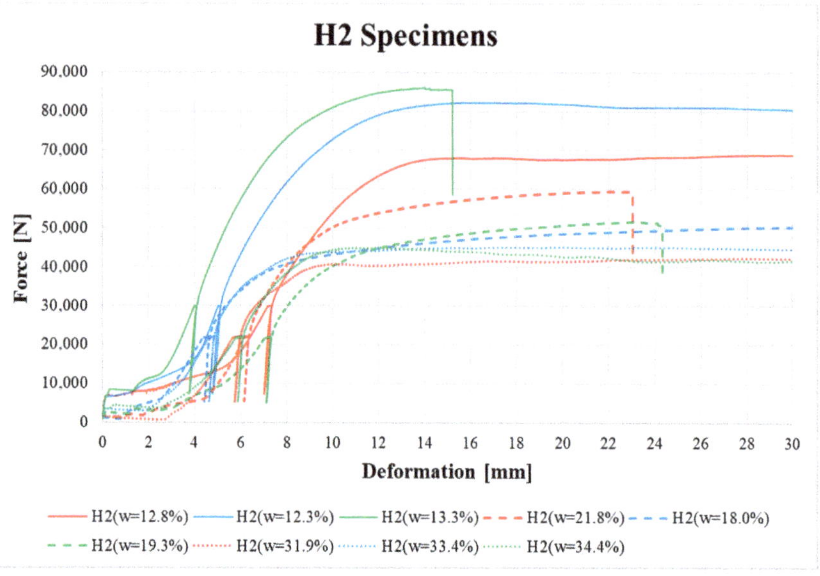

Figure 10. Load–deformation curves of H2 specimens.

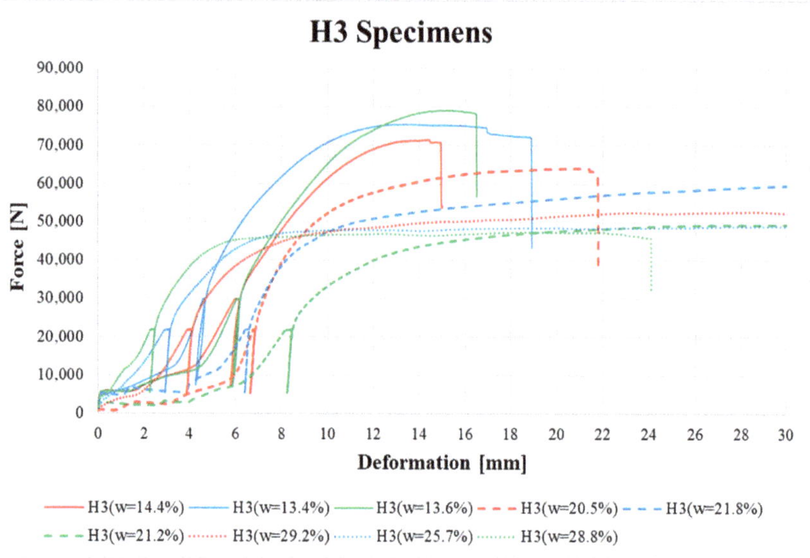

Figure 11. Load–deformation curves of H3 specimens.

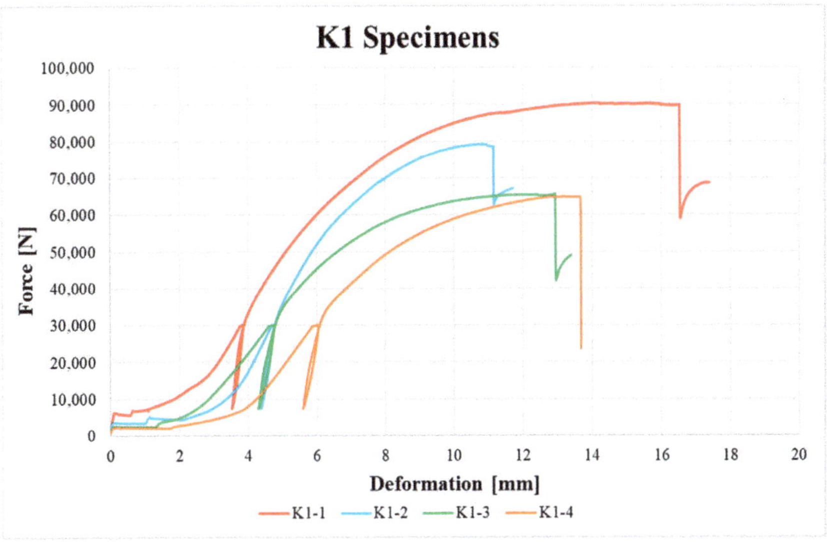

Figure 12. Load–deformation curves of K1 specimens.

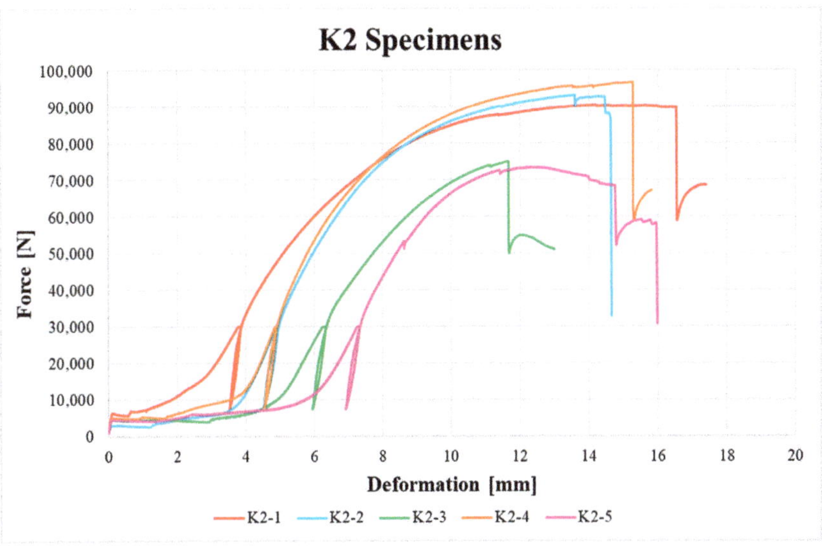

Figure 13. Load–deformation curves of K2 specimens.

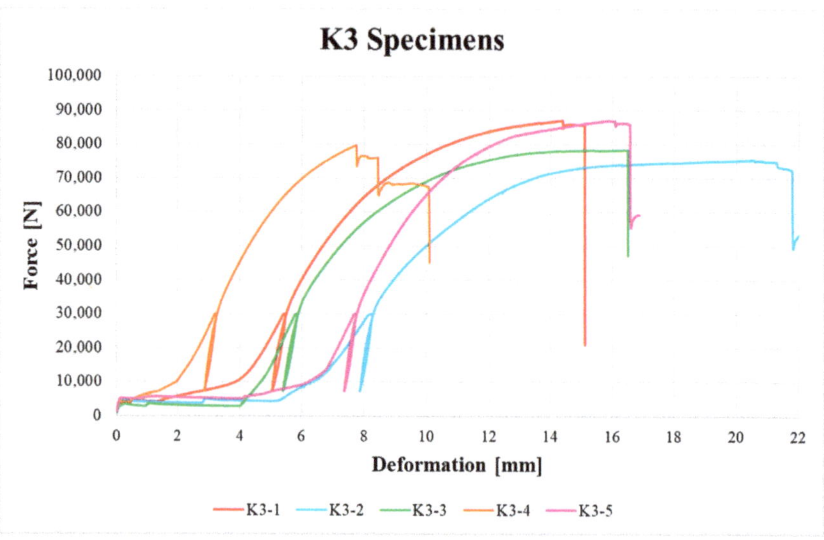

Figure 14. Load–deformation curves of K3 specimens.

Figures 15 and 16 show the probability density function and the variance in load-carrying capacity of squared timber and round timber specimens with an equilibrium moisture content of about 12%. The statistical evaluation uses a normal distribution with input parameters, which were evaluated from the results of the experiment (see the average value and standard deviation in Table 2).

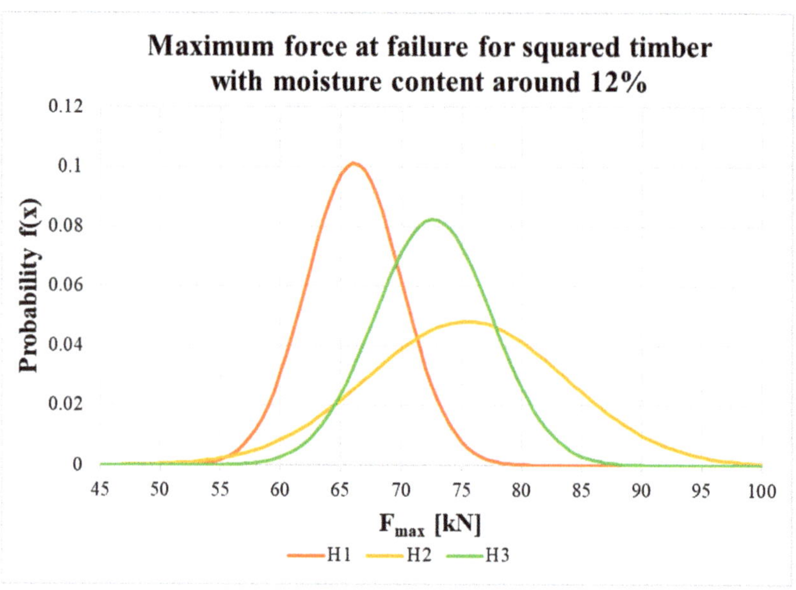

Figure 15. Probability density function of maximum force at failure for H1, H2 and H3.

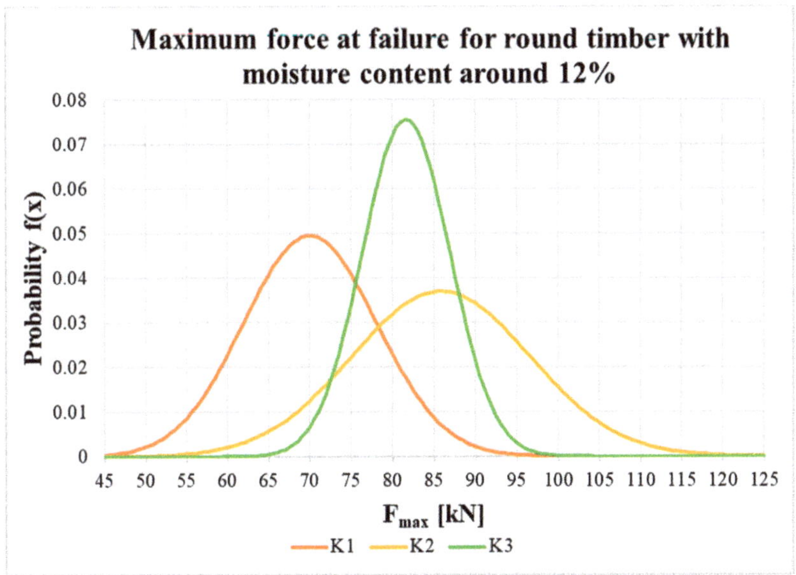

Figure 16. Probability density function of maximum force at failure for K1, K2 and K3.

Table 2 shows statistically evaluated values of bulk density and moisture content for the timber material (the values given are for timber specimens during their manufacturing), the average maximum load-carrying capacity with standard deviation, the coefficient of variation and characteristic value based on testing, the calculated characteristic load-carrying capacity without contribution from the rope effect according to Eurocode 5 [27] (see the calculation below) and the values of the initial slip modulus and the slip moduli according to the above-mentioned formulas for all types (H1, H2, H3, K1, K2 and K3). The results were not normalized for the density data due to low variability in density. A normal distribution of test values was assumed in the statistical evaluation in accordance with [42,43].

Table 2. Values of bulk density, moisture content, maximum force, characteristic force and slip moduli for all types of specimens.

Comparison	H1	H2	H3	K1	K2	K3
ρ (AVG) [kg/m^3]	417	411	414	458	455	431
ρ (SD) [kg/m^3]	13.4	20.0	20.6	34.7	45.1	34.3
ρ (COV) [%]	3.2	4.9	5.0	7.8	9.9	8.1
w (AVG) [%]	12.8	12.6	13.0	13.2	12.8	12.8
w (SD) [%]	0.2	0.6	0.8	0.6	0.7	0.4
$F_{max,test}$ (AVG) [kN]	66.10	75.43	72.63	69.98	85.81	81.62
$F_{max,test}$ (SD) [kN]	3.94	8.26	4.84	8.04	10.79	5.28
$F_{max,test}$ (COV) [%]	6.0	11.0	6.7	11.5	12.5	6.5
$F_{k,test}$ [kN]	56.41	55.10	60.72	50.20	59.39	68.65
$F_{k,EC5}$ [kN]			46.53			
k_i (AVG) [kN/mm]	5852	5931	5530	5889	5804	5519
k_s (AVG) [kN/mm]	6817	5644	6446	11,298	11,342	11,129

here,

ρ is the bulk density of timber specimens;
w is the moisture content of timber specimens;
$F_{max,test}$ is the maximum force at failure based on test data (average value and standard deviation);
$F_{k,test}$ is the characteristic force at failure based on test data (determined according to [42,43]);
$F_{k,EC5}$ is the characteristic load-carrying capacity according to the standard [27];
k_i is the initial slip modulus of connections based on test data;
k_s is the slip modulus of connections based on test data.

In accordance with Eurocode 5 [27], a double shear steel-to-timber connection was considered (steel plate of any thickness as the central member). The characteristic value of the load-carrying capacity, without the contribution from the rope effect for the given type of connection, according to the above-mentioned standard is $F_{v,Rk} = 46{,}528$ N (twice 23,264 N). In order to calculate the load-carrying capacity, the necessary input values for C24 strength timber (determined on the basis of the standard [42]) and M20 grade 8.8 bolts were used. The calculation according to Eurocode 5 is given below in Table 3, where the failure mode *b* is decisive.

Table 3. Calculation of characteristic load-carrying capacity according to Eurocode 5.

Calculation according to Eurocode 5		
Fastener diameter [mm]		20
Characteristic fastener tensile strength [N/mm^2]		800
Characteristic fastener yield moment [N/mm^2]		579,281
Thickness of the timber side member [mm]		60
Characteristic timber density [kg/m^3]		350
Characteristic embedment strength in timber [N/mm^2]		22.96
Characteristic load-carrying capacity in single shear [N]	failure mode *a*	27,552
	failure mode *b*	23,264
	failure mode *c*	37,512

To obtain the design values, it is necessary to recalculate the characteristic value using the modified factor for the duration of load and moisture content k_{mod} ($k_{mod} = 1.1$ for the 1st/2nd service class and $k_{mod} = 0.9$ for the 3rd service class, both for instantaneous load duration) and the partial factor for the material properties γ_M ($\gamma_M = 1.3$ for connections).

The effect of moisture content on the real load-carrying capacity of the connection is shown in Figures 17–19. The obtained data are complemented with regression line equations and correlation coefficients. The graphs are also complemented with the characteristic load-carrying capacity F_k and design load-carrying capacities for 1st/2nd service class F_d *SC1/SC2* (design values for both classes were the same) and 3rd service class F_d *SC3* for instantaneous load duration based on calculation according to the standard (see the calculation below).

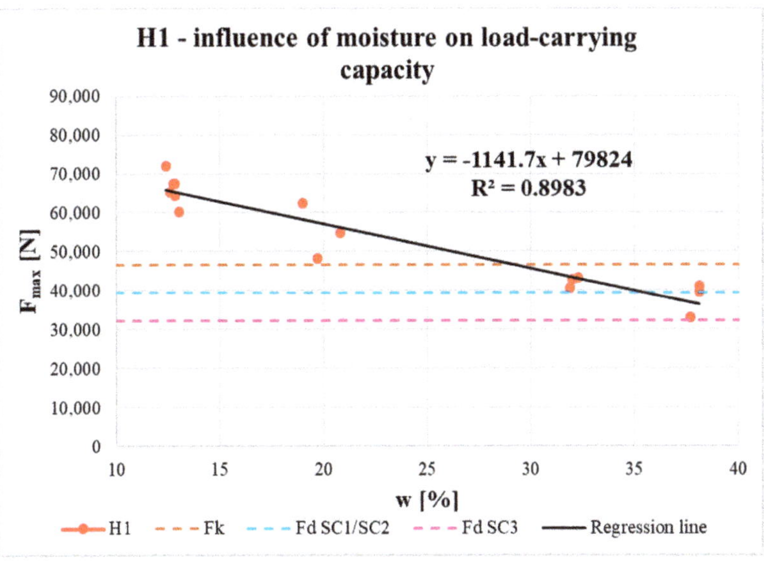

Figure 17. Influence of moisture content on load-carrying capacity of H1 specimens.

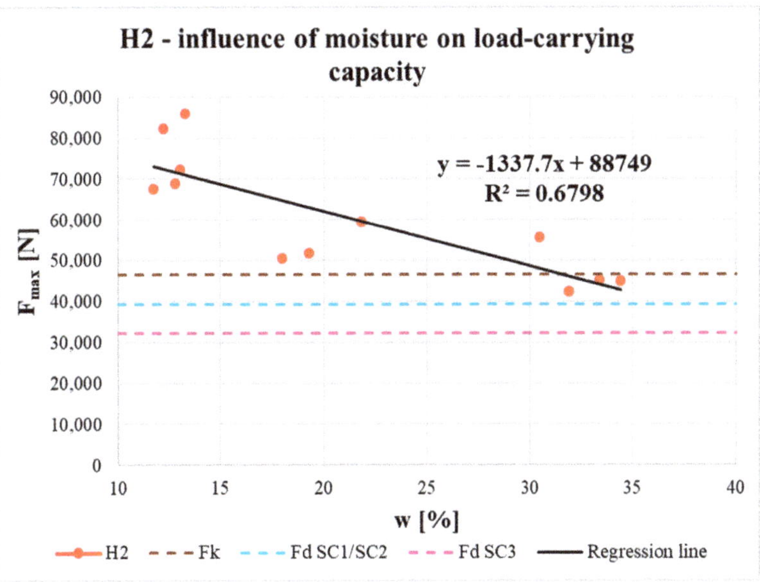

Figure 18. Influence of moisture content on load-carrying capacity of H2 specimens.

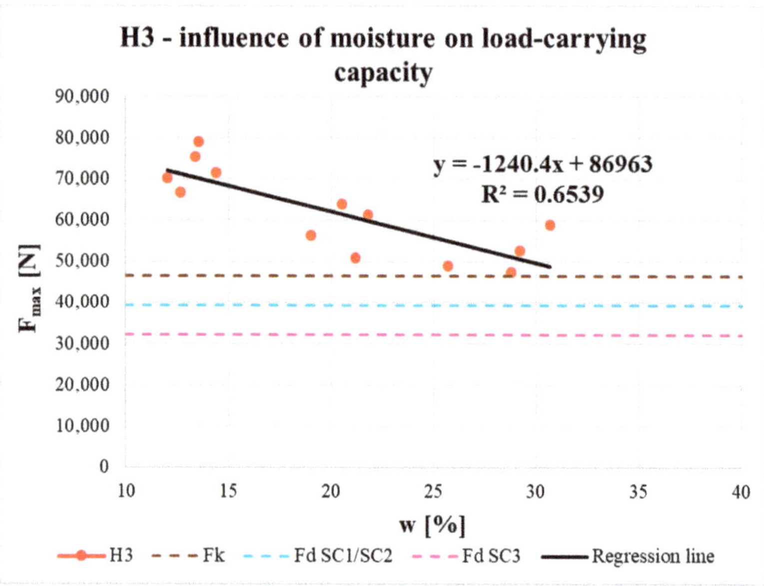

Figure 19. Influence of moisture content on load-carrying capacity of H3 specimens.

Figure 20 shows the load-carrying capacities of all three types of connections in squared timber at different moisture content levels, which were obtained by putting those levels into the regression line from Figures 17–19.

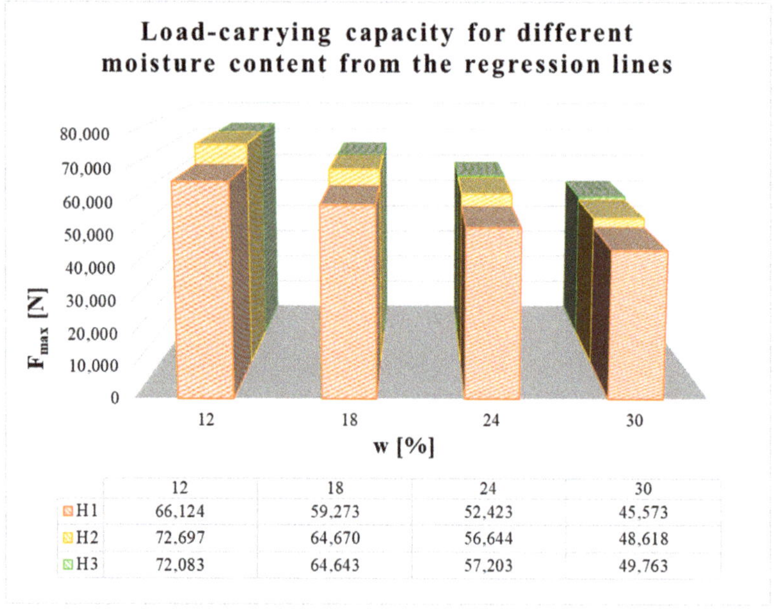

	12	18	24	30
H1	66,124	59,273	52,423	45,573
H2	72,697	64,670	56,644	48,618
H3	72,083	64,643	57,203	49,763

Figure 20. Load-carrying capacity for different moisture content from the regression lines.

6. Discussion

From the initial loading, deformation of holes in timber members occurred until the specimens failed due to tension perpendicular to the grain. Failure always occurred by timber splitting under the bolt, when the tensile strength perpendicular to the grain was exceeded and cross-links between individual fibers in timber were broken (see Figure 8). This phenomenon can be attributed to the fact that timber under tension has little plasticity and is broken by brittle fractures. The failure in most specimens occurred after a visible embedment (ductile failure usually preceded brittle failure). The splitting crack spread very quickly in one half of a timber specimen (the other half remained only embedded) under one of the bolts and in some cases also between the bolts. Brittle failure indicated that higher capacities could be obtained and were not limited to the ductile failure modes from Eurocode 5 (depending on the embedment strength of the timber and yield moment of the fastener, see Section 3).

The maximum force values at specimen failure with moisture content around 12% (see Table 2) showed that, at the shortest distance between the fastener and loaded end (i.e., $7d$), the average values were lowest for both the squared timber and round timber. For specimens with longer distances (i.e., $9d$ and $11d$), the average load-carrying capacity was higher, and the highest value was surprisingly for middle-length specimens (H2 and K2 specimens). Higher load-carrying capacities were measured for round timber specimens than for squared specimens, probably due to the higher average bulk density of the timber (higher by between 4.1% and 11.8%). The results also show that there was no increase in real load-carrying capacity above the end distance $9d$, so the use of a higher end distance has no practical significance.

However, in construction, useful values are so called characteristic values, evaluated as the 5% quantile of the approximated probability distribution of the measured values. Characteristic values are significantly influenced by the number of specimens (reduction in the number of test specimens, $k_s(n)$, according to [43]) and variability in the measured values (sample standard deviation). The characteristic load-carrying capacities for all types of specimens were higher than the standard value for the double shear steel-to-timber connection (steel plate of any thickness as the central member). The highest values were found for H3 and K3 specimens (the longest distance of the bolt from the loaded end); for H3 specimens it was higher by 30.5% and for K3 specimens it was higher by 47.5%. Nevertheless, it is not possible to conclude there was a significant influence exerted by the distance between the fastener and the loaded end on the characteristic load-carrying capacity of the connection. This is complicated by the small number of tested specimens, which considerably affects the statistical evaluation of specimens with higher variance (especially specimens with coefficients of variation higher than 10%).

In addition, considerable variability can be seen in the slip moduli (see Table 2). The slip modulus k_s was higher for round timber than for squared timber (almost twice as high), which is probably, again, due to the higher bulk density of round timber.

It is known that with increasing moisture content in timber elements (related to service class assignment), the load-carrying capacity of elements, and their connections, is reduced by the modification factor k_{mod}. This decrease in real load-carrying capacity with increasing moisture content was also confirmed experimentally. The measured data could be approximated by a straight line using regression analysis (see Figures 17–19). Correlation coefficients were evaluated (R = −0.95 for H1 specimens, R = −0.82 for H2 specimens, R = −0.81 for H3 specimens) and they indicated significant correlation between the examined parameters. The slopes of all the regression lines were quite similar, which indicates a similar effect of moisture content on load-carrying capacity for all types of specimens. The small differences were mainly due to the end distances.

Regression lines were then used to determine the load-carrying capacities for different moisture contents. The results show that at the shortest distance between the fastener and loaded end (i.e., $7d$) the load-carrying capacity was lowest. For specimens with longer distances (i.e., $9d$ and $11d$), the average load-carrying capacities were higher and the values for both distances were almost comparable. It can therefore be observed that with increasing distance between the fastener and the loaded end, there was a slight increase in the real load-carrying capacity (from the minimum standard value of $7d$ to the value of $9d$). As the distance increased further, the increase in load-carrying capacity was practically negligible (from $9d$ to $11d$).

To assign the structure to service class 1, the moisture content of the built-in timber elements needs to be around the equilibrium moisture content of 12% (only for interiors). To assign the structure to service class 2, this moisture content must not exceed 20%. In cases where the moisture content is more than 20%, it is necessary to assign the structure to service class 3 [27]. Figures 17–19 show that for all types of specimens with moisture contents around 12% (corresponding to service class 1) and 20% (corresponding to service class 2), the real load-carrying capacities were higher than the design value for service classes 1 or 2, even higher than the characteristic value according to the standard [27]. Furthermore, for specimens with the highest moisture content (corresponding to service class 3), the real load-carrying capacities were higher than the design value for service class 3 according to the standard [27]. For H2 specimens, the real load-carrying capacities were higher than the design value for service classes 1 or 2 and for H3 specimens they were even higher than the characteristic value. The values of real load-carrying capacities for H1 specimens with the highest moisture content point to the importance of carefully designing timber structures, especially exterior details (connections), because of the significant influence from atmospheric conditions, which can lead to rainwater flowing into improperly designed details and, subsequently, significantly decrease the real load-capacity capacity to the limit according to the standard [27].

With increasing moisture content in the squared timber specimens, a slight increase in the slip modulus can be observed, which appears to be due to the swelling of timber when it is saturated with water. However, no significant correlation was found between the slip modulus and moisture content dependence.

According to Figures 10 and 11, specimens H2 and H3 also showed higher ductility in the final phase of loading before total brittle failure compared to specimen H1.

7. Conclusions

Experimental testing is the most appropriate way to verify the load-carrying capacity and stiffness of timber structures. Therefore, tests on bolted connections in squared and round timber with inserted steel plates, with different distances between the fastener and the loaded end, were performed.

Connections with inserted steel plates are also widely used in exterior timber structures (e.g., lookout towers [1], bridges [2], etc.) that are exposed to significant temperature and moisture content changes. Therefore, the effect of increased moisture content in timber on the load-carrying capacity of the connections was also analyzed in the experiment.

The following partial conclusions can be drawn:

- Connection failure always occurred by timber splitting under the bolt, when the tensile strength perpendicular to the grain was exceeded.
- The average load-carrying capacity was lowest for the shortest fastener distance for both the squared timber and round timber. For specimens with higher distances, the average load-carrying capacity was higher.
- Considerable variability in the slip modulus can be seen in the evaluation. The slip modulus is higher for the round timber than for the squared timber, which is probably due to the higher bulk density of round timber.
- The decrease in real load-carrying capacity with increasing moisture content was also confirmed experimentally. The measured data could be approximated by a straight line using regression analysis.

The limitations of this research include the use of only one fastener in a row (bolt M12 8.8) and only one species of timber (spruce, C24 strength class), the limited number of test specimens (which affects the statistical evaluation) and the monotonic static testing.

Exterior structures are also predominantly loaded by wind, which has the characteristics of an alternating dynamic load [46]. On the basis of these facts, future research on this issue should incorporate experiments with cyclic and dynamic tests [8,47].

Author Contributions: Conceptualization, P.D. and A.L.; Data curation, P.D.; Formal analysis, P.D.; Funding acquisition, P.D.; Investigation, P.D. and A.L.; Methodology, P.D., O.S. and A.L.; Supervision, A.L.; Validation, P.D. and A.L.; Visualization, O.S.; Writing—original draft, P.D. All authors have read and agreed to the published version of the manuscript.

Funding: This paper was financially supported by the Ministry of Education, specifically by the Student Research Grant Competition of the Technical University of Ostrava in 2020 under identification number SP2020/138.

Conflicts of Interest: The authors confirm that there are no known conflict of interest associated with this paper.

References

1. Šmak, M.; Barnat, J.; Straka, B.; Kotásková, P.; Havířová, Z. Dowelled joints in timber structures experiment-design-realization. *Wood Res.* **2016**, *61*, 651–662.
2. Fojtik, R.; Lokaj, A.; Gabriel, J. *Wooden Bridges and Footbridges*; Czech Chamber of Authorized Engineers and Technicians in Construction (CKAIT): Prague, Czech Republic, 2017; p. 158. ISBN 978-80-88265-04-7.
3. Bergman, R.; Puettmann, M.; Taylor, A.; Skog, K.E. The carbon impacts of wood products. *For. Prod. J.* **2014**, *64*, 220–231. [CrossRef]
4. Blass, H.J.; Sandhaas, C. *Timber Engineering—Principles for Design*; Karlsruher Institut für Technologie: Karlsruhe, Czech Republic, 2017; ISBN 978-3-7315-0673-7.
5. Borgström, E.; Karlsson, R. *Design of Timber Engineering—1*; Swedish Wood: Stockholm, Sweden, 2016.
6. Sandhaas, C.; Munch-Andersen, J.; Dietsch, P. *Design of Connections in Timber Structures*; Shaker-Verlag GmbH: Düren, Germany, 2018; ISBN 978-3-8440-6144-4.
7. Que, Z.; Hou, T.; Gao, Y.; Teng, Q.; Chen, Q.; Wang, C.; Chang, C. Influence of Different Connection Types on Mechanical Behavior of Girder Trusses. *J. Bioresour. Bioprod.* **2019**, *4*, 89–98.
8. Lokaj, A.; Klajmonova, K. Carrying capacity of round timber bolted joints with steel plates under cyclic loading. *Adv. Mater. Res.* **2013**, *838–841*, 634–638. [CrossRef]
9. Lokaj, A.; Klajmonova, K. Round timber bolted joints exposed to static and dynamic loading. *Wood Res.* **2014**, *59*, 439–448.
10. Lokaj, A.; Klajmonova, K.; Mikolasek, D.; Vavrusova, K. Behavior of round timber bolted joints under tension load. *Wood Res.* **2016**, *61*, 819–830.
11. Lokaj, A.; Klajmonova, K. Comparison of behaviour of laterally loaded round and squared timber bolted joints. *Frat. ed Integrita Strutt.* **2017**, *11*, 56–61. [CrossRef]
12. Lokaj, A.; Klajmonova, K. Round timber bolted joints reinforced with self-drilling screws. *Procedia Eng.* **2015**, *114*, 263–270.
13. Kiwelu, H.M. Experimental study on the effect of moisture on bolt embedment and connection loaded parallel to grain for timber structures. *Tanzan. J. Eng. Technol.* **2019**, *38*, 47–59.
14. Blass, H.J.; Schädle, P. Ductility aspects of reinforced and non-reinforced joints. *Eng. Struct.* **2011**, *33*, 3018–3026. [CrossRef]
15. Sandhaas, C. Mechanical Behaviour of Timber Joints with Slotted-in Steel Plates. Ph.D. Thesis, TU Delft, Delft, The Netherlands, 2012.
16. Cabrero, J.M.; Yurita, M.; Quenneville, P. Brittle failure in the parallel-to-grain direction of multiple shear softwood timber connections with slotted-in steel plates and dowel-type fasteners. *Constr. Build. Mater.* **2019**, *216*, 296–313.
17. Sawata, K. Strength of bolted timber joints subjected to lateral force. *J. Wood Sci.* **2015**, *61*, 221–229. [CrossRef]
18. Glisovic, I.; Stevanovic, B.; Kocetov-Misulic, T. Embedment test of wood for dowel-type fasteners. *Wood Res.* **2012**, *57*, 639–650.
19. Jockwer, R.; Fink, G.; Kohler, J. Assessment of the failure behaviour and reliability of timber connections with multiple dowel-type fasteners. *Eng. Struct.* **2018**, *172*, 76–84. [CrossRef]
20. Rahim, N.L.; Raftery, G.M.; Quenneville, P. Stiffness of bolted timber connection. In Proceedings of the World Conference on Timber Engineering, Seoul, Korea, 20–23 August 2018.

21. Awaludin, A.; Saputro, D.N. Bolt spacing and end distance of bolted connection of laminated veneer lumber (LVL) sengon. *Civ. Eng. Dimens.* **2017**, *19*, 1–6.

22. Salenikovich, A.; Legras, B.; Mohammad, M.; Quenneville, P. Effect of moisture on the performance of bolted connections in timber structures. In Proceedings of the 11th Conference on Timber Engineering, Trentino, Italy, 20–24 June 2010; Volume 2, pp. 960–969.

23. Carll, C. Moisture durability for wood products. *VTT Symp.* **2010**, *263*, 142–147.

24. Roszyk, E.; Stachowska, E.; Majka, J.; Mania, P.; Broda, M. Moisture-dependent strength properties of thermally-modified fraxinus excelsior wood in compression. *Materials* **2020**, *13*, 1647. [CrossRef]

25. Sjödin, J.; Johansson, C.J. Influence of initial moisture induced stresses in multiple steel-to-timber dowel joints. *Holz Roh Werkst.* **2007**, *65*, 71–77. [CrossRef]

26. Rammer, D.R.; Winistorfer, S.G. Effect of moisture content on dowel-bearing strength. *Wood Fiber Sci.* **2001**, *33*, 126–139.

27. EN 1995-1-1. *Eurocode 5: Design of Timber Structures—Part 1-1: General—Common Rules and Rules for Buildings*; Czech Standards Institute: Praha, Czech Republic, 2006.

28. Zhu, E.; Pan, J.; Zhou, X.; Zhou, H. Experiments of load-carrying capacity of bolted connections in timber structures and determination of design value. *J. Build. Struct.* **2016**, *37*, 54–63.

29. Jorissen, A.J.M. Double Shear Timber Connections with Dowel Type Fasteners. Ph.D. Thesis, TU Delft, Delft, The Netherlands, 1998.

30. Yurrita, M.; Cabrero, J.M. New design model for splitting in timber connections with one row of fasteners loaded in the parallel-to-grain direction. *Eng. Struct.* **2020**, *223*, 111155. [CrossRef]

31. Quenneville, P.; Mohammad, M. On the failure modes and strength of steel-wood-steel bolted timber connections loaded parallel-to-grain. *Can. J. Civ. Eng.* **2011**, *27*, 761–773. [CrossRef]

32. Johansen, K.W. Theory of timber connections. *Int. Assoc. Bridg. Struct. Eng.* **1949**, *9*, 249–262.

33. Jorissen, A.J.M.; Fragiacomo, M. General notes on ductility in timber structures. *Eng. Struct.* **2011**, *33*, 2987–2997. [CrossRef]

34. Cabrero, J.M.; Yurita, M. Performance assessment of existing models to predict brittle failure modes of steel-to-timber connections loaded parallel-to-grain with dowel-type fasteners. *Eng. Struct.* **2018**, *171*, 895–910. [CrossRef]

35. Vavrusova, K.; Mikolasek, D.; Lokaj, A.; Klajmonova, K.; Sucharda, O. Determination of carrying capacity of steel-timber joints with steel rods glued-in parallel to grain. *Wood Res.* **2016**, *61*, 733–740.

36. Franke, B.; Quenneville, P. Prediction of the load capacity of dowel-type connections loaded perpendicular to grain for solid wood and wood products. In Proceedings of the World Conference on Timber Engineering, Auckland, New Zealand, 15–19 July 2012.

37. Dobes, P.; Lokaj, A.; Sucharda, O. Test results of connections of timber structures. In Proceedings of the 204th IASTEM International Conference, Amsterdam, The Netherlands, 10 September 2019; Volume 204, pp. 1–4.

38. Dobes, P.; Lokaj, A.; Sucharda, O. Load-carrying capacity of bolted connections of round timber with different distances between the fastener and the loaded end. In Proceedings of the 3rd International Conference on Sustainable Development in Civil, Urban and Transportation Engineering, Ostrava, Czech Republic, 21–23 October 2020; Volume 3, pp. 109–112.

39. Awaludin, A.; Hirai, T.; Hayashikawa, T.; Sasaki, Y. Load-carrying capacity of steel-to-timber joints with a pretensioned Bolt. *J. Wood Sci.* **2008**, *54*, 362–368. [CrossRef]

40. Chen, J.; Wang, H.; Yu, Y.; Liu, Y.; Jiang, D. Loosening of bolted connections under transverse loading in timber structures. *Forests* **2020**, *11*, 816. [CrossRef]

41. EN 73 2824-1. Strength grading of wood—Part 1: Coniferous sawn timber. In *Metrology and Testing*; Czech Office for Standards: Praha, Czech Republic, 2015.

42. EN 384+A1. *Structural Timber—Determination of Characteristic Values of Mechanical Properties and Density*; Czech Office for Standards, Metrology and Testing: Praha, Czech Republic, 2019.

43. EN 14358. *Timber Structures—Calculation of Characteristic 5—Percentile Values and Acceptance Criteria for a Sample*; Czech Office for Standards, Metrology and Testing: Praha, Czech Republic, 2017.

44. EN 13183-2. *Moisture Content of a Piece of Sawn Timber—Part 2: Estimation by Electrical Resistance Method*; Czech Office for Standards, Metrology and Testing: Praha, Czech Republic, 2002.

45. EN 26891. *Timber Structures. Joints Made with Mechanical Fasteners. General Principles for the Determination of Strength and Deformation Characteristics*; Czech Office for Standards, Metrology and Testing: Praha, Czech Republic, 1994.

46. Solarino, F.; Giresini, L.; Chang, W.S.; Huang, H. Experimental Tests on a Dowel-Type Timber Connection and Validation of Numerical Models. *Buildings* **2017**, *7*, 116. [CrossRef]

47. Piazza, M.; Polastri, A.; Tomasi, R. Ductility of timber joints under static and cyclic loads. *Proc. Inst. Civ. Eng.-Struct. Build.* **2011**, *164*, 79–90. [CrossRef]

Publisher's Note: MDPI stays neutral with regard to jurisdictional claims in published maps and institutional affiliations.

 *materials*

Article

Dynamic Testing of Lime-Tree (*Tilia Europoea*) and Pine (*Pinaceae*) for Wood Model Identification

Anatoly Bragov [1], Leonid Igumnov [1,2,*], Francesco dell'Isola [1], Alexander Konstantinov [1,2], Andrey Lomunov [1] and Tatiana Iuzhina [1]

[1]   Research Institute for Mechanics, National Research Lobachevsky State University of Nizhny Novgorod, 603950 Nizhny Novgorod, Russia; bragov@mech.unn.ru (A.B.); fdellisola@gmail.com (F.d.); constantinov.al@yandex.ru (A.K.); lomunov@mech.unn.ru (A.L.); yuzhina_tatiana@mech.unn.ru (T.I.)

[2]   Research and Education Mathematical Center "Mathematics for Future Technologies", 603950 Nizhny Novgorod, Russia

*   Correspondence: igumnov@mech.unn.ru; Tel.: +7-910-792-7826

Received: 22 October 2020; Accepted: 19 November 2020; Published: 20 November 2020

Abstract: The paper presents the results of dynamic testing of two wood species: lime-tree (*Tilia europoea*) and pine (*Pinaceae*). The dynamic compressive tests were carried out using the traditional Kolsky method in compression tests. The Kolsky method was modified for testing the specimen in a rigid limiting holder. In the first case, stress–strain diagrams for uniaxial stress state were obtained, while in the second, for uniaxial deformation. To create the load a gas gun was used. According to the results of the experiments, dynamic stress–strain diagrams were obtained. The limiting strength and deformation characteristics were determined. The fracture energy of lime and pine depending on the type of test was also obtained. The strain rates and stress growth rates were determined. The influence of the cutting angle of the specimens relative to the grain was noted. Based on the results obtained, the necessary parameters of the wood model were determined and their adequacy was assessed by using a special verification experiment.

Keywords: timber; natural composite; Kolsky method; deformation diagrams; wood species; energy absorption; wood model; verification

1. Introduction

Wood is a complex natural composite. It is widely used as a material for damping intensive dynamic loads of shock or explosive nature. In order to conduct reliable numerical analysis of the designs of containers for transporting hazardous substances, using wood as shock damping components, reliable mathematical models are required that take into account its complex multicomponent structure of wood. The actively developing models of wood deformation and destruction can be used in various design complexes to simulate the behavior of technically complex structures that incorporate wood elements. However, the complex nature of wood means that not a single model can be used for all purposes, thus different models must be used to solve various specific problems. In order to reliably describe the behavior of wood in a dynamically loaded structure, its model should include a sufficiently large set of parameters that take into account the deformation anisotropy and the dependence of the strength properties on the strain rate, density, temperature and moisture content. For example, in the manual for the LS-DYNA software package, the wood model No. 143 has 29 parameters: five modules for transversely isotropic constitutive equations, six tensile strengths for yield criteria, four hardening parameters to peak stresses, eight softening parameters after peak, and six parameters speed effect [1]. The LS-DYNA calculation complex library contains model parameters for only two wood species: southern yellow pine and Douglas fir. This necessitates detailed studies of various wood species for various types of stress–strain states in a wide range of strain rates and temperatures, in order to equip mathematical models of wood with parameters (identification) that can adequately describe the behavior of the engineering structure containing wood components under shock wave loads.

The first detailed review of using wood as an element of damping structures was given by Johnson [2]. In the report, he noted that the dynamic properties of wood are not well understood. In the past three decades, quite a lot of works have appeared that explored various aspects of the high-speed deformation and destruction of wood, including the use of wood as a damping material in containers for transportation of radioactive materials by air, road and rail transport [3]. A large amount of research into the dynamic properties of wood was performed by Reid et al. [4–6]. In these works, the dependences of destructive stress and energy absorption on the impact velocity were obtained.

It was noted that dynamic destructive stresses are several times higher than static ones. The authors also postulated that failure stresses of specimens fabricated along the grain are an order of magnitude higher than that of the specimens cut transverse to the grain. The relationship between mechanical parameters of selected wood species (*Carya* sp., *Fagus sylvatica* L., *Acer platanoides* L., *Fraxinus excelsior* L., *Ulmus minor* Mill.) and the grain orientation angle concern loading direction was investigated in [7]. It was observed that as the angle between fibers and loading direction increases whereas the mechanical characteristics of all wood species decrease.

In the works of Bragov et al., using the Kolsky method, dynamic diagrams of birch, aspen, and sequoia were obtained at strain rates of ~10^3 s^{-1} with different direction of cutting specimens relative to the direction of timber fibers [8,9]. The deformation diagrams of specimens under loading along the grain are much higher than those under loading transverse to the grain. The limiting deformative characteristics have the opposite direction.

In recent years, interest in studying the effect of moisture, density and cutting angle on the mechanical properties of wood has increased significantly. Adalian et al. [10] and Eisenacher et al. [11] carried out a cycle of tests on wood in the range of strain rate from 10^{-3} to 10^3 s^{-1}. Wouts et al. [12], Ha et al. [13,14] and Hao et al. [15] considered the damping ability of various biological materials of plant origin from the point of view of improving artificial damping structures. Mach et al. [16] calculated the hysteresis losses (dissipated energy) as the area bounded by the loading and unloading curves, whereas stored energy (recoverable energy) is defined as the area under the unloading curve. Thus, the total energy is calculated by summing the area of the hysteresis loss and the stored energy. In the work of Novikov et al. [17], a large number of tests of sequoia, birch, aspen, and pine was carried out at different cutting angles, temperatures ranged from −30 °C to +65 °C, and at the moisture content of 5%, 20% and 30%. As a result, the dependence of strength on moisture and cutting angle was determined.

The mechanical properties of wood strongly depend on the place of growth, its age, the place of cutting and the type of stress–strain state. Therefore, the results obtained by different authors may differ significantly from each other. The purpose of this work is to conduct a detailed study of the effect of strain rate and type of stress–strain state on the mechanical properties of wood as well as to identify and verify the wood model.

2. Test Methods, Materials and Specimens

For dynamic compressive tests of wood, the Kolsky method and split Hopkinson pressure bar (SHPB) technology were used [18]. Figure 1 shows the installation diagram, as well as the main formulas for determining the parametric dependencies of deformation, stress and strain rate of the sample under compression. Two measuring bars were made of high-strength aluminum alloy and had a diameter of 20 mm and length of 1.5 m. During testing, a striker accelerated in the barrel of a gas gun impacts the SHPB and excites an elastic compression wave in the loading measuring bar, which, upon reaching the specimen, deforms it. The second (support) measuring bar acts as a dynamometer-waveguide and allows you to register the transmitted strain pulse $\varepsilon^T(t)$ and then to determine the process of stress developing in the specimen. The pulse $\varepsilon^R(t)$ reflected from the specimen in the loading bar makes it possible to determine the process of strain rate change in the specimen, and its integration allows to determine the process of the specimen deformation development. Small-length foil strain gages were used for registration elastic strain pulses in the bars. Based on these pulses, the parametric processes of stress, strain and strain rate over time are calculated using the formulae shown in the lower part of Figure 1. Then, excluding time as a parameter, a dynamic stress–strain curve is constructed with the known law of variation of the strain rate [19].

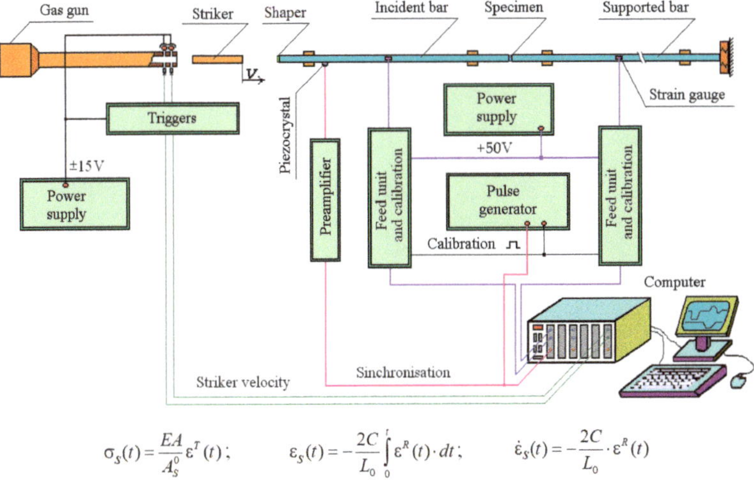

$$\sigma_s(t) = \frac{EA}{A_s^0}\varepsilon^T(t); \qquad \varepsilon_s(t) = -\frac{2C}{L_0}\int_0^t \varepsilon^R(t) \cdot dt; \qquad \dot{\varepsilon}_s(t) = -\frac{2C}{L_0} \cdot \varepsilon^R(t)$$

Figure 1. Installation scheme and basic dependencies.

Dynamic tests were carried out with specimens of lime-tree (*Tilia europoea*) and pine (*Pinaceae*) with a diameter of 20 mm and a length of 10 mm, cut from solid wood at an angle of 0° and 90° relative to the axis of the tree trunk. The flat ends of the samples were carefully hand-grinded before testing. The physical parameters of tested materials are shown in the Table 1.

Table 1. The physical parameters of tested materials.

Wood Species	Lime-Tree		Pine	
Parameter	Along the Grain	Transverse to the Grain	Along the Grain	Transverse to the Grain
Density, g/cm^3	0.505 ± 0.02	0.512 ± 0.015	0.435 ± 0.01	0.484 ± 0.015
Moisture content, %	7.5 ± 0.2	7.9 ± 0.3	6.8 ± 0.3	7.1 ± 0.2

The total amount of tested samples of each wood species amounted 50 pieces: 40 pieces were used for mechanical testing and 10 pieces were used for subsequent wood model verification. In each test mode during mechanical testing (strain rate and loading direction), 4–5 experiments were carried out, the results of which were then averaged. All the experiments were carried out in laboratory conditions at room temperature and 50% air humidity.

3. Experimental Results

Using the above methods, the dynamic tests of pine and lime-tree were carried out. As a result, their dynamic stress–strain curves, the ultimate strength and deformative characteristics of the specimens cut along and transverse to the grain were obtained. Dynamic strain diagrams for pine and lime-tree under uniaxial stress state are presented in Figures 2 and 3. The figures show the averaged stress–strain curves and appropriate strain rate-strain curves. The spread of the obtained properties was no more than 6%. In all figures solid lines show the dependences of the true stress of the specimen on its deformation σ~ε, and the dotted lines (in the lower part of the diagrams) correspond to the history of the strain rate change ɛ̇~ε (the appropriate axis is on the right).

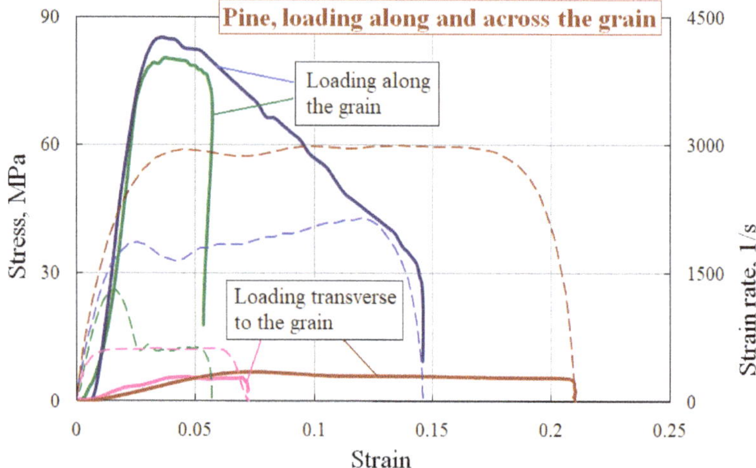

Figure 2. Deformation diagrams of pine under loading along and across the grain.

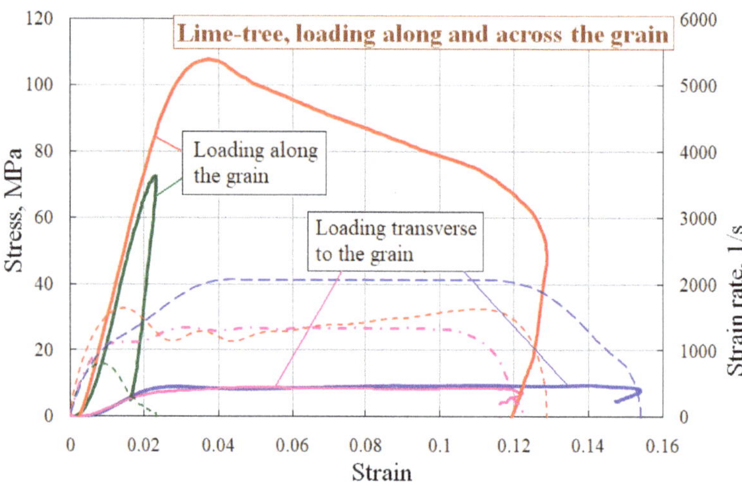

Figure 3. Deformation diagrams of lime-tree under loading along and across the grain.

Two test modes on strain rate were chosen: with non-destructive and destructive results. From the obtained stress–strain diagrams, it follows that for both wood species the specimens cut and loaded along the grain have the largest modulus of the load branch in the diagram, as well as ultimate stress and energy absorption.

Wood endurance can be indicated in Figure 4, where diagrams of fivefold loading of a pine specimen with its minor damages (curves 1–5) and a single independent loading of another similar specimen with its complete failure (curve 6) are presented. The diagrams are located on the deformation axis conditionally in order to make it easier to assess the effect of multiple loading on the steepness of the load sections of the stress–strain curves. The average strain rate at repeated loads was ~600–800 s^{-1}, and in the case of specimen failure, the strain rate was about 2200 s^{-1}. One can clearly see a decrease in the steepness of the load section of stress–strain curve during repeated loads by a factor of 2–3, which is associated with partial destruction of the specimen during each loading and violation of the flatness of its ends.

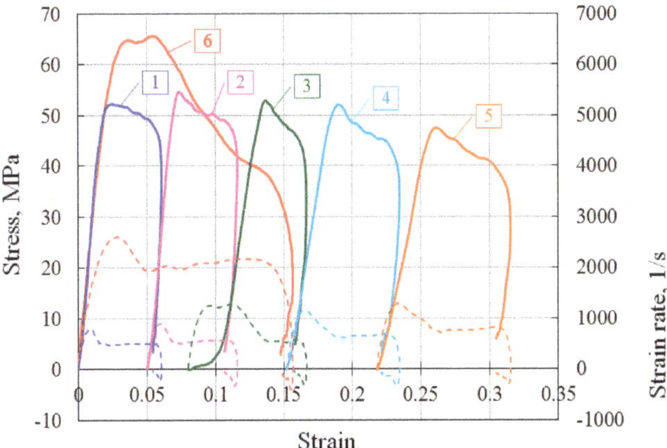

Figure 4. An example of multiply repeated loading of the specimen with its minor damage.

As an important characteristic of the timber damping capacity, the energy absorption of the tested wood species was estimated under loading along and transverse to the grain by calculating the area under the curve σ~ε (Figure 5).

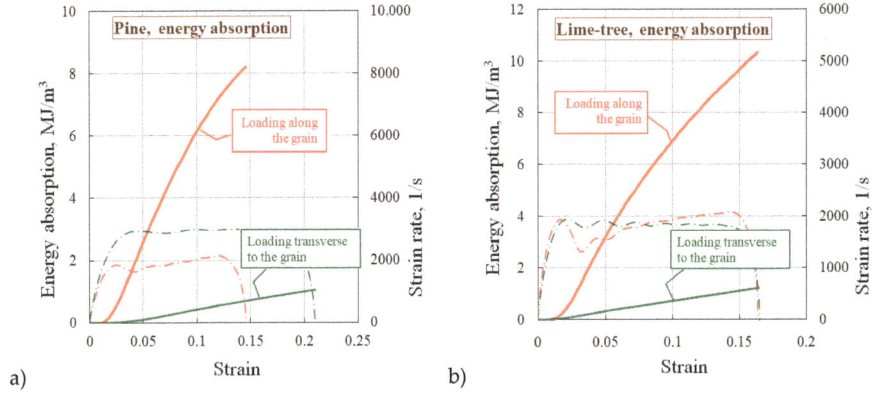

Figure 5. Energy absorption of the tested wood species: (**a**) for pine, (**b**) for lime-tree.

For both wood species, there is a significant excess of energy absorption by the specimens cut and tested along the grain, compared to the specimens cut and tested transverse to the grain.

It is interesting to compare the results on the damping ability of wood-based materials obtained by other researchers. In the work [12] the damping capacity of two types of deciduous and coniferous wood: beech and spruce under loading in the longitudinal, tangential and radial directions were compared. The highest specific energy absorbed was noted for specimens under longitudinal loading, and the smallest—under tangential loading. This is traditional tendency for wood materials.

While in [14], the damping ability of the mesocarp layer in durian shell (tropical fruit) was investigated and, as a result of the research, it was found the inverse effect: specific energy absorption of the mesocarp layer under lateral loading is higher than that under axial one. This may be due to the fact that the durian shell does not have the same clearly pronounced fiber structure as in wood, therefore the authors' accentuation on the lower strength and damping ability of the material in the axial direction of load application compared to lateral loading refers to the radial and tangential strength of shell of this fruit.

4. Identification and Verification of the Wood Model

To describe the behavior of wood under dynamic loads in the library of the LS-DYNA calculation complex, there is model No. 143 MAT_WOOD. This is a model of a transversely isotropic material for solid elements. It is possible to set material properties or use a predefined set of constants, but only for two species of pine growing in the USA: southern yellow pine and Douglas fir.

The primary features of the model are:

- Transverse isotropy for the elastic constitutive equations (different properties are modeled parallel and perpendicular to the grain).
- Yielding with associated plastic flow formulated with separate yield (failure) surfaces for the parallel- and perpendicular-to-the-grain modes.
- Hardening in compression formulated with translating yield surfaces.
- Post-peak softening formulated with separate damage models for the parallel and perpendicular-to-the-grain modes.
- Strength enhancement at high strain rates.

Model 143 for pine contains a predefined set of 29 parameters depending on moisture content and temperature. According to the results of the pine tests, some parameters of this model were adjusted for a particular batch of samples under study. These parameters are shown in Table 2.

Table 2. Main mechanical properties of pine for model identification.

Designation	Value
Density	450 kg/m^3
Modulus of elasticity in the direction of the grain	10,500 MPa
Flow stress during compression in the direction of the grain	80 MPa
The modulus of elasticity in the direction perpendicular to the grain	246 MPa
Flow stress during compression in the direction perpendicular to the grain	10 MPa

It should be noted that the elastic moduli of pine under loading along and across the grain were obtained as a result of quasi-static tests, since the Kolsky method, in principle, does not allow constructing a stress–strain curve in which the slope dσ/dε of the elastic loading section would be equal to the static modulus of elasticity. Usually the steepness of this section is several times less than the static Young's modulus. The reasons for this are as follows:

- The difference in the cross-sectional areas of the measuring bars and the sample (some part of the incident wave is reflected from the free surface around the sample), causing an apparent deformation of the sample,
- The presence of gaps between the ends of the measuring bars and the sample (due to the non-parallelism of its ends and their roughness or insufficient "pressing" of the sample during the preparation of the test),
- Dispersion during the propagation of waves in the bars (decrease in the steepness of the transmitted pulse during the propagation time to the recording strain gauge).

In order to confirm the adequacy of the model parameters determined from experiments, it is necessary to verify it, but in experiments other than those in which these parameters were obtained. For this purpose, we used a model experiment in full-scale and numerical implementation. The simulation of the indentation process was performed using the finite element method. The explicit time integration scheme was used for solving the equations of motion in time. Modeling was performed using the Dynamics-2® software package [20]. The numerical experiment was simulated in an axisymmetric formulation, and its design corresponded to a similar natural test.

To verify the model of material behavior, we used an experimental scheme for dynamic indentation by using the system of a split Hopkinson bar. The indicated procedure is schematically shown in Figure 6. The sample 5 and a removable indenter 4 with a hemispherical head are sandwiched between the measuring bars 2 and 6. Upon impact loading by the striker 1, a compression impulse is generated in the bar 2 the duration of which depends on the length of the striker and the amplitude on the speed of the striker. In the case of hemispherical indenter, the contact area at the initial moment of indentation is very small, therefore, the amplitude of the reflected pulse is very significant, and so the sample is loaded several times. For reliable recording of additional loading cycles, it is necessary to increase the length of the supporting bar in comparison with the length of the loading bar as much as required to register loading cycles [21].

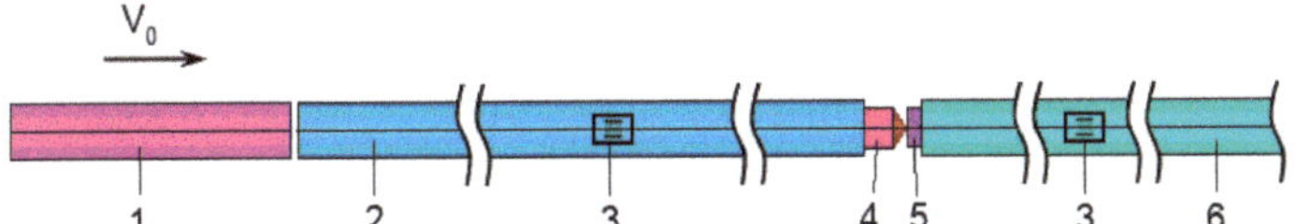

Figure 6. High speed indentation experiment scheme.

Using strain gauges 3, elastic strain pulses are recorded in the measuring bars (Figure 7). The following pulses are indicated by numbers: 1—incident (loading) strain pulse $\varepsilon^I(t)$, 2—reflected pulse in the first loading cycle $\varepsilon_1^R(t)$, 3—incident (loading) strain pulse $\varepsilon_2^I(t)$ in the second loading cycle, 4—reflected strain pulse in the second loading cycle $\varepsilon_2^R(t)$, 5—transmitted pulse (first cycle) $\varepsilon_1^T(t)$, 6—transmitted pulse (second cycle) $\varepsilon_2^T(t)$.

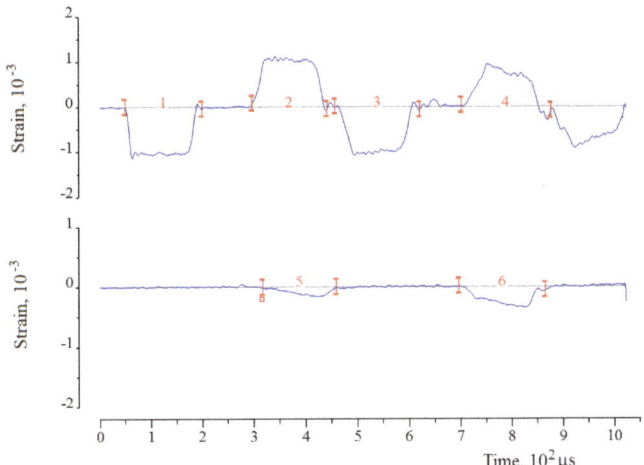

Figure 7. Typical waveform obtained in the experiment for high-speed indentation, taking into account additional cycles: **upper beam**—data from bar 2; **lower beam**—data from bar 6.

The loading of the SHPB system in the numerical experiment (Figure 8) is performed in the same way as in the natural experiment—with the help of a striker having an initial velocity V_0. In the calculation process, the incident and reflected strain pulses are calculated in element 2 and the past—in element 6. The time dependences obtained during the simulation are compared with the corresponding values recorded during the experiment.

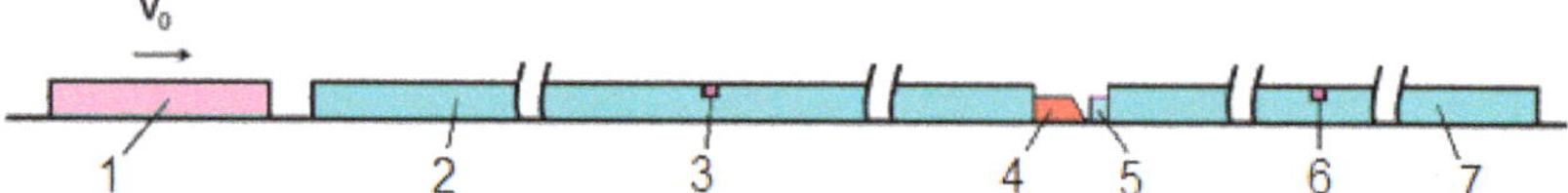

Figure 8. High speed indentation experiment simulation scheme.

The indenter is made of high-strength tungsten–cobalt alloy and is modeled by a rigid non-deformable material.

To assess the processes occurring in the sample, data obtained from measuring bars is used. In accordance with the Kolsky formulas, one can calculate the law of change of the indenter penetration rate into the sample $V(t)$ and the penetration resistance force $F(t)$:

$$V(t) = C_1 \cdot (\varepsilon^I(t) + \varepsilon^R(t)) - C_2 \cdot \varepsilon^T(t)$$
$$F(t) = E_2 \cdot S_2\, \varepsilon^T(t).$$

(1)

here $\varepsilon^I(t)$, $\varepsilon^R(t)$ and $\varepsilon^T(t)$ denote the incident, reflected, and transmitted strain pulses in the measuring bars, respectively, E is the elastic modulus, and C is the bar velocity of sound. The subscripts 1 and 2 refer to the first (loading) and second (supporting) measuring bars.

When modeling the process of dynamic indentation into wood samples, the following scheme was used: a sample and an indenter were considered (Figure 9). The spatial discretization of the indenter and the sample was done by using a solid three-dimensional finite element with one integration point. Due to the presence of symmetry, a quarter of the geometric model was considered. On the planes of symmetry, the corresponding boundary conditions were specified: at the nodes lying on the plane with the normal in the direction of the oY axis, zero velocities in the oY direction and zero velocities of rotation about the oX and oZ axes were set; at the nodes lying on the plane with the normal in the direction of the oZ axis, zero velocities in the oZ direction and zero velocities of rotation about the oX and oY axes were set.

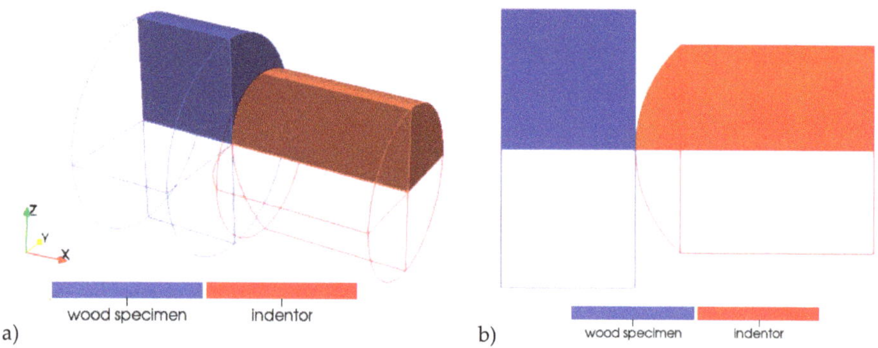

Figure 9. Geometric statement of the problem of numerical simulation: (**a**) 3D configuration, (**b**) plane configuration.

Zero velocities in the direction of the oX axis were set at the nodes belonging to the surface of the sample, which in full-scale experiments rested against the transmitted measuring bar.

The indenter was modeled by an absolutely rigid undeformable body. For the indenter, the law of the change in the speed of its movement in the axial direction was set. The time dependence of indenter's axial velocity for a particular experiment was calculated using the Equation (1) based on the signals recorded in the measuring bars.

The "surface–surface" contact interaction was set between the indenter and the sample.

In natural experiments on the indentation the samples of pine were used in the form of tablets with a length of 10 mm and a diameter of 20 mm. Some of the samples were cut in the direction of the wood grain, while another part of the samples was cut in the perpendicular to the grain direction.

As mentioned above, in the process of dynamic indentation, the sample is subjected to multiple loading in the experiment (see Figure 7), thus undergoing a certain deformation in each loading cycle. High-speed film recording of the indentation process makes it possible to estimate a number of loading cycle and indentation depth at which the destruction of the material occurs. Figure 10 shows the frames of such registration, made using a high-speed camera HSFC Pro.

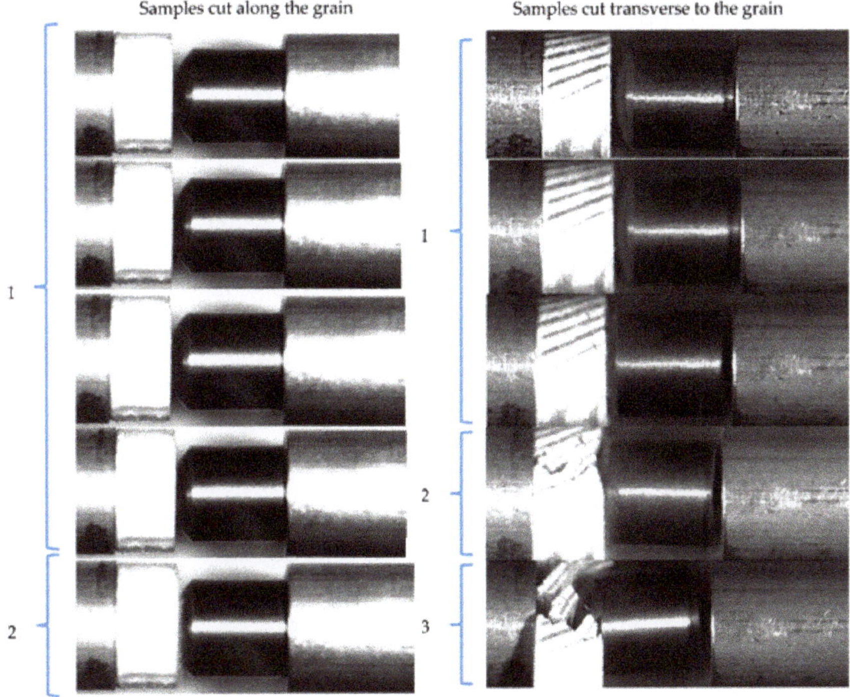

Figure 10. High-speed registration of dynamic indentation.

The left part of the figure corresponds to the experiment with the sample cut in the direction along the grain, whereas the right part corresponds to the experiment with the sample cut in the direction transverse to the grain. The numbers on the left of the high-speed registration images indicate the number of loading cycles (running number of impulse in the loading measuring bar) which one can see in Figure 7. It can be seen in Figure 10 that the samples cutting off parallel to the grain remain intact throughout the experiment, which is confirmed by the final state of such samples. Samples obtained by cutting in the direction perpendicular to the grain retain their integrity during the first loading cycle, however, cracks and gaps between the layers of wood appear in the second cycle, which grow and progress in subsequent loading cycles, leading to the separation of the sample into parts.

It can be noted that for the same load amplitude, the samples obtained by cutting in the direction of the fiber remain intact, while the samples cut perpendicular to the fiber exhibit failure already in the first loading cycle.

A comparison of the shape of the imprint on the pine sample after the experiment on the dynamic indentation of a hemispherical indenter is shown in Figure 11.

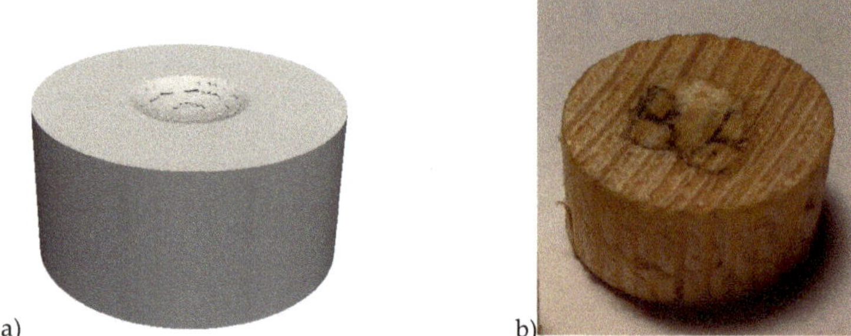

Figure 11. The shape of the imprints obtained by numerical modeling (**a**) and in the natural test (**b**).

Figure 12 shows a comparison of the time dependences of the indentation resistance of pine sample recorded in the experiments (solid colored lines) with the data obtained by numerical modeling (black dashed lines) during dynamic indentation of hemispherical indenter into specimen both along and across the grain.

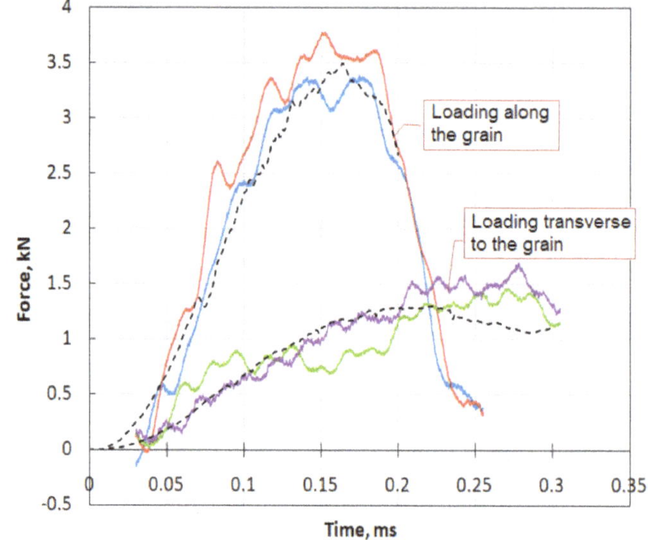

Figure 12. Comparison of experimental and calculated pulses in the supporting bar.

It can be seen that the model used allows us to accurately reproduce the features of the deformation of real material, namely, the difference in its deformation behavior in different directions with respect to grain orientation.

The results obtained are in qualitative agreement with the data from previous studies of high strain rate behavior of wood [8,17]. Correction of some parameters of the MAT_WOOD model made it possible to describe the deformation behavior of the studied wood species with sufficient accuracy. A further increase in the quality of the mathematical model is possible by taking into account the effect of the strain rate in a wide range of its change, as well as taking into account the fracture of the material.

In addition, it is necessary to conduct more complex model experiments to verify the fracture criteria, for example, high-speed penetration and perforation of wood plates.

5. Conclusions

Dynamic tests of lime-tree and pine were conducted. There is a strong anisotropy of the properties of the tested materials: the specimens exhibit the greatest strength under load applied along the grain, while the lowest strength is observed under loading transverse to the grain. The load branch module is non-linear and, as a rule, smaller than the unloading branch module (while maintaining the specimen integrity). At the specimen cutting angle of 90° with respect to the direction of the fibers, the stress–strain diagram after reaching a certain threshold stress value (about 8 MPa for lime-tree and about 10 MPa for pine) is close to the ideal plastic diagram. At 0° cutting angle, the initial section of the diagrams is close to linear. That is, an elastic deformation occurs. However, after reaching an ultimate stress value (yield strength), the diagram becomes nonlinear with stress relaxation. Especially such behavior is observed in the experiments in which the failure of specimens occurs.

Using a modification of the SHPB method, experiments were performed on the dynamic indentation of indenters with a hemispherical head into the samples parallel and perpendicular to the direction of the fibers. Dependences of resistance to penetration on time are constructed. In the direction of the fiber (along the grain), the resistance force was more than three-times greater than in the transverse direction. The results of the natural experiment were compared with the results of numerical modeling, in which the wood behavior was described by the MAT_WOOD model from the LS-DYNA software, in which some of the parameters were obtained experimentally. A good coincidence was observed.

Author Contributions: Conceptualization, A.B. and L.I.; data curation, T.I.; formal analysis, A.K.; funding acquisition, L.I.; investigation, A.L.; methodology, A.K. and A.L.; project administration, A.B.; resources, L.I.; software, A.K. and T.I.; supervision, F.d.; validation, F.d. and A.K.; visualization, A.K.; writing—original draft, A.L.; writing—review and editing, A.B. All authors have read and agreed to the published version of the manuscript.

Funding: Experimental investigations of wood were performed with the financial support of the Ministry of Science and Higher Education of the Russian Federation (task 0729-2020-0054). Numerical simulation of wood behavior was supported by the grant of the Government of the Russian Federation (contract No.14.Y26.31.0031).

Conflicts of Interest: The authors declare no conflict of interest. The funders had no role in the design of the study; in the collection, analyses, or interpretation of data; in the writing of the manuscript, or in the decision to publish the results.

References

1. Murray, Y.D. *Manual for LS-DYNA Wood Material Model 143*; Publication No. FHWA-HRT-04-097; Federal Highway Administration: Washington, DC, USA, 2007.
2. Johnson, W. Historical and present-day references concerning impact on wood. *Int. J. Impact Eng.* **1986**, *4*, 161–174. [CrossRef]
3. Neumann, M. Investigation of the Behavior of Shock-Absorbing Structural Parts of Transport Casks Holding Radioactive Substances in Terms of Design Testing and Risk Analysis. Ph.D. Thesis, Bergische Universität Wuppertal, Wuppertal, Germany, 2009.
4. Reid, S.R.; Reddy, T.Y.; Peng, C. Dynamic compression of cellular structures and materials. In *Structural Crashworthiness and Failure*; Jones, N., Wierzbicki, T., Eds.; Taylor & Francis Publication: London, UK; New York, NY, USA, 1993; pp. 257–294.
5. Reid, S.R.; Peng, C. Dynamic Uniaxial Crushing of Wood. *Int. J. Impact Eng.* **1997**, *19*, 531–570. [CrossRef]
6. Harrigan, J.J.; Reid, S.R.; Tan, P.J.; Reddy, T.Y. High rate crushing of wood along the grain. *Int. J. Mech. Sci.* **2005**, *47*, 521–544. [CrossRef]
7. Mania, P.; Siuda, F.; Roszyk, E. Effect of Slope Grain on Mechanical Properties of Different Wood Species. *Materials* **2020**, *13*, 1503. [CrossRef] [PubMed]
8. Bragov, A.M.; Lomunov, A.K. Dynamic properties of some wood species. *J. Phys. IV* **1997**, *7*, 487–492. [CrossRef]

9. Bragov, A.M.; Lomunov, A.K.; Sergeichev, I.V.; Gray, I.I.I.G.T. Dynamic behaviour of birch and sequoia at high strain rates. "Shock Compression of Condensed Matter". In Proceedings of the Conference of the American Physical Society Topical Group "APS-845", Baltimore, MD, USA, 31 July–5 August 2005; American Institute of Physics: Melville, NY, USA, 2006; pp. 1511–1514.

10. Adalian, C.; Morlier, P. Modeling the behaviour of wood during the crash of a cask impact limiter. In Proceedings of the PATRAM'98 Conference Proceedings, Paris, France, 10–15 May 1998; Volume 1.

11. Eisenacher, G.; Scheidemann, R.; Neumann, M.; Wille, F.; Droste, B. Crushing characteristics of spruce wood used in impact limiters of type B packages. Packaging and transportation of radioactive materials. In Proceedings of the PATRAM 2013, San Francisco, CA, USA, 18–23 August 2013.

12. Wouts, J.; Haugou, G.; Oudjene, M.; Coutellier, D.; Morvan, H. Strain rate effects on the compressive response of wood and energy absorption capabilities. Part A: Experimental investigation. *Compos. Struct.* **2016**, *149*, 315–328. [CrossRef]

13. Ha, N.S.; Lu, G. A review of recent research on bio-inspired structures and materials for energy absorption applications. *Composites* **2020**, *181*, 107496. [CrossRef]

14. Ha, N.S.; Lu, G.; Shu, D.W.; Yu, T.X. Mechanical properties and energy absorption characteristics of tropical fruit durian (Durio zibethinus). *J. Mech. Behav. Biomed. Mater.* **2020**, *104*, 103603. [CrossRef] [PubMed]

15. Hao, J.; Wu, X.; Oporto, G.; Wang, J.; Dahle, G.; Nan, N. Deformation and Failure Behavior of Wooden Sandwich Composites with Taiji Honeycomb Core under a Three-Point Bending Test. *Materials* **2018**, *11*, 2325. [CrossRef] [PubMed]

16. Mach, K.J.; Hale, B.B.; Denny, M.W.; Nelson, D.V. Death by small forces: A fracture and fatigue analysis of wave-swept macroalgae. *J. Exp. Biol.* **2007**, *210*, 2231–2243. [CrossRef] [PubMed]

17. Bol'Shakov, A.P.; Balakshina, M.A.; Gerdyukov, N.N.; Zotov, E.V.; Muzyrya, A.K.; Plotnikov, A.F.; Novikov, S.A.; Sinitsyn, V.A.; Shestakov, D.I.; Shcherbak, Y.I. Damping properties of sequoia, birch, pine, and aspen under shock loading. *J. Appl. Mech. Tech. Phys.* **2001**, *42*, 202–210. [CrossRef]

18. Kolsky, H. An investigation of the mechanical properties of materials at very high rates of loading. *Proc. Phys. Soc. Lond.* **1949**, *62*, 676–700. [CrossRef]

19. Bragov, A.M.; Lomunov, A.K. Methodological aspects of studying dynamic material properties using the Kolsky method. *Int. J. Impact Eng.* **1995**, *16*, 321–330. [CrossRef]

20. Bazhenov, V.G.; Zefirov, S.V.; Kochetkov, A.V.; Krylov, S.V.; Feldgun, V.R. The Dynamica-2 software package for analyzing plane and axisymmetric nonlinear problems of nonstationary interaction of structures with compressible media. *Mat. Modelirovanie* **2000**, *12*, 67–72.

21. Bragov, A.M.; Lomunov, A.K.; Sergeichev, I.V. Modification of the Kolsky method for studying properties of low-density materials under high-velocity cyclic strain. *J. Appl. Mech. Tech. Phys.* **2001**, *42*, 1090–1094. [CrossRef]

Publisher's Note: MDPI stays neutral with regard to jurisdictional claims in published maps and institutional affiliations.

 *materials*

Article

Effect of Membrane Components of Transverse Forces on Magnitudes of Total Transverse Forces in the Nonlinear Stability of Plate Structures

Zbigniew Kołakowski *  **and Jacek Jankowski**

Department of Strength of Materials, Faculty of Mechanical Engineering, Lodz University of Technology, Stefanowskiego 1/15, PL-90-924 Lodz, Poland; jacek.jankowski@p.lodz.pl
* Correspondence: zbigniew.kolakowski@p.lodz.pl

Received: 30 September 2020; Accepted: 16 November 2020; Published: 20 November 2020

Abstract: For an isotropic square plate subject to unidirectional compression in the postbuckling state, components of transverse forces in bending, membrane transverse components and total components of transverse forces were determined within the first-order shear deformation theory (FSDT), the simple first-order shear deformation theory (S-FSDT), the classical plate theory (CPT) and the finite element method (FEM). Special attention was drawn to membrane components of transverse forces, which are expressed with the same formulas for the first three theories and do not depend on membrane deformations. These components are nonlinearly dependent on the plate deflection. The magnitudes of components of transverse forces for the four theories under consideration were compared.

Keywords: nonlinear stability; square plate; shear forces; components of transverse forces in bending; membrane components of transverse forces; 4 methods (CPT, FSDT, S-FSDT, FEM)

---

1. Introduction

In the mid-20th century, Reissner [1,2] presented a plate theory accounting for the transverse shear deformation effect. This is a stress-based approach. Mindlin [3] offered a theory based on a displacement approach, where transverse shear stress was assumed to be the same through the plate thickness, and the shear correction factor k^2 (the so-called Mindlin correction factor) appeared. For transversely inextensible plates and $k^2 = 5/6$, values of stresses are equal in the Reissner and Mindlin plate theories [4]. The theories are characterised by an equivalent approximation degree known as the Reissner–Mindlin plate theory. Their comparison is discussed in [5–8], etc. Theoretical considerations can be found in [9–15] for higher-order shear deformation theories as well.

In [4,5], the equations of Reissner and Mindlin plates, including the parameter, which allows for an interpretation of these theories for transversally inextensible plates, were derived. In [6], for Reissner, Mindlin and Reddy plate models, a solution to the rectangular transverse plate sinusoidally loaded and freely supported along all the edges is given. Vibrations in the plate–beam system, in which the Reissner–Mindlin plate model related to the Timoshenko beam model was applied, were analysed in [7]. In [11,16], it is shown that when the plate thickness is around zero, the solution for the Reissner–Mindlin plate becomes close to the solution within the Kirchhoff–Love plate theory (the so-called classical plate theory (CPT)). In [17], formulations of the mixed finite element were given based on the mechanism of the shear locking phenomenon and the general variation method for the Reissner theory, including the Lagrange multiplier method. An extensive literature survey devoted to the two primary plate theories, i.e., the Kirchhoff plate theory and the Reissner–Mindlin plate theory, is found in [18]. The main purpose of [18] was to present a history of refinement in the Reissner–Mindlin theory, showing the up-to-date state of knowledge in this field of research.

Within the plate theories accounting for the shear deformation effect (FSDT), one can find other approaches, established, for instance, on superposition of bending and shear deflections (i.e., a two-variable refined plate theory [19–30] or a single-variable refined theory [31]).

Endo and Kimura [19] first suggested the simple first-order shear deformation theory (S-FSDT). Employing Hamilton's principle, an alternative formula in which a deflection in bending is the primary variable instead of the angle of rotation in bending and, at the same time, some limitations on neglecting the Reissner boundary effects are imposed, was given [2,32]. In the S-FSDT, three equations in the original formulation are reduced to two and the boundary conditions are also subject to respective modifications; however, the way the system is modelled remains unaltered (for a more detailed analysis, see Appendix A). Moreover, in [26–28], when two independent variables φ and w_s are considered, two differential equations with boundary conditions are attained. In the static analysis, two differential equations are uncoupled. The boundary conditions should be uncoupled as well, which is not possible in general. By introducing a bending relationship between the quantities, differences between the Reissner and Mindlin plate theories were investigated in [8]. In [24], the first two-variable shear deformation theory (FSDT) considering in-plane rotation, which allows one to correctly predict the response of plates for arbitrary boundary conditions in the analysis of buckling and vibrations of isotropic plates, was presented.

Interesting critical remarks to the above-mentioned plate theories can be found in [32–36]. To consider the Reissner boundary effect, a rotary potential, which is a fast-varying solution to the boundary layer, should be applied apart from the function φ. The boundary effect covers only some boundary conditions (e.g., pure tension or contact problems). The author of [32] suggested to refer to the presented version of the theory as a modern form of the CPT.

The finite element method employs the FSDT [9,18,33,37–41]. The effect of shear locking in finite shell elements and a loss in accuracy was explained in terms of the occurrence of solutions to the boundary layer. A shear locking problem occurs in the FEM, as shape functions cannot approximate a fast-variable solution to the boundary layer. [33,39]. Shear locking does not cause membrane deformations. In the majority of cases, the effect of shear deformation on displacements should be considered only. Solutions to the boundary layer are neglected.

In composites widely applied nowadays, the behaviour of individual layers can be affected considerably by transverse shear deformation [42]. These materials show low shear characteristics beyond the plane, which should be accounted for in numerous analyses (e.g., [15,20,21,43]).

In the works devoted to the plate dynamics, the effect of transverse rigidity on shear deformation and the influence of rotary inertia on frequencies and modes of plate vibrations were investigated (e.g., [3,5,15,17,27,30,31]). In [44], for the plate jointly supported along the whole circumference and subjected to free vibrations, it was shown that deflections in bending and shear deformation vibrated in phase (i.e., the total deflection is equal to their sum) in the first branch, whereas in the second branch, shear deformation and bending deflections vibrate in antiphase, where the deflection in shear deformation is predominant, i.e., the total deflection is in phase along with the deflection in shear deformation.

In the literature, apart from vibrations, the stability of individual rectangular plates is analysed (e.g., [15,16]). In [22], a buckling analysis of isotropic and orthotropic plates employing a two-variable refined plate theory is presented, whereas in [43], the S-FSDT for composite plates with four unknowns is discussed. The authors do not know any works devoted to the nonlinear stability of plates accounting for the transverse shear deformation effect.

All the works under discussion refer to the cases when there are no membrane forces in the plate structure. These forces appear in thin-walled structures (where $h/a < 0.05$) for loads exceeding the critical loads, that is to say, for postbuckling equilibrium paths. In the literature known to the authors of the present paper, there is a lack of works devoted to the nonlinear stability of thin-walled plates, in which transverse shear deformation is considered.

In the classical theory of thin plates (CPT), total equivalent transverse Kirchhoff forces were introduced only in [36,45] employing the variational method as far as the literature known to the authors is concerned. In the CPT, a notion of equivalent Kirchhoff transverse forces is introduced to satisfy a proper number of boundary conditions. In the variational approach to the CPT, there is no need to introduce the notion of Kirchhoff forces. These forces "emerge themselves" from the theory in such an approach. As shown in [46], it is necessary to introduce the notion of total equivalent Kirchhoff transverse forces, which results from Stokes' theorem concerning a change of the surface integral for the equilibrium equations into a plate circumference-oriented integral, that is to say, for the boundary conditions. In total Kirchhoff forces, two components of transverse forces appear; one of them is a derivative of internal moments, and the second is a projection of membrane forces on the transverse direction. The membrane forces also appear in other nonlinear problems, such as the deflection of thin-walled transversely loaded plates.

In the present work, the authors have decided to deal with the influence of these additional membrane components on the magnitudes of total Kirchhoff forces within the CPT for isotropic square plates subject to compression in the postbuckling state. These limitations were taken to facilitate an interpretation of the obtained results. For verification purposes, solutions to the Reissner theory (FSDT) and the Mindlin theory within the S-FSDT approach, i.e., after the introduction of two independent functions of displacements along the z-axis (i.e., the total lateral displacement w and the bending deflection ϕ), are presented. In the FSDT and the S-FSDT, the Reissner boundary condition was neglected. The governing equations within the three theories under consideration were derived with variational methods, allowing one to indicate two different components of transverse forces resulting from internal moments and membrane forces. For transversally inextensible plates, the membrane shear forces are independent of membrane deformation. For these three theories, the results for membrane forces and total forces were presented.

In composite materials, transverse shear deformation substantially affects the delamination of composites. In the failure criteria of composites, the impact of transverse components of membrane forces (i.e., in compression) is neglected. In the authors' opinion, these components are predominant in the postbuckling state and should be considered in composite failure criteria. The main aim of the paper is to draw attention to the theoretical background for membrane components of transverse forces in the expressions for transverse forces in the theory of thin plates, which are not accounted for in FEM shell elements.

2. Formulation of the Problem

The nonlinear stability of a square isotropic plate freely supported along the whole circumference and subject to compression along the x-axis (Figure 1) is analysed. The plate material is assumed to obey Hooke's law.

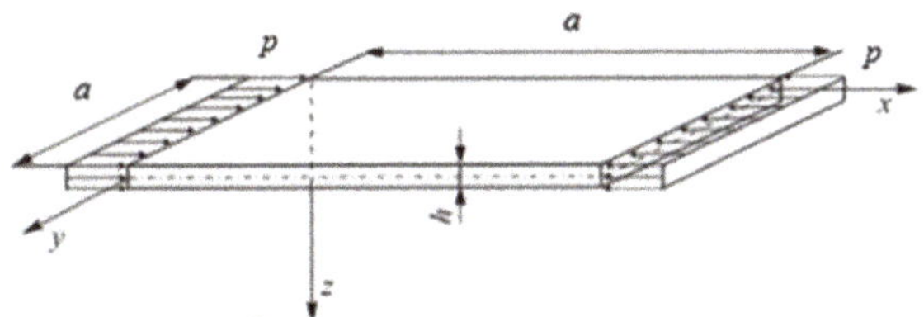

Figure 1. Square plate freely supported along all edges under compression.

In this study, for postbuckling equilibrium paths, transverse shear forces are analysed in detail for the transversally inextensible plate. The analysis is conducted within three theories of thin plates, namely the classical plate theory CPT (i.e., the Kirchhoff plate theory), the simple first-order shear deformation theory (S-FSDT) in a two-variable refined plate version and the Reissner plate theory (FSDT).

The governing equations of the three theories under consideration are presented in Appendix A. The equations were derived within a variational approach, which allows the equilibrium equations and the boundary conditions to be expressed explicitly. The solutions to the nonlinear problem of stability of the square plate for the three theories are presented in Appendix A. Instead of a system of two equations of equilibrium in the central plate plane (i.e., after an introduction of the function of Airy forces F, the system is satisfied identically), an equation of inseparability of deformations was derived (Appendix A).

According to the considerations presented in the Appendix A, transverse shear forces have two components (compare: the FSDT (A13), the S-FSDT (A24), the CPT (A33), respectively).

The first components depend on the derivatives of internal moments on the plate. Thus, the components can be referred to as transverse shear forces in bending. These forces are expressed with the following relationships: (A45) for the CPT and (A55) for the S-FSDT, correspondingly. The forces have a very similar structure. A difference lies only in the reduction factor $1/(1 + \eta)$ in (A55). Moreover, for the FSDT in (A64), a difference with respect to the S-FSDT occurs in the numerical coefficient two instead of (3-v) for the S-FSDT. A change in the numerical coefficient results from different boundary conditions for the FSDT and the S-FSDT. A more detailed analysis can be found in the Appendix A.

The second components depend on projections of membrane transverse forces on the direction perpendicular to the central plate plane. These components can be referred to as transverse shear forces in compression. For the three theories under consideration, these forces are expressed with identical formulas (compare (A46) for the CPT, (A56) for the S-FSDT and (A65) for the FSDT in the Appendix A). It is caused by the fact that the effect of shear deformation is not accounted for, as the forces are determined on the basis of the displacement w and the function of Airy forces F.

3. Analysis of the Calculation Results

A detailed analysis was conducted for a steel square plate (Figure 1) of the following dimensions: $a = 100$ mm, $h = 1$ mm and the material constants: $E = 200$, GPa, $v = 0.3$.

The ideal plate is supported freely along all edges and subjected to uniform compression with the stress p along the x-axis. The boundary conditions for the three theories under consideration (i.e., the FSDT, the S-FSDT and the CPT) are given in detail in the Appendix. The analytical results attained were verified with the commercial ANSYS software [47] employing the FEM (details to be found in Appendix A).

In the detailed analysis, the postbuckling state (or the so-called postbuckling equilibrium path) was dealt with, as only then plate deflections appear for the perfect plate. It is accompanied by the appearance of two transverse components of shear forces, that is to say, in bending and compression (the so-called membrane components).

The following index symbols are introduced in the study: C for the CPT, S for the S-FSDT, F for the FSDT, and A for ANSYS (FEM), respectively.

Firstly, for the three theories, the corresponding bifurcation loads (or the so-called critical loads), listed in detail in Table 1, were determined. According to the Appendix, values of the bifurcation loads for the FSDT (A62) and the S-FSDT (A54) are identical and slightly lower by the factor $1/(1 + \eta)$ than the CPT. For the data assumed in the analysis, according to (A63), we have $\eta = 0.000564$, which corresponds to $1/(1 + \eta) = 0.9994$. As can be seen, according to (A63), corrections for the S-FSDT and the FSDT are very inconsiderable when compared to the CPT for the assumed ratio of ($h/a = 0.01$). The results obtained within the three theories are in conformity with the FEM outcomes.

Table 1. Values of critical stresses according to the classical plate theory (CPT), the simple first-order shear deformation theory (S-FSDT), the first-order shear deformation theory (FSDT) and ANSYS.

Critical Stresses in MPa			
CPT	**S-FSDT**	**FSDT**	**ANSYS (FEM)**
p_{cr}^C	p_{cr}^S	p_{cr}^F	p_{cr}^A
72.30	72.26	72.26	72.30

The determined value of the critical stress (A43) for the CPT was introduced into the relationship for the total equivalent Kirchhoff force $\hat{Q}_x^C$ (A47a) and then the component $\hat{Q}_x^C$ when $p = p_{cr}^C$, is equal to

$$
\begin{aligned}
\hat{Q}_x^C\left(p = p_{cr}^C\right) &= \left[(3-v)D\left(\tfrac{\pi}{a}\right)^3 W - p_{cr}^C hW\tfrac{\pi}{a}\right]\cos\tfrac{\pi x}{a}\sin\tfrac{\pi y}{a} \\
&\quad - \tfrac{EhW^3}{8}\left(\tfrac{\pi}{a}\right)^3 \cos\tfrac{\pi x}{a}\sin\tfrac{\pi y}{a}\cos\tfrac{2\pi y}{a} \\
&= -(1+v)\left(\tfrac{\pi}{a}\right)^3 DW\cos\tfrac{\pi x}{a}\sin\tfrac{\pi y}{a} - \tfrac{EhW^3}{8}\left(\tfrac{\pi}{a}\right)^3 \cos\tfrac{\pi x}{a}\sin\tfrac{\pi y}{a}\cos\tfrac{2\pi y}{a}
\end{aligned}
\tag{1}
$$

As can be easily noticed, in (1) there is a minus sign at both terms of the right-hand side. However, mutual relations depend on the relationships of products of trigonometric functions. The first term $\hat{Q}_x^C\left(p = p_{cr}^C\right)$ in (1) attains extreme values for $x = 0$; a and $y = a/2$ and the second respectively minimum for $x = 0$; a and $y = a/2$. Attention should be drawn to the fact that when $x = 0$ and $y = a/2$, the first term has a minus sign, and the second term has a plus sign. The opposite situation takes place when $x = a$ and $y = a/2$, i.e., a plus sign is in the first term and a minus sign is in the second. The extreme values $\hat{Q}_x^C$ are attained inside the square plate.

A further analysis dealt with postbuckling states. According to (A39c), the force component N_{xy} equals zero, and one of two membrane components of transverse forces (according to (A12), (A23) and (A32)) vanishes as well.

Next, maximal absolute values of components of transverse forces in bending, membrane components (or in compression) or total forces for the three theories, determined according to the formulas given in Appendix A for five overload values of critical load, i.e., $1.2 \leq p^\theta/p_{cr}^\theta \leq 2.0$ (where the index $\theta = C, S, F$), are listed in Table 2. In this table, values of the dimensionless deflection W/h and $1/(1 + \eta)$ are also presented.

For the CPT, the equivalent Kirchhoff forces Q_x^C, Q_y^C are equal according to (A45). On the other hand, values of membrane components of the transverse forces $\overline{Q}_x^C, \overline{Q}_y^C$ differ depending on the overload p^θ/p_{cr}^θ. For the overload equal to 1.2, the ratio of maximal absolute components $\overline{Q}_x^C/\overline{Q}_y^C$ equals almost 5, whereas for the overload equal to 2, the ratio of membrane components is 1.4. The membrane forces $\overline{Q}_x^C, \overline{Q}_y^C$ are independent of membrane deformation. It results from the fact that the membrane force Q_x^C has a term linearly dependent on deflection and in the third power, which for Q_y^C is in the third power only. A detailed analysis can be found in Appendix A. Components of the total equivalent Kirchhoff force $\hat{Q}_y^C$ are always higher for the range of loads under analysis than $\hat{Q}_x^C$. It follows from the term that is linear with respect to W, dependent on the overload $\hat{Q}_x^C$.

In Figure 2, the maximal absolute values of components of transverse Kirchhoff forces for the CPT versus p^C/p_{cr}^C, listed in Table 2, are presented.

For the S-FSDT, the maximal absolute values of components of the transverse forces $\left|\hat{Q}_x^S\right|_{max}$ and $\left|\hat{Q}_y^S\right|_{max}$ are the same in practice as for the CPT, which results from a very low value of the correction η. For the FSDT, force components in bending are 1.35 times lower for the CPT and the S-FSDT (cf. Formulas (A64) and (A45)). Similarly as for the S-FSDT, components of the total transverse force $\hat{Q}_y^F$ are always larger than $\hat{Q}_x^F$. When total transverse forces are accounted for in the CPT, the S-FSDT and the FSDT, they yield higher values than the equivalent Kirchhoff force by approx. 1.5 times for the component with respect to the x-axis (i.e., with a lower index x) and more than 2 times for the component with respect to the y-axis.

Table 2. Values of components of transverse forces for the CPT, the S-FSDT, the FSDT and ANSYS.

Theory	Symbol	Unit	p^θ/p_{cr}^θ (Where the Index $\theta=C,S,F,A$)				
			1.2	**1.4**	**1.6**	**1.8**	**2.0**
CPT (index C)	W/h	-	0.7656	1.083	1.326	1.531	1.712
	$1/(1+\eta)$	-			1.0		
	$\left\|Q_x^C\right\|_{max}$	N/mm	1.17	1.66	2.03	2.34	2.62
	$\left\|Q_y^C\right\|_{max}$	N/mm	1.17	1.66	2.03	2.34	2.62
	$\left\|\overline{Q}_x^C\right\|_{max}$	N/mm	1.73	2.55	3.44	4.42	5.49
	$\left\|\overline{Q}_y^C\right\|_{max}$	N/mm	0.347	0.983	1.80	2.78	3.88
	$\left\|\hat{Q}_x^C\right\|_{max}$	N/mm	0.651	1.22	1.96	2.80	3.73
	$\left\|\hat{Q}_y^C\right\|_{max}$	N/mm	1.52	2.64	3.84	5.13	6.51
S-FSDT (index S)	W/h	-	0.7651	1.0823	1.325	1.530	1.711
	$1/(1+\eta)$	-			0.9994		
	$\left\|\hat{Q}_x^S\right\|_{max}$	N/mm	0.652	1.22	1.96	2.80	3.73
	$\left\|\hat{Q}_y^S\right\|_{max}$	N/mm	1.52	2.64	3.83	5.12	6.51
FSDT (index F)	W/h	-	0.7651	1.0823	1.325	1.530	1.711
	$1/(1+\eta)$	-			0.9994		
	$\left\|\hat{Q}_x^F\right\|_{max}$	N/mm	0.898	1.54	2.27	3.16	4.13
	$\left\|\hat{Q}_y^F\right\|_{max}$	N/mm	1.21	2.21	3.31	4.52	5.83
FEM (index A)	W/h	-	0.76	1.07	1.31	1.50	1.68
	$\left\|Q_x^A\right\|_{max}$	N/mm	1.00	1.52	1.98	2.41	2.83
	$\left\|Q_y^A\right\|_{max}$	N/mm	0.96	1.37	1.67	1.91	2.08

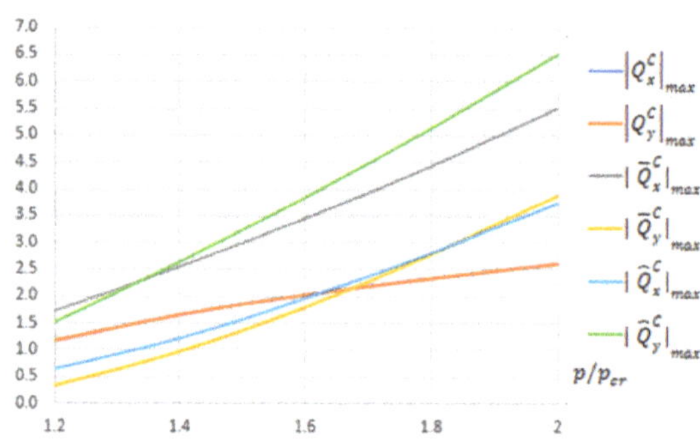

Figure 2. Maximal absolute values of components of the transverse forces $\left|Q_x^C\right|_{max}$, $\left|Q_y^C\right|_{max}$, $\left|\overline{Q}_x^C\right|_{max}$, $\left|\overline{Q}_y^C\right|_{max}$, $\left|\hat{Q}_x^C\right|_{max}$, $\left|\hat{Q}_y^C\right|_{max}$ in N/mm for the CPT.

For the FEM, the values of components of the transverse forces $\left|Q_x^A\right|_{max}$ are higher than $\left|Q_y^A\right|_{max}$. At the overload equal to 1.2, the ratio $\left|Q_x^A\right|_{max}/\left|Q_x^A\right|_{max}$ is 1.05, but for the overload of 2.0, it is equal to 1.36, respectively. The values $\left|Q_x^A\right|_{max}$ and $\left|Q_y^A\right|_{max}$ are closest to the equivalent Kirchhoff force Q_x^C, Q_y^C. Thus, the transverse forces $\left|Q_x^A\right|_{max}$, $\left|Q_y^A\right|_{max}$ determined within the FEM have a different character than the total transverse forces for the CPT, the S-FSDT and the FSDT, determined on the basis of components in bending and compression. It can originate from the fact that membrane components were neglected in the FEM analysis.

In Figure 3, the maximal absolute values of transverse resultant forces for the CPT, the S-FSDT and the FSDT (of which the values are listed in Table 2) versus overload are collected. The results for ANSYS are shown as well.

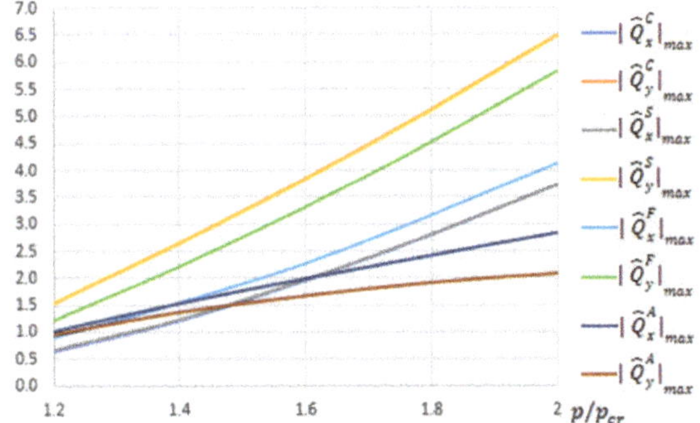

Figure 3. Maximal absolute values of the total transverse forces, $\left|Q_x^S\right|_{max}, \left|Q_y^S\right|_{max}, \left|Q_x^F\right|_{max},$ $\left|Q_y^F\right|_{max}, \left|Q_x^A\right|_{max}, \left|Q_y^A\right|_{max}$ in N/mm for the CPT, the S-FSDT, the FSDT and the finite element method (FEM).

The present study was primarily aimed at drawing attention to a necessity to consider the effect of membrane components on total transverse forces for $1.2 \le p^\theta / p_{cr}^\theta \le 2.0$.

To illustrate the effect of components of transverse Kirchhoff forces for the CPT, their distributions are presented in subsequent figures (Figures 4–9). The components are shown in the contour drawings (denoted as a) and plane drawings (denoted as b) for the whole square plate and the overload $p^C / p_{cr}^C = 2$ in the following sequence: Q_x^C (Figure 4), $\overline{Q}_x^C$ (Figure 5), $\hat{Q}_x^C$ (Figure 6), Q_y^C (Figure 7), $\overline{Q}_y^C$ (Figure 8) and $\hat{Q}_y^C$ (Figure 9).

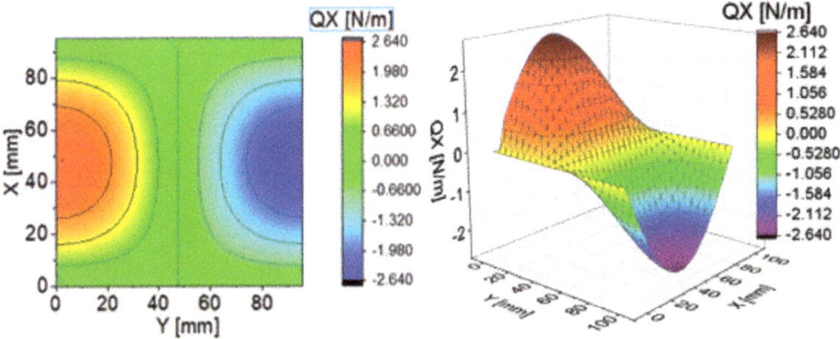

Figure 4. Contour-surface chart of Q_x^C

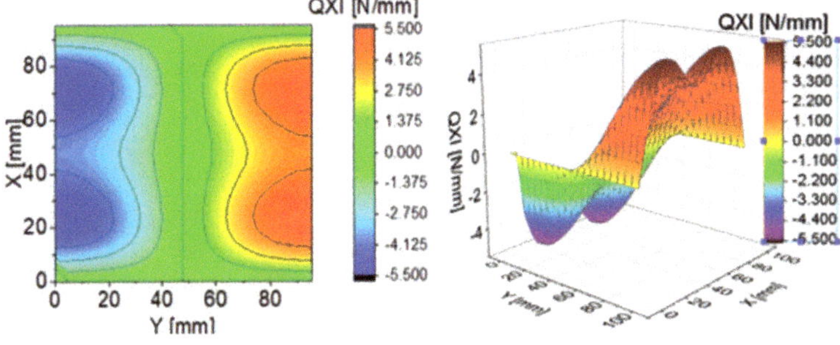

Figure 5. Contour-surface chart of $\overline{Q}_x^C$

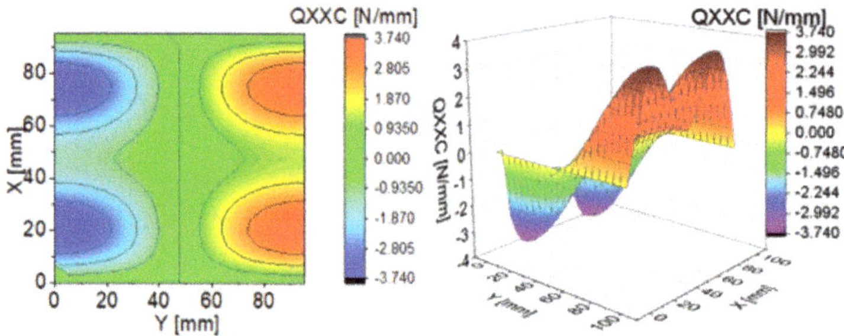

Figure 6. Contour-surface chart of $\hat{Q}_x^C$

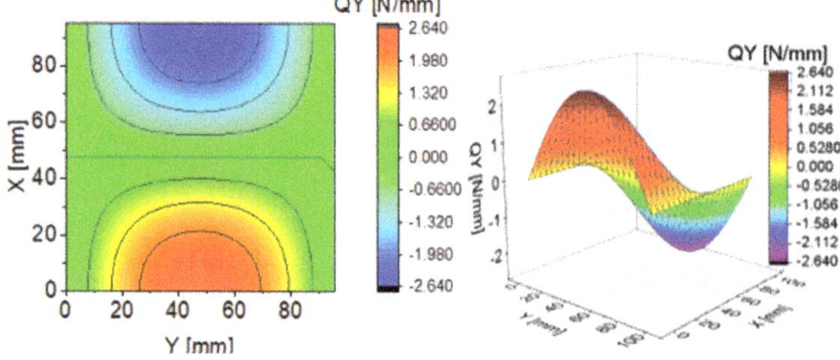

Figure 7. Contour-surface chart of Q_y^C

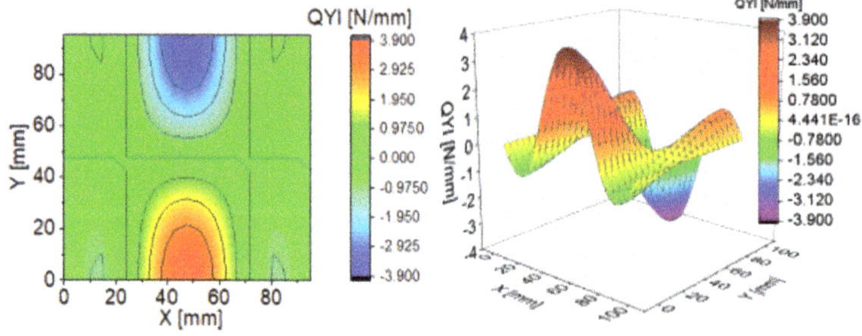

Figure 8. Contour-surface chart of Q_y^C

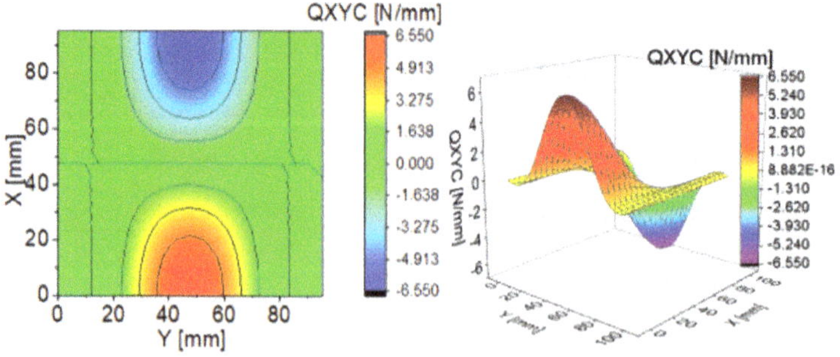

Figure 9. Contour-surface chart of $\hat{Q}_y^C$

The distributions of Q_x^C and Q_y^C presented in Figures 4 and 7 are the same according to (A45). When the distributions of membrane forces are compared, the distribution for $\overline{Q}_x^C$ (Figure 5) is more complex than $\overline{Q}_y^C$ (Figure 8). However, in the authors' opinion, the total transverse forces $\hat{Q}_x^C$ (Figure 6) and $\hat{Q}_y^C$ (Figure 9), which should be employed in failure criteria, are the most important. The components $\hat{Q}_y^C$ are larger than $\hat{Q}_x^C$, opposite to what happens in the FEM (Table 2).

It should be underlined once again that for the square plate under analysis, one of two membrane force components, which is dependent nonlinearly on the deflection W, equals zero (i.e., for $N_{xy} = 0$).

4. Conclusions

The effect of membrane components of transverse forces on total transverse forces within the three theories: the CPT, the S-FSDT and the FSDT, was discussed. When membrane components are taken into consideration, an increase can be observed in transverse forces equal to 1.5 times, at least for the square plate, freely supported along the whole circumference under consideration. It results from the fact that membrane components of transverse forces depend nonlinearly on the plate deflection. The results were compared to the FEM. The membrane transverse forces are independent of membrane deformation.

In composite materials, a failure of the structure resulting from delamination exerts a considerable effect on its integrity and load-carrying capacity. Transverse shear effects significantly influence the behaviour of composites. In the composite failure criteria, the impact of transverse force components in compression is neglected. From the authors' viewpoint, these components prevail in the postbuckling state, which was proven in this study and should be considered in the failure criteria of composites, for instance, the Hashin failure criterion for 3D, LaRC04(3D), Matrix Failure under the additional condition that $\sigma_{33} = 0$.

Author Contributions: Conceptualization, Z.K; methodology, Z.K. and J.J.; software, Z.K. and J.J.; validation, Z.K. and J.J.; formal analysis, Z.K.; writing—original draft preparation, Z.K.; writing—review and editing, J.J. All authors have read and agreed to the published version of the manuscript.

Funding: This research received no external funding.

Appendix A

Appendix A.1. Governing Equations in the FSDT, the S-FSDT and the CPT

In this Appendix A, governing equations for three theories, namely the first-order shear deformation plate theory (FSDT), the simple first-order shear deformation theory (S-FSDT) and the classical plate theory (CPT), are presented. The equilibrium equations and the boundary conditions are attained within a variational approach.

The following geometrical relationships for the plate component are assumed [36,44,46]

$$\varepsilon_x = u_{,x} + \frac{1}{2}w_{,x}^2 \tag{A1a}$$

$$\varepsilon_y = v_{,y} + \frac{1}{2}w_{,y}^2 \tag{A1b}$$

$$2\varepsilon_{xy} = \gamma_{xy} = u_{,y} + v_{,x} + w_{,x}w_{,y} \tag{A1c}$$

And

$$\kappa_x = -\psi_{x,x} \quad \kappa_y = -\psi_{y,y} \quad \kappa_{xy} = -\left(\psi_{x,y} + \psi_{y,x}\right) \tag{A2}$$

where u, v are components of the plate displacement vector along the axis x, y, correspondingly, w is the total displacement vector along the z-axis and ψ_x, ψ_y are the rotation angles of a transverse normal due to bending of the axes x, y, respectively, whereas the plane $x - y$ overlaps the central plane before its buckling.

In this approach to the plate theory when transverse shear is accounted for, it is assumed that the total rotation angles of the normal to the central plane in two planes are, respectively [19,44]

$$w_{,x} = \psi_x + \beta_x \quad w_{,y} = \psi_y + \beta_y \tag{A3}$$

where β_x, β_y are the transverse shear angles.

Internal cross-sectional forces can be expressed in the form [19,44]

$$N_x = \frac{Eh}{1 - v^2}\left(\varepsilon_x + v\varepsilon_y\right) \tag{A4a}$$

$$N_y = \frac{Eh}{1 - v^2}\left(\varepsilon_y + v\varepsilon_x\right) \tag{A4b}$$

$$N_{xy} = \frac{Eh}{1 - v^2}\frac{1 - v}{2}\gamma_{xy} \tag{A4c}$$

$$M_x = -D\left(\psi_{x,x} + v\psi_{y,y}\right) \tag{A5a}$$

$$M_y = -D\left(\psi_{y,y} + v\psi_{x,x}\right) \tag{A5b}$$

$$M_{xy} = -D\frac{1 - v}{2}\left(\psi_{x,y} + \psi_{y,x}\right) \tag{A5c}$$

$$Q_x = k^2 Gh(w_{,x} - \psi_x) \tag{A6a}$$

$$Q_y = k^2 Gh\left(w_{,y} - \psi_y\right) \tag{A6b}$$

The coefficient k^2 occurring in Formulas (A6) is known as the Mindlin correction factor.

The total potential energy Π of a thin rectangular plate of the following dimensions: $\ell \times b \times h$ can be written as [36]

$$\begin{aligned}
\Pi = U - W = {}& \tfrac{1}{2}\int_0^\ell \int_0^b \left(N_x\varepsilon_x + N_y\varepsilon_y + N_{xy}\gamma_{xy}\right)dxdy \\
& -\tfrac{1}{2}\int_0^\ell \int_0^b \left[M_x\psi_{x,x} + M_y\psi_{y,y} + M_{xy}\left(\psi_{x,y} + \psi_{y,x}\right)\right]dxdy \\
& +\tfrac{1}{2}\int_0^\ell \int_0^b \left[Q_x(w_{,x} - \psi_x) + Q_y\left(w_{,y} - \psi_y\right)\right]dxdy - \int_0^b hp^0(y)u\big|_{x=0}^{x=\ell}dy
\end{aligned} \tag{A7}$$

where U is the elastic strain internal energy, W is the work of external forces and $p^0(y)$ is the plate prebuckling external load in the central plane along the x-axis.

Appendix A.1.1. FSDT

To determine a variation of the total energy Π (A7), the following relations were substituted: (A1), (A2) and (A4)–(A6). Having grouped the terms including the same variations and summed each group of the terms to zero (due to mutual independence of the variations), the following equations are obtained

- Equations of equilibrium:

$$\int_0^\ell \int_0^b \left[N_{x,x} + N_{xy,y}\right]\delta u\,dxdy = 0 \tag{A8a}$$

$$\int_0^\ell \int_0^b \left[N_{xy,x} + N_{y,y}\right]\delta v\,dxdy = 0 \tag{A8b}$$

$$\int_0^\ell \int_0^b \left[Q_{x,x} + Q_{y,y} + \left(N_x w_{,x} + N_{xy}w_{,y}\right)_{,x} + \left(N_{xy}w_{,x} + N_y w_{,y}\right)_{,y}\right]\delta w\,dxdy = 0 \tag{A8c}$$

$$\int_0^\ell \int_0^b \left[M_{x,x} + M_{xy,y} - Q_x \right] \delta\psi_x \, dx\, dy = 0 \tag{A8d}$$

$$\int_0^\ell \int_0^b \left[M_{xy,x} + M_{y,y} - Q_y \right] \delta\psi_y \, dx\, dy = 0 \tag{A8e}$$

- Boundary conditions for $x = const$

$$\int_0^b \left[N_x - hp^0(y) \right] \delta u \, dy \big| x = const = 0 \tag{A9a}$$

$$\int_0^b N_{xy} \delta v \, dy \big| x = const = 0 \tag{A9b}$$

$$\int_0^b \left[N_{xy} w_{,y} + N_x w_{,x} + Q_x \right] \delta w \, dy \big| x = const = 0 \tag{A9c}$$

$$\int_0^b \left[M_x \right] \delta\psi_x \, dy \big| x = const = 0 \tag{A9d}$$

$$\int_0^b \left[M_{xy} \right] \delta\psi_y \, dy \big| x = const = 0 \tag{A9e}$$

for $y = const$

$$\int_0^\ell N_y \delta v \, dx \big| y = const = 0 \tag{A10a}$$

$$\int_0^\ell N_{xy} \delta u \, dx \big| y = const = 0 \tag{A10b}$$

$$\int_0^\ell \left[N_{xy} w_{,x} + N_y w_{,y} + Q_y \right] \delta w \, dx \big| y = const = 0 \tag{A10c}$$

$$\int_0^\ell \left[M_{xy} \right] \delta\psi_x \, dx \big| y = const = 0 \tag{A10d}$$

$$\int_0^\ell \left[M_y \right] \delta\psi_y \, dx \big| y = const = 0 \tag{A10e}$$

Equation (A8) is a system of equilibrium equations and relationships, and (A9) and (A10) are boundary conditions for $x = const$ and $y = const$, respectively. Equations (A8)–(A10) hold for the FSDT. The above equations were determined from variational methods for which, according to Stokes' theorem, the surface integral can be transformed into a circumference-oriented integral. The form the equations are presented follows from it.

According to (A6) and (A8d,e), transverse forces are expressed with the relations

$$Q_x^F = M_{x,x} + M_{xy,y} \tag{A11a}$$

$$Q_y^F = M_{y,y} + M_{xy,x} \tag{A11b}$$

where the upper index F was introduced for the FSDT. These are components of transverse forces dependent on derivatives of internal moments.

In Conditions (A9c) and (A10c) for the components of transverse forces, forces of the same characters are added, i.e., additional components of transverse forces that depend on projections of membrane forces, that is to say [36,45]

$$\overline{Q}_x^F = N_x w_{,x} + N_{xy} w_{,y} \tag{A12a}$$

$$\overline{Q}_y^F = N_y w_{,y} + N_{xy} w_{,x} \tag{A12b}$$

These forces do not affect membrane deformations.

When (A11) and (A12) are considered, according to (A9c) and (A10c), the total transverse forces $\hat{Q}_x^F$ and $\hat{Q}_y^F$ are introduced

$$\hat{Q}_x^F = Q_x^F + \overline{Q}_x^F = M_{x,x} + M_{xy,y} + N_x w_{,x} + N_{xy} w_{,y} \tag{A13a}$$

$$\hat{Q}_y^F = Q_y^F + \overline{Q}_y^F = M_{y,y} + M_{xy,x} + N_y w_{,y} + N_{xy} w_{,x} \tag{A13b}$$

Appendix A.1.2. S-FSDT

Next, the independent functions ψ_x, ψ_y (A2) were substituted with a potential function $\phi(x, y)$ [19,32–34,44] such that

$$\phi_{,x} = \psi_x \quad \phi_{,y} = \psi_y \tag{A14}$$

This approach consists of the fact that the function ϕ is treated as deflections in bending. Because the function ϕ is introduced, the Reissner boundary conditions [32], not considered in the present study, are ignored.

If $\beta_x = w_{s,x}$, $\beta_y = w_{s,y}$ hold for (A3), the shear deflection is obtained from the relationship $w_s = w - \phi$. From (A3) and (A14), it follows that

$$w_{,x} = \phi_{,x} + \beta_x = \phi_{,x} + w_{s,x} \quad w_{,y} = \phi_{,y} + \beta_y = \phi_{,y} + w_{s,y} \tag{A15}$$

When (A15) is accounted for, the internal forces (A5) and (A6) are written as

$$M_x = -D\left(\phi_{x,x} + v\phi_{y,y}\right) \tag{A16a}$$

$$M_y = -D\left(\phi_{y,y} + v\phi_{x,x}\right) \tag{A16b}$$

$$M_{x,y} = -D\left(1 - v\right)\phi_{x,y} \tag{A16c}$$

$$Q_x = k^2 Gh(w_{,x} - \phi_{,x}) \tag{A17a}$$

$$Q_y = k^2 Gh\left(w_{,y} - \phi_{,y}\right) \tag{A17b}$$

From the variation of the total energy Π (A7) with respect to the displacement variable components u, v, w, ϕ, we obtain

- Equations of equilibrium

$$\int_0^\ell \int_0^b \left[N_{x,x} + N_{xy,y}\right]\delta u\, dxdy = 0 \tag{A18a}$$

$$\int_0^\ell \int_0^b \left[N_{xy,x} + N_{y,y}\right]\delta v\, dxdy = 0 \tag{A18b}$$

$$\int_0^\ell \int_0^b \left[Q_{x,x} + Q_{y,y} + \left(N_x w_{,x} + N_{xy} w_{,y}\right)_{,x} + \left(N_{xy} w_{,x} + N_y w_{,y}\right)_{,y}\right]\delta w\, dxdy = 0 \tag{A18c}$$

$$\int_0^{\ell} \int_0^{b} \left[M_{x,xx} + 2M_{xy,xy} + M_{y,yy} - Q_{x,x} - Q_{y,y} \right] \delta\phi \, dx \, dy = 0 \tag{A18d}$$

- Boundary conditions for $x = const$

$$\int_0^{b} \left[N_x - hp^0(y) \right] \delta u \, dy \big|_{x = const} = 0 \tag{A19a}$$

$$\int_0^{b} N_{xy} \delta v \, dy \big|_{x = const} = 0 \tag{A19b}$$

$$\int_0^{\ell} \left[N_{xy} w_{,y} + N_x w_{,x} + Q_x \right] \delta w \, dy \big|_{x = const} = 0 \tag{A19c}$$

$$\int_0^{b} \left[2M_{xy,y} + M_{x,x} - Q_x \right] \delta\phi \, dy \big|_{x = const} = 0 \tag{A19d}$$

$$\int_0^{b} \left[M_x \right] \delta\phi_{,x} \, dy \big|_{x = const} = 0 \tag{A19e}$$

for $y = const$

$$\int_0^{\ell} N_y \delta v \, dx \big|_{y = const} = 0 \tag{A20a}$$

$$\int_0^{\ell} N_{xy} \delta u \, dx \big|_{y = const} = 0 \tag{A20b}$$

$$\int_0^{\ell} \left[N_{xy} w_{,x} + N_y w_{,y} + Q_y \right] \delta w \, dx \big|_{y = const} = 0 \tag{A20c}$$

$$\int_0^{\ell} \left[2M_{xy,x} + M_{y,y} - Q_y \right] \delta\phi \, dx \big|_{y = const} = 0 \tag{A20d}$$

$$\int_0^{\ell} \left[M_y \right] \delta\phi_{,y} \, dx \big|_{y = const} = 0 \tag{A20e}$$

for the plate corner, i.e., for $x = const$ and $y = const$

$$2M_{xy} \delta\phi \big|_{x = const} \big|_{y = const} = 0 \tag{A21}$$

Equations (A18)–(A21) correspond to the S-FSDT, i.e., to a two-variable refined plate theory.

According to (A6), (A19d) and (A20d), the transverse forces dependent on the variable ϕ are expressed as

$$Q_x^S = M_{x,x} + 2M_{xy,y} \tag{A22a}$$

$$Q_y^S = M_{y,y} + 2M_{xy,x} \tag{A22b}$$

where the upper index S was introduced for the S-FSDT.

The membrane components of the transverse forces dependent on the variable w, according to (A19c) and (A20c), take the form

$$\overline{Q}_x^S = N_x w_{,x} + N_{xy} w_{,y} \tag{A23a}$$

$$\overline{Q}_y^S = N_y w_{,y} + N_{xy} w_{,x} \tag{A23b}$$

Analogously to (A13), the total transverse forces $\hat{Q}_x^S$ and $\hat{Q}_y^S$ for the S-FSDT are as follows

$$\hat{Q}_x^S = Q_x^S + \overline{Q}_x^S = M_{x,x} + 2M_{xy,y} + N_x w_{,x} + N_{xy} w_{,y} \tag{A24a}$$

$$\hat{Q}_y^S = Q_y^S + \overline{Q}_y^S = M_{y,y} + 2M_{xy,x} + N_y w_{,y} + N_{xy} w_{,x} \tag{A24b}$$

Comparing the formulas for transverse forces in pairs of (A11) and (A22), as well as (A13) and (A24), it can be easily noticed that in the case of the S-FSDT, we have a coefficient two at a derivative of the torque M_{xy}, which for the FSDT is equal to one.

Appendix A.1.3. CPT

In the classical theory of plates (CPT), transverse forces are neglected (A6) and, moreover, in (A2) and (A3), it should be [36]

$$w_{,x} = \psi_x \quad w_{,y} = \psi_y \tag{A25}$$

Taking into consideration (A25) in (A5), we have [46]

$$M_x = -D\left(w_{,xx} + v w_{,yy}\right) \tag{A26a}$$

$$M_y = -D\left(w_{,yy} + v w_{,xx}\right) \tag{A26b}$$

$$M_{xy} = -D(1-v)w_{,xy} \tag{A26c}$$

When the above-mentioned relations are accounted for and it is assumed that $Q_x = Q_y = 0$ in (A7), the following equations are obtained [46]

- Equations of equilibrium

$$\int_0^\ell \int_0^b \left[N_{x,x} + N_{xy,y}\right]\delta u \, dx \, dy = 0 \tag{A27a}$$

$$\int_0^\ell \int_0^b \left[N_{xy,x} + N_{y,y}\right]\delta v \, dx \, dy = 0 \tag{A27b}$$

- Boundary conditions for $x = const$

$$\int_0^b \left[N_x - h p^0(y)\right]\delta u \, dy |x = const = 0 \tag{A28a}$$

$$\int_0^b N_{xy} \delta v \, dy |x = const = 0 \tag{A28b}$$

$$\int_0^\ell \left[M_{x,x} + 2M_{xy,y} + N_{xy} w_{,y} + N_x w_{,x}\right]\delta w \, dy |x = const = 0 \tag{A28c}$$

$$\int_0^b M_x \delta w_{,x} \, dy |x = const = 0 \tag{A28d}$$

for $y = const$

$$\int_0^\ell N_y \delta v \, dx \Big| y = const = 0 \tag{A29a}$$

$$\int_0^\ell N_{xy} \delta u \, dx \Big| y = const = 0 \tag{A29b}$$

$$\int_0^\ell \left[M_{y,y} + 2M_{xy,x} + N_{xy} w_{,x} + N_y w_{,y}\right]\delta w \, dx \Big| y = const = 0 \tag{A29c}$$

$$\int_0^\ell \left[M_y\right]\delta w_{,y}dx\Big|y = const = 0 \tag{A29d}$$

for the plate corner, i.e., for $x = const$ and $y = const$

$$2M_{xy}\delta w|x = const\big|y = const = 0 \tag{A30}$$

In the history of the CPT, for the first two components in (A28c) and (A29c), a term of equivalent Kirchhoff transverse forces, was introduced and defined as

$$Q_x^C = M_{x,x} + 2M_{xy,y} \tag{A31a}$$

$$Q_y^C = M_{y,y} + 2M_{xy,x} \tag{A31b}$$

where the upper index C was used for the CPT.

By analogy to the FSDT and the S-FSDT, according to (A28c) and (A29c), the following components were assumed

$$\overline{Q}_x^C = N_x w_{,x} + N_{xy} w_{,y} \tag{A32a}$$

$$\overline{Q}_y^C = N_y w_{,y} + N_{xy} w_{,x} \tag{A32b}$$

The above-mentioned components of transverse forces result from the projection of membrane forces. Thus, they can be referred to as equivalent Kirchhoff membrane forces.

Taking into account the two above-mentioned systems, the total equivalent Kirchhoff transverse forces $\hat{Q}_x^C$ and $\hat{Q}_y^C$ for the CPT were written as [36,45,46]

$$\hat{Q}_x^C = Q_x^C + \overline{Q}_x^C = M_{x,x} + 2M_{xy,y} + N_x w_{,x} + N_{xy} w_{,y} \tag{A33a}$$

$$\hat{Q}_y^C = Q_y^C + \overline{Q}_y^C = M_{y,y} + 2M_{xy,x} + N_y w_{,y} + N_{xy} w_{,x} \tag{A33b}$$

Formulas (A24) and (A33) have the same structure. It should be remembered that for the S-FDST, these equations are for the two variables w, ϕ, whereas for the CPT, only for the variable w.

Appendix A.1.4. Shear Forces

When we compare relationships (A13) for the FSDT, (A24) for the S-FDST and (A32) for the CPT, we can see that the expressions for total transverse forces are identical in practice. In the FSDT, at the derivative of the torque M_{xy} there is a coefficient equal to one and not two, as it takes place for the S-FSDT and the CPT. It is caused by additional boundary conditions (A9c) and (A10c) for the FSDT.

For postbuckling states, there are membrane forces in the central plane that yield simultaneously projections for additional components of transverse forces (A12), (A23) and (A32) for the FSDT, the S-FSDT and the CPT, correspondingly. In the case of postbuckling equilibrium paths, these additional components are larger than the components of transverse forces (A11), (A22) and (A31).

Appendix A.2. Solution to Governing Equations within the FSDT, the S-FSDT and the CPT

A square isotropic plate freely supported along all edges, compressed along the x-axis (Figure 1), was analysed. The plate of the dimensions a and the thickness h was assumed to have the following material constants: Young's modulus E and Poisson's ratio v. The considerations were limited to an elastic range.

Appendix A.2.1. Continuity Equation of Deformations (the So-Called Inseparability Equation of Deformations)

On the assumption that relations (A1) and (A4) hold for the three theories under consideration (i.e., the FSDT, the S-FSDT and the CPT), two first equations of equilibrium for each theory are the same (cf. (A8a,b), (A18a,b) and (A27a,b)). To solve them, a function of Airy forces F was introduced. It is defined as follows [36,45]

$$N_x = \sigma_x h = F_{,yy} \tag{A34a}$$

$$N_y = \sigma_y h = F_{,xx} \tag{A34b}$$

$$N_{xy} = \tau_{xy} h = -F_{,xy} \tag{A34c}$$

Taking into consideration (A34) in (A8a,b), (A18a,b) and (A26a,b), it was found that both the equations were identically zero. That system of equations was substituted by one equation referred to as a continuity equation of deformations or an inseparability equation of deformations.

For this purpose, a system of Equations (A1) was rewritten as

$$\varepsilon_{x,yy} + \varepsilon_{y,xx} - \gamma_{xy,xy} = w_{,xy}^2 - w_{,xx}w_{,xy} \tag{A35}$$

When we take account of relations (A1), (A4), (A34) in the above-mentioned system, we obtain a continuity equation of deformations in the form [36,45]

$$VVF \equiv F_{,xxxx} + 2F_{,xxyy} + F_{,yyyy} = E\left(w_{,xy}^2 - w_{,xx}w_{,xy}\right) \tag{A36}$$

The equation is linear with respect to F and nonlinear with respect to w.

The deflection of a square plate freely supported is approximated in the following way [36]

$$w = W\sin\frac{\pi x}{a}\sin\frac{\pi y}{a} \tag{A37}$$

which satisfies the following boundary conditions

$$w(x = 0) = w(x = a) = w(y = 0) = w(y = a) = 0 \tag{A38}$$

When (A37) is substituted into (A36), a function of the membrane forces F is determined, and then components of the membrane forces are defined as [36]

$$N_x = F_{,yy} = -\frac{EhW^2}{8}\left(\frac{\pi}{a}\right)^2\cos\frac{2\pi y}{a} - ph \tag{A39a}$$

$$N_y = F_{,xx} = -\frac{EhW^2}{8}\left(\frac{\pi}{a}\right)^2\cos\frac{2\pi x}{a} \tag{A39b}$$

$$N_{xy} = -F_{,xy} = 0 \tag{A39c}$$

where p is the stress along the x-axis (Figure 1).

The functions of forces (A39) fulfil the following boundary conditions [36,46]

$$u(x = 0) = u(x = a) = const \quad N_{xy}(x = 0) = N_{xy}(x = a) = 0 \tag{A40a}$$

$$v(y = 0) = v(y = a) = const \quad N_{xy}(y = 0) = N_{xy}(y = a) = 0 \tag{A40b}$$

Appendix A.2.2. Solution to the Nonlinear Problem of Stability for the CPT

Within the CPT, a solution to the nonlinear problem after an introduction of the force function (A34) is composed of two nonlinear Equations (A36) and (A27c) with respect to F and w, correspondingly.

One of the solutions is Relationship (A39). The second solution is the equation of Equilibrium (A27c) with respect to w, which was solved with the Galerkin–Bubnov method [36].

The assumed function of deflection in (A37), according to (A26), fulfils the boundary conditions of free support

$$M_x(x=0) = M_x(x=a) = M_y(y=0) = M_y(y=a) = 0 \tag{A41}$$

To solve the problem, Relationships (A37) and (A39) are introduced into Equation (A27c). Finally, we obtain an equation that describes the postbuckling equilibrium path for the CPT

$$p^C = p_{cr}^C + \frac{E\pi^2}{8a^2}W^2 \tag{A42}$$

where

$$p_{cr}^C = \frac{4D\pi^2}{ha^2} \tag{A43}$$

Relationship (A42) can also be expressed as [36]

$$\left(1 - \frac{p^C}{p_{cr}^C}\right) + 0.34125\left(\frac{W}{h}\right)^2 = 0 \tag{A44}$$

Transverse forces in the CPT are referred to as equivalent Kirchhoff transverse forces (A31) and, according to (A37), expressed with the following relations

$$Q_x^C = M_{x,x} + 2M_{xy,y} = (3-v)D\left(\frac{\pi}{a}\right)^3 W\cos\frac{\pi x}{a}\sin\frac{\pi y}{a} \tag{A45a}$$

$$Q_y^C = M_{y,y} + 2M_{xy,x} = (3-v)D\left(\frac{\pi}{a}\right)^3 W\sin\frac{\pi x}{a}\cos\frac{\pi y}{a} \tag{A45b}$$

From (A32), for equivalent Kirchhoff forces in compression, taking account of (A3) and (A39), we have

$$\overline{Q}_x^C = N_x w_{,x} + N_{xy}w_{,y} = -\frac{EhW^3}{8}\left(\frac{\pi}{a}\right)^3 \cos\frac{\pi x}{a}\sin\frac{\pi y}{a}\cos\frac{2\pi y}{a} - phW\frac{\pi}{a}\cos\frac{\pi x}{a}\sin\frac{\pi y}{a} \tag{A46a}$$

$$\overline{Q}_y^C = N_y w_{,y} + N_{xy}w_{,x} = -\frac{EhW^3}{8}\left(\frac{\pi}{a}\right)^3 \sin\frac{\pi x}{a}\cos\frac{\pi y}{a}\cos\frac{2\pi x}{a} \tag{A46b}$$

For transversally inextensible membrane plates, the shear forces are independent of membrane deformation.

The relations for total equivalent Kirchhoff transverse forces result from the two above-mentioned systems of equations and (A33)

$$\hat{Q}_x^C = (3-v)D\left(\frac{\pi}{a}\right)^3 W\cos\frac{\pi x}{a}\sin\frac{\pi y}{a} - phW\frac{\pi}{a}\cos\frac{\pi x}{a}\sin\frac{\pi y}{a} - \frac{EhW^3}{8}\left(\frac{\pi}{a}\right)^3 \cos\frac{\pi x}{a}\sin\frac{\pi y}{a}\cos\frac{2\pi y}{a} \tag{A47a}$$

$$\hat{Q}_y^C = (3-v)D\left(\frac{\pi}{a}\right)^3 W\sin\frac{\pi x}{a}\cos\frac{\pi y}{a} - \frac{EhW^3}{8}\left(\frac{\pi}{a}\right)^3 \sin\frac{\pi x}{a}\cos\frac{\pi y}{a}\cos\frac{2\pi x}{a} \tag{A47b}$$

As one can easily notice, the first terms on right-hand sides in (A47) are positive, whereas the remaining terms are negative. The two first terms in (A47a) and the first term in (A47b) are a linear function of W, whereas the remaining terms are nonlinear with respect to W. To solve the problem, the deflection W should be determined from Equation (A44) for the given load p, and next the values of transverse forces (A45)–(A47).

The first component $\hat{Q}_x^C$ in (A47) depends formally on the value of the compressive load p, whereas the second $\hat{Q}_y^C$ does not. However, it should be remembered that for the given load $p > p_{cr}^C$, we have the deflection W, and thus, indirectly, the equivalent Kirchhoff transverse forces $\hat{Q}_x^C$ and $\hat{Q}_y^C$ depend on p and W.

Appendix A.2.3. Solution to the Nonlinear Problem of Stability for the S-FSDT

A solution to the nonlinear stability problem for the S-FSDT consists of three equations: (A36) and (A18c,d). The solutions to (A36) are relationships (A39). The solution to the remaining two equations, when we take into account (A37), is predicted in the form

$$\phi = \Phi sin\frac{\pi x}{a}sin\frac{\pi y}{a} \tag{A48}$$

which, according to (A16) fulfils, the boundary conditions of free support

$$M_x(x=0) = M_x(x=a) = M_y(y=0) = M_y(y=a) = 0 \tag{A49}$$

Having substituted (A37) and (A48) into (A18d), a linear relation between amplitudes of the functions W and Φ was determined as

$$W = \Phi(1+\eta) \tag{A50}$$

where

$$\eta = \frac{2D}{k^2Gh}\left(\frac{\pi}{a}\right)^2 \tag{A51}$$

Nonlinear Equation (A18c) was solved within the Galerkin–Bubnov method, as for the CPT. For this purpose, Relations (A16), (A17), (A39) and (A50) were introduced into the above-mentioned equation, and after some transformations, an equation describing the postbuckling equilibrium path was obtained

$$p^S = p_{cr}^S + \frac{EW^2}{8}\left(\frac{\pi}{a}\right)^2 \tag{A52a}$$

where

$$p_{cr}^S = \frac{4D\pi^2}{ha^2}\frac{1}{1+\eta} \tag{A52b}$$

Relation (A52) can be expressed in an analogous way to (A44) as

$$\left(1-\frac{p^s}{p_{cr}^S}\right) + 0.34125(1+\eta)\left(\frac{W}{h}\right)^2 = 0 \tag{A53}$$

From the formal point of view, Relationships (A42) and (A52) have an identical structure due load and the deflection W, and a difference lies in critical forces only. From a comparison of (A43) and (A53), we have

$$p_{cr}^S = \frac{4D\pi^2}{ha^2}\frac{1}{1+\eta} = p_{cr}^C\frac{1}{1+\eta} \tag{A54}$$

The transverse components of forces in the S-FSDT, according to (A16), (A22) and (A50), are in the form

$$Q_x^S = (3-v)D\left(\frac{\pi}{a}\right)^3\frac{W}{1+\eta}cos\frac{\pi x}{a}sin\frac{\pi y}{a} \tag{A55a}$$

$$Q_y^S = (3-v)D\left(\frac{\pi}{a}\right)^3\frac{W}{1+\eta}sin\frac{\pi x}{a}cos\frac{\pi y}{a} \tag{A55b}$$

and, according to (A23), (A37), (A39) and (A48), we have

$$\overline{Q}_x^S = -\frac{EhW^3}{8}\left(\frac{\pi}{a}\right)^3cos\frac{\pi x}{a}sin\frac{\pi y}{a}cos\frac{2\pi y}{a} - phW\frac{\pi}{a}cos\frac{\pi x}{a}sin\frac{\pi y}{a} \tag{A56a}$$

$$\overline{Q}_y^S = -\frac{EhW^3}{8}\left(\frac{\pi}{a}\right)^3 sin\frac{\pi x}{a}cos\frac{\pi y}{a}cos\frac{2\pi x}{a} \tag{A56b}$$

Systems of equations (A46) and (A56) have an identical structure.

From the above two systems of equations and (A24), we obtain

$$\hat{Q}_x^S = (3-v)D\left(\frac{\pi}{a}\right)^3\frac{W}{1+\eta}cos\frac{\pi x}{a}sin\frac{\pi y}{a} - phW\frac{\pi}{a}cos\frac{\pi x}{a}sin\frac{\pi y}{a}$$
$$- \frac{EhW^3}{8}\left(\frac{\pi}{a}\right)^3 cos\frac{\pi x}{a}sin\frac{\pi y}{a}cos\frac{2\pi y}{a} \tag{A57a}$$

$$\overline{Q}_y^S = (3-v)D\left(\frac{\pi}{a}\right)^3\frac{W}{1+\eta}sin\frac{\pi x}{a}cos\frac{\pi y}{a} - \frac{EhW^3}{8}\left(\frac{\pi}{a}\right)^3 sin\frac{\pi x}{a}cos\frac{\pi y}{a}cos\frac{2\pi x}{a} \tag{A57b}$$

Comparing relations (A47) for the CPT and (A57) for the S-FSDT, it can be easily seen that the formulas differ only by the factor $1/(1+\eta)$ in the first term of right-hand sides of expressions in (A57). Moreover, the first two components of the right-hand side of (A57a) and the first term in (A57b) depend linearly on W, whereas the remaining ones are in the third power for W.

Appendix A.2.4. Solution to the Nonlinear Problem of Stability for the FSDT

For the FSDT, a solution to the nonlinear stability problem consists of a function of Airy forces F (A39) and a system of equilibrium equations (A8c–e). The solutions to these equations are the function w (A37) and the functions ψ_x, ψ_y, which were assumed in the form

$$\psi_x = \Psi_x cos\frac{\pi x}{a}sin\frac{\pi y}{a} \tag{A58a}$$

$$\psi_y = \Psi_y sin\frac{\pi x}{a}cos\frac{\pi y}{a} \tag{A58b}$$

The function w (A37) fulfils the boundary conditions (A38), whereas the functions ψ_x, ψ_y satisfy the conditions of free support

$$\psi_x(y=0) = \psi_x(y=a) = 0 \tag{A59a}$$

$$\psi_y(x=0) = \psi_y(x=a) = 0 \tag{A59b}$$

$$M_x(x=0) = M_x(x=a) = 0 \tag{A59c}$$

$$M_y(y=0) = M_y(y=a) = 0 \tag{A59d}$$

After the substitution of (A58) into two equations (A8d,e), the following linear dependencies between the functions W, Ψ_x, Ψ_y are attained

$$\Psi_x = \Psi_y = \frac{W}{1+\eta}\left(\frac{\pi}{a}\right) \tag{A60}$$

where relation (A51) holds.

A further step was a solution to the nonlinear Equation (A51), which describes the postbuckling equilibrium path. This solution was obtained employing, similarly as for the CPT and the S-FSDT, the Galerkin–Bubnov method. The postbuckling equation of equilibrium has the form

$$p^F = p_{cr}^F + \frac{EW^2}{8}\left(\frac{\pi}{a}\right)^2 \tag{A61a}$$

or

$$\left(1-\frac{p^F}{p_{cr}^F}\right) + 0.34125(1+\eta)\left(\frac{W}{h}\right)^2 = 0 \tag{A61b}$$

where

$$p_{cr}^F = \frac{4D\pi^2}{ha^2}\frac{1}{1+\eta} = p_{cr}^S = p_{cr}^C\frac{1}{1+\eta} \tag{A62}$$

As can be easily seen in (A62), the values of critical loads for the S-FSDT and the FSDT are identical. The postbuckling equilibrium paths for the S-FSDT (A52), (A52a) and the FSDT (A61), (A61a) are the same, and, additionally, they differ from the CPT only by the factor $1/(1+\eta)$.

In a comparison of the Kirchhoff theory (i.e., CPT) with the Mindlin (the S-FSDT version) and Reissner (FSDT) theories, the reduction factor η plays a very important role (A51).

For the CPT, the factor η, according to [3,46], can be written for the case when $G \to \infty$, which leads to the relation $\eta = 0$.

The reduction factor η (A51) can be transformed to

$$\eta = \frac{2D}{k^2Gh}\left(\frac{\pi}{a}\right)^2 = \frac{\pi^2}{3k^2(1-v)}\left(\frac{h}{a}\right)^2 \tag{A63}$$

For $k^2 = 5/6$, $v = 0.3$ from (A63), we obtain $\eta = 5.64(h/a)^2$. When $h/a = 0.1$, $\eta = 0.0564$; whereas when $h/a = 0.04$, $\eta = 0.009$. That means that for the plate of medium thickness (when $h/a = 0.1$), the correction for the S-FSDT and the FSDT when compared to the CPT is lower than 6%. When $h/a = 0.04$, it is less than 1%. For thin plates (i.e., when h/a<0.05), the reduction factor $\eta < 0.014$.

In the last stage, the determined components of transverse forces in the FSDT, according to (A60), (A37) and (A8), were written as

$$Q_x^F = 2D\left(\frac{\pi}{a}\right)^3\frac{W}{1+\eta}\cos\frac{\pi x}{a}\sin\frac{\pi y}{a}$$
$$Q_y^F = 2D\left(\frac{\pi}{a}\right)^3\frac{W}{1+\eta}\sin\frac{\pi x}{a}\cos\frac{\pi y}{a} \tag{A64}$$

and additionally, according to (A9), (A10), (A39) and (A58), we have

$$\overline{Q}_x^F = -\frac{EhW^3}{8}\left(\frac{\pi}{a}\right)^3\cos\frac{\pi x}{a}\sin\frac{\pi y}{a}\cos\frac{2\pi y}{a} - phW\frac{\pi}{a}\cos\frac{\pi x}{a}\sin\frac{\pi y}{a}$$
$$\overline{Q}_y^F = -\frac{EhW^3}{8}\left(\frac{\pi}{a}\right)^3\sin\frac{\pi x}{a}\cos\frac{\pi y}{a}\cos\frac{2\pi x}{a} \tag{A65}$$

Dependencies (A46), (A56) and (A65) have the same structure for the CPT, the S-FSDT and the FSDT. These are transverse force components dependent on membrane forces.

From the system of the above two equations and (A24), we obtain

$$\hat{Q}_x^F = 2D\left(\frac{\pi}{a}\right)^3\frac{W}{1+\eta}\cos\frac{\pi x}{a}\sin\frac{\pi y}{a} - phW\frac{\pi}{a}\cos\frac{\pi x}{a}\sin\frac{\pi y}{a}$$
$$- \frac{EhW^3}{8}\left(\frac{\pi}{a}\right)^3\cos\frac{\pi x}{a}\sin\frac{\pi y}{a}\cos\frac{2\pi y}{a} \tag{A66a}$$

$$\overline{Q}_y^F = 2D\left(\frac{\pi}{a}\right)^3\frac{W}{1+\eta}\sin\frac{\pi x}{a}\cos\frac{\pi y}{a} - \frac{EhW^3}{8}\left(\frac{\pi}{a}\right)^3\sin\frac{\pi x}{a}\cos\frac{\pi y}{a}\cos\frac{2\pi x}{a} \tag{A66b}$$

When Relations (A47) for the CPT and (A57) for the S-FSDT are compared, one can easily see that they differ in the factor $1/(1+\eta)$ only in the first term of right-hand sides of expressions (A57).

Like for the CPT and the S-FSDT, the first two terms of the right-hand side (A66a) and the first term in (A66b) depend linearly on W, whereas the remaining terms are in the third power for W.

Appendix A.2.5. Solution to the Nonlinear Problem of Stability in the FEM (ANSYS)

In the given system of coordinates (Figure A1), a square plate jointly supported was investigated. The support was achieved taking the degrees of freedom away along the following directions: x for

$x = 0$ (displacement $u = 0$), y for $y = 0$ (displacement $v = 0$), z for $z = 0$ (displacement $w = 0$, perpendicular to the central plane of the plate).

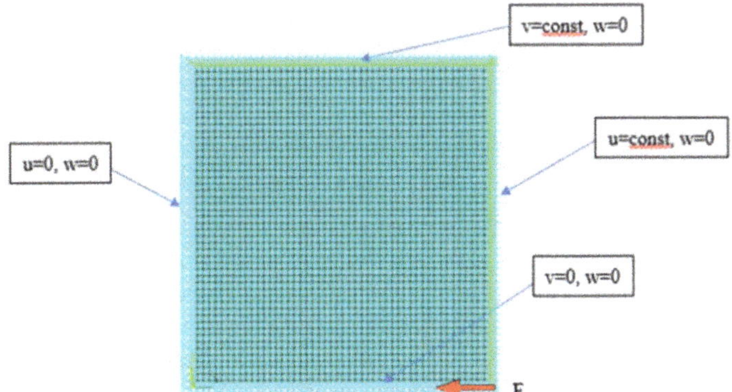

Figure A1. FEM model of the plate.

Numerical calculations were performed with the commercial ANSYS® software [47] based on the FEM. In Figure A1, a compression force along the x-direction, applied to the node, is indicated. The displacements u and v for x = a and y = a (a-plate length/width), respectively, were assumed to be constant, using the *Couple Dof's* function, which allows for controlling a group of nodes (the so-called *Slaves*) with one main node (the so-called *Master*). In the numerical model, a Shell181 finite element with six degrees of freedom was used. The FEM model had 2500 elements, 15,606 degrees of freedom, and the element size was 2 mm. In the nonlinear problem, an option of *Update geom*, i.e., a possibility to impose an initial shape imperfection, which is a solution to the plate postbuckling state, was involved. In the computations, the imperfection amplitude equal to $w0 = 0.0001$ h was assumed. The problem was solved within the Newton–Raphson method for the system of equilibrium equations, which is a system of algebraic equations, for 1000 load steps within the range $1.001 \leq P/P_{cr} \leq 2.000$.

In the Shell181 element description [46], the transverse forces are denoted as Q_{13} and Q_{23}. For uniform notations in the presented study, it is assumed that

$$Q_x^A \equiv Q_{13} \quad Q_y^A \equiv Q_{23} \tag{A67}$$

where the index A refers to ANSYS® [47] based on the FEM.

In FEM procedures, the manufacturer issues certain supplements and corrections with respect to the traditional version of the assumed theory, for instance, for the FSDT.

References

1. Reissner, E. On the theory of bending of elastic plates. *J. Math. Phys.* **1944**, *23*, 184–191. [CrossRef]
2. Reissner, E. The effect of transverse shear deformation on the bending of elastic plates. *ASME J. Appl. Mech.* **1945**, *12*, A69–A77.
3. Mindlin, R.D. Influence inertia and shear on flexural motions of isotropic, elastic plates. *ASME J. Appl. Mech.* **1951**, *18*, 31–38.
4. Baptista, M. An elementary derivation of basic equations of the Reissner and Mindlin plate theories. *Eng. Struct.* **2010**, *32*, 906–909. [CrossRef]
5. Baptista, M. Refined Mindlin-Reissner theory of forced vibrations of shear deformable plates. *Eng. Struct.* **2011**, *33*, 265–272. [CrossRef]
6. Baptista, M. Comparison of Reissner, Mindlin and Reddy plate models with exact three dimensional solution for simply supported isotropic and transverse inextensible rectangular plate. *Meccanica* **2012**, *47*, 257–268. [CrossRef]

7. Labuschagne, A.; van Resenburg, N.F.J.; van der Merwe, A.J. Vibration of a Reissner-Mindlin-Timoshenko plate beam system. *Math. Comput. Model.* **2009**, *50*, 1033–1044. [CrossRef]
8. Wang, C.M.; Lim, G.T.; Reddy, J.N.; Lee, K.H. Relationships between bending solutions of Reissner and Mindlin plate theories. *Eng. Struct.* **2001**, *23*, 838–849. [CrossRef]
9. Chróścielewski, J.; Makowski, J.; Pietraszkiewicz, W. *Statyka i Dynamika Powłok Wielopłatowych*; IPPT PAN: Warsaw, Poland, 2004. (In Polish)
10. Woźniak, C. *Mechanics of Elastic Plates and Shells*; PWN: Warsaw, Poland, 2001; p. 766. (In Polish)
11. Wu, S.R. Reissner-Mindlin plate theory for elastodynamics. *J. Appl. Math.* **2004**, *3*, 179–189. [CrossRef]
12. Lo, K.H.; Christensen, R.M.; Wu, E.M. A high-order theory of plate deformation, Part 1: Homogeneous plates. *ASME J. Appl. Mech.* **1977**, *44*, 663–668. [CrossRef]
13. Lo, K.H.; Christensen, R.M.; Wu, E.M. A high-order theory of plate deformation. Part 2: Laminated plates. *ASME J. Appl. Mech.* **1977**, *44*, 669–676. [CrossRef]
14. Reddy, J.N. A general non-linear third-order theory of plates with moderate thickness. *Int. J. Non-Linear Mech.* **1990**, *25*, 677–686. [CrossRef]
15. Reddy, J.N.; Phan, N.D. Stability and vibration of isotropic, orthotropic and laminated plates according to a higher-order shear deformation theory. *J. Sound Vib.* **1985**, *98*, 157–170. [CrossRef]
16. Volmir, A.S. *Flexible Plates and Shells*; State Publishing House of Technical-Theoretical Literature: Moscow, Russia, 1956. (In Russian)
17. Cai, L.; Rong, T.; Chen, D. Generalized mixed variational methods for Reissner plate and its applications. *Comput. Mech.* **2002**, *30*, 29–37. [CrossRef]
18. Cen, S.; Shang, Y. Developments of Mindlin-Reissner plate elements. *Math. Probl. Eng.* **2015**, *2015*, 456740. [CrossRef]
19. Endo, M.; Kimura, N. An alternative formulation of the boundary value problem for the Timoshenko beam and Mindlin plate. *J. Sound Vib.* **2007**, *301*, 355–373. [CrossRef]
20. Ghugal, Y.M.; Shimpi, R.P. A review of refined shear deformation theories of isotropic and anisotropic laminated plates. *J. Reinf. Plast. Compos.* **2002**, *21*, 775–813. [CrossRef]
21. Kim, S.E.; Thai, H.-T.; Lee, J. A two variable refined plate theory for laminated composite plates. *Compos. Struct.* **2009**, *89*, 197–205. [CrossRef]
22. Kim, S.-E.; Thai, H.-T.; Lee, J. Buckling analysis of plates using the two variable refined plate theory. *Thin-Walled Struct.* **2009**, *47*, 455–462. [CrossRef]
23. Nelson, R.B.; Lorch, D.R. A refined theory for laminated orthotropic plates. *J. Appl. Mech.* **1974**, *41*, 177–183. [CrossRef]
24. Park, M.; Choi, D.-H. A two-variable first-order shear deformation theory considering in-plane rotation for bending, buckling and free vibration analyses of isotropic plates. *Appl. Math. Model.* **2018**, *61*, 49–71. [CrossRef]
25. Reddy, J.N. A refined nonlinear theory of plates with transverse shear deformation. *Int. J. Solids Struct.* **1984**, *20*, 881–896. [CrossRef]
26. Shimpi, R.P.; Patel, H.G. A two variable refined plate theory for orthotropic plate analysis. *Int. J. Solids Struct.* **2006**, *43*, 6783–6799. [CrossRef]
27. Shimpi, R.P.; Patel, H.G. Free vibrations of plate using two variable refined plate theory. *J. Sound Vib.* **2006**, *296*, 979–999. [CrossRef]
28. Shimpi, R.P. Refined plate theory and its variants. *AIAA J.* **2002**, *40*, 137–146. [CrossRef]
29. Soldatos, K.P. On certain refined theories for plate bending. *ASME J. Appl. Mech.* **1988**, *55*, 994–995. [CrossRef]
30. Thai, H.-T.; Kim, S.-E. Free vibration of laminated composite plates using two variable refined plate theory. *Int. J. Mech. Sci.* **2010**, *52*, 626–633. [CrossRef]
31. Shimpi, R.P.; Shetty, R.A.; Guha, A. A single variable refined theory for free vibrations of a plate using inertia related terms in displacements. *Eur. J. Mech. A Solids* **2017**, *65*, 136–148. [CrossRef]
32. Vasiliev, V.V. Modern conceptions of plate theory. *Compos. Struct.* **2000**, *48*, 39–48. [CrossRef]
33. Taylor, M.W.; Vasiliev, V.V.; Dillard, D.A. On the problem of shear-locking in finite elements based on shear deformable plate theory. *Int. J. Solids Struct.* **1997**, *34*, 859–875. [CrossRef]
34. Vasiliev, V.V. The theory of thin plates. *Mech. Solids* **1992**, *27*, 22–42. (In Russian)
35. Vasiliev, V.V. A discussion on classical plate theory. *Mech. Solids* **1995**, *30*, 130–138. (In Russian)

36. Vasiliev, V.V.; Lure, S.A. On refined theories of beams, plates, and shells. *J. Comput. Math.* **1992**, *26*, 546–557. [CrossRef]

37. Allman, D.J. A compatible triangular element including vertex rotations for plane elasticity analysis. *Comput. Struct.* **1984**, *19*, 1–8. [CrossRef]

38. Averill, R.C.; Reddy, J.N. Behaviour of plate elements based on the first order shear deformation theory. *Eng. Comput.* **1990**, *7*, 57–74. [CrossRef]

39. Bathe, K.J. *Finite Element Procedures*; Prentice-Hall International, Inc.: Upper Saddle River, NJ, USA, 1996.

40. Kant, T. Numerical analysis of thick plates. *Comput. Methods Appl. Mech. Eng.* **1982**, *31*, 1–18. [CrossRef]

41. Zienkiewicz, O.C.; Taylor, R.L. *The Finite Element Method*, 4th ed.; McGraw-Hill Book Co.: London, UK, 1991; Volume 2.

42. Jones, R.M. *Mechanics of Composite Materials*, 2nd ed.; Taylor and Francis, Inc.: Philadelphia, PA, USA, 1999.

43. Thai, H.-T.; Choi, D.-H. A simple first-order shear deformation theory for laminated composite plates. *Compos. Struct.* **2013**, *106*, 754–763. [CrossRef]

44. Manevich, A.; Kolakowski, Z. To the theory of transverse vibration of plates regarding shear deformation. *Int. Appl. Mech.* **2014**, *50*, 196–205. [CrossRef]

45. Bilstein, W. *Anwendung der Nichtlinearen Beultheorie auf Vorverformte, Mit Diskreten Laengssteifen Verstaerkte Rechteckplatten unter Laengsbelastung*; Veroeffentlichungen des Institutes fuer Statik and Stahlbau der Technichen Hochschule Darmstadt: Darmstadt, Germany, 1974; Heft 25.

46. Kolakowski, Z.; Krolak, M. Modal coupled instabilities of thin-walled composite plate and shell structures. *Compos. Struct.* **2006**, *76*, 303–313. [CrossRef]

47. *User's Guide ANSYS®*; Academic Research, Help System, (n.d.): Minneapolis, MN, USA, 2019.

Publisher's Note: MDPI stays neutral with regard to jurisdictional claims in published maps and institutional affiliations.

Article

Modeling Tensile Damage and Fracture Behavior of Fiber-Reinforced Ceramic-Matrix Minicomposites

Zhongwei Zhang [1,*], Yufeng Liu [2,3], Longbiao Li [4,*] and **Daining Fang [1]**

[1] Institute of Advanced Structure Technology, Beijing Institute of Technology, Beijing 100081, China; fangdn@pku.edu.cn
[2] Science and Technology of Advanced Functional Composite Materials Laboratory, Aerospace Research Institute of Materials & Processing Technology, Beijing 100076, China; liuyf123@csu.edu.cn
[3] Powder Metallurgy Research Institute, Central South University, Changsha 410083, China
[4] College of Civil Aviation, Nanjing University of Aeronautics and Astronautics, No.29, Yudao St., Nanjing 210016, China
* Correspondence: 6120190123@bit.edu.cn (Z.Z.); llb451@nuaa.edu.cn (L.L.)

Received: 19 August 2020; Accepted: 25 September 2020; Published: 27 September 2020

Abstract: Evolution of damage and fracture behavior of fiber-reinforced mini ceramic-matrix composites (mini-CMCs) under tensile load are related to internal multiple damage mechanisms, i.e., fragmentation of the brittle matrix, crack defection, and fibers fracture and pullout. In this paper, considering multiple micro internal damage mechanisms and related models, a micromechanical constitutive stress–strain relationship model is developed to predict the nonlinear mechanical behavior of mini-CMCs under tensile load corresponding to different damage domains. Relationships between multiple micro internal damage mechanisms mentioned above and tensile micromechanical multiple damage parameters are established. Experimental tensile nonlinear behavior, internal damage evolution, and micromechanical tensile damage parameters corresponding to different damage domains of two different types of mini-CMCs are predicted. The effects of constitutive properties and damage-related parameters on nonlinear behavior of mini-CMCs are discussed.

Keywords: ceramic-matrix composites (CMCs); minicomposite; tensile; damage; fracture

---

1. Introduction

With the development of thermodynamics, the advancement of component integrated design technology, the weight reduction brought by structural simplification and the comprehensive development of material technology, the thrust-to-weight ratio of aeroengines has gradually increased. However, on the premise of maintaining the engine layout and not changing the conventional metal materials, the improvement of aerodynamic, thermal, component design and structural weight reduction techniques can only increase the ratio between the thrust and weight of the aeroengines to approximately 14. For aeroengines with ratios between the thrust and a weight of approximately 12–15 or higher, more breakthroughs must be made in new materials, new process applications and new structural design, such as polymer-matrix composites (PMCs) or metal-matrix composites (MMCs) for cold section components in aeroengine and ceramic-matrix composites (CMCs) for hot section components in aeroengines. Two main types of CMCs are used in aeroengines, including, C/SiC and SiC/SiC. For aeroengines, the operating temperatures of C/SiC and SiC/SiC are approximately 1650 °C and 1450 °C, respectively. Increasing the operating temperature of SiC fibers can raise the operating temperature of SiC/SiC to 1650 °C. Results show that the application of CMCs in the hot section components of the combustion chamber, turbine, and nozzle can increase the operating temperature of the aeroengine by 300–500 °C, reduce the weight by 50–70% and increase the thrust by 30–100% [1–4].

In order to safely and confidently implement CMCs in different thermo-structural component applications, one should understand the complexity of designing with such brittle composite systems. The mechanical behavior of minicomposites can be used to analyze the internal complex damage evolution of woven composites [5–8]. Many in situ monitoring techniques, including acoustic emission (AE), X-ray microtomography, and acousto-ultrasonics (AU), have been used to detect and reveal internal damage evolution mechanisms in minicomposites [9–12]. Under tensile load, the damages in the minicomposites can be divided into different stages and the damage stages are affected by the fiber type and interphases [13–16]. Heat-treatment at elevated temperature causes damages in the fiber and interphase, leading to the degradation of tensile performance of mini-CMCs [17–21].

The objective in the present analysis is to build the relationship between tensile nonlinear damage behavior and multiple micro internal damage mechanisms of mini-CMCs. A micromechanical nonlinear stress–strain relationship model is given to predict the tensile stress–strain relationship of mini-CMCs when either partial or complete interfacial debonding occurs. Experimental tensile damage evolution and related multiple microdamage parameters of two different mini-CMCs are predicted. Effects of constitutive properties and damage-related parameters on damage and fracture of mini-CMCs are analyzed.

2. Theoretical Analysis

The tensile nonlinear behavior of mini-CMCs is mainly attributed to multiple microdamage mechanisms. Through in situ experimental observations and monitoring on internal damage state [12,13], damage evolution and fracture of mini-CMCs under tensile load can be divided into three main domains, including:

(1) Domain I, the linear domain.
(2) Domain II, the nonlinear domain due to microdamages of matrix fragmentation.
(3) Domain III, the secondary linear and final fracture domain due to gradual fibers fracture and pullout.

In this section, the micromechanical nonlinear constitutive relationship of three damage domains mentioned above is developed and related to the microdamage state of matrix fragmentation and fiber fractures and pullouts inside of mini-CMCs. A micromechanical parameter of interface debonding fraction η is adopted to describe composite internal damage evolution.

2.1. Domain I

In Domain I, no damages occur in mini-CMCs, and the linear relationship between applied stress σ and composite stress–strain ε_c can be determined by Equation (1) [4].

$$\varepsilon_c = \frac{\sigma}{E_c} \tag{1}$$

where E_c is the elastic modulus of the undamaged composite.

2.2. Domain II

When multiple microdamage mechanisms occur in mini-CMCs, the micro stress field of the damaged composite is affected by composite constitutive properties (i.e., volume of the fiber and matrix V_f and V_m, and fiber radius r_f) and damage state (i.e., fragmentation length of the brittle matrix l_c, and crack defection length at the interface l_d). The fiber axial stress distribution can be determined by Equation (2) [4].

$$\sigma_f(x) = \begin{cases} \frac{\sigma}{V_f} - \frac{2\tau_i}{r_f}x, & x \in [0, l_d] \\ \sigma_{fo} + \left(\frac{V_m}{V_f}\sigma_{mo} - 2\frac{l_d}{r_f}\tau_i\right)\exp\left(-\rho\frac{x-l_d}{r_f}\right), & x \in \left[l_d, \frac{l_c}{2}\right] \end{cases} \tag{2}$$

where τ_i the interface frictional shear stress, σ_{fo} and σ_{mo} the axial stress of the fiber and matrix in the interface bonding region, and ρ the shear–lag model parameter. The stochastic model in Equation (3) can be used to determine the fragmentation length of the brittle matrix [22], and the relationship between the applied stress and debonding length at the interface is obtained by Equation (4) [23].

$$l_c = l_{sat}\left\{1 - \exp\left[-\left(\frac{\sigma_m}{\sigma_R}\right)^m\right]\right\}^{-1} \tag{3}$$

$$l_d = \frac{r_f}{2}\left(\frac{V_m E_m \sigma}{V_f E_c \tau_i} - \frac{1}{\rho}\right) - \sqrt{\left(\frac{r_f}{2\rho}\right)^2 + \frac{r_f V_m E_m E_f}{E_c \tau_i^2}\zeta_d} \tag{4}$$

where l_{sat} is the saturation matrix fragmentation length, σ_m is the stress carried by the matrix, σ_R is the characteristic stress for matrix fragmentation, and ζ_d is the debonding energy of the interface.

The composite strain ε_c is equal to the strain of intact fibers, and the nonlinear stress–strain constitutive relationship can be determined by Equation (5).

$$\varepsilon_c(\sigma) = \begin{cases} \frac{\sigma}{V_f E_f}\eta - \frac{\tau_i}{E_f}\frac{l_d}{r_f}\eta + \frac{\sigma_{fo}}{E_f}(1-\eta) - \frac{1}{\rho E_f}\frac{2 r_f}{l_c}\left(\frac{V_m}{V_f}\sigma_{mo} - 2\frac{l_d}{r_f}\tau_i\right) \\ \times\left[\exp\left(-\frac{\rho}{2}\frac{l_c}{r_f}(1-\eta)\right)-1\right] - (\alpha_c - \alpha_f)\Delta T, \eta < 1 \\ \frac{\sigma}{V_f E_f} - \frac{\tau_i}{2 E_f}\frac{l_c}{r_f} - (\alpha_c - \alpha_f)\Delta T, \eta = 1 \end{cases} \tag{5}$$

where E_f is the elastic modulus of the fiber, η is the ratio between debonding length l_d and half of fragmentation length of the brittle matrix l_c, α_f and α_c are the axial thermal expansion coefficient of the fiber and the composite, and ΔT is the temperature difference between testing and fabricated temperature.

2.3. Domain III

Upon approaching saturation of fragmentation of the brittle matrix, some fibers gradually fracture and pullout inside of the minicomposite. Considering intact and broken fiber in the different damage regions, the axial stress of the fiber can be determined by Equation (6) [4].

$$\sigma_f(x) = \begin{cases} \Phi - \frac{2\tau_i}{r_f}x, x \in [0, l_d] \\ \sigma_{fo} + \left(\Phi - \sigma_{fo} - 2\frac{l_d}{r_f}\tau_i\right)\exp\left(-\rho\frac{x-l_d}{r_f}\right), x \in \left[l_d, \frac{l_c}{2}\right] \end{cases} \tag{6}$$

where Φ is the stress of unbroken fiber, which can be determined by Equation (7) [24].

$$\frac{\sigma}{V_f} = \Phi(1 - P) + \frac{2\tau_i}{r_f}\langle L \rangle P \tag{7}$$

where $\langle L \rangle$ is the average pullout length of a broken fiber, and P the probability of fiber being broken.

Considering fibers fracture inside of mini-CMCs, the nonlinear stress–strain relationship at damage Domain III is acquired by Equation (5).

$$\varepsilon_c(\sigma) = \begin{cases} \frac{\Phi}{E_f}\eta - \frac{\tau_i}{E_f}\frac{l_d}{r_f}\eta + \frac{\sigma_{fo}}{E_f}(1-\eta) - \frac{1}{\rho E_f}\frac{2 r_f}{l_c}\left(\Phi - \sigma_{fo} - 2\frac{l_d}{r_f}\tau_i\right) \\ \times\left[\exp\left(-\frac{\rho}{2}\frac{l_c}{r_f}(1-\eta)\right)-1\right] - (\alpha_c - \alpha_f)\Delta T, \eta < 1 \\ \frac{\Phi}{E_f} - \frac{\tau_i}{E_f}\frac{l_c}{2 r_f} - (\alpha_c - \alpha_f)\Delta T, \eta = 1 \end{cases} \tag{8}$$

3. Experimental Comparison

Nonlinear tensile damage and fracture behavior of two different types of mini-CMCs are predicted. The effect of heat treatment at 1300 and 1500 °C on tensile nonlinear damage development behavior of mini-CMCs at room temperature is also analyzed. The material properties of different mini-CMCs are listed in Table 1.

Table 1. Material properties of unidirectional SiC/SiC and C/SiC minicomposite.

Items	Hi-Nicalon™ SiC/SiC [13]	Hi-Nicalon™ Type S SiC/SiC [13]	Tyranno™ ZMI SiC/SiC [13]	SiC/SiC [18]	T300™ C/SiC [14]
$r_f/(\mu m)$	7	6	5.5	6.5	3.5
$V_f/(\%)$	25.8	22.8	27.5	23	30
$E_f/(GPa)$	270	400	170	122	120
$E_m/(GPa)$	350	350	350	303	150
$\alpha_f/(10^{-6}/^{\circ}C)$	3.5	4.5	4.0	3.1	−0.38
$\alpha_m/(10^{-6}/^{\circ}C)$	4.6	4.6	4.6	4.6	2.8
$\sigma_R/(MPa)$	420	350	280	220	180
m	6	6	8	3	5
$l_{sat}/(\mu m)$	564	667	667	780	400
m_f	5	5	5	5	5
$\sigma_{uts}/(MPa)$	644	488	498	399	348
$\varepsilon_f/(\%)$	0.45	0.31	0.36	0.88	0.67

3.1. SiC/SiC Minicomposite

Figure 1 shows the experimental data and theoretical predicted results of tensile nonlinear curves and fragmentation density of brittle matrix versus applied stress curves of Hi-Nicalon™ SiC/SiC mini-CMC. The predicted results of tensile curves and fragmentation density of brittle matrix evolution curves both agreed well with experimental data. Three main damage domains are obtained from the nonlinear tensile curve, including:

(1) Domain I, the linear domain. The domain starts from the initial loading to the first matrix fragmentation stress σ_{mc} = 200 MPa, and the first fragmentation density of the matrix is approximately λ_{mc} = 0.02/mm, and the corresponding composite tensile strain is approximately ε_c = 0.06%.

(2) Domain II, the nonlinear region due to matrix fragmentation. The domain starts from σ_{mc} = 200 MPa to the saturation matrix fragmentation stress σ_{sat} = 560 MPa, and the saturation fragmentation density of the matrix is approximately λ_{sat} = 1.76/mm, and the composite tensile strain is approximately ε_c = 0.34%.

(3) Domain III, the secondary linear and final fracture domain. The domain starts from σ_{sat} = 560 MPa to composite tensile strength σ_{uts} = 644 MPa.

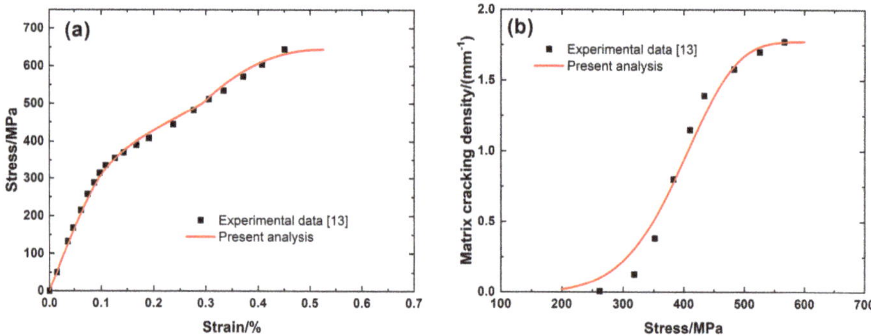

Figure 1. (a) Tensile nonlinear stress–strain curves; and, (b) matrix fragmentation density versus applied stress curves of Hi-Nicalon™ SiC/SiC mini ceramic-matrix composites (mini-CMCs).

Figure 2 shows experimental data and theoretical predicted results of tensile nonlinear curves and fragmentation density of the brittle matrix versus applied stress curves of Hi-Nicalon™ Type S SiC/SiC minicomposite. The predicted results of tensile curves and fragmentation density of brittle matrix evolution curves agree with experimental data. Three main damage domains are obtained from the nonlinear tensile curve, including:

(1)   Domain I, the linear domain. The domain starts from initial loading to $\sigma_{mc} = 200$ MPa with $\lambda_{mc} = 0.05$/mm and $\varepsilon_c = 0.08\%$.

(2)   Domain II, the nonlinear region due to matrix fragmentation. The domain starts from $\sigma_{mc} = 200$ MPa to $\sigma_{sat} = 470$ MPa with $\lambda_{sat} = 1.49$/mm and $\varepsilon_c = 0.3\%$.

(3)   Domain III, the secondary linear and final fracture region. The domain starts from $\sigma_{sat} = 470$ MPa to $\sigma_{uts} = 488$ MPa.

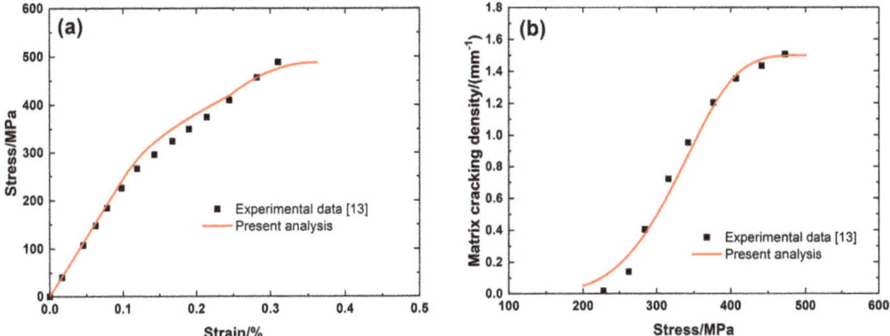

Figure 2. (**a**) Tensile nonlinear stress–strain curves; and, (**b**) matrix fragmentation density versus applied stress curves of Hi-Nicalon^{TM} Type S SiC/SiC mini-CMC.

Figure 3 shows experimental data and theoretical predicted results of tensile nonlinear curves and fragmentation density evolution of brittle matrix curves versus applied stress of Tyranno^{TM} ZMI SiC/SiC minicomposite. Predicted results of tensile curves and fragmentation density of brittle matrix evolution curves agree well with experimental data. Three main damage domains are obtained from the nonlinear tensile curve, including:

(1)   Domain I, the linear region. The domain starts from initial loading to $\sigma_{mc} = 150$ MPa with $\lambda_{mc} = 0.01$/mm and $\varepsilon_c = 0.05\%$.

(2)   Domain II, the nonlinear region due to matrix fragmentation. The domain starts from $\sigma_{mc} = 150$ MPa to $\sigma_{sat} = 350$ MPa with $\lambda_{mc} = 1.49$/mm and $\varepsilon_c = 0.198\%$.

(3)   Domain III, the secondary linear and final fracture region. The domain starts from $\sigma_{sat} = 350$ MPa to $\sigma_{uts} = 498$ MPa.

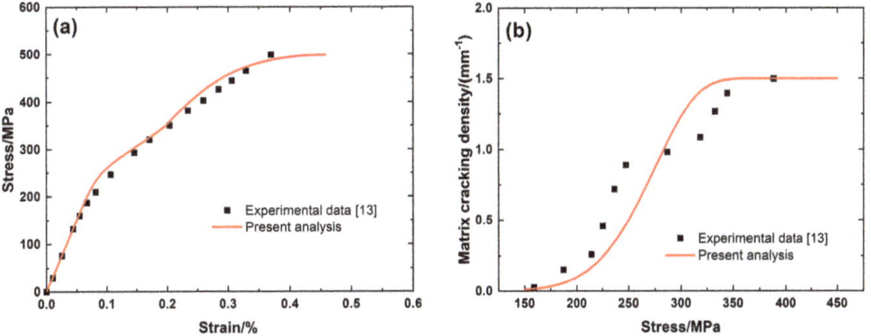

Figure 3. (**a**) Tensile nonlinear stress–strain curves; and, (**b**) matrix fragmentation density versus applied stress curves of Tyranno^{TM} ZMI SiC/SiC mini-CMC.

Figure 4 shows experimental data and theoretical predicted results of tensile nonlinear curves and fragmentation density evolution of the brittle matrix versus applied stress of SiC/SiC minicomposite without heat treatment. The predicted results of tensile curves and fragmentation density evolution of the brittle matrix curves agree well with experimental data. Three main damage domains are obtained from the nonlinear tensile curve, including:

(1) Domain I, the linear region. The domain starts from initial loading to σ_{mc} = 35 MPa with λ_{mc} = 0.04/mm and ε_c = 0.013%.

(2) Domain II, the nonlinear region. The domain starts from the σ_{mc} = 35 MPa to σ_{sat} = 340 MPa with λ_{sat} = 1.27/mm and ε_c = 0.6%.

(3) Domain III, the secondary linear and final fracture domain. The domain starts from σ_{sat} = 340 MPa to σ_{uts} = 399 MPa.

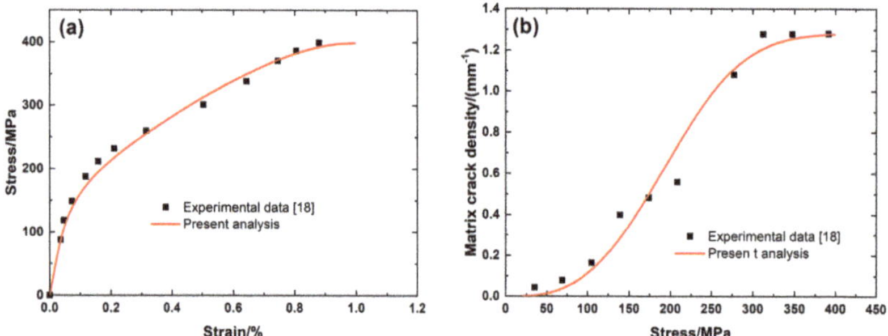

Figure 4. (**a**) Tensile nonlinear stress–strain curves; and, (**b**) matrix fragmentation density versus applied stress curves of SiC/SiC mini-CMC without heat treatment.

Figure 5 shows experimental data and theoretical predicted results of tensile nonlinear curves and fragmentation density evolution of brittle matrix versus applied stress of SiC/SiC minicomposite after heat treatment at 1300 °C. The predicted results of tensile curves and fragmentation density evolution of brittle matrix versus applied stress curves agree well with experimental data. Three main damage domains are obtained from the nonlinear tensile curve, including:

(1) Domain I, the linear region. The domain starts from initial loading to σ_{mc} = 50 MPa with λ_{mc} = 0.02/mm and ε_c = 0.012%.

(2) Domain II, the nonlinear region due to matrix fragmentation. The domain starts from σ_{mc} = 50 MPa to σ_{sat} = 210 MPa with λ_{sat} = 0.28/mm and ε_c = 0.13%.

(3) Domain III, the secondary linear and final fracture region. The domain starts from σ_{sat} = 210 MPa to σ_{uts} = 362 MPa.

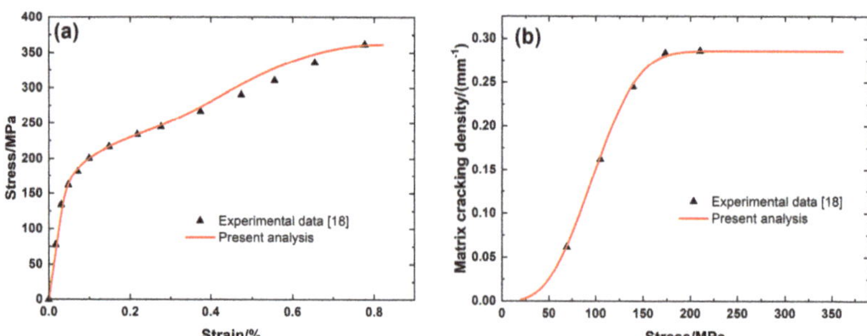

Figure 5. (**a**) Tensile nonlinear stress–strain curves; and, (**b**) matrix fragmentation density versus applied stress curves of SiC/SiC mini-CMC after heat treatment at 1300 °C in Ar atmosphere.

Figure 6 shows experimental data and theoretical predicted results of tensile nonlinear curves and fragmentation density evolution of brittle matrix versus applied stress of SiC/SiC minicomposite after heat treatment at 1500 °C. The predicted results of tensile curves and fragmentation density of brittle matrix versus applied stress curves both agree well with experimental data. Three main damage domains are obtained from the nonlinear tensile curve, including:

(1)   Domain I, the linear region. The domain starts from initial loading to σ_{mc} = 50 MPa with λ_{mc} = 0.005/mm and ε_c = 0.017%.

(2)   Domain II, the nonlinear region. The domain starts from σ_{mc} = 50 MPa to σ_{sat} = 220 MPa with λ_{sat} = 0.58/mm and ε_c = 0.39%.

(3)   Domain III, the secondary linear and final fracture domain. The domain starts from σ_{sat} = 220 MPa to σ_{uts} = 246 MPa.

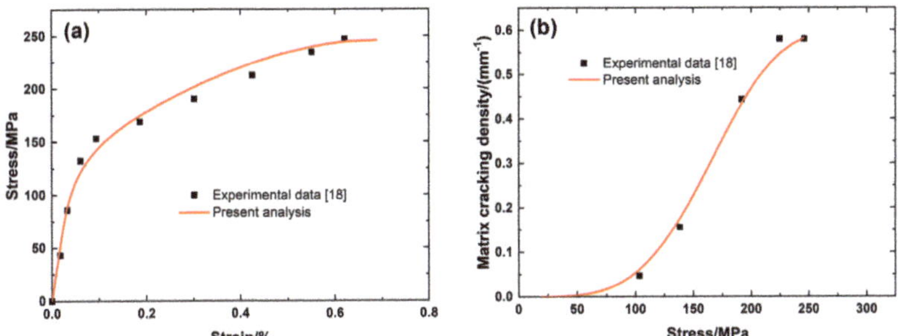

Figure 6. (**a**) Tensile nonlinear stress–strain curves; and, (**b**) matrix fragmentation density versus applied stress curves of SiC/SiC mini-CMC after heat treatment at 1500 °C in Ar atmosphere.

3.2. C/SiC Minicomposite

Figure 7 shows experimental data and theoretical predicted results of tensile curves and the interface debonding fraction versus applied stress curves of unidirectional C/SiC minicomposite. The predicted results of tensile curves agree with experimental data, as shown in Figure 7a, and the composite tensile fractures at the condition of partial interface debonding, as shown in Figure 7b. Three main damage domains are obtained from the nonlinear tensile curve, including:

(1)   Domain I, the linear region. The domain starts from initial loading to σ_{mc} = 130 MPa.

(2)   Domain II, the nonlinear region. The domain starts from σ_{mc} = 130 MPa to σ_{sat} = 250 MPa.

(3)   Domain III, the secondary linear and final fracture domain. The domain starts from σ_{sat} = 250 MPa to σ_{uts} = 348 MPa. The interface debonding fraction increases to $2l_d/l_c$ = 0.78 at tensile strength σ_{uts} = 348 MPa.

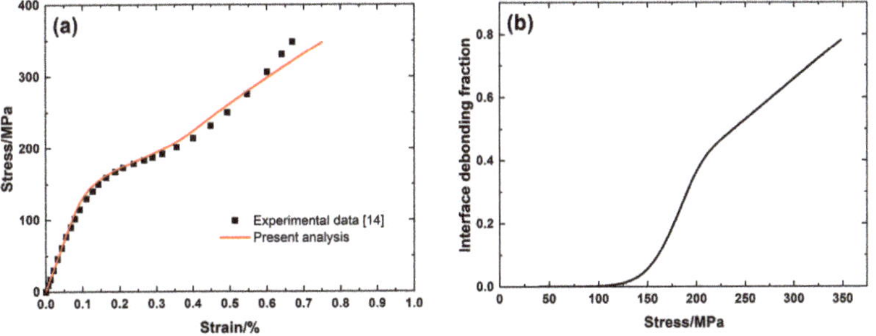

Figure 7. (**a**) Tensile nonlinear stress–strain curves; and, (**b**) the interface debonding fraction versus applied stress curves of unidirectional T–300™ C/SiC mini-CMC.

4. Discussion

In this section, the constitutive properties and damage-related parameters on tensile nonlinear stress–strain curves of SiC/SiC mini-CMC are discussed.

4.1. Effect of Fiber Volume

Figure 8 shows the tensile nonlinear curves and the evolution of interface debonding fraction for different fiber volumes. For low fiber volume $V_f = 25\%$, the debonding fraction increases from η ($\sigma_d = 166$ MPa) = 0 to η ($\sigma_{uts} = 521$ MPa) = 0.33, and the failure strain of the composite is ε_f ($\sigma_{uts} = 521$ MPa) = 0.6%. For high fiber volume $V_f = 30\%$, the debonding fraction increases from η ($\sigma_d = 204$ MPa) = 0 to η ($\sigma_{uts} = 626$ MPa) = 0.32, and the failure strain of the composite is $\varepsilon_f = 0.64\%$. At high fiber volume, the debonding fraction at the same applied stress decreases, and the composite strain at nonlinear Domain II decreases, and the composite tensile strength and failure strain increase.

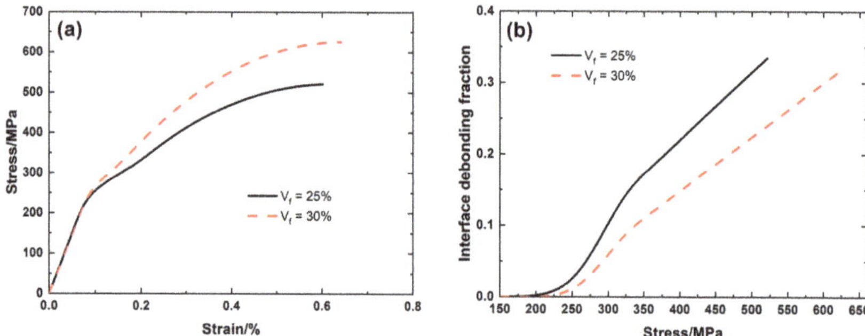

Figure 8. (**a**) Tensile nonlinear stress–strain curves; and, (**b**) the debonding fraction versus applied stress curves of SiC/SiC mini-CMC for different fiber volumes.

4.2. Effect of Interface Properties

Figure 9 shows the tensile nonlinear curves and the evolution of the interface debonding fraction for different interface shear stresses. For low interface shear stress $\tau_i = 20$ MPa, the debonding fraction increases from η ($\sigma_d = 166$ MPa) = 0 to η ($\sigma_{uts} = 521$ MPa) = 0.5, and the failure strain of the composite is $\varepsilon_f = 0.73\%$. For high interface shear stress $\tau_i = 40$ MPa, the debonding fraction increases from η ($\sigma_d = 168$ MPa) = 0 to η ($\sigma_{uts} = 521$ MPa) = 0.25; and the failure strain of the composite is $\varepsilon_f = 0.25\%$. At high interface shear stress, the debonding fraction at the same applied stress decreases, and the composite strain at nonlinear Domain II decreases, and the composite failure strain decreases.

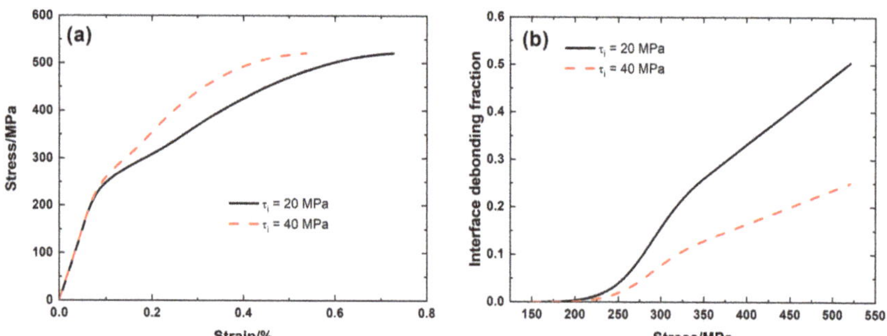

Figure 9. (**a**) Tensile nonlinear stress–strain curves; and, (**b**) the fraction of the interface debonding versus applied stress curves of SiC/SiC mini-CMC for different interface shear stress.

Figure 10 shows the tensile nonlinear curves and the evolution of the interface debonding fraction for different interface debonding energy. For low interface debonding energy $\zeta_d = 4\,J/m^2$, the debonding fraction increases from η ($\sigma_d = 192$ MPa) = 0 to η ($\sigma_{uts} = 521$ MPa) = 0.31, and the failure strain of the composite is $\varepsilon_f = 0.59\%$. For high interface debonding energy $\zeta_d = 6\,J/m^2$, the debonding fraction increases from η ($\sigma_d = 234$ MPa) = 0 to η ($\sigma_{uts} = 521$ MPa) = 0.57, and the failure strain of the composite is $\varepsilon_f = 0.57\%$. At high interface debonding energy, the interface debonding stress increases, the debonding fraction at the same applied stress decreases, and the composite strain during nonlinear Domain II decreases, and the failure strain of the composite decreases.

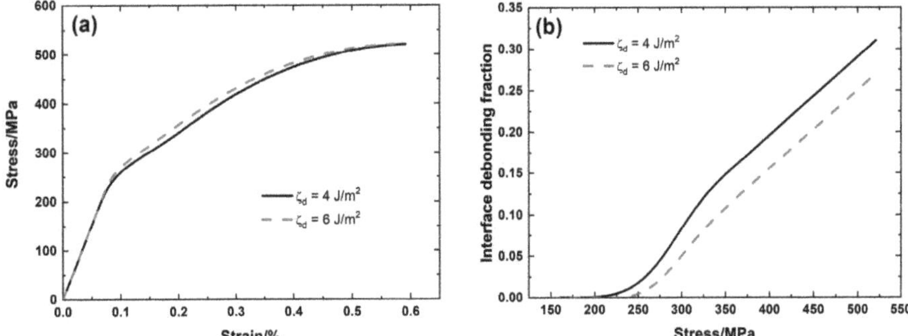

Figure 10. (**a**) Tensile nonlinear stress–strain curves; and, (**b**) the fraction of the interface debonding versus applied stress curves of SiC/SiC mini-CMC for different interface debonding energy.

5. Conclusions

In this paper, the tensile damage and fracture behavior of two different mini-CMCs are investigated. Multiple microdamage mechanisms of matrix fragmentation, fibers failure and pullout are considered in the analysis of tensile nonlinear curves. Experimental tensile nonlinear curves and internal damage evolution of SiC/SiC and C/SiC mini-CMCs are predicted. The effects of material properties and damage-related parameters on tensile damage and fracture at different domains are analyzed.

(1) Predicted tensile nonlinear curves and matrix fragmentation density evolution curves agree with experimental data, and the matrix fragmentation approaches saturation before tensile fracture, and the interface partial debonding remains till tensile fracture.

(2) Microdamage parameters of first matrix fragmentation stress, saturation matrix fragmentation stress and density, composite tensile strength and failure strain are obtained from tensile stress–strain curves and can be used to characterize tensile nonlinear behavior of mini-CMCs.

(3) At higher fiber volume, the debonding fraction at the same applied stress decreases, and the composite strain at nonlinear Domain II decreases, and the composite tensile strength and failure strain increase.

(4) At higher interface shear stress and interface debonding energy, the debonding fraction at the same applied stress decreases, and the composite strain at nonlinear Domain II decreases, and the failure strain of the composite decreases.

Author Contributions: Conceptualization: Z.Z. and L.L.; methodology: L.L.; writing—original draft preparation: Z.Z., Y.L. and L.L.; writing—review and editing: D.F. All authors have read and agreed to the published version of the manuscript.

Funding: This research was funded by Fundamental Research Funds for the Central Universities of China, grant number NS2019038.

Acknowledgments: The authors also wish to thank four anonymous reviewers and editors for their helpful comments on an earlier version of the paper.

Conflicts of Interest: The authors declare no conflict of interest. The funders had no role in the design of the study; in the collection, analyses, or interpretation of data; in the writing of the manuscript, or in the decision to publish the results.

References

1. Padture, N.P.; Gell, M.; Jordan, E.H. Thermal barrier coatings for gas-turbine engine applications. *Science* **2002**, *296*, 280–284. [CrossRef] [PubMed]
2. Naslain, R.R. SiC-matrix composites: Nonbrittle ceramics for thermo-structural application. *Int. J. Appl. Ceram. Technol.* **2005**, *2*, 75–84. [CrossRef]
3. Zok, F.W. Ceramic-matrix composites enable revolutionary gains in turbine efficiency. *Am. Ceram. Soc. Bull.* **2016**, *95*, 22–28.
4. Li, L.B. *Damage, Fracture and Fatigue of Ceramic-Matrix Composites*; Springer Nature Singapore Pte Ltd.: Singapore, 2018.
5. Morscher, G.N.; Martinez-Fernandez, J. Determination of interfacial properties using a single-fiber microcomposite test. *J. Am. Ceram. Soc.* **1996**, *79*, 1083–1091. [CrossRef]
6. Li, H.; Morscher, G.; Lee, J.; Lee, W. Tensile and stress-rupture behavior of SiC/SiC minicomposite containing chemically vapor deposited zirconia interphase. *J. Am. Ceram. Soc.* **2004**, *87*, 1726–1733. [CrossRef]
7. Yu, H.; Zhou, X.; Zhang, W.; Peng, H.; Zhang, C. Mechanical behavior of SiC_f/SiC composites with alternating PyC/SiC multilayer interphases. *Mater. Des.* **2013**, *44*, 320–324. [CrossRef]
8. Krishnan, S.; Vijay, V.; Siva, S.; Painuly, A.; Naithani, N.; Devasia, R. A comparative study on C_f/PyC/SiC minicomposites prepared via CVI process for hypersonic engine application. *Int. J. Appl. Ceram. Technol.* **2018**, *15*, 1110–1123. [CrossRef]
9. Mamiya, T.; Kagawa, Y. Tensile fracture behavior and strength of surface-modified SiTiCO fiber SiC-matrix minicomposites fabricated by the PIP process. *J. Am. Ceram. Soc.* **2000**, *83*, 433–435. [CrossRef]
10. Martinez-Fernandez, J.; Morscher, G.N. Room and elevated temperature tensile properties of single tow Hi-Nicalon, carbon interphase, CVI SiC matrix minicomposites. *J. Eur. Ceram. Soc.* **2000**, *20*, 2627–2636. [CrossRef]
11. Chateau, C.; Gelebart, L.; Bornert, M.; Crepin, J.; Boller, E.; Sauder, C.; Ludwig, W. In situ X-ray microtomography characterization of damage in SiC/SiC minicomposites. *Compos. Sci. Technol.* **2011**, *71*, 916–924. [CrossRef]
12. Maillet, E.; Godin, N.; R'Mili, M.; Reynaud, P.; Fantozzi, G.; Lamon, J. Damage monitoring and identification in SiC/SiC minicomposites using combined acousto-ultrasonics and acoustic emission. *Compos. Part A* **2014**, *57*, 8–15. [CrossRef]
13. Almansour, A.S. Use of Single-Tow Ceramic Matrix Minicomposites to Determine Fundamental Room and Elevated Temperature Properties. Ph.D. Thesis, University of Akron, Akron, OH, USA, 2017.
14. Wang, Y.; Cheng, L.; Zhang, L. Prediction of the mechanical behavior of a minicomposite based on grey Verhulst models. *Ceram. Silik.* **2017**, *61*, 372–377. [CrossRef]
15. Zhang, S.; Gao, X.; Chen, J.; Dong, H.; Song, Y. Strength model of the matrix element in SiC/SiC composites. *Mater. Des.* **2016**, *101*, 66–71. [CrossRef]
16. Zhang, S.; Gao, X.; Chen, J.; Dong, H.; Song, Y.; Zhang, H. Effect of micro-damage on the nonlinear constitutive behavior of SiC/SiC minicomposites. *J. Ceram. Sci. Technol.* **2016**, *7*, 341–348.
17. Ochiai, S.; Kimura, S.; Tanaka, H.; Tanaka, M.; Hojo, M.; Morishita, K.; Okuda, H.; Nakayama, H.; Tamura, M.; Shibata, K.; et al. Residual strength of PIP-processed SiC/SiC single-tow minicomposite exposed at high temperatures in air as a function of exposure temperature and time. *Compos. Part A* **2004**, *35*, 41–50. [CrossRef]
18. Han, X.; Gao, X.; Song, Y. Effect of heat treatment on the microstructure and mechanical behavior of SiC/SiC mini-composites. *Mater. Sci. Eng. A* **2019**, *746*, 94–104. [CrossRef]
19. Guo, S.; Kagawa, Y. Tensile fracture behavior of continuous SiC fiber-reinforced SiC matrix composites at elevated temperatures and correlation to in situ constituent properties. *J. Eur. Ceram. Soc.* **2002**, *22*, 2349–2356. [CrossRef]
20. Yang, C.; Zhang, L.; Wang, B.; Huang, T.; Jiao, G. Tensile behavior of 2D-C/SiC composites at elevated temperatures: Experiment and modeling. *J. Eur. Ceram. Soc.* **2017**, *37*, 1281–1290. [CrossRef]

21. Li, L.B. Modeling matrix multicracking development of fiber-reinforced ceramic-matrix composites considering fiber debonding. *Int. J. Appl. Ceram. Technol.* **2019**, *16*, 97–107. [CrossRef]
22. Curtin, W.A. Multiple matrix cracking in brittle matrix composites. *Acta Met. Mater.* **1993**, *41*, 1369–1377. [CrossRef]
23. Gao, Y.; Mai, Y.; Cotterell, B. Fracture of fiber-reinforced materials. *J. Appl. Math. Phys.* **1988**, *39*, 550–572. [CrossRef]
24. Curtin, W.A. Theory of mechanical properties of ceramic-matrix composites. *J. Am. Ceram. Soc.* **1991**, *74*, 2837–2845. [CrossRef]

 *materials*

Article

The Usefulness of Pine Timber (*Pinus sylvestris* L.) for the Production of Structural Elements. Part II: Strength Properties of Glued Laminated Timber

Radosław Mirski [1,*], Dorota Dziurka [1,*], Monika Chuda-Kowalska [2], Jakub Kawalerczyk [1], Marcin Kuliński [1] and Karol Łabęda [3]

1   Department of Wood Based Materials, Faculty of Wood Technology, Poznań University of Life Sciences, Wojska Polskiego 38/42, 60-627 Poznań, Poland; jakub.kawalerczyk@up.poznan.pl (J.K.); kmarcin97@gmail.com (M.K.)
2   Institute of Structural Analysis, Faculty of Civil and Transport Engineering, Poznan University of Technology, pl. Sklodowskiej-Curie 5, 60-965 Poznań, Poland; monika.chuda-kowalska@put.poznan.pl
3   Department of Furniture, Faculty of Wood Technology, Poznań University of Life Sciences, Wojska Polskiego 38/42, 60-627 Poznań, Poland; karol.labeda@up.poznan.pl
*   Correspondence: rmirski@up.poznan.pl (R.M.); dorota.dziurka@up.poznan.pl (D.D.); Tel.: +48-61-848-7416 (R.M.); +48-61-848-7619 (D.D.)

Received: 14 August 2020; Accepted: 8 September 2020; Published: 11 September 2020

Abstract: The paper assessed the feasibility of manufacturing glued structural elements made of pine wood after grading it mechanically in a horizontal arrangement. It was assumed that the pine wood was not free of defects and that the outer lamellas would also be visually inspected. This would result in only rejecting items with large, rotten knots. Beams of the assumed grades GL32c, GL28c and GL24c were made of the examined pine wood. Our study indicated that the expected modulus of elasticity in bending was largely maintained by the designed beam models but that their strength was connected with the quality of the respective lamellas, rather than with their modulus of elasticity. On average, the bending strength of the beams was 44.6 MPa. The cause of their destruction was the individual technical quality of a given item of timber, which was loosely related to its modulus of elasticity, assessed in a bending test. Although the modulus of elasticity of the manufactured beam types differed quite significantly (11.45–14.08 kN/mm^2), the bending strength for all types was similar. Significant differences occurred only during a more detailed analysis because lower classes were characterized by a greater variation of the bending strength. In this case, beams with a strength of 24 MPa to 50 MPa appeared.

Keywords: beams; glued laminated timber; modulus of elasticity; pine wood; laboratory tests

---

1. Introduction

Developments in the construction industry and searching for ways to use conventional and alternative structural materials have provided new materials: EWPs (Engineering Wood Products). In the case of EWPs, the idea is to obtain a full-quality product from a material that was originally not suitable for specific uses due to its size or insufficient quality [1,2]. Nowadays, Europe and the world have seen developments in the technology of the manufacturing and application of glued timber, mainly GLT (Glued Laminated Timber). This material fits very well with the EWP technology trend. GLT has the typical features of solid timber: light weight, good strength, elasticity, durability, easy processing and a unique feature, i.e., it is readily shaped into cross-sections. Its cross-section has a layered structure, enabling the manufacturing of components with variable cross-sectional heights, as needed [3–6].

Wooden components are glued with binding agents that guarantee a high strength under static and dynamic loads. Structural timber layers, especially those for making load-bearing structures, are nearly always combined by means of resorcinol-phenol-formaldehyde (PRF) or melamine-urea-formaldehyde (MUF) glues. Apart from these, polyurethane-based glue is becoming more and more popular. All these adhesives should provide a high durability in variable environmental conditions [7–11].

Studies on the durability of straight and bent beams composed of wooden layers glued together with synthetic resins indicate that such elements have comparable load-bearing capacities to those made of solid timber. Moreover, the layered timber structure has resulted in an improved quality of the material [3,12–21].

Recent research studies on glued laminated timber have focused, among other things, on materials glued lengthwise and crosswise. Test results for high-grade solid timber and glued laminated timber, even that obtained from lower-quality timber after grading, indicate that the latter has better strength properties. This is mostly attributed to the distribution of defects and to the gluing process. For instance, the studies focused on the shaping of cross-sections of items made out of thin timber that was comprised of different grade materials with respect to their strength [22–30].

According to JCR (Journal Citation Reports), more than 700 scientific papers on glulam timber were published in the last 20 years. They considered, among other things, the component's reinforcement methods by using glass or carbon fibers [31–34], the effect of different factors on the behavior of steel-timber composites (STC) [35–37] and glulam-concrete beams [38–40].

As mentioned earlier, in the case of glued laminated timber, the cross-sections of the obtained elements can be shaped as required. However, what is important is that the strength is also improved, and it is generally higher than the combined elements. The coefficient of variation for the bending test is also improved [41,42]. In the works of Tomasi [43] and Gonzales [37], a significant improvement in the mechanical properties of glued laminated beams reinforced with steel rods was indicated. However, they pointed out that the quality of the steel-wood joint was of significant importance in these systems. The abovementioned authors draw attention to systems made of different lamella qualities; however, there is the possibility of having the lamellas be joined in their width [44,45] or having them made of different species [46,47]. The created systems result from certain fixed concepts, as described in appropriate standard or the model described in the works of Fink [48], Foschi and Barrett [49], or Hernandez et al. [50] may be the basis for their creation. Predicting the future quality of the produced GLT elements is an overriding task for engineers who support this branch of the wood industry because, in contrast to the mass production of wood-based panels with stabilized parameters, testing the produced GLT is expensive and difficult.

The parameter of GLT which determines its suitability is, first of all, the strength of lamellas with different parameters and properties, as defined in applicable normative documents, taking into consideration the acceptable classes of glued timber [51–53]. The current strength classification system (EN 338) [54] for structural timber enables the use of a single-strength class in the range of C16–C30 for beams of uniform structure and in combination with lower classes C16 and C18 in the case of a nonuniform structure (EN 384) [55]. The class C is related to the static bending strength of manufactured material. The strength of defect-free pine wood is usually in the range of 90–110 MPa. However, because of its natural features, which from a technological perspective are often recognized as defects, the mechanical properties deteriorate considerably. The strength characteristics of the majority of timbers are often below 20 MPa, which results mostly from the occurrence of knots. They appear in various sections with an interval of 40 to 60 cm. The timber properties are affected not only by the number and size of the knots, but also by their soundness. Low-quality knots are usually cut, and the obtained elements are combined with the use of finger joints. This technique is known and has been constantly developed since the Second World War [56]. The knots are an important type of defect and due to their dimensions affect in particular the pine wood (Figure 1a,b). For comparison, a spruce timber is characterized by the presence of significantly smaller knots (Figure 1c). Thus, their removal contributes to an improvement in both the technical quality and visual aspects of each piece of pine timber. However, it seems like, in the case of glued components, the occurrence of knots has a smaller effect and the visual side is less important.

 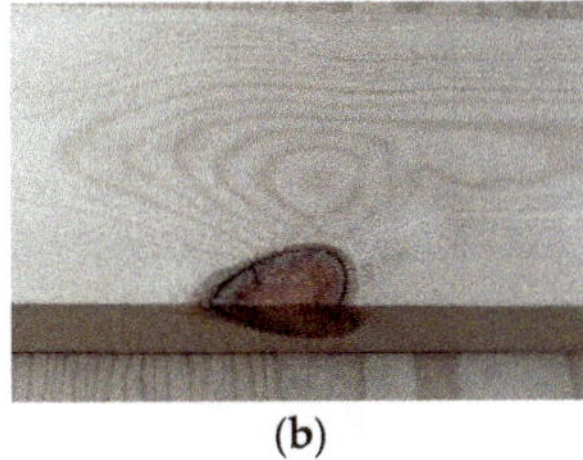 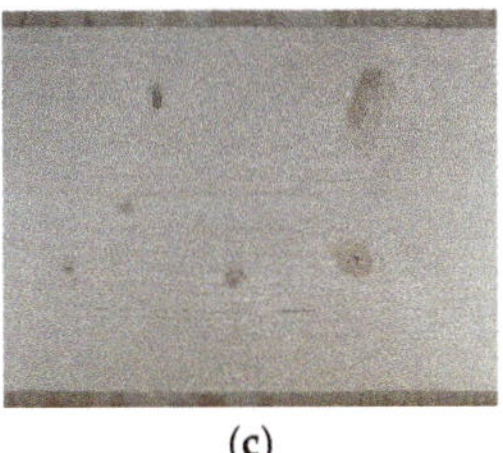

(a) (b) (c)

Figure 1. The examples of knots from: (**a**,**b**) Scots pine and (**c**) Norway spruce.

Consequently, the aim of the presented work was to investigate the possibility of using pine timber sorted solely on the basis of mechanical properties, with the exception of the outer layers. The outer lamellas of the eight-layer beams were also assessed visually. During the assessments, the pieces of timber having edge knots or large rotten knots were classified as unsuitable for the outer layers.

2. Materials and Methods

The research material was pine wood with the following dimensions: 137 mm wide × 39.50 mm thick × 3485 mm long. The average density of the timber items was 571 kg/m^3 (average moisture 8.98%). The pine wood was obtained by sawing timber in the form of logs having round cross sections and originating from the Forest Division Olesno (50°52′30″ N 18°25′00″ E). The obtained sawn timber was dried to a moisture content of 10% ± 2%. After drying, the sawn timber was organized so as to obtain a uniform thickness of all the lamellas. The preliminary assessment was performed in accordance with EN 338. The detailed description of the modulus of elasticity assessments is included in the first part of the research. Selected timber items were used for the preparation, in semi-industrial conditions, of glued beams with a diameter of 137 mm × 300 mm, i.e., comprising eight layers. With the exception of the outer layers, the choice of the lamellas for making the beams depended only on the determined value of the modulus of elasticity. The outer layers, with the exception of the required value of the modulus of elasticity, were required to have no edge knots. The raw material originating from that region is characterized by a higher percentage of timber, whose physico-mechanical parameters enable a considerable portion of it (45%) to be classified into higher classes than C24 (details will follow in the next chapter). Therefore, it was assumed that the respective beam models would satisfy the conditions for a modulus of elasticity set out for grades GL24c, GL28c and GL32c according to EN 14,080 [57]. The elastic properties of the beams layers were determined according to Bodig and Jayne [58], assuming that the beam was symmetric and contained eight lamellas (1):

$$E_{ef} = \frac{1}{J_y} \sum_{i=1}^{4} E_i \left[J_{yi} + A_i (d_i)^2 \right] \tag{1}$$

where:

E_{ef}—effective/substitute modulus of elasticity, N/mm^2,

J_y—area moment of inertia, mm^4,

E_i—modulus of elasticity of layer, N/mm^2,

A_i—cross-sectional area, mm^2,

d—distance from the neutral axis, mm.

The adopted values of the modulus of elasticity for various types of beams are shown in Table 1.

Just before being used for the preparation of glued beams, the timber items were further processed via a plan to improve their surfaces before gluing them together. The effective thickness of individual lamellas was 37.5 mm. The resulting surface was covered with an amount of glue of 220–250 g/m^2. Melamine-urea-formaldehyde resin (MUF 1247) and its dedicated hardener (2526), both from Akzo Nobel (Amsterdam, Netherlands), were used as the binding agent. The mixture was prepared while taking into account the conditions prevailing in the laboratory room. The hardener was used at 20 g per each 100 g of resin, as recommended by Akzo Nobel for that resin. The glue was applied using a roller coating machine. The beams were manufactured at a room temperature range between 20 °C and 24 °C. The press loading time was around 12–15 min. Four beams were pressed at the same time under a pressure of 0.48 MPa for 20 h. Four beams were manufactured each day. Pressing was conducted with the use of an industrial press equipped with hydraulic cylinders dedicated to the production of glued, structural elements (FOST, Czersk, PL). After production, the beams were air-conditioned in the laboratory for min. four weeks. The conditions in the laboratory were controlled: the temperature was 21 ± 2 °C, and the air humidity was 55–65%. After the period of air conditioning, the beams were assessed for their mechanical properties. Due to the weight of the beams, they were not planed. Excess glue was manually removed immediately prior to testing the mechanical properties.

Table 1. Elastic properties of designed beams.

Beam Type	Number of Samples	E_{mean}	E_{min}	E_{max}	$E_{mean\ 1st\ layer}$	$E_{mean\ 4st\ layer}$	$MOR_{declaration}$ *
					kN/mm^2		N/mm^2
GL24c	12	11.71	11.25	11.93	12.53/1.42 **	8.48/1.88	24
GL28c	14	12.82	11.98	13.50	13.96/5.44	8.08/11.69	28
GL32c	22	14.84	14.13	16.52	16.45/8.64	8.58/11.06	32

* Modulus of rupture (MOR)—characteristic value according to EN 14,080 [47], ** CoV (coefficient of variation).

The resulting beams were evaluated for their 4-point bending strength, in accordance with the diagram shown in Figure 2. Figure 3 shows the appearance of the test stand. It was equipped with: a hydraulic cylinder (50 Mg, Hi-Force, Daventry, UK), hydraulic pump (50 Mg, Hi-Force, Daventry, UK), oil flow rate regulator (Hi-Force, Daventry, UK), force sensor (CL 16 tm 500 kN, ZEPWN, Marki, PL) and deformation sensor (KTC-600-P, Variohm Eurosensor, Towcester, UK).

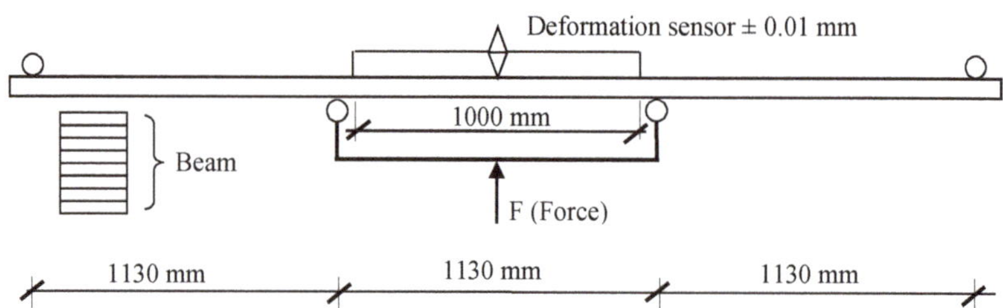

Figure 2. The scheme of the tested eight-layered beam.

In order to take into account the influence of the moisture content on the modulus of elasticity, the obtained results were calculated in accordance with Bauschinger's Equation (2):

$$E_{12} = E_{MC}[1 + \alpha_{MC} \cdot (MC - 12)] \tag{2}$$

where:

E_{12}—modulus of elasticity of wood for a moisture content of 12%, N/mm^2,

E_{MC}—modulus of elasticity of wood for a moisture content of 4% < w < 20%,

α_{MC}—coefficient of variation of the modulus of elasticity of wood after its moisture content changed by 1%—assumed to be 0.02,

MC—absolute moisture content of wood, %.

The destructive test included the assessment of the point and cause of failure for each specific beam.

The results of the experimental measurements were analyzed using the STATISTICA 13.0 package (Version 13.0, StatSoft Inc., Tulsa, OK, USA).

Figure 3. Test stand for the evaluations of the bending strength and modulus of elasticity.

3. Results and Discussion

The mean values of the modulus of elasticity are shown in Table 2. The values shown therein indicate that the prepared beams, with the exception of grade GL32c, exhibited a low variability of the modulus of elasticity in bending. Moreover, the obtained values were close to or only slightly higher than the assumed ones (negative value of δ). Since the moisture of the beams during the test differed considerably from 12% (the average moisture for all the beams was 8.83%), the outcomes were recalculated using Bauschinger's Equation (2). With the exception of grade GL32c beams, the calculated values of the modulus of elasticity were only slightly lower than the assumed ones. For GL32c, the relative difference was 5.1%. Assuming that the values of the modulus of elasticity calculated for 12% MC are appropriate, it should be expected that the beams satisfy the assumptions in this regard.

Table 2. Elastic properties of the designed beams.

Beam Type	Assumed Values		Determined Values		δ * (%)	$E_{mean\ per\ 12\%MC}$ (kN/mm²)	$E_{5percentyl\ per\ 12\%MC}$ (kN/mm²)
	E_{meanZ} (kN/mm²)	CoV (%)	E_{meanP} (kN/mm²)	CoV (%)			
GL24c	11.71	1.81	12.79	6.42	−9.21	11.45	10.43
GL28c	12.82	3.83	13.63	6.84	−6.31	12.78	11.78
GL32c	14.84	4.00	14.94	14.1	−0.68	14.08	11.68

*—Relative change: $\delta = (E_{meanZ} - E_{meanP})/E_{meanZ} \times 100\%$.

It is assumed that the prepared beams should have a static bending strength that is not lower than 24 N/mm², 28 N/mm² and 32 N/mm², respectively, for beam types GL24c, GL28c and GL32c. The lowest strength for all the prepared beams was 29.97 N/mm², and the highest was 55.38 N/mm². However, the static bending strength of the beams had a normal distribution (Figure 4), and, importantly, its standard deviation was only 6.45 N/mm² and its variation coefficient was 14.5%, even though they were designed for different values of the modulus of elasticity. This means that the strength of the obtained beams was characterized by a low variability and was not strongly correlated with the designed system.

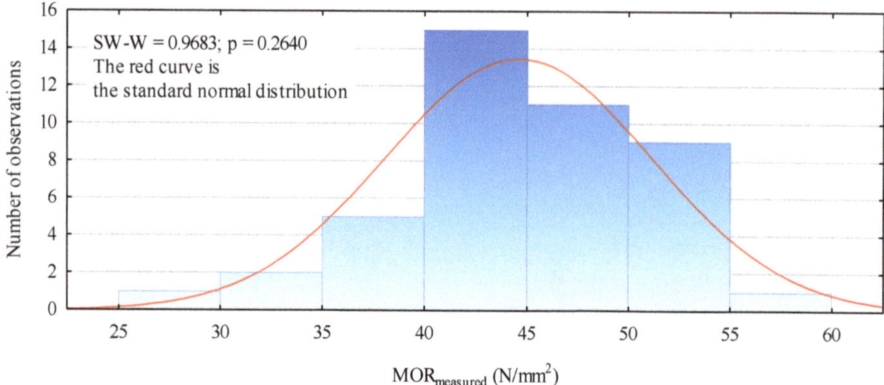

Figure 4. A histogram of the static bending strength for the glued beams made of mechanically graded timber.

Hence, the static bending strength is not correlated with the grade of the designed beams.

The data in Figure 5 show that all the models are characterized by a similar strength of around 44.5 N/mm^2, regardless of the assumed timber grade, whereas an analysis of the modulus of elasticity shows the presence of two clearly different groups.

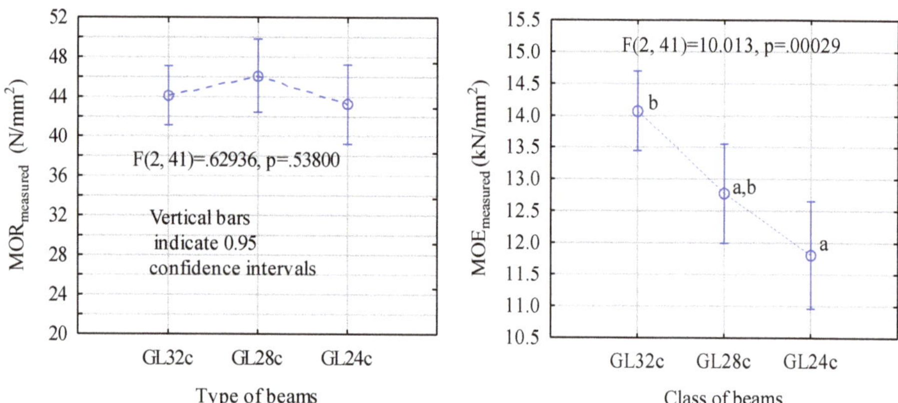

Figure 5. A one-factor ANOVA for the system: glued beam's grade—static bending strength; and beam's grade—modulus of elasticity. The letters denote uniform groups for Tukey's test.

The values of strength obtained in the bending test were also recalculated with Bauschinger's formula, using a factor of $\alpha = 0.04$ this time. The results obtained with that factor are shown in Figure 5. The mean values calculated for all the beams were thus reduced from 44.5 N/mm^2 to 38.6 N/mm^2, which is still rather high. However, the assignment to the specific grade GL is based on a 5-percentile value of strength. For the represented number of samples, this value is the lowest or close to the lowest.

The values shown in Figure 6 indicate that the beams assigned to the groups GL32c and GL24c satisfied the strength requirement, reaching the following values: 32.5 N/mm^2 for grade GL32c beams and 24.4 N/mm^2 for grade GL24c beams. The batch of beams modeled to be assigned to grade GL28c did not satisfy requirements and should have been assigned to grade GL24c, even though it had the highest mean value. What is important, in the case of that group, is that its assignment to the specific grade was attributed to a value regarded as being a statistical extreme. Moreover, the second lowest value of the static bending strength reached in that group was as high as 36.8 N/mm^2. Without taking into account the strength of the three beams with the lowest values, the 5-percentile value would be 32.5 N/mm^2.

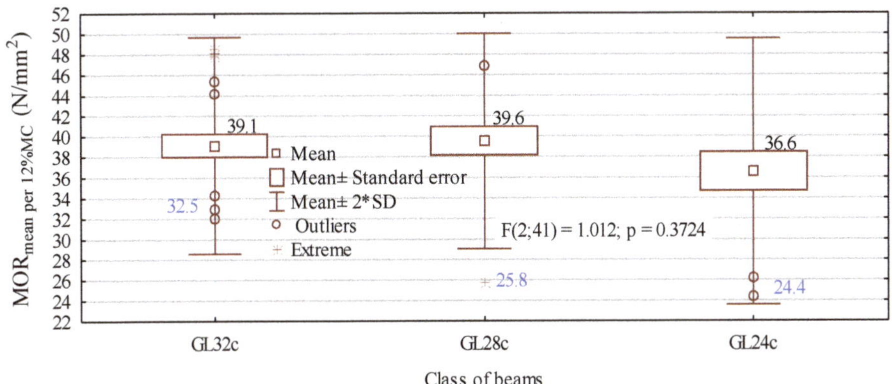

Figure 6. Characteristic values of the static bending strength for the prepared beam (samples). The numbers in blue denote the 5-percentile value.

It is hard to predict the exact point of failure and the potential strength in some cases. For Beam 41 (Figure 7a), the cause of destruction was found, as expected, in the second and third lamellas and was due to the presence of large rotten knots in the pure bending zone. On the other hand, the beam demonstrated a strength that was nearly twice as high as expected. In the second case, failure occurred in the middle zone, for the lamellas 3/4/5 from the top, in a practically knotless zone, at a strength of about 98 kN (Figure 7b).

(a)   (b)

Figure 7. Beam failure images: (a) GL24c—MOR(MOR$_{12\%}$)—48.5(43.4) N/mm^2—the strongest in the group, (b) GL28c—MOR(MOR$_{12\%}$)—30.4(25.8) N/mm^2—the weakest in the group.

Obviously, the presence of knots is the main cause of the beams' destruction. On the other hand, nearly 60% of the beams failed because of damage of the outer lamellas, and some 34% failed because of damage of the middle lamellas. For three beams, the exact starting point of destruction could not be identified (Figure 8).

The type of destruction propagating from the beam's middle zone was only dominant for grade GL32c beams. In the other cases, more than 70% involved destruction in the outer layer. It would be unjustified to reject the zero hypothesis that states that the strength of the prepared beams depends on the starting point of the propagation of the destruction (Figure 9). The average static bending strength for the beams destroyed as a consequence of damage to the outer lamellas was 39.6 N/mm^2, and the value was 37.3 N/mm^2 for those where the destruction originated in the middle layer. A different situation was observed for the modulus of elasticity. In this case, the differences were statistically significant and beams with a higher MOE value were destroyed mainly in the middle layer. This is probably attributable to the fact that the beams with higher moduli of elasticity had higher-quality outer lamellas and were capable of withstanding the arising stress, whereas lower-grade timber, though located deeper, was exposed to critical/damaging stress.

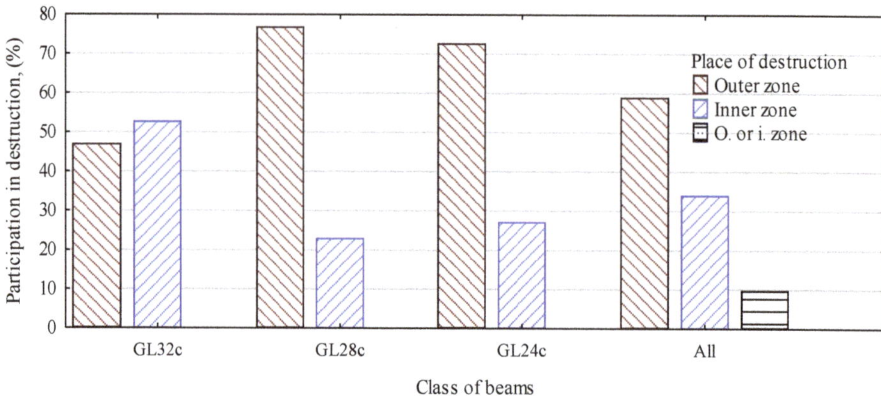

Figure 8. Impact of the point of destruction for various beam grades.

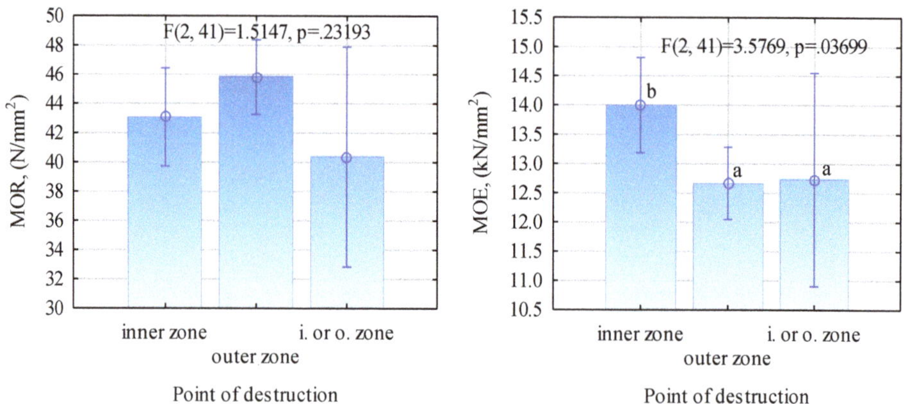

Figure 9. ANOVA of the assessment of the static bending strength and modulus of elasticity relative to the point of destruction (Letters mark uniform groups determined with the Tukey HSD test).

In order to determine the effect of the relation between the supports' spacing (l) and the height of the manufactured beams (h), a model system with the cross-sections' dimensions of 138 × 300 mm and a strength of 32 MPa was adopted. The J_y value for the adopted system was 31,050 cm⁴. According to EN 408 [59], the l/h relation should be 18 ± 3; however, in the conducted research, the beams were characterized by a relation of 13.3. The shear forces diagrams and bending moments diagrams of the beams, with a support spacing of 18 × h = 5400 mm (in compliance with the standard) and with a support spacing of 3390 mm for experimental beams, are presented in Figure 10. Moreover, the significant physical quantities' characteristics for the bend test are included in Table 3.

On the basis of the results presented in Figure 10 and Table 3, it can be concluded that the significant reduction in the beams' length in comparison with the regulations of EN 408 [59] showed an increase in the shear stresses by 60%. The considerable increase in the shear force value can lead to the beams' destruction in the inner zone, more precisely in the inner lamellas. The assumed timber length (and consequently also the beam length) derived from the most efficient breakdown of the 14-m long logs into 3.5-m long sections. This type of division ensured less material wastes and made it easier to move the research materials; however, this length can influence the obtained results. The average shear strength of pine wood is around 10 N/mm² (it ranges between 6 and 14 N/mm²). However, our observations show that this did not have a significant impact on the obtained research results. Most of the beams were destroyed between the pressures. In exceptional cases, the beams were damaged outside the pressures, but mainly in the tension zone.

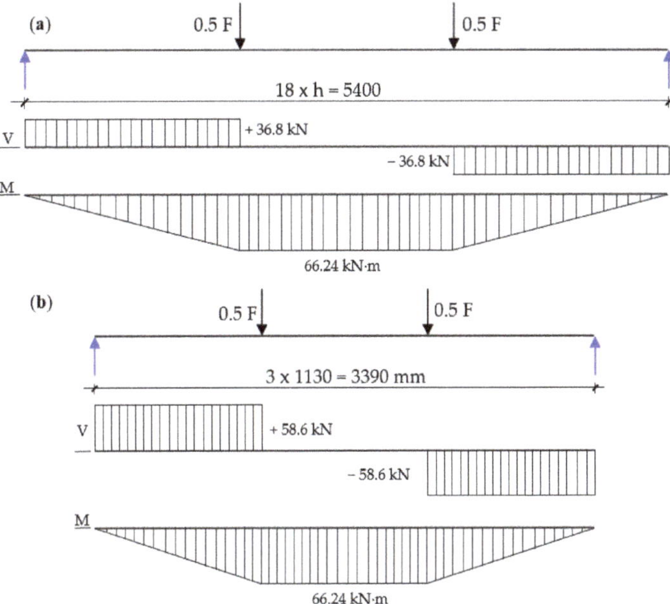

Figure 10. Shear forces diagrams and bending moments diagrams for (a) beams consistent with the dimensional requirements of EN 408 and (b) experimental beams used in the conducted research.

Table 3. Physical quantities for the bend test—No. 1 and 2: theoretical beams, No. 3: a beam with the maximum strength obtained during the tests.

No.	Type of Beam (mm)	Class of MOR (N/mm²)	F (N)	V (kN)	M (kN·m)	τ_{xz} (N/mm²)
1	18 × h = 5400	32	73,600	36.8	66.2	1.3
2	11.3 × h = 3390	32	117,240	58.6	66.2	2.1
3	11.3 × h = 3390	48.6	154,770	77.4	66.2	2.8

For solid structures or homogeneous glued laminated timber, the stress distribution will be linear over the entire height of the section (GL32h—Table 4). Composite beams are constructed from more than one type of material so as to increase stiffness or strength (or to reduce cost). In the analyzed case (e.g., GL32c—Table 4), the layers are glued together. Therefore, it should be assumed that the deformations at the interface of the layers are the same. In the elastic range at the height of each layer, the stress distribution will be linear. However, due to the variable E modulus for each layer, we observe stress jumps at the layers' borders.

It appears that stress jumps at the layers' borders, although not large, may contribute to the destruction of the beams in the deeper lamellas. Deeper lamellas were clearly damaged where there were wood defects. In the case of very high-quality external lamellas, the second and third layers were

Materials **2020**, *13*, 4029

responsible for the quality of the beam. It should be remembered that the lamellas of these layers were only assessed in terms of the linear elasticity modulus.

Table 4. Modules of elasticity of individual layers for beams of the Gl32h (homogeneous) and GL32c (combined) type and their stress diagrams.

Beams	GL32h		GL32c		
	E (N/mm^2)	σ_x (N/mm^2)	E * (N/mm^2)	σ_x (N/mm^2)	
		32.2			35.7
	14,200		16,450	21.6	26.8
	14,200		13,270	12.4	14.4
	14,200		11,420	4.66	6.20
	14,200	0	8580	0	
	14,200		8580		0
	14,200		11,420	6.20	4.66
	14,200		13,270	14.4	12.4
	14,200		16,450	26.8	21.6
		32.2			35.7

*—laboratory beam GL32c.

4. Conclusions

For eight-layered beams made of grade GL24c, GL28c and GL32c timber, the modulus of elasticity was only slightly different from its assumed values, and the obtained beams satisfied the requirements of the standard EN-14080 [57] in this regard.

The static bending strength obtained from the 4-point bending test was not related to the class of the designed beam models. Regardless of the assumed class, the average beam strength was above 36.6 N/mm^2.

We adopted a procedure for preparing timber before making glued components, which provided the designed systems with satisfactory modulus of elasticity values and a static bending strength that was essentially higher than its assumed value. However, the scope of the study needs to be extended to include beams with other cross-sections so that the objective of this research work can be fully met. Nonetheless, at this point in our research, it seems that the visual grading of timber can be limited to only the timber items that are intended for use as outer layers.

Author Contributions: Conceptualization, R.M.; Methodology, R.M and M.K.; Validation, D.D., M.C.-K., J.K., and K.Ł.; Formal analysis, R.M. and M.C.-K.; Investigation, D.D., J.K., and K.Ł.; Resources, R.M., D.D., M.K. and K.Ł.; Writing—Original Draft Preparation, R.M., D.D., and M.C.-K.; Writing—Review and Editing, R.M. M.K. and D.D.; Visualization, R.M.; Project Administration, R.M. and D.D.; Funding Acquisition, R.M. All authors have read and agreed to the published version of the manuscript.

Funding: This research was funded by the National Centre for Research and Development, grant number BIOSTRATEG3/344303/14/NCBR/2018. The authors are grateful for the support of the Ministry of Science and Higher Education program "Regional Initiative of Excellence" in the years 2019–2022, Project No. 005/RID/2018/19.

Conflicts of Interest: The authors declare no conflict of interest.

References

1. Borysiuk, P. Veneer Laminated Wood LVL meets the global markets for wood-based construction materials. *Biul. Inf. OBRPPD* **2007**, *1–2*, 63–71. (In Polish)

2. Borysiuk, P.; Kowaluk, G. Types and examples of applications of wood-based construction materials. *Biul. Inf. OBRPPD* **2007**, *3–4*, 115–119. (In Polish)

3. Sterr, R. Untersuchungen zur Dauerfestigkeit von Schichtholzbalken. Mitteilung aus dem Institut für Holzforschung und Holztechnik der Universität München (Investigations on the fatigue resistance of laminated wood beams). *Holz Roh Werkstoff* **1963**, *21*, 7–61. [CrossRef]

4. Gerhards, C.C. Effect of early wood and latewood on stress-wave measurements parallel to the grain. *Wood Sci. Technol.* **1975**, *11*, 69–72.

5. Krzosek, S. *Strength Grading of Polish Pine Structural Sawn Timber*; Wydawnictwo SGGW: Warsaw, Poland, 2009. (In Polish)

6. Wieruszewski, M.; Gołuński, G.; Hruzik, G.J.; Gotych, V. Glued elements for construction. *Ann. Warsaw Univ. Life Sci. SGGW For. Wood Technol.* **2010**, *72*, 453–458.

7. Serrano, E. Glued-in rods for timber structures—A 3D model and finite element parameter studies. *Int. J. Adh. Adhesiv.* **2001**, *21*, 115–127. [CrossRef]

8. Kawalerczyk, J.; Siuda, J.; Mirski, R.; Dziurka, D. Hemp flour as a formaldehyde scavenger for melamine-urea-formaldehyde in plywood production. *BioResources* **2020**, *15*, 4052–4064.

9. Kägi, A.; Niemz, P.; Mandallaz, D. Einfluss der Holzfeuchte und ausgewählter technologischer Parameter auf die Verklebung mit 1K-PUR Klebstoffen unter extremen klimatischen Bedingungen. *Holz Roh Werkstoff* **2006**, *64*, 261–268. [CrossRef]

10. Krystofiak, T.; Proszyk, S.; Lis, B. Adhesives for large-scale wooden construction elements for buildings. *Drewno Wood* **2008**, *51*, 61–79. (In Polish)

11. Krystofiak, T.; Proszyk, S.; Lis, B.; Wieruszewski, M.; Gotych, V. Studies upon durability of glue lines in production of laminated wood for skeleton building prepared in industrial conditions. In Proceedings of the XXth Symposium Adhesives in Woodworking Industry, Zvolen, Slovakia, 29 June–1 July 2011; pp. 169–176.

12. Bos, H. The Potential of Flax Fibres as Reinforcement for Composite Materials. Ph.D. Thesis, Technische Universiteit Eindhoven, Eindhoven, The Netherlands, 2004. [CrossRef]

13. Brol, J. Effectiveness of reinforcing wooden beams with CFRP bands. In Proceedings of the VII Konferencja Naukowa "Drewno i Materiały Drewnopochodne w Konstrukcjach Budowlanych", Międzyzdroje, Poland, 12–13 May 2006. (In Polish).

14. Burawska, I. Studies on Changes in Strength of Weakened Wooden Bending Beams When Reinforced with Carbon Fibre. Master's Thesis, Szkoła Główna Gospodarstwa Wiejskiego w Warszawie, Warsaw, Poland, 2012. (In Polish).

15. Burawska-Kupniewska, I.; Krzosek, S.; Mankowski, P.; Grzeskiewicz, M.; Mazurek, A. The influence of pine logs (*Pinus sylvestris* L.) quality class on the mechanical properties of timber. *BioResources* **2019**, *14*, 9287–9297.

16. Burawska, I.; Tomusiak, A.; Beer, P. Influence of the length of CFRP tape reinforcement adhered to the bottom part of the bent element on the distribution of normal stresses and on the elastic curve. *Ann. Warsaw Univ. Life Sci. SGGW. For. Wood Technol.* **2011**, *73*, 186–191.

17. Burawska, I.; Zbieć, M.; Beer, P. Study of peel and shear strength of adhesive joint between pine wood and CFRP tape.Chip and chipless woodworking processes. In Proceedings of the 8th International Science Conference, Technical University in Zvolen, Zvolen, Slovakia, 6–8 September 2012; pp. 35–40.

18. Burawska, I.; Zbieć, M.; Tomusiak, A.; Beer, P. Local reinforcement of timber with composite and lignocellulosic materials. *BioResources* **2015**, *10*, 457–468. [CrossRef]

19. Orłowicz, R.; Gil, Z.; Fanderejewska, E. Extension capacity of spiral bars used in joints of wooden structures. In Proceedings of the VII Konferencja Naukowa: Drewno i Materiały Drewnopochodne w Konstrukcjach Budowlanych, Szczecin-Międzyzdroje, Poland, 27–29 May 2004; pp. 161–167. (In Polish).

20. Ritter, M.A.; Williamson, T.G.; Moody, R.C. Innovations in glulam timber bridge design. In *Structures Congress 12: Proceedings of Structures Congress '94, Atlanta, GA, USA, 24–28 April 1994*; Baker, N.C., Goodno, B.J., Eds.; American Society of Civil Engineers: New York, NY, USA, 1994; Volume 2, pp. 1298–1303.

21. Rapp, P.; Lis, Z. Testing of wooden beams reinforced with carbon fibre bands. *Inżynieria Budownictwo* **2001**, *7*, 390–392. (in Polish).

22. Czuczeło, K. Testing of Glued Laminated Timber from Strength Sorted Pine Wood. Ph.D. Thesis, Katedra Nauki o Drewnie i Ochrony Drewna, Wydział Technologii Drewna SGGW w Warszawie, Warsaw, Poland, 2005. (In Polish).

23. Brandner, R. Production and technology of cross laminated timber (CLT): State-of-the-art report. In *Focus Solid Timber Solutions, European Conference on Cross Laminated Timber, Graz, Austria, 21–22 May 2013*, 2nd ed.; COST Action: Graz, Austria, 2014; pp. 3–36.

24. Bejder, A.K.; Kirkegaard, P.H.; Wraber, I.K.; Falk, A. The Materiality of Novel Timber Architecture—Developing a Model for Analysing and Evaluating Materials in Architecture. 2016. Available online: https://vbn.aau.dk/en/publications/the-materiality-of-novel-timber-architecture-developing-a-model-f (accessed on 6 July 2020).

25. Bejder, A.K.; Kirkegaard, P.H.; Wraber, I.K.; Falk, A. The Materiality of Novel Timber Architecture—Based on a Case Study Analysis of Cross-Laminated Timber. 2011. Available online: https://vbn.aau.dk/en/publications/the-materiality-of-novel-timber-architecture-based-on-a-casestudy (accessed on 6 July 2020).

26. Bejder, A.K.; Kirkegaard, P.H.; Fisker, A.M. *On the Architectural Qualities of Cross Laminated Timber*; Cruz, P.J.S., Ed.; Taylor & Francis Group: London, UK, 2010; pp. 119–121. [CrossRef]

27. Augustin, M.; Blaß, H.J.; Bogensperger, T.; Ebner, H.; Ferk, H.; Fontana, M.; Frangi, A.; Hamm, P.; Jöbstl, R.A.; Moosbrugger, T.; et al. *BSP Handbuch. Holz-Massivbauweise in Brettsperrholz—Nachweise auf Basis des Neuen Europäischen Normenkonzepts*; Verlag der Technischen Universität Graz: Graz, Austria, 2010.

28. Falk, A. Cross-laminated timber plate tensegrity and folded roofs. In Proceedings of the Wood for Good: Symposium on Innovation in Timber Design and Research, Copenhagen, Danmark, 20 September 2010.

29. Espinoza, O.; Trujillo, V.R.; Laguarda, M.F.; Buehlmann, U. Cross-laminated timber: Status and research needs in Europe. *BioResources* **2016**, *11*, 281–295. [CrossRef]

30. Krauss, A.; Fabisiak, E.; Szymański, W. The ultrastructural determinant of radial variability of compressive strength among the grain of Scots pine wood. *Ann. Warsaw Univ. Life Sci. SGGW For. Wood Techno.* **2009**, *68*, 431–435.

31. Alfred Franklin, V.; Christopher, T. Fracture energy estimation of DCB specimens made of glass/epoxy: An experimental study. *Adv. Mater. Sci. Eng.* **2013**, *2013*, 7. [CrossRef]

32. Kumar, K.; Rao, S.; Gopikrishna, N. Evaluation of strain energy release rate of epoxy glass fibre laminate (mode—I). *Int. Educ. Res. J.* **2017**, *3*, 44–46.

33. Blackman, B.R.K.; Kinloch, A.J.; Paraschi, M. The determination of the mode II adhesive fracture resistance, GIIC, of structural adhesive joints: An effective crack length approach. *Eng. Fract. Mech.* **2005**, *72*, 877–897. [CrossRef]

34. De Moura, M.; Campilho, R.; Gonçalves, J.P.M. Pure mode II fracture characterization of composite bonded joints. *Int. J. Solids Struct.* **2009**, *46*, 1589–1595. [CrossRef]

35. Hu, Q.; Gao, Y.; Meng, X.; Diao, Y. Axial compression of steel–timber composite column consisting of H-shaped steel and glulam. *Eng. Struct.* **2020**, *216*, 110561. [CrossRef]

36. Le, T.D.H.; Tsai, M.-T. Experimental assessment of the fire resistance mechanisms of timber–steel composites. *Materials* **2019**, *12*, 4003. [CrossRef] [PubMed]

37. Gonzales, E.; Tannert, T.; Vallee, T. The impact of defects on the capacity of timber joints with glued-in rods. *Int. J. Adhes. Adhes.* **2016**, *65*, 33–40. [CrossRef]

38. Otero-Chans, D.; Estévez-Cimadevila, J.; Suárez-Riestra, F.; Martín-Gutiérrez, E. Experimental analysis of glued-in steel plates used as shear connectors in Timber-Concrete-Composites. *Eng. Struct.* **2018**, *170*, 1–10. [CrossRef]

39. Schänzlin, J.; Fragiacomo, M. Analytical derivation of the effective creep coefficients for timber-concrete composite structures. *Eng. Struct.* **2018**, *172*, 432–439. [CrossRef]

40. Estévez-Cimadevila, J.; Otero-Chans, D.; Martín-Gutiérrez, E.; Suárez-Riestra, F. Testing of different non-adherent tendon solutions to reduce short-term deflection in full-scale timber-concrete-composite T-section beams. *J. Build. Eng.* **2020**, *31*, 101437. [CrossRef]

41. Toratti, T.; Schnabl, S.; Turk, G. Reliability analysis of aglulam beam. *Struct. Saf.* **2007**, *29*, 279–293. [CrossRef]

42. Falk, R.H.; Colling, F. Laminating effects in glued-laminated timber beams. *J. Struct. Eng.* **1995**, *121*, 1857–1863. [CrossRef]

43. Tomasi, R.; Parisi, M.A.; Piazza, M. Ductile design of glued-laminated timber beams. *Pract. Period. Struct. Design Construct.* **2009**, *14*, 113–122. [CrossRef]

44. Hiramatsu, Y.; Fujimoto, K.; Miyatake, A.; Shindo, K.; Nagao, H.; Kato, H.; Ido, H. Strength properties of glued laminated timber made from edge-glued laminae II: Bending, tensile, and compressive strength of glued laminated timber. *J. Wood Sci.* **2011**, *57*, 66–70. [CrossRef]

45. Fujimoto, K.; Hiramatsu, Y.; Miyatake, A.; Shindo, K.; Karube, M.; Harada, M.; Ukyo, S. Strength properties of glued laminated timber made from edge-glued laminae I: Strength properties of edge-glued karamatsu (Larix kaempferi) laminae. *J. Wood Sci.* **2010**, *56*, 444–451. [CrossRef]

46. Castro, G.; Paganini, F. Mixed glued-laminated timber of Poplar and Eucalyptus grandis clones. *Holz Roh Werkst.* **2003**, *61*, 291–298. [CrossRef]

47. Teles, R.F.; Cláudio, H.S.; Menezzi, D.; De Souza, F.; De Souza, M.R. Theoretical and experimental deflections of glued laminated timber beams made from a tropical hardwood. *Wood Mater. Sci. Eng.* **2013**, *8*, 89–94. [CrossRef]

48. Fink, G.; Frangi, A.; Kohler, J. Probabilistic approach formodelling the load-bearing capacity of glued laminatedtimber. *Eng. Struct.* **2015**, *100*, 751–762. [CrossRef]

49. Foschi, R.O.; Barrett, J.D. Glued-laminated beam strength. *J. Struct. Div. ASCE* **1980**, *106*, 1735–1754.

50. Hernandez, R.; Bender, D.; Richburg, B.; Kline, K. Probabilistic modeling of gluedlaminated timber beams. *Wood Fiber Sci.* **1992**, *24*, 294–306.

51. Jöbstl, R.A.; Schickhofer, G. Comparative examination of creep of GLT and CLT slabs in bending. In Proceedings of the 40th Meeting of CIB-W18, Bled, Slovenia, 28–31 August 2007.

52. Thiel, A. ULS and SLS Design of CLT and its implementation in the CLT designer. In *Solid Timber Solutions, European Conference on Cross Laminated Timber, Graz, Austria, 21–22 May 2013*, 2nd ed.; COST Action: Graz, Austria, 2014; pp. 77–102.

53. Fragiacomo, M. Seismic behavior of cross-laminated timber buildings: Numerical modeling and design provisions. In *Solid Timber Solutions, European Conference on Cross Laminated Timber, Graz, Austria, 21–22 May 2013*, 2nd ed.; COST Action: Graz, Austria, 2014; pp. 3–36.

54. European Committee for Standardization. *Structural Timber—Strength Classes*; EN 338; European Committee for Standardization: Brussels, Belgium, 2016.

55. European Committee for Standardization. *Structural Timber—Determination of Characteristic Values of Mechanical Properties and Density*; EN 384; European Committee for Standardization: Brussels, Belgium, 2018.

56. Pereira, M.C. de M.; Calil Neto, C.; Icimoto, F.H.; Calil Junior, C. Evaluation of tensile strength of a *Eucalyptus grandis* and *Eucalyptus urophyla* hybrid in wood beams bonded together by means of finger Joints and polyurethane-based glue. *Mater. Res.* **2016**, *19*, 1270–1275. [CrossRef]

57. European Committee for Standardization. *Timber Structures—Glued Laminated Timber and Glued Solid Timber—Requirements*; EN 14080; European Committee for Standardization: Brussels, Belgium, 2013.

58. Bodig, J.; Jayne, B.A. *Mechanics of Wood and Wood Composites*; Van Nostrand Reinhold: Hoboken, NJ, USA, 1982; p. 376.

59. European Committee for Standardization. *Timber Structures—Structural Timber and Glued Laminated Timber—Determination of Some Physical and Mechanical Properties*; EN 408; European Committee for Standardization: Brussels, Belgium, 2012.

 *materials*

Article

The Usefulness of Pine Timber (*Pinus sylvestris* L.) for the Production of Structural Elements. Part I: Evaluation of the Quality of the Pine Timber in the Bending Test

Radosław Mirski [1,*], Dorota Dziurka [1,*], Monika Chuda-Kowalska [2], Marek Wieruszewski [1], Jakub Kawalerczyk [1] and Adrian Trociński [1]

[1] Department of Wood Based Materials, Faculty of Wood Technology, Poznań University of Life Sciences, Wojska Polskiego 38/42, 60-627 Poznań, Poland; marek.wieruszewski@up.poznan.pl (M.W.); jakub.kawalerczyk@up.poznan.pl (J.K.); adrian.trocinski@up.poznan.pl (A.T.)

[2] Institute of Structural Analysis, Faculty of Civil and Transport Engineering, Poznan University of Technology, pl. Sklodowskiej-Curie 5, 60-965 Poznań, Poland; monika.chuda-kowalska@put.poznan.pl

* Correspondence: rmirski@up.poznan.pl (R.M.); dorota.dziurka@up.poznan.pl (D.D.); Tel.: +48-61-848-7416 (R.M.); +48-61-848-7619 (D.D.)

Received: 22 July 2020; Accepted: 2 September 2020; Published: 7 September 2020

Abstract: The study assessed the quality of pine lumber by marking the modulus of elasticity in the horizontal system. The research material was a plank with the following dimensions: 137 mm wide × 39.50 mm thick × 3485 mm long. The pine wood was obtained by sawing timber in the form of logs with round cross sections and originating from the Forest Division Olesno (50°52′30′′ N, 18°25′00′′ E). Each long log was sawn to provide four logs of about 3.5 m, which were marked as butt-end logs (O), middle logs (S)—2 items, and top logs (W). The origin of the logs from the trunk (*Pinus sylvestris* L.) has a significant impact on the physical and mechanical properties of the wood from which they are made. Only butt-end logs (log type O) allows for the production of high-quality timber elements. The pine timber that was evaluated in this paper had a high density of about 570 kg/m^3 and a high percentage of timber items were assigned to class C24 and higher (above 50%). The adopted horizontal model of evaluation of the modulus of elasticity gave similar results to those obtained in an evaluation according to the EN-408.

Keywords: modulus of elasticity; pine wood; wood defects; knots; laboratory tests

1. Introduction

To improve the applicability of timber in the construction industry, manufacturers are required to provide components with as low a number of defects as possible. Such components are made both as solid and glued semi-products [1–3]. It is the top priority to obtain dimensionally stable components with very good mechanical properties from readily available pine wood (*Pinus sylvestris* L.) originating from Polish forests [4–10]. Glued laminated timber has typical features of a solid timber: light weight, good strength, elasticity, durability, easy processing, and a unique feature (i.e., is readily shaped cross-sections). Its cross-section has a layered structure, enabling the manufacturing of components with variable cross-sectional height as needed [11–14].

The complete and practical application of pine wood for the manufacturing of glued components requires timber strength grading [2,15–19]. The procedure is based on the fundamental knowledge of the structure of wood and the defects it may have [20–24]. The preferred timber grading techniques are based on the use of machines in accordance with EN 338, EN 408, and EN 14081-1 [25–27]. If the machines are unavailable, visually graded timber is used. In Poland, the method is based on the Polish Standard PN-D-94021-10P [28].

The wood of the pine tree is valued for its high availability, good strength characteristics, and high productivity of each long log, which is due to the specific anatomical characteristics of this species. As previous studies have shown, not only the geographical region, but also the location of the log in a given tree zone influences the quality of the sawn timber harvested. It was shown that the middle and butt logs had a lower number of defects. Moreover, the proportion of raw material without defects increases as the distance from the core increases. The occurrence of a large number of knots in the Polish raw material is considered a negative quality characteristic. It is known that the number and intensity of defects affects the static bending strength of lumber.

The most frequently determined strength parameter of materials is the modulus of elasticity in static bending. The parameter is a basis for the comparison of the technical value of timber: the higher the modulus, the better the technological quality of structural materials. As the structure of the wood is anisotropic, and to ensure some flaws can be masked, the value of the modulus of elasticity does not always translate directly into proportional bending strength. This problem makes it necessary to check properties, identified in non-destructive testing, with real strength. Since the evaluation of strength requires the destruction of the tested material, it is more advantageous to determine the relationship between the modulus and the strength of the laboratory test [2,14,17].

The numerous advantages of glued laminated timber enable its use in practically every type of building regardless of its intended application. This assumption is the basis for undertaking the quality testing of glued semi-products and components, taking into consideration their use for construction purposes.

Assessing the impact of the quality of the various classes of timber originating from Polish forests to be used as sawn timber or glued components may provide valuable information to wood technology specialists in Poland, contributing to developments in the production of wood for construction applications in the form GLT (glued laminated timber).

The aim of the study was to assess the quality of pine timber in a horizontal bending test. Pieces of timber intended for testing were obtained by sawing logs from model trees from a 125-year-old stand.

2. Materials and Methods

The research material was pine wood with the following dimensions: 137 mm wide × 39.50 mm thick × 3485 mm long. The pine wood was obtained by sawing timber in the form of logs with round cross sections and originating from the Forest Division Olesno (50°52′30″N, 18°25′00″E). The age of the tree stand was about 125 years, and the average tree height was 25 m. Butt-end long logs of about 14 m were obtained from every timber item. Each long log was sawn to provide four logs of about 3.5 m, which were marked as butt-end logs (O), middle logs (S)—2 items, and top logs (W). Each log was processed to obtain a flitch from which the main timber intended for manufacturing structural beams was obtained. Each timber item was measured to determine its linear dimensions, density, and modulus of elasticity (MOE). MOE was found by determining the deflection for a given load in accordance with the diagram shown in Figure 1. Figure 2 shows the appearance of the test stand. The assumed preliminary load was 134.9 N (13.75 kg). At that value, the deformation sensor was reset before increasing the load to 517.5 N (52.75 kg). The timber was deflected eight times, but the deflection values were recorded for the final five measurements only.

The modulus of elasticity (MOE) of lumber is calculated from Equation (1):

$$\text{MOE} = \frac{F \times a \times l^2}{8 \times f \times J} \tag{1}$$

where F is the force, N; a is the distance from the applied force to the support, mm; l is the length of the deflection measuring section, mm; f is the level of deformation, mm; and J is the moment of inertia, mm^4.

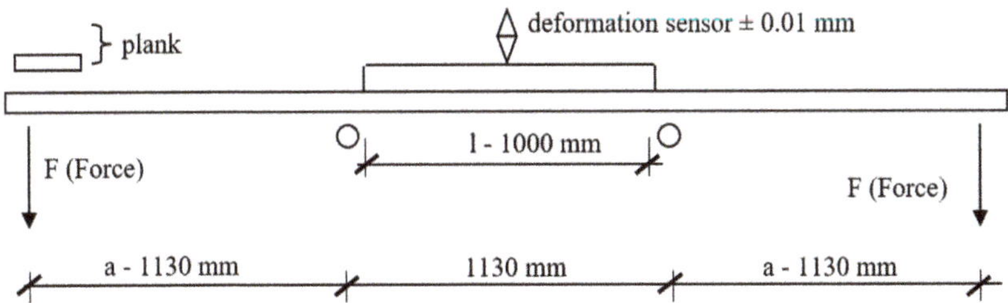

Figure 1. The experimental setup used to identify MOE parameter.

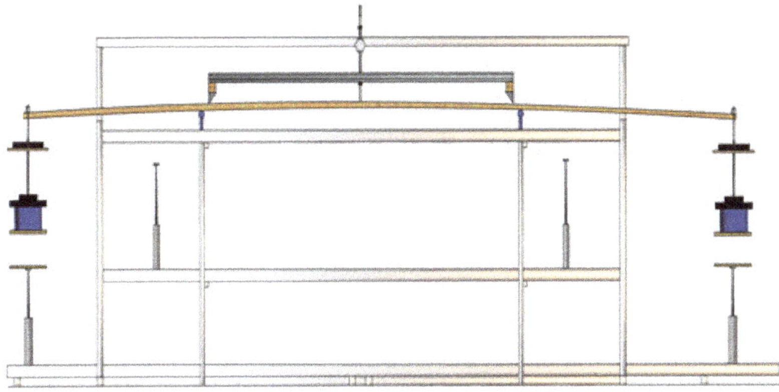

Figure 2. Lumber test stand (Grzegorz Zmyślony).

SYLVAC callipers with a measurement accuracy of 0.01 mm were used to determine the sawn timber section (thickness and width). To determine the length, a linear gauge with 1 mm graduation was used. Weight was evaluated using the Radwag tensometer scale. Moreover, the moisture content of each piece was determined using a Tanel HIT-1 hammer moisture meter.

Since the moisture of the timber during the test differed considerably from 12% (between 10.2% and 13.5%), the outcomes were recalculated using Bauschinger's Equation (2):

$$E_{12} = E_{MC}[1 + \alpha_{MC} \times (MC - 12)] \tag{2}$$

where E_{12} is the modulus of elasticity of wood for a moisture content of 12%, N/mm^2; E_{MC} is the modulus of elasticity of wood for a moisture content of 4% < w < 20%; α_{MC} is the coefficient of variation of the modulus of elasticity of wood after its moisture content changed by 1%, which is assumed to be 0.02; and MC is the absolute moisture content of wood, %.

The assessment covered 486 timber items (samples). Each piece was given an individual number so that it was easy to prepare the lumber sets for the production of beams.

The results of the experimental measurements were analyzed using the STATISTICA 13.0 package (Version 13.0, StatSoft Inc., Tulsa, OK, USA). The obtained results were evaluated using ANOVA (analysis of variance) variance analysis, whereas Tukey's test was most frequently used to determine homogeneous groups.

3. Results and Discussion

In the evaluation of structural timber, density is one of the crucial or most important parameters of interest. Average density of the timber items was 571 kg/m^3 (average moisture 8.98%). The items obtained from the top and middle logs were characterized by similar densities. For those from the butt-end logs, the density was much higher (50 kg/m^3) (Figure 3).

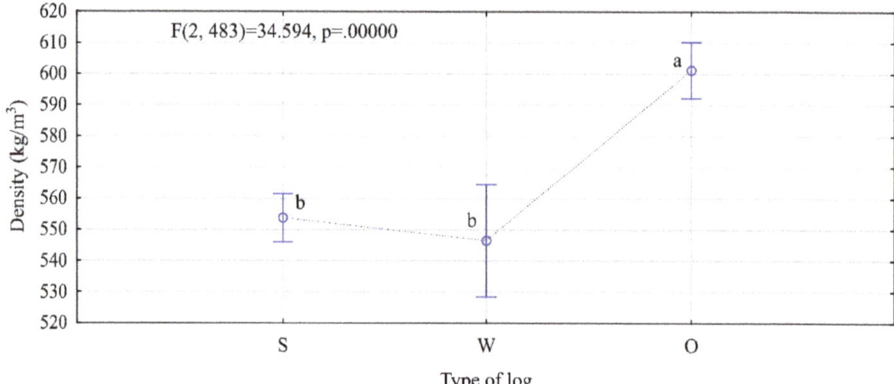

Figure 3. Density of analyzed timber (W—top logs, S—middle logs, O—butt-end logs. The letters denote uniform groups, as determined by Tukey's HSD (honest significant difference) test).

The tested timber did not follow a normal density distribution (Figure 4). Its density distribution was characterized by a skewness of 0.343 and kurtosis of 0.361. Hence, the timber's density distribution had a positive skew, which means that the most of analyzed timber items were characterized by densities below 600 kg/m^3. This was confirmed by previous observations and data concerning the number of items originating from the various locations along the tree length. It should be noted that the number of butt-end logs was half that of the middle logs.

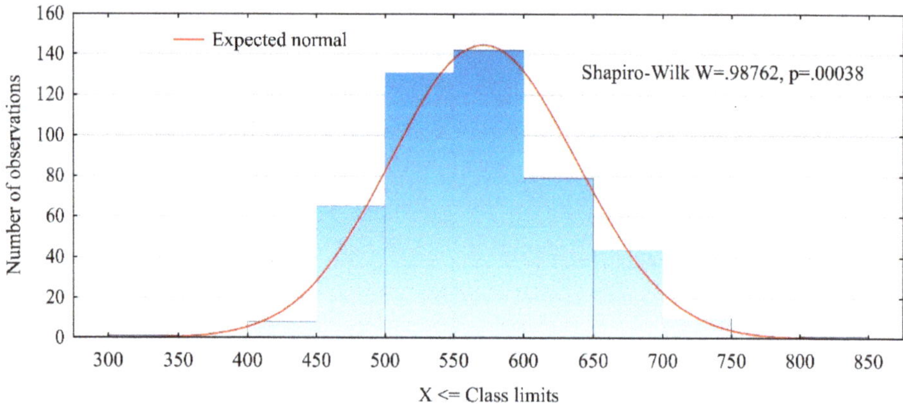

Figure 4. Density distribution for the timber.

When analyzing density in accordance with EN 338 as an indicator of the strength class of coniferous timber, classes C24, C27, and C30 would comprise only one timber item each, C45 with nine items and C50 the remaining ones. This means that for this batch of timber, its strength is determined mainly by its modulus of elasticity. Statistically, significant differences between the moduli of elasticity were observed for timber items located in different sections along the tree length. The highest moduli were observed for the butt-end logs and the lowest or the top logs. In the first case, the mean value of the modulus of elasticity was 12.4 kN/mm^2 and in the second was 9.10 kN/mm^2. The middle-section timber items were characterized by an average modulus of elasticity of 11.3 kN/mm^2 (Figure 5). Timber from the top section represented classes below C20, timber from the middle section classes C24–C27, and the butt-end timber represents classes above C30. What is important, even though the adopted model of assessing the modulus of elasticity is very different from the assumptions of EN-408, it does provide values close to those stated in that standard [29]. A study has shown that the average modulus of elasticity of timber from the region of interest was 11,915 N/mm^2 for a SD (standard deviation) = 2440 N/mm^2, whereas in the test sample of timber, the average modulus was around 11,556 N/mm^2 for a SD of 2554 N/mm^2. This indicates that the adopted method of evaluation of the modulus of elasticity in this case provided convergent results for the modulus of elasticity.

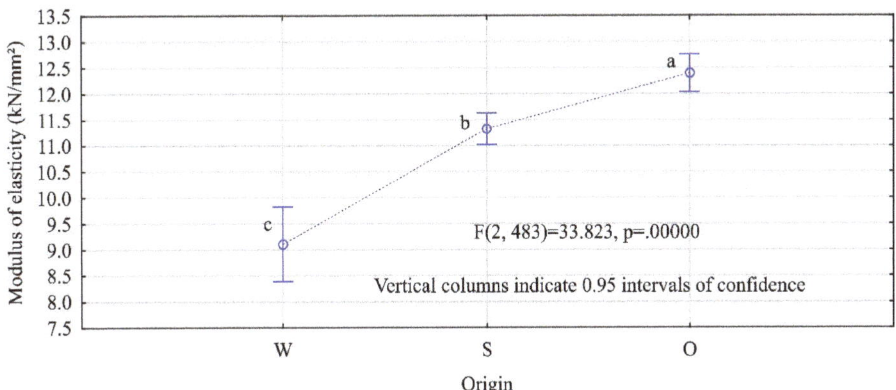

Figure 5. MOE of timber depending on the location of the log along the tree length (W—top logs, S—middle logs, O—butt-end logs. The letters denote uniform groups, as determined by Tukey's HSD test).

A more detailed analysis of the distribution of MOE values for the timber items of interest is illustrated in Figure 6.

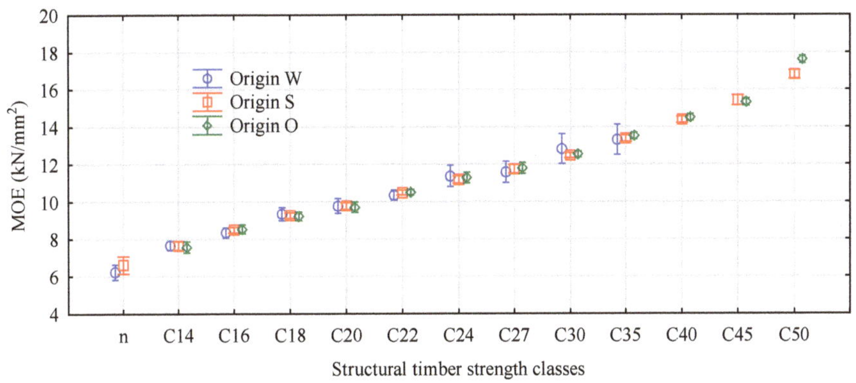

Figure 6. Timber strength classes by location. Legend: W—top logs, S—middle logs, O—butt-end logs.

The broadest range was represented by the timber obtained by sawing the mid-section logs, which comprised both items classified as class C50 and classes below C14. The butt-end log timber, even though its modulus of elasticity was very high, comprised items of high average strength classes ranging from C14 to C50. The top log timber was represented both by unclassified items and those graded between C14 and C35.

Figure 7 shows the number of items representing the respective classes of timber strength and the cumulative percentage of the number of unclassified timber items relative to class C50.

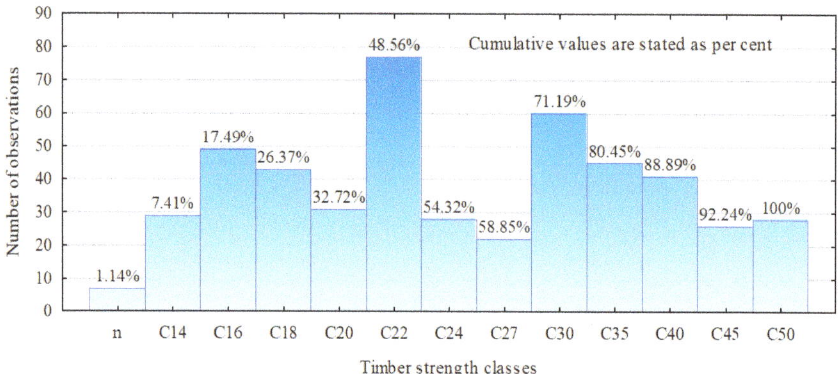

Figure 7. Number of items in strength classes for timber from the Forest Division Olesno.

C24 is the timber class that is essentially used for roof truss systems and other load-bearing components. This class was represented by a rather low number of items, merely 5.7% of the analyzed timber items. On the other hand, there were only 48.6% of timber items below class C24. The region thus provides timber with exceptionally high mechanical strength, especially with regard to the parameter of interest, modulus of elasticity, which is loosely correlated with bending strength. Since the importance of the quality of timber, in this case, its modulus of elasticity, increases with a squared distance between the timber item's axis and the beam's neutral axis, then the higher the beam, the lower the grade of the timber that may be positioned nearer the neutral axis. This means that the higher the beam, the higher number of low grades of timber can be compensated with good grade timber positioned nearer the outer beams. Therefore, the authors chose to prepare such technical models of beams as could be obtained from the available timber.

4. Conclusions

The origin of the logs from the trunk (*Pinus sylvestris* L.) has a significant impact on the physical and mechanical properties of the wood from which they are made. Only butt-end logs (log type O) allows for the production of high-quality timber elements.

The pine timber from *Pinus sylvestris* L., which was evaluated in this paper, had a high density of about 570 kg/m^3 and a high percentage of timber items assigned to class C24 and higher (above 50%), as found in the 4-point bending test to assess its modulus of elasticity.

We used the horizontal model of evaluation of the modulus of elasticity of which the results, for the batch of interest, were similar to those obtained in an evaluation according to EN-408 [25,28].

Author Contributions: Conceptualization, R.M.; Methodology, R.M.; Validation, D.D., M.C.-K., A.T., and M.W.; Formal analysis, R.M. and M.C.-K.; Investigation, D.D., A.T., and M.W.; Resources, R.M., D.D., J.K., and M.W.; Writing—Original Draft Preparation, R.M., D.D., and M.C.-K.; Writing—Review and Editing, R.M., J.K., and D.D.; Visualization, J.K. and R.M.; Project Administration, R.M. and D.D.; Funding Acquisition, R.M. All authors have read and agreed to the published version of the manuscript.

Funding: This research was funded by the National Center for Research and Development, grant number BIOSTRATEG3/344303/14/NCBR/2018. The authors are grateful for the support of the Ministry of Science and Higher Education Program "Regional Initiative of Excellence" in the years 2019–2022, Project No. 005/RID/2018/19.

Acknowledgments: We would like to thank Grzegorz Zmyślony for his help in conducting the research.

Conflicts of Interest: The authors declare no conflict of interest.

References

1. Czuczeło, K.; Dzbeński, W. *Influence of the Mechanical Quality of the Timber on the Properties of Glued Laminated Elements*; VI Konferencja Naukowa Wydziału Budownictwa i Architektury; Wydawnictwo Uczelniane Politechniki Szczecińskiej: Szczecin, Poland, 2004; pp. 35–42. (In Polish)
2. Czuczeło, K. Testing of glued laminated timber from strength sorted pine wood. Ph.D. Thesis, Warsaw University of Life Sciences—SGGW, Warszawa, Poland, 2005. (In Polish)
3. Baltrušaitis, A. Investigation of influence of knots on short-and long-term strength of glulam beams. In Proceedings of the 6th World Conference on Timber Engineering, British Columbia, BC, Canada, 31 July–3 August 2000.
4. Buchholz, J.; Hruzik, J.G.; Meyer, B. Investigations of selected properties of degradated pine sawntimber. *Folia Forestalia Polonica* **1990**, *21*, 17–31. (In Polish)
5. Dzbeński, W. *Polish Construction Timber against European Requirements*; IX Konferencja Naukowa Wydziału Technologii Drewna SGGW: Warszawa, Poland, 1995. (In Polish)
6. Hruzik, J.G. Consumption of wood raw material and wood materials in products of the sawmill industry. *Drewno-Wood* **2006**, *49*, 25–44. (In Polish)
7. Rapp, P. Experimental testing of glued longitudinal wooden beam joints. In Proceedings of the V Konferencja Naukowa, Drewno i Materiały Drewnopochodne w Konstrukcjach Budowlanych, Szczecin, Poland, 17–18 May 2002; pp. 137–148. (In Polish)
8. Rapp, P. Large-span wooden beams glued together from thicker elements. In Proceedings of the IV Konferencja Naukowa, Drewno i Materiały Drewnopochodne w Konstrukcjach Budowlanych, Szczecin-Świnoujście, Poland, 27–28 September 1999; pp. 291–296. (In Polish).
9. Neuhaus, H. *Wood Construction, Engineer's Handbook*; Polskie Wydawnictwo Techniczne: Rzeszów, Poland, 2006.
10. Jeitler, G.; Katzengruber, R.; Schickhofer, G. Quality control of wedge-jointed sawn timber. *Gazeta Drzewna* **2007**, *10*, 38. (In Polish)
11. Sterr, R. Untersuchungen zur Dauerfestigkeit von Schichtholzbalken. Mitteilung aus dem Institut für Holzforschung und Holztechnik der Universität München (Investigations on the fatigue resistance of laminated wood beams). *Holz als Roh und Werkstoff* **1963**, *21*, 7–61. [CrossRef]
12. Gerhards, C.C. Effect of early wood and latewood on stress-wave measurements parallel to the grain. *Wood Sci. Technol.* **1975**, *11*, 69–72.
13. Krzosek, S. *Strength Grading of Polish Pine Structural Sawn Timber*; Wydawnictwo SGGW: Warsaw, Poland, 2009. (In Polish)
14. Wieruszewski, M.; Gołuński, G.; Hruzik, G.J.; Gotych, V. Glued elements for construction. *Ann. Warsaw Univ. Life Sci.—SGGW For. Wood Technol.* **2010**, *72*, 453–458.
15. Krzosek, S. Strenght grading of structural lumber. *Lekkie Budownictwo Szkieletowe* **1998**, *4*, 10–11. (In Polish)
16. Krzosek, S. Strenght grading of sawn timber in the production of glued laminated timber components. *Przemysł Drzewny* **2002**, *3*, 32–33.
17. Krzosek, S.; Grześkiewicz, M. Strenght grading Polish-grown Pinus silvestris L. structural sawn timber using Timber Grader MTG and visual method. *Ann. Warsaw Univ. Life Sci.—SGGW For. Wood Technol.* **2008**, *66*, 26–31.
18. Pajchrowski, G.; Jabłoński, L.; Szumiński, G. *National Sawmill Base for the Production of Structural Timber That Meets the Requirements of the Construction Products Directive (89/106/EEC). STAGE VII. Analysis of Development Directions of Equipment for Strength Classification of Structural Timber*; Instytut Technologii Drewna, Zakład Badania i Zastosowania Drewna: Poznań, Poland, 2009. (In Polish)
19. Kotwica, E. Solid Structural Timber Certification. Available online: http://www.fachowydekarz.pl/index.php/na-rynku/81-certyfikacja-drewna-konstrukcyjnego-litego.html (accessed on 6 July 2020). (In Polish).
20. Diebold, R.; Schleifer, A.; Glos, P. Machine grading of structural sawn timber from various softwood and hardwood species. In Proceedings of the 12th Internat. Symp. Nondestructive Testing of Wood, Sopron, Hungary, 13–15 September 2000; pp. 139–146.

21. Duju, A.; Nakai, T.; Nagao, H.; Tanaka, T. Nondestructive evaluation of mechanical strength of sarawak timbers. In Proceedings of the 12th Internat. Symp. Nondestructive Testing of Wood, Sopron, Hungary, 13–15 September 2000; pp. 131–137.

22. Mańkowski, P.; Krzosek, S. Application of scanners in the wood industry. *Przemysł Drzewny* **2001**, *10*, 20–23. (In Polish)

23. Noskowiak, A.; Szumiński, G. *Investigations of Mechanical Properties of Construction Pine Lumber Obtained from the Silesian Land*; Typescript Instytut Technologii Drewna: Poznań, Poland, 2005. (In Polish)

24. Kozakiewicz, P.; Krzosek, S. *Wood Materials Engineering*; Wydawnictwo SGGW: Warszawa, Poland, 2013. (In Polish)

25. *EN 338. Structural Timber—Strength Classes*; European Committee for Standardization: Brussels, Belgium, 2016.

26. *EN 408. Timber structures—Structural Timber and Glued Laminated Timber—Determination of Some Physical and Mechanical Properties*; European Committee for Standardization: Brussels, Belgium, 2012.

27. *EN 14081-1. Timber Structures—Strength Graded Structural Timber with Rectangular Cross Section-Part. 1: General Requirements*; European Committee for Standardization: Brussels, Belgium, 2016.

28. *PN-D-94021-10—Constructional Softwood Sorted by Strength Grading Methods*; Polish Committee for Standardization: Warsaw, Poland, 2013. (In Polish)

29. Burawska-Kupniewska, I.; Krzosek, S.; Mankowski, P.; Grzeskiewicz, M.; Mazurek, A. The influence of pine logs (*Pinus sylvestris* L.) quality class on the mechanical properties of timber. *BioRes* **2019**, *14*, 9287–9297.

materials

Article

Dynamic Pulse Buckling of Composite Stanchions in the Sub-Cargo Floor Area of a Civil Regional Aircraft

Andrea Sellitto [1],*[ID], Francesco Di Caprio [2][ID], Michele Guida [3][ID], Salvatore Saputo [1][ID] and Aniello Riccio [1][ID]

[1] Department of Engineering, University of Campania Luigi Vanvitelli, via Roma 29, 81031 Aversa, Italy; salvatore.saputo@unicampania.it (S.S.); aniello.riccio@unicampania.it (A.R.)
[2] CIRA—Italian Aerospace Research Centre, via Maiorise snc, 81043 Capua, Italy; f.dicaprio@cira.it
[3] Department of Industrial Engineering, University of Naples Federico II, p.le Tecchio 80, 80125 Naples, Italy; michele.guida@unina.it
* Correspondence: andrea.sellitto@unicampania.it

Received: 15 July 2020; Accepted: 11 August 2020; Published: 14 August 2020

Abstract: This work is focused on the investigation of the structural behavior of a composite floor beam, located in the cargo zone of a civil aircraft, subjected to cyclical low-frequency compressive loads with different amplitudes. In the first stage, the numerical models able to correctly simulate the investigated phenomenon have been defined. Different analyses have been performed, aimed to an exhaustive evaluation of the structural behavior of the test article. In particular, implicit and explicit analyses have been considered to preliminary assess the capabilities of the numerical model. Then, explicit non-linear analyses under time-dependent loads have been considered, to predict the behavior of the composite structure under cyclic loading conditions. According to the present investigation, low-frequency cyclic loads with peak values lower than the static buckling load value are not capable of triggering significant instability.

Keywords: dynamic pulse buckling; composite stanchion; FE analysis; nonlinear analysis; crashworthiness

---

1. Introduction

Buckling is an instability phenomenon typical of "thin" structures (characterized by at least one very small dimension compared to the others). Usually, buckling has been considered a purely static phenomenon. The classic example is that of the Euler beam in which a beam stuck at one end is loaded from the tip to the other, with a compressive load lower than the limit of elasticity of the material. Theoretically, if the force is perfectly centered and the beam is free of imperfections, the latter would remain in equilibrium under the action of any load. In order for the instability to occur, it is mandatory to destabilize the beam by means of an external action, immediately removed. After the perturbation, three cases can occur, depending on the applied load: The beam returns to the initial equilibrium configuration (stable equilibrium); the beam moves to a new equilibrium condition different from the initial one (indifferent equilibrium); the beam moves away indefinitely from the initial equilibrium configuration (unstable equilibrium). The so-called buckling load is the lowest of the loads for which equilibrium is indifferent. However, buckling can also be caused by loads that vary over time [1–3]. The application of a time-dependent axial load to a beam, which then induces lateral vibrations and can eventually lead to instability, is something that has been investigated by many authors [4–10].

Generally, dynamic buckling depends on prescribed dynamic loads, and is distinguished between buckling from oscillatory loads and buckling from transient loads, so two forms of dynamic buckling are classified: Vibration and pulse buckling.

Vibration buckling is characterized by oscillating loads producing amplification of the vibrations up to unacceptable values, and, in the absence of damping or nonlinear effects, a phenomenon similar to vibration resonance is predicted. In the vibration buckling, the force producing the instability is considered as a parameter multiplying the displacement, while in the vibration resonance, the load exciting the motion is evaluated as an applied force.

In pulse buckling, the presence of applied transient loads generates much larger and unacceptable amplitudes than the static buckling load of the bar as a result of the plastic response or large-deformation post-buckled state. The pulse buckling is characterized by buckling increasing rather than oscillating, and the modes produce wavelengths larger than the Euler wavelength for a given load.

The dynamic buckling of the column in an aircraft landing stanchion is a great deal and, in this work, the structural behavior of a composite floor beam subjected to low-frequency cyclic load conditions has been investigated. Indeed, the structural limits of beam-like structures is driven by their stability, rather than by their strength. The loss of stability induced by static loads is a well-known phenomenon, extensively investigated [11]. However, the loads that actually act on the structures are characterized by dynamic behavior, which can result in dynamic instability [12–15]. For these reasons, more detailed studies on the dynamic buckling phenomenon became mandatory in order to accurately predict the structural limits of the components subjected to realistic loads [16].

Dynamic buckling has a relatively recent history. One of the first research studies on dynamic buckling can be found in [17], where a theoretical solution considering a simply supported rectangular plate subjected to varying floor loads over time has been developed. In [18], a criterion that relates dynamic buckling to the duration of the load has been introduced. The effects of high-intensity and short-duration loads have been studied in [19]. According to the investigation, long-lasting critical dynamic buckling loads may be of less intensity than the corresponding static buckling loads.

In [20], a numerical and experimental investigation was performed on composite laminated shells by correlating their natural vibration frequencies and mode shapes with the buckling loads and modes. Moreover, parametric studies on the dynamic instability of carbon-fiber-reinforced plastic (CFRP) cylindrical shells were introduced in [21]. Different parameters were considered, such as the shape and the duration of pulse loading, and the initial geometric imperfection of the structure. According to the study, the dynamic load factor (DLF), defined as the ratio between the dynamic and the static buckling loads, is higher than that if short-pulse-duration loadings are used. Then, as the pulse duration increases, the DLF decreases, being lower than that in the vicinity of the natural frequency of the structure.

Other works have focused on the investigation of dynamic buckling on damaged composite structures. In [22], the nonlinear dynamic pulse buckling of a plate with embedded delamination was numerically investigated, and the DLF was computed. Different types of pulse loadings (sinusoidal, exponential, and rectangular) and different delamination sizes and depths were considered. Moreover, in [23], the dynamic instability of damaged composite components was studied, and the effects of the damage and its position were assessed. The damage was numerically simulated by reducing the stiffness of the components.

Different studies on the dynamic buckling induced by impact loading can be found in the literature [24–26]. In [27], the dynamic buckling of an elastic cylindrical shell subjected to axial impact was presented, and the influence of the boundary condition on the critical velocity of the impactor was assessed. In [28], the dynamic pulse buckling of composite laminated beams subjected to impact in the axial direction was investigated. In particular, the influence of different beams parameters (transverse and axial inertias, shear deformation, cross-section's rotational inertia, and axial shortening) on the beams' pulse buckling response was assessed.

Interesting insight on the dynamic instability can be found in [29]. A very detailed study was presented by the authors, which categorized the phenomenon according to the nature of the external load, whose direction can vary depending on the deformation of the structure, classifying the dynamic buckling problems as flutter or pulse buckling.

In this work, detailed numerical analyses have been presented, aimed to investigate the dynamic buckling response of the composite stanchion, in the finite element code Abaqus environment (release 2019, Dassault Systèmes Simulia Corp: Providence, Rhode Island, RI, SUA). Preliminary finite element models have been presented and compared, considering different in-plane and through-the-thickness element dimensions, different element types, and different material formulations, in order to find the model that more faithfully reproduces the investigated phenomenon. The latter has been used for the subsequent numerical investigation: The static stability limit of the composite stanchion (identified by its static critical buckling load) has been found; then, three different low-frequency cyclic load conditions have been considered (below, close, and above the static critical buckling load). In Section 2, the test cases are described, while in Section 3, the numerical analyses are introduced, including preliminary analyses aimed to determine the most suitable numerical model able to describe the dynamic phenomenon, and the results are critically discussed.

2. Test Case Description

The investigated test case, which has been previously validated in [30–32], is representative of the stanchion of the cargo floor of a regional aircraft of general aviation [33]. Images of the stanchions and their location in a typical fuselage section are reported in Figure 1.

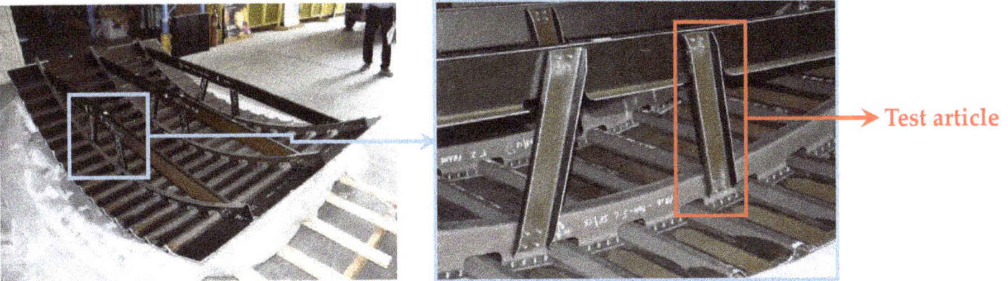

Figure 1. Location of the stanchion in a typical fuselage section.

Two different test articles, with different geometrical characteristics, have been considered. In particular, Test Article 1 is 380 mm long, while the length of Test Article 2 is 315 mm. More details on the test cases' geometry and boundary conditions can be found in Figure 2. According to Figure 2, both sides of the specimens have been potted. A compressive displacement has been applied on one side, while the opposite side has been fixed.

The stanchion material is a composite made of carbon fibers and epoxy matrix at high polymerization temperature. The mechanical properties of the composite lamina are reported in Table 1, where ρ is the nominal density, th is the lamina thickness, E is Young's modulus, G is the shear modulus, υ is Poisson's ratio, X_t and X_c are respectively the longitudinal tensile and compressive strengths, Y_t and Y_c are respectively the transversal tensile and compressive strengths, and S_c is the shear strength. The stacking sequence of the laminate is [45; −45; 90; −45; 45; 0; 0; 0; 0; 0; 45; −45; 90; −45; 45].

Table 1. Mechanical properties of the lamina.

ρ [g/cm^3]	th [mm]	E_{11} [MPa]	E_{22} [MPa]	G_{12} [MPa]	G_{13} [MPa]	G_{23} [MPa]	ν_{12} [-]	X_t [MPa]	X_c [MPa]	Y_t [MPa]	Y_c [MPa]	S_c [MPa]
1.6	0.186	135,000	8430	4160	4160	3328	0.26	2257	800	75	171	85

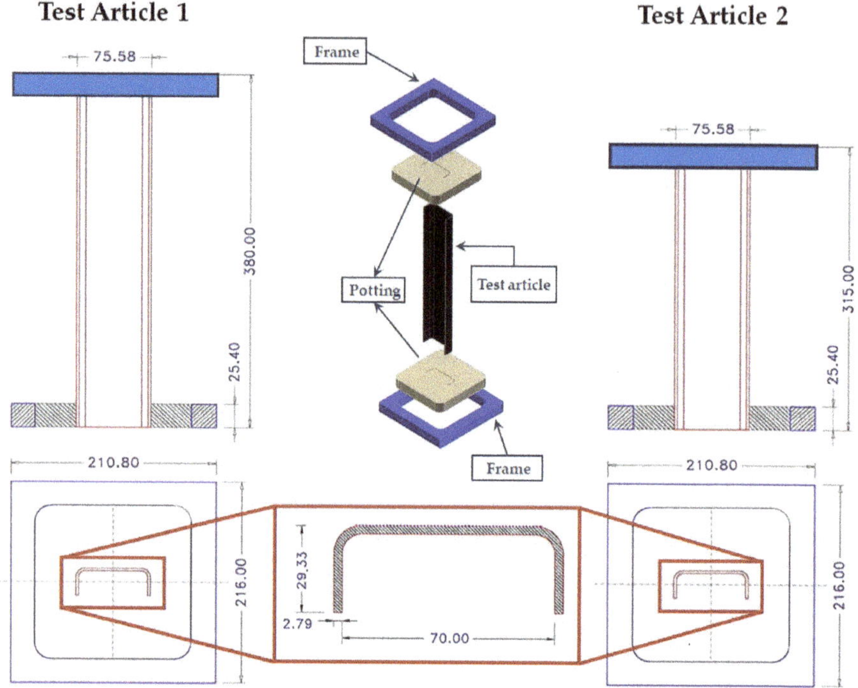

Figure 2. Geometrical description and boundary conditions.

3. Results and Discussion

3.1. Mesh Convergence Analysis

A preliminary mesh convergence analysis has been carried out to determine the best compromise between computational costs and accuracy of the results in terms of predicted stiffness. Hence, different static linear analyses have been performed, varying the in-plane and through-the-thickness element sizes. In particular, three different mesh element sizes were considered: Coarser (8 mm), intermediate (4 mm), and finer (2 mm). Moreover, for each in-plane element size, three through-the-thickness mesh configurations have been investigated, resulting in the nine mesh configurations that have been analyzed and compared to each other. In particular, Figures 3–5 report the model configuration considering, respectively, one, three, and five elements through the thickness. Eight-node Abaqus Continuum shell elements SC8R [34], with 3 degrees of freedom (DoF) per node and a reduced integration scheme, have been used to discretize the model.

The different composite layups, which depend on the number of elements in the thickness direction, are shown in Figure 6.

A linear static compression test has been performed on the configurations, to compare the results in terms of stiffness and computational time (16-core Intel Xeon E5-2687W 3.10 GHz processor, 64 GB RAM). The results are summarized in Table 2.

According to the results reported in Table 2, the configuration 3B represents the best compromise between accuracy of the results and computational time. Hence, this model will be used in all the subsequent analyses.

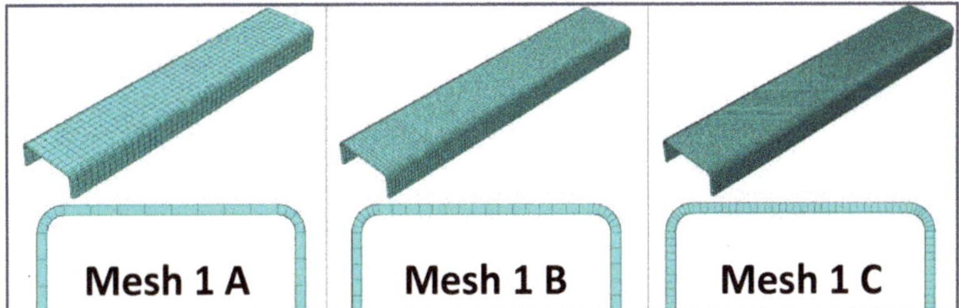

Figure 3. One element in the thickness direction: Coarser mesh (Mesh 1A), intermediate mesh (Mesh 1B), and finer mesh (Mesh 1C).

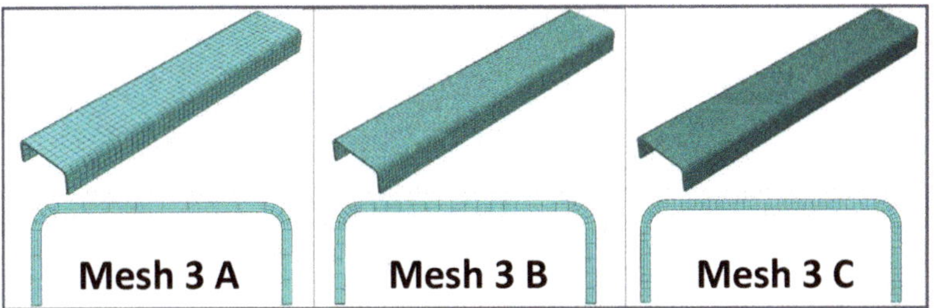

Figure 4. Three elements in the thickness direction: Coarser mesh (Mesh 3A), intermediate mesh (Mesh 3B), and finer mesh (Mesh 3C).

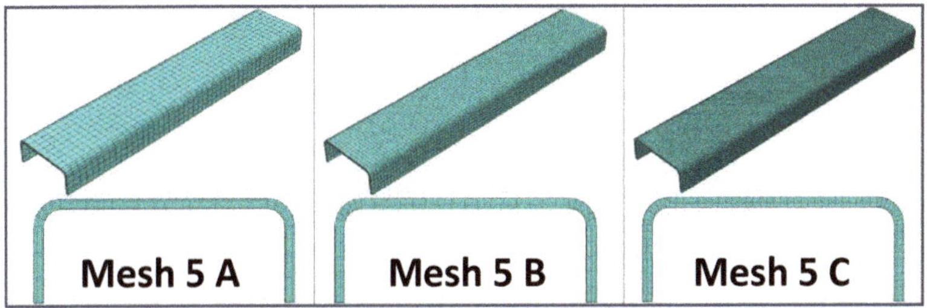

Figure 5. Five elements in the thickness direction: Coarser mesh (Mesh 5A), intermediate mesh (Mesh 5B), and finer mesh (Mesh 5C).

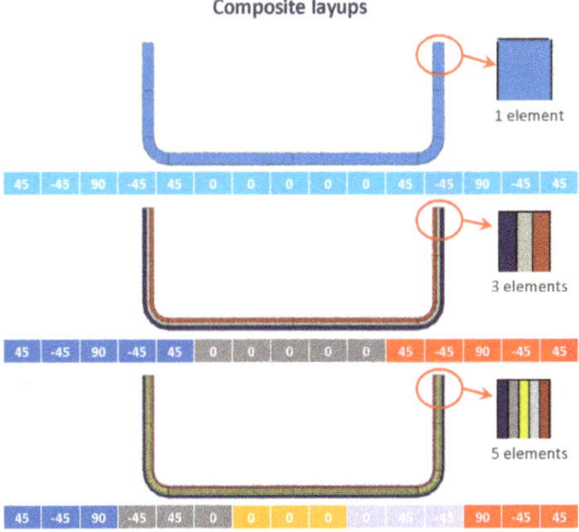

Figure 6. Composite layups.

Table 2. Mesh convergence analysis results.

Number of Elements in the Thickness	Element Size	Model Name	Number of Elements (Total)	Stiffness [kN/mm]	Computational Time
1	8 mm	Mesh 1A	1056	56.220	10 s
	4 mm	Mesh 1B	3572	56.317	15 s
	2 mm	Mesh 1C	12,920	56.340	21 s
3	8 mm	Mesh 3A	3168	56.215	13 s
	4 mm	Mesh 3B	10,716	56.305	18 s
	2 mm	Mesh 3C	38,760	56.327	33 s
5	8 mm	Mesh 5A	5280	56.097	16 s
	4 mm	Mesh 5B	17,860	56.180	21 s
	2 mm	Mesh 5C	64,600	56.192	52 s

3.2. Validation of the Numerical Model

The selected numerical model has been validated by comparison with experimental test campaign results, in terms of stiffness and failure. In particular, two experimental tests have been performed:

- Experimental Test T1: Compressive test aimed to determine the stiffness of the structure;
- Experimental Test T2: Compressive test up to the total failure.

In Figure 7 and Table 3, it is possible to appreciate the good agreement between the numerical predicted solutions and the results of the experimental campaign tests in terms of stiffness and failure load. Moreover, the failure mode of the test article, computed by using the Hashin's failure criteria, has also been predicted with a high level of accuracy, as highlighted in Figure 8. No buckling phenomena occur up to the failure load both experimentally and numerically.

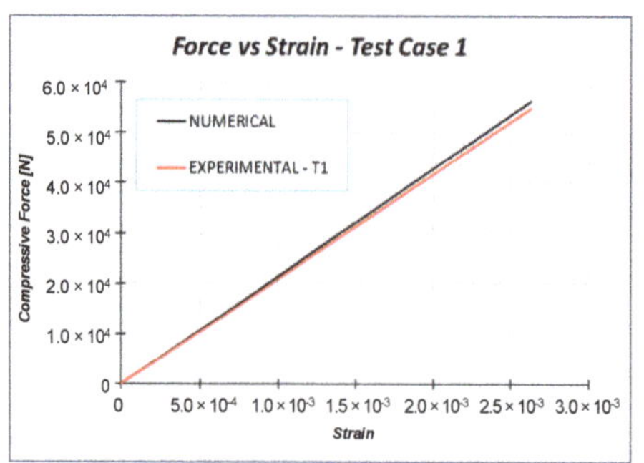

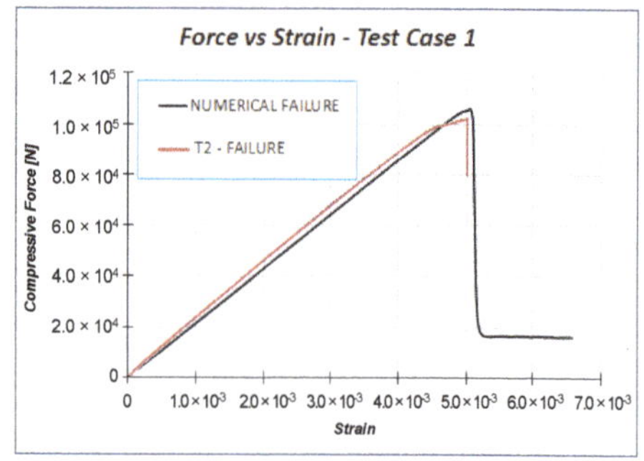

Figure 7. Numerical-experimental comparisons: Load vs. strain.

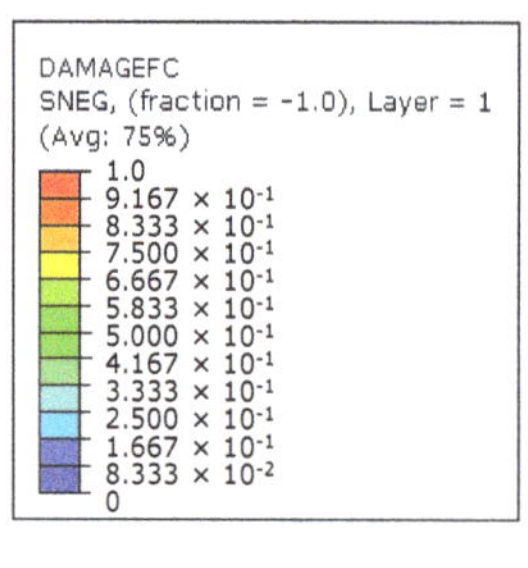

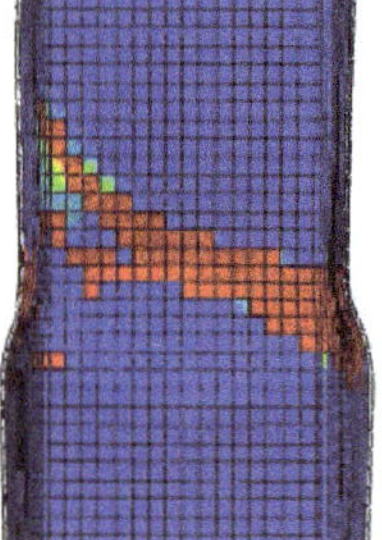

Figure 8. Numerical-experimental comparisons: Failures.

Table 3. Numerical-experimental comparisons: Stiffness and failure load.

	Stiffness	**Failure Load**
Experimental	54.8 kN/mm	103.7 kN
Numerical	56.3 kN/mm	106.0 kN
Error	2.6%	2.2%

3.3. Dynamic Buckling Analysis

Once the numerical model has been validated, explicit numerical analyses [35,36] have been performed on Test Case 2 to investigate the arising of the dynamic buckling phenomenon. However, as a preliminary step, compressive tests have been performed on Test Case 2 by using implicit and explicit formulations, and the results have been compared in order to assess the robustness of the numerical model and the influence of the solver's scheme. Experimental data are available only for quasi-static loading conditions and, therefore, for implicit solutions. As the numerical investigations presented in the next sections have been performed with an explicit solver, the implicit-explicit comparison aims to investigate the possible influence of the solution scheme on the results. In the explicit analysis, the load is applied by a smoothed step in order to minimize the inertial effect (related to dynamic simulation). Thus, no significant differences are expected; anyway, in order to highlight this aspect, the implicit–explicit comparison is presented, to demonstrate that differences in structural response, in the explicit model, are related only to different loading conditions and cannot be addressed to the adopted solution scheme. Both models were discretized by means of continuum shell elements (SC8R). All explicit analyses have been performed setting a structural damping equal to 2%. Good agreement has been found between the solutions of both formulations, as confirmed by the load–strain chart in Figure 9, the element failures in Figure 10, and the failure displacements and loads in Table 4.

According to Figures 9 and 10, no relevant differences between the implicit and explicit solution can be found. The slight deviation in the failure path is related to small differences in the load distribution, due to the difference in the calculation of nodal displacements and forces between the two solution schemes. Indeed, the results must not be focused on the number of failed elements (Figure 10) but, instead, on the global behavior, which is reported in the graph of Figure 9. The buckling load of the structure is very close to the failure load, so no post-buckling regime is expected. The slight deviation between the two curves (at about 3.0×10^{-3} strain) is related to the damage propagation, not to the post-buckling regime. In the explicit analysis, the elements reach the total failure state faster with respect to the implicit one, leading to an earlier and smaller loss in stiffness.

Table 4. Implicit–explicit formulation comparisons: Failure displacement and failure load.

	Failure Displacement	**Failure Load**
Implicit	1.59 mm	107.3 kN
Explicit	1.60 mm	104.0 kN
Error	0.06%	3.1%

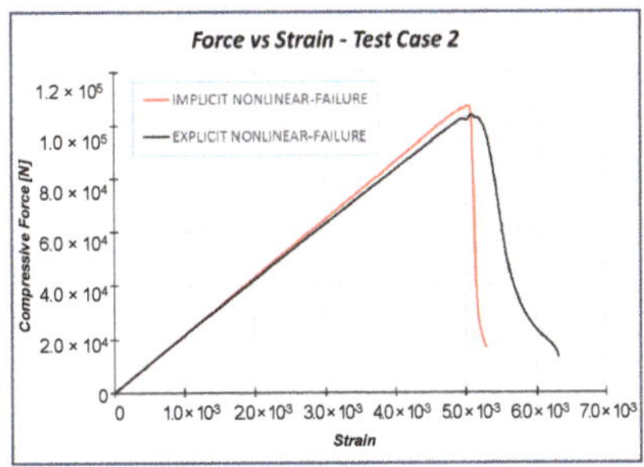

Figure 9. Implicit–explicit formulation comparisons: Load vs. strain.

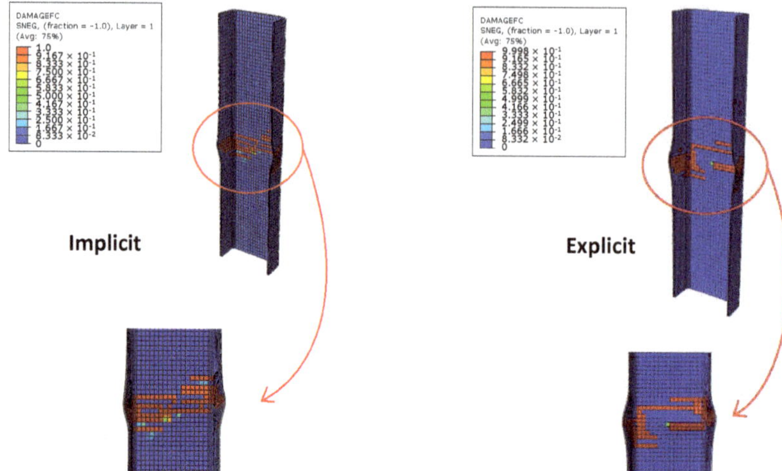

Figure 10. Implicit–explicit formulation comparisons: Failures.

In the dynamic explicit analyses, a rigid plate has been used to apply the time-dependent load onto the structure as a controlled displacement, as shown in Figure 11, which reports the control points where the outputs have been monitored as well. Seventy-two control points have been used to evaluate the occurrence of the buckling phenomena; they are placed along the entire stanchion and in a different section location in the back-to-back configuration.

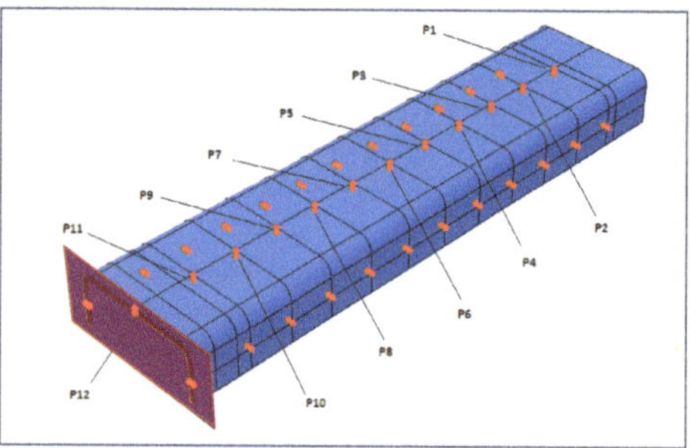

Figure 11. Explicit analysis: Rigid plate and control points.

Furthermore, several numerical element formulations have been investigated, to assess the capability of different numerical formulations to simulate the dynamic phenomenon. Hence, alongside the aforementioned continuum shell (SC8R) element discretization with three elements in the thickness direction (Model 3B), the following models, summarized in Table 5, have been introduced:

- Model S: This model has been discretized by using four-node linear shell (S4) elements [34]. The same in-plane element size of Model 3B has been considered; however, obviously, no division in the thickness direction has been employed. The equivalent properties of the entire laminate, reported in Table 6, have been used for each element.

- Model 1R: This model has been discretized by using eight-node linear solid elements with a reduced integration scheme (C3D8R) [34]. The same in-plane mesh discretization used for Model 3B and Model S has been considered. One element has been placed in the thickness direction, in order to avoid the use of the layered option. The equivalent mechanical properties of the entire laminate (Table 6) have been assigned to the elements.

- Model 3R: This model is similar to model 1R but differs from the latter as three elements in the thickness direction have been considered. Hence, the properties of the equivalent five-plies sublaminates corresponding to those shown in Figure 6 (Table 7) have been properly assigned to each element.

Table 5. Investigated element formulations.

Model Name	Element Type	# of Elements in the Thickness	Laminate Type
Model 3B	Continuum Shell (SC8R)	3	Composite Layup
Model S	Shell (S4)	1	Composite Layup
Model 1R	3D Stress (C3D8R)	1	Equivalent Laminate
Model 3R	3D Stress (C3D8R)	3	Equivalent Laminate

Table 6. Equivalent laminate mechanical properties.

th [mm]	E_{11} [MPa]	E_{22} [MPa]	E_{33} [MPa]	G_{12} [MPa]	G_{13} [MPa]	G_{23} [MPa]	ν_{12} [-]
2.79	60,267	37,845	8430	20,559	20,559	3328	0.43

Table 7. Equivalent sublaminate mechanical properties.

Stacking Sequence	th [mm]	E_{11} [MPa]	E_{22} [MPa]	E_{33} [MPa]	G_{12} [MPa]	G_{13} [MPa]	G_{23} [MPa]	ν_{12} [-]	ν_{13} [-]	ν_{23} [-]
$[0]_5$	0.93	135,000	8430	8430	4160	4160	3328	0.26	0.3	0.3
$[-45, 45, 90, 45, -45]$	0.93	22,674	39,437	8430	28,759	28,759	3328	0.44	0.3	0.3

The numerical models have been subjected to a triangular load with a maximum compressive amplitude equal to 0.75 mm (smaller than critical displacement estimated by numerical analysis, which is about 1.53 mm), shown in Figure 12. Then, the numerical models have been compared in terms of the free side displacements and fixed side reactions reported in Figure 13.

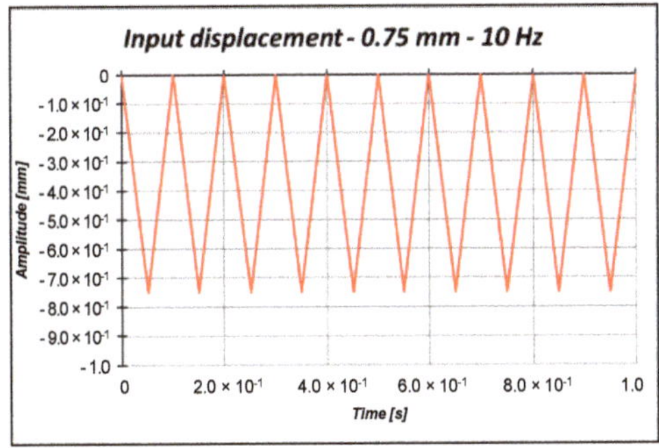

Figure 12. Input triangular displacement.

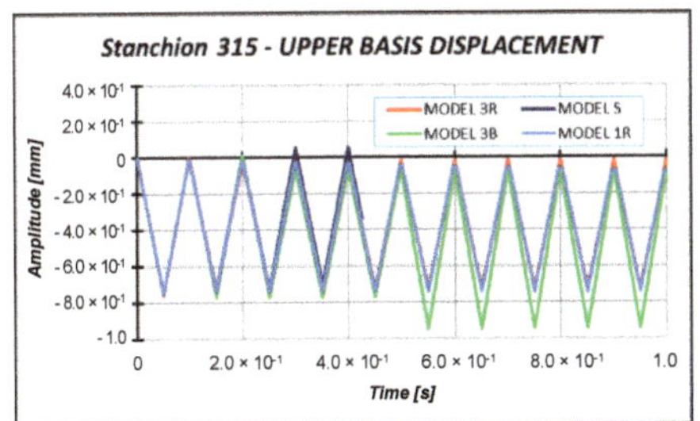

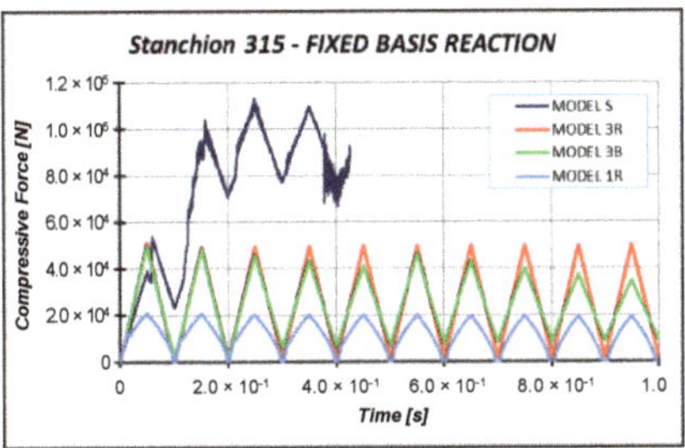

Figure 13. Free side displacement and fixed side reaction comparisons.

The results reported in Figure 13 demonstrated that the most suitable numerical model able to simulate a dynamic compressive analysis is model 3R, which has been selected in the subsequent analyses. Indeed, in model 3B, the displacement of the specimen is equal to the compressive displacement imposed on the plate only in the initial stage of the load cycle, up to t = 0.25 s. Then, the displacements start to differ from each other. This phenomenon is even more clear starting from t = 0.5 s. Moreover, the specimen discretized by model 3B is not able to completely be unloaded after 0.2 s, according to the results of the fixed side reaction. Model S experienced convergence issues, as highlighted by the results in terms of fixed side reactions, which are very irregular. Model 1R experiences issues in terms of free side displacement very similar to those encountered by Model 3B. Moreover, Model 1R is strongly affected by hourglass, as highlighted by fixed side reactions.

Sinusoidal cyclic displacements have then been applied on model 3R. According to preliminary eigenvalue buckling analyses, not reported here for the sake of brevity, the critical Test Case 2 displacement is 1.53 mm. Hence, three input displacements, characterized by a sinusoidal shape with a 10 Hz frequency, have been considered:

- Applied displacement = 1.0 mm (below the critical displacement);
- Applied displacement = 1.53 mm (equal to the critical displacement);
- Applied displacement = 2.0 mm (above the critical displacement).

Figures 14–16 report the displacements, the reactions, and the energies, as a function of time, for each load case.

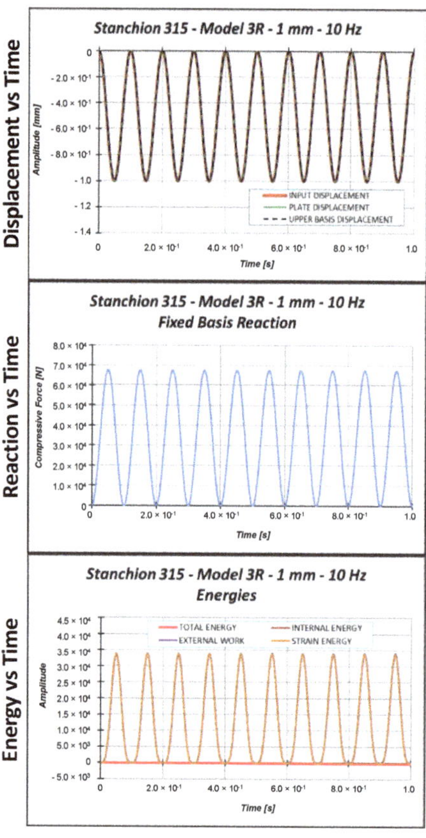

Figure 14. Displacement, reaction, and energies as a function of time: 1.0 mm applied displacement.

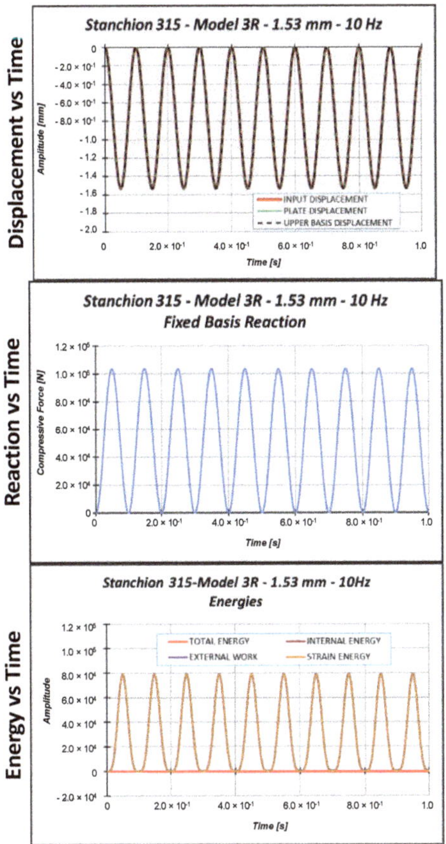

Figure 15. Displacement, reaction, and energies as a function of time: 1.53 mm applied displacement.

More detailed information on the dynamic behavior of the investigated test case can be obtained from Figure 17, which compares the stiffness corresponding to each load amplitude.

According to Figure 17, the stiffness of the model loaded by a maximum displacement equal to 1 mm is almost constant, meaning that no buckling occurs. The stiffness of the specimen loaded with a maximum amplitude equal to 1.53 mm (static buckling load) experiences a clear variation in correspondence to the maximum compression. However, this variation is almost instantaneous; hence, it cannot be considered representative of the buckling phenomenon. These considerations cannot be drawn for the 2 mm applied displacement. Indeed, in this last load case, the stiffness variation is more pronounced and lasts longer in time. Considering the provided results, it is quite clear that the applied load velocity (about 2 × maximum displacement × frequency) is too small to generate any resonance issue, which could lead to instability phenomena. Further, a constant structural damping (0.2%) has been adopted; hence, small slowdown effects on the elastic wave propagation can be observed with respect to the quasi-static case.

Other information can be taken considering the out-of-plane displacements U2 of the central points of the specimen (points P3, P6, and P9 in Figure 11), reported in Figures 18–20.

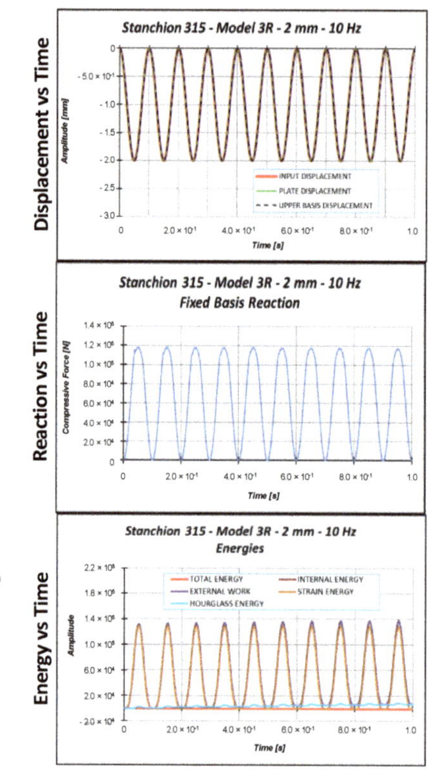

Figure 16. Displacement, reaction, and energies as a function of time: 2.0 mm applied displacement.

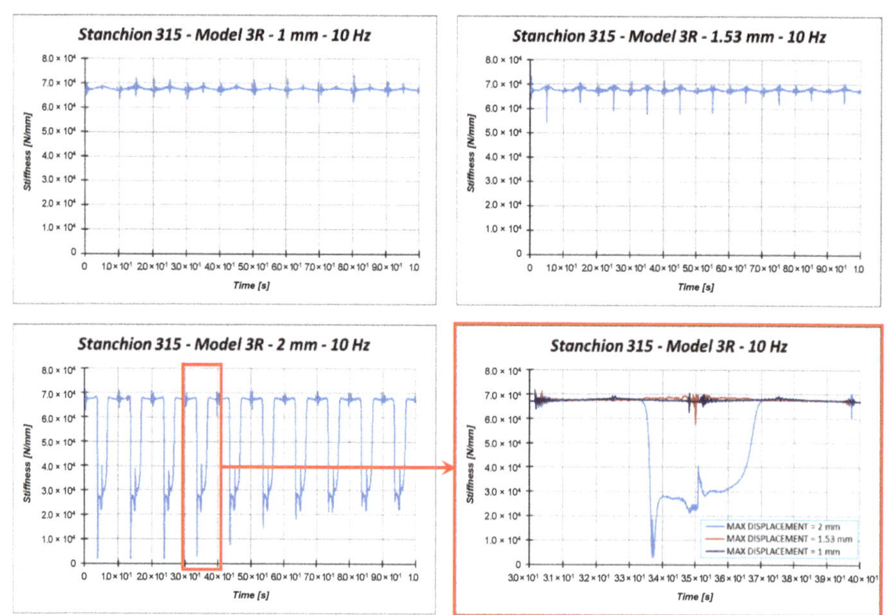

Figure 17. Model 3R: Stiffness comparisons.

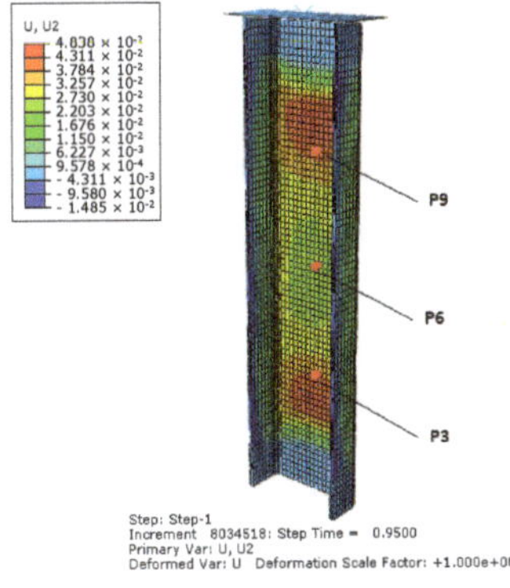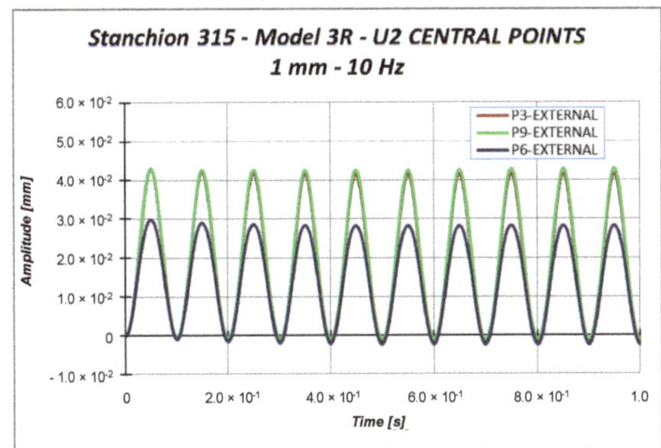

Figure 18. Out-of-plane displacement: 1.0 mm applied displacement.

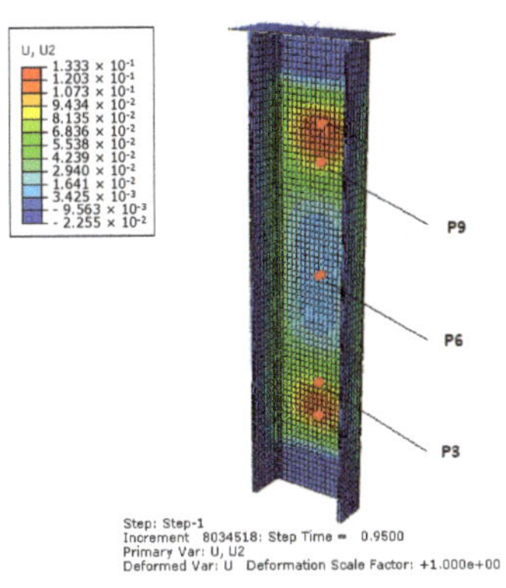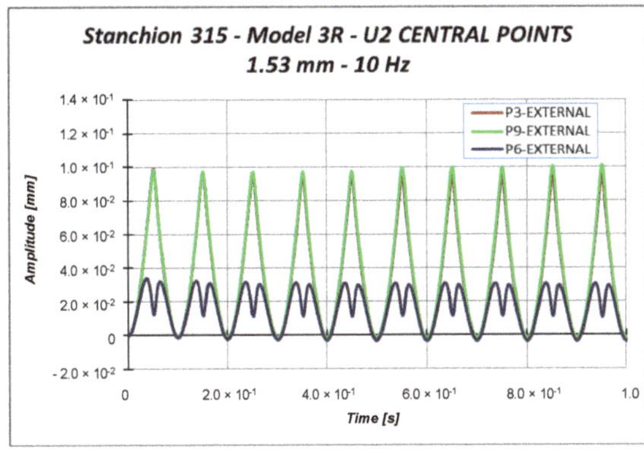

Figure 19. Out-of-plane displacement: 1.53 mm applied displacement.

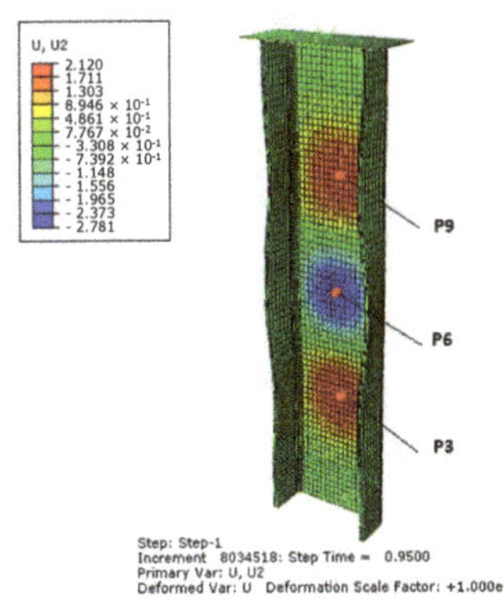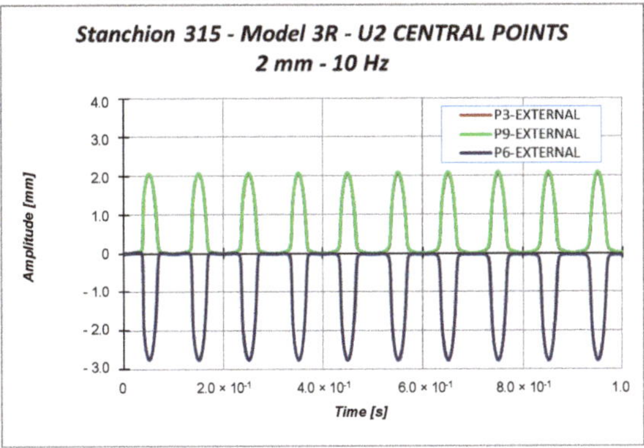

Figure 20. Out-of-plane displacement: 2.0 mm applied displacement.

Indeed, Figures 18–20 highlight the symmetrical out-of-plane displacement patterns resulting from the structural response of these symmetric systems (due to the adopted material formulation), as suggested in [37].

Considering a maximum applied displacement equal to 1 mm (Figure 18), no buckling occurs and all the points experience minimal out-of-plane displacements, which follows a sinusoidal and regular behavior. In the case of 1.53 mm maximum applied displacement (Figure 19), a more irregular sinusoidal behavior can be observed. Moreover, point P6 experiences a double peak, corresponding to the maximum compressive displacement, which can be considered representative of the buckling onset. Figure 20 reports the out-of-plane displacements in the case of 2 mm compression. In this case, point P6 experiences a shift in the value of the displacement, due to the buckling. This can be confirmed by monitoring the strains of the internal and external points considering a maximum applied displacement of 1.53 mm (Figure 21) and 2.0 mm (Figure 22).

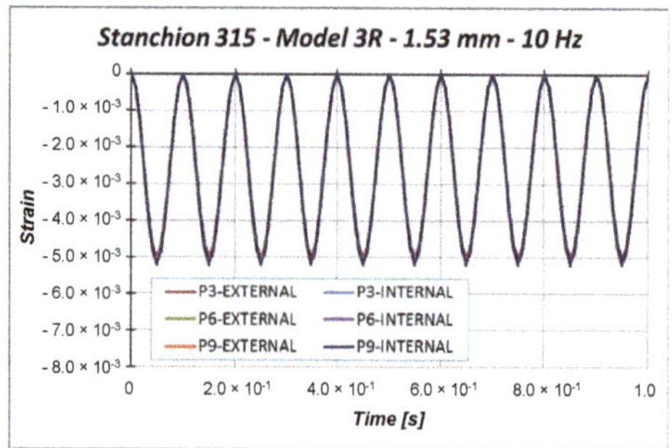

Figure 21. Strain: 1.53 mm applied displacement.

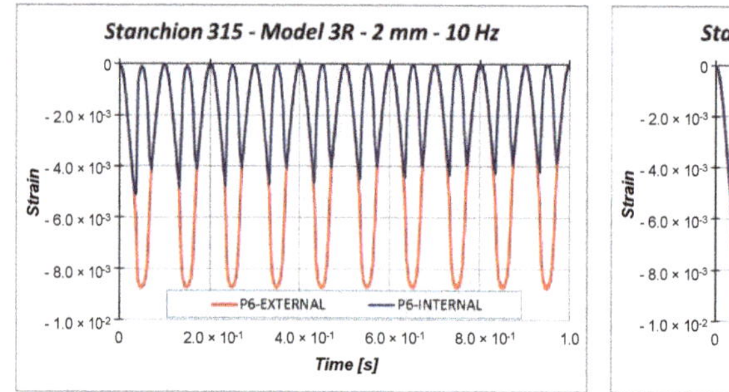
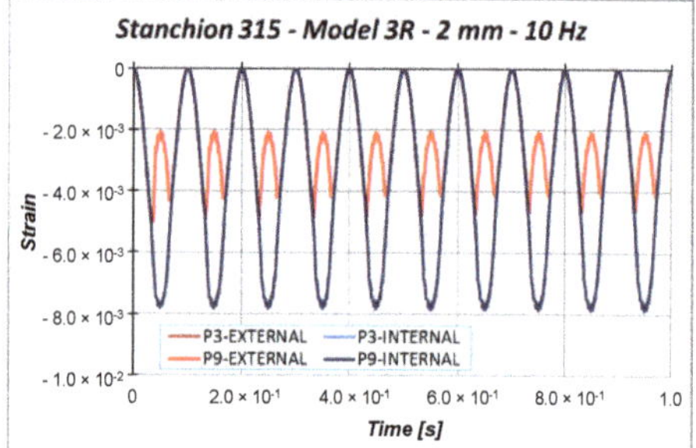

Figure 22. Strain: 2.0 mm applied displacement.

Indeed, the strains in Figure 21 are almost coincident, proving that compressive strain is mainly due to the compressive applied displacement and is not representative of the buckling phenomenon. On the other side, Figure 22 shows that the external point P6 experiences a compressive strain variation, while the internal point P6 experiences a tensile strain variation, which clearly indicated that buckling occurs. The strains' divergence (outer/inner control point), which indicates the buckling onset, occurs at 76.6% of the maximum amplitude, which corresponds to a displacement very close to the critical one. Further, with the increment in cycles, the structural behavior is very repetitive, but it is possible to see a small and slow reduction in the critical displacement.

4. Conclusions

In this work, the structural behavior of a composite floor beam subjected to cyclical low-frequency compressive loads has been presented. Sensitivity analyses have been carried out in order to find the most suitable numerical model able to simulate the dynamic analyses, considering different mesh discretization, element, and section formulations. The numerical model has been validated with respect to experimental data obtained under quasi-static loading conditions, thus adopting the implicit formulation. In order to investigate the influence of the solution scheme on the numerical results, an explicit–implicit comparison has been performed. The results highlighted that the adopted technique (smoothed step) allows us to minimize the inertial effects and, thus, to have a good agreement between the two solver schemes. Therefore, the explicit model can also be considered indirectly validated. Further, considering that the applied loading velocity, in dynamic cyclic loading conditions, is very small (lower than 0.1 m/s), the previous consideration can be extended to the final model.

Explicit non-linear analyses have been performed, applying dynamic 10 Hz frequency compressive loads. Load cycles with different amplitudes have been considered: 1 mm (about 70% of the static buckling value), 1.53 mm (close to the static buckling value), and finally 2 mm (above the static buckling value).

The results showed that low-frequency cyclic loads (compared to the test article's own frequencies, that are in the order of hundreds Hz) with peak values lower than the static buckling load value are not capable of triggering significant instability. This behavior is confirmed by comparing the stiffness of the models as a function of time for the different load levels applied at the same 10 Hz frequency.

In particular, all reported results highlight that the applied load velocity (obtained from the load frequency) is small enough to not generate any resonance issue; therefore, the deformation elastic wave is able to cover the entire specimen length in a time smaller than that related to a single load cycle. Thus, considering the applied load frequency, no significant differences were observed with respect to the static case. All of this will be confirmed by future experimental activities in which the frequency-damping dependency will also be investigated (also beyond the applied frequency). This parameter, assumed constant in this work, could generate a slowdown in the propagation of the elastic wave and, thus, anticipate resonance effects.

Author Contributions: All authors equally contribute to this work. All authors have read and agreed to the published version of the manuscript.

Funding: This research received no external funding.

References

1. Box, F.; Kodio, O.; O'Kiely, D.; Cantelli, V.; Goriely, A.; Vella, D. Dynamic Buckling of an Elastic Ring in a Soap Film. *Phys. Rev. Lett.* **2020**, *124*, 198003. [CrossRef]
2. Kodio, O.; Goriely, A.; Vella, D. Dynamic buckling of an inextensible elastic ring: Linear and nonlinear analyses. *Phys. Rev. E* **2020**, *101*, 053002. [CrossRef]
3. Ali, A.Y.; Hasan, H.M. Nonlinear dynamic stability of an imperfect shear deformable orthotropic functionally graded material toroidal shell segments under the longitudinal constant velocity. *Proc. Inst. Mech. Eng. Part C J. Mech. Eng. Sci.* **2019**, *233*, 6827–6850. [CrossRef]
4. Amabili, M.; Païdoussis, M.P. Review of studies on geometrically nonlinear vibrations and dynamics of circular cylindrical shells and panels, with and without fluid-structure interaction. *Appl. Mech. Rev.* **2003**, *56*, 349–381. [CrossRef]
5. Alijani, F.; Amabili, M. Non-linear vibrations of shells: A literature review from 2003 to 2013. *Int. J. Non-Linear Mech.* **2014**, *58*, 233–257. [CrossRef]
6. Kubiak, T. Static and dynamic buckling of thin-walled plate structures. In *Static and Dynamic Buckling of Thin-Walled Plate Structures*; Springer: London, UK, 2013.
7. Bhimaraddi, A.; Chandrashekhara, K. Nonlinear vibrations of heated antisymmetric angle-ply laminated plates. *Int. J. Solids Struct.* **1993**, *30*, 1255–1268. [CrossRef]
8. Kumar, P.; Srinivas, J. Vibration, buckling and bending behavior of functionally graded multi-walled carbon nanotube reinforced polymer composite plates using the layer-wise formulation. *Compos. Struct.* **2017**, *177*, 158–170. [CrossRef]
9. Kumar, L.R.; Datta, P.K.; Prabhakara, D.L. Vibration and stability behavior of laminated composite curved panels with cutout under partial in-plane loads. *Int. J. Struct. Stab. Dyn.* **2005**, *5*, 75–94. [CrossRef]
10. Kolahchi, R.; Zhu, S.-P.; Keshtegar, B.; Trung, N.-T. Dynamic buckling optimization of laminated aircraft conical shells with hybrid nanocomposite martial. *Aerosp. Sci. Technol.* **2020**, *98*, 105656. [CrossRef]

11. Bigoni, D. *Nonlinear Solid Mechanics: Bifurcation Theory and Material Instability*; Cambridge University Press: New York, NY, USA, 2012.

12. Javani, M.; Kiani, Y.; Eslami, M.H. Dynamic snap-through of shallow spherical shells subjected to thermal shock. *Int. J. Press. Vessel. Pip.* **2020**, *179*, 104028. [CrossRef]

13. Shokravi, M.; Jalili, N. Thermal dynamic buckling of temperature-dependent sandwich nanocomposite quadrilateral microplates using visco-higher order nonlocal strain gradient theory. *J. Therm. Stress.* **2019**, *42*, 506–525. [CrossRef]

14. Zhang, J.; Chen, S.; Zheng, W. Dynamic buckling analysis of functionally graded material cylindrical shells under thermal shock. *Contin. Mech. Thermodyn.* **2019**, *32*, 1095–1108. [CrossRef]

15. Yasunaga, J.; Uematsu, Y. Dynamic buckling of cylindrical storage tanks under fluctuating wind loading. *Thin-Walled Struct.* **2020**, *150*, 106677. [CrossRef]

16. Kubiak, T. Estimation of dynamic buckling for composite columns with open cross-section. *Comput. Struct.* **2011**, *89*, 2001–2009. [CrossRef]

17. Zizicas, G.A. Dynamic buckling of thin elastic plates. *Trans. ASME* **1952**, *74*, 1257.

18. Budiansky, B.; Roth, R.S. Axisymmetric dynamic buckling of clamped shallow spherical shells. *NASA TN* **1962**, *1510*, 597–606.

19. Lindberg, H.E.; Florence, A.L. *Dynamic Pulse Buckling*; Springer: Dordrecht, The Netherlands, 1987.

20. Labans, E.; Abramovich, H.; Bisagni, C. An experimental vibration-buckling investigation on classical and variable angle tow composite shells under axial compression. *J. Sound Vib.* **2019**, *449*, 315–329. [CrossRef]

21. Zaczynska, M.; Abramovich, H.; Bisagni, C. Parametric studies on the dynamic buckling phenomenon of a composite cylindrical shell under impulsive axial compression. *J. Sound Vib.* **2020**, *482*, 11546. [CrossRef]

22. Mondal, S.; Ramachandra, L. Nonlinear dynamic pulse buckling of imperfect laminated composite plate with delamination. *Int. J. Solids Struct.* **2020**, *198*, 170–182. [CrossRef]

23. Karthick, S.; Datta, P.K. Dynamic instability characteristics of thin plate like beam with internal damage subjected to follower load. *Int. J. Struct. Stab. Dyn.* **2015**, *15*, 1450048. [CrossRef]

24. Wang, Y.-H.; Han, Z.-J. Buckling study of composite plates subjected to impact loading. *IOP Conf. Ser. Earth Environ. Sci.* **2019**, *267*, 022022.

25. Wang, P.; Li, S.-Q.; Yu, G.-J.; Wu, G.-Y. Dynamic Crushing Behavior of Graded Hollow Cylindrical Shell under Axial Impact Loading. *Chin. J. High Press. Phys.* **2017**, *31*, 778–784.

26. Gui, Y.; Ma, J. Buckling of step cylindrical shells under axial impact load. *J. Vib. Shock* **2019**, *38*, 200–205, 228.

27. Gui, Y.; Xu, J.; Ma, J. Dynamic Buckling of a Cylindrical Shell with a General Boundary Condition under an Axial Impact. *J. Appl. Mech. Tech. Phys.* **2019**, *60*, 712–723. [CrossRef]

28. Zhang, Z.; Taheri, F. Numerical studies on dynamic pulse buckling of FRP composite laminated beams subject to an axial impact. *Compos. Struct.* **2002**, *56*, 269–277. [CrossRef]

29. Bažant, Z.P.; Cedolin, L. *Stability of Structures*; Dover Publication, Inc.: Mineola, NY, USA, 2010.

30. Riccio, A.; Raimondo, A.; Di Caprio, F.; Fusco, M.; Sanitá, P. Experimental and numerical investigation on the crashworthiness of a composite fuselage sub-floor support system. *Compos. Part B Eng.* **2018**, *150*, 93–103. [CrossRef]

31. Di Caprio, F.; Ignarra, M.; Marulo, F.; Guida, M.; Lamboglia, A.; Gambino, B. Design of composite stanchions for the cargo subfloor structure of a civil aircraft. *Procedia Eng.* **2016**, *167*, 88–96. [CrossRef]

32. Guida, M.; Marulo, F.; Abrate, S. Advances in crash dynamics for aircraft safety. *Prog. Aerosp. Sci.* **2018**, *98*, 106–123. [CrossRef]

33. Marulo, F.; Guida, M.; Di Caprio, F.; Ignarra, M.; Lamboglia, A.; Gambino, B. Fuselage Crashworthiness Lower Lobe Dynamic Test. *Procedia Eng.* **2016**, *167*, 120–128. [CrossRef]

34. Smith, M. *ABAQUS/Standard User's Manual, Version 2019*; Dassault Systèmes Simulia Corp: Providence, RI, USA, 2019.

35. Riccio, A.; Saputo, S.; Sellitto, A.; Raimondo, A.; Ricchiuto, R. Numerical investigation of a stiffened panel subjected to low velocity impacts. *Key Eng. Mater.* **2015**, *665*, 277–280. [CrossRef]

36. Riccio, A.; Cristiano, R.; Saputo, S.; Sellitto, A. Numerical methodologies for simulating bird-strike on composite wings. *Compos. Struct.* **2018**, *202*, 590–602. [CrossRef]
37. Zucco, G.; Weaver, P.M. The role of symmetry in the post-buckling behaviour of structures. *Proc. R Soc. A Math. Phys. Eng. Sci.* **2020**, *476*, 20190609. [CrossRef] [PubMed]

 *materials*

Article

Catastrophic Influence of Global Distortional Modes on the Post-Buckling Behavior of Opened Columns

Andrzej Teter [1] and **Zbigniew Kolakowski** [2,*]

[1] Department of Applied Mechanics, Faculty of Mechanical Engineering,
Lublin University of Technology (LUT), Nadbystrzycka 36, PL–20-618 Lublin, Poland; a.teter@pollub.pl
[2] Department of Strength of Materials, Faculty of Mechanical Engineering, Lodz University of Technology,
Stefanowskiego 1/15, PL-90-924 Lodz, Poland
* Correspondence: zbigniew.kolakowski@p.lodz.pl

Received: 2 July 2020; Accepted: 21 July 2020; Published: 25 July 2020

Abstract: The multimodal buckling of thin-walled isotropic columns with open cross-sections under uniform compression is discussed. Column lengths were selected to enable strong interactions between selected eigenmodes. In the case of short columns or very long ones subjected to compression, single-mode buckling can be observed only and the effect under discussion does not occur. In the present study, the influence of higher global modes on the load-carrying capacity and behavior in the post-buckling state of thin-walled structures with open cross-sections is analyzed in detail. In the literature known to the authors, higher global modes are always neglected practically in the analysis due to their very high values of bifurcation loads. However, the phenomenon of an unexpected loss in the load-carrying capacity of opened columns can be observed in the experimental investigations. It might be explained using multimode buckling when the higher global distortional-flexural buckling modes are taken into account. In the conducted numerical simulations, a significant influence of higher global distortional-flexural buckling modes on the post-buckling equilibrium path of uniformly compressed columns with C- and TH-shaped (the so-called "top-hat") cross-sections was observed. The columns of two lengths, for which strong interactions between selected eigenmodes were seen, were subject to consideration. Two numerical methods were applied, namely, the semi-analytical method (SAM) using Koiter's perturbation approach and the finite element method (FEM), to solve the problem. The SAM results showed that the third mode had a considerable impact on the load-carrying capacity, whereas the FEM results confirmed a catastrophic effect of the modes on the behavior of the structures under analysis, which led to a lack of convergence of numerical calculations despite an application of the Riks algorithm. All elastic-plastic effects were neglected.

Keywords: C-section; TH-section; distortional mode; medium length; interactive buckling; compression; Koiter's theory; FEM

1. Introduction

Cold-formed isotropic columns under compression are commonly used as load-carrying elements in structures of any type. Their behavior under increasing loading depends on numerous factors, such as geometrical dimensions, kind of load, initial imperfections, and material properties. In the majority of cases, the performance of thin-walled structures results not from their strength but predominantly from their stability. In the general case, different buckling modes occur. Short columns and very long ones are subject to single-mode buckling and it is difficult to observe the effects of interactions of various buckling modes. These are local and global buckling modes, respectively. Only for medium-long columns, multimodal buckling could be observed. The key problem consists of a selection of eigenmodes, which are accounted for in the multimodal analysis. In the literature, it is easy to find many monographs, e.g., [1,2], and papers, e.g., [3–5], presenting the state of the art in the field of the nonlinear stability of thin-walled structures. Despite this fact, these issues have not been well recognized so far and are the subject of interest for numerous researchers. Numerical methods and theories, which are applied in the nonlinear analysis of stability and load-carrying capacity, have been constantly developed and improved. Among them, one can mention those widely used, namely: The Generalized Beam Theory (GBT), DSM (Direct Strength Method), and its modifications, e.g., the GBTUL (Generalized Beam Theory University of Lisbon), FSM (Finite Strip Method), FEM (Finite Element Method), and cFEM (constrained Finite Element Method). We focus only on recent publications devoted to the interactive buckling of thin-walled beams with open cross-sections by Ádány and Schafer [6–9], Davison [10], Hancock [11], Rasmussen [12,13], Silvestre [14], Szymczak [15], and Camotim et al. [16–26], and many more can be enumerated.

Development of the theory of the interactive buckling of thin-walled structures was discussed in the survey paper [11]. Martins, Camotim, Gonçalves, and Dinis [20–23] investigated interactive buckling and its effect on the post-buckling behavior and failure in columns [20] and beams [21–23] of various cross-sections. In [21], it is noticed that "secondary global-bifurcation D-G interaction—SGI" or "secondary distortional-bifurcation D-G interaction SDI" can occur when the beam geometry has an initial imperfection corresponding to three buckling modes (one distortional and two global).

The phenomenon of an unexpected loss in the load-carrying capacity of C-beams under bending was observed in the experimental investigations presented in [27,28]. In [28], an unexpected behavior of the holders used to support and load the models, which consisted of a rotation of the holders from their plane of bending, was observed. That behavior was interpreted as an effect of the secondary global distortional-lateral buckling mode on the interactive buckling of the C-section under bending, in the web plane. The results of further theoretical investigations were published in [29,30]. In [30], it is analyzed at what length of the steel C-beam under bending the effect of the second global mode on the load-carrying capacity was most considerable. The conclusions coming from those analyses pointed to beams of medium length. In [28–30], the problem was solved with the semi-analytical method (SAM) developed by the authors and based on Koiter's asymptotic theory [31,32]. In the SAM, all analyzed buckling modes corresponding to bifurcation loads and the number of halfwaves along the longitudinal direction are calculated based on equilibrium equations and boundary conditions with the modified numerically strict transfer matrix method. This method is very useful for the interpretation of interactions of various buckling modes in the full range of loading. Moreover, it facilitates the understanding of the phenomena that occur during coupled buckling. In [33,34], this subject was further analyzed but extended to composite beams. Additionally, FEM analyses were conducted to validate the proposed SAM model and to verify the post-buckling equilibrium paths and load-carrying capacity attained. Such a comparison enables the scope to which the method can be applied to be determined, when the secondary global mode should be considered regarding the secondary global buckling mode and the range of beam lengths. The examples of computations cover the post-buckling equilibrium paths of C- and top-hat (TH)-section beams with given lengths, made of steel and laminates with different arrangements of plies.

The present paper is a continuation and expansion of the above-mentioned investigations for steel C- and TH-section columns with medium length and under uniform compression. The phenomenon of an unexpected loss in the load-carrying capacity of opened columns can be observed in the experimental investigations. It is worth mentioning that higher buckling loads are huge, but considering it in SAM leads to the significant decrease in the post-buckling equilibrium path and load-carrying capacity. It might be explained using multimode buckling when the higher global distortional-flexural buckling modes are taken into account. The results of computations were obtained with two numerical methods, the SAM and the FEM, where the second one was used for verification purposes. The considerations were limited to isotropic structures due to easier interpretations of the results than for composite structures, including FGMs and so on [35,36].

2. Formulation of the Problem

The multimodal elastic buckling of C-section (Figure 1a) and TH-section (the so-called "top-hat") (Figure 1b) columns with medium lengths and under compression, made of isotropic materials like steel, was analyzed. The material the columns were made of satisfies Hooke's law. The column lengths were chosen to ensure strong interactions between the selected buckling eigenmodes. The numerical simulations were conducted with two methods, namely the semi-analytical method (SAM) and the finite element method (FEM), to discuss the effect of global distortional-flexural modes on the post-buckling behavior of the columns under compression in detail.

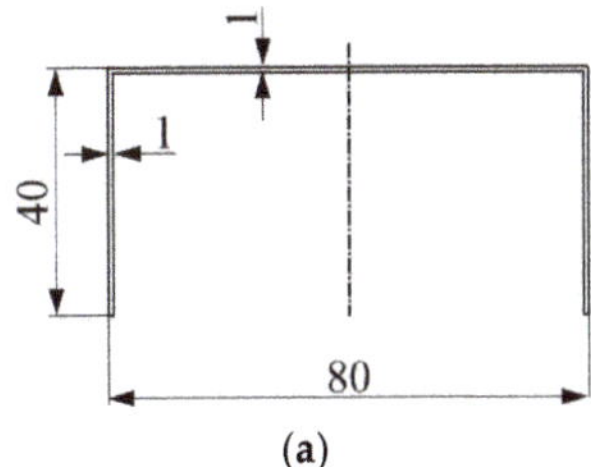

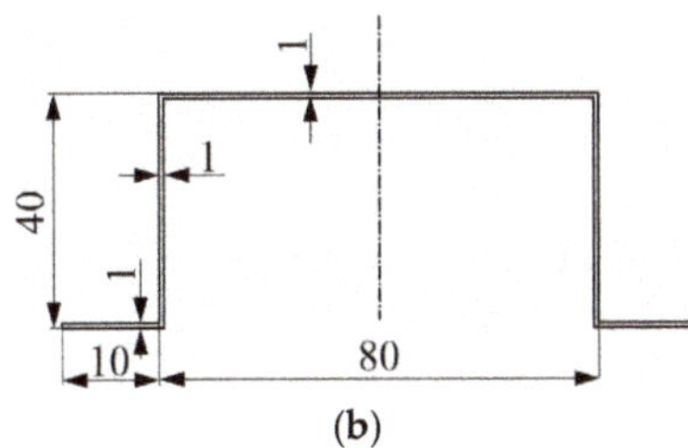

Figure 1. Outlines of the cross-sections and their dimensions: C-section (**a**) and top-hat (TH)-section (**b**).

2.1. Semi-Analytical Method (SAM)

The analytical-numerical method (ANM) [31] and the semi-analytical method (SAM) [32,37] based on Koiter's asymptotical perturbation theory within Byskov–Hutchinson's formulation had been developed for more than 30 years by the authors [28–30,33,34]. The methods allow one to solve the nonlinear stability problem of thin-walled structures made of isotropic, orthotropic, and composite materials. The structures under consideration are prismatic thin-walled columns built of plates connected along longitudinal edges and subjected to complex longitudinal loading. To account for all possible modes of global, local, and coupled buckling, a plate model (i.e., 2D) of thin-walled structures is applied. The columns are freely supported at their ends. For each plate, the exact geometrical relationships (i.e., full Green's strain tensor) are assumed:

$$\begin{aligned}
\varepsilon_x &= u_{,x} + \tfrac{1}{2}(w_{,x}^2 + v_{,x}^2 + u_{,x}^2) \\
\varepsilon_y &= v_{,y} + \tfrac{1}{2}(w_{,y}^2 + u_{,y}^2 + v_{,y}^2) \\
2\varepsilon_{xy} &= \gamma_{xy} = u_{,y} + v_{,x} + w_{,x}w_{,y} + u_{,x}u_{,y} + v_{,x}v_{,y}
\end{aligned} \tag{1}$$

and

$$\kappa_x = -w_{,xx} \quad \kappa_y = -w_{,yy} \quad \kappa_{xy} = -2w_{,xy} \tag{2}$$

where u, v, w—components of the displacement vector of the plate along the x-, y-, and z-axis direction, respectively.

According to Koiter's theory, the fields of displacement and the fields of sectional forces are distributed into a series with respect to the dimensionless deflection amplitude [31,32,37]:

$$\begin{aligned}
U &\equiv (u,v,w) = \lambda U_0 + \zeta_r U_r + \zeta_r^2 U_{rr} + \ldots \\
N &\equiv (N_x, N_y, N_{xy}) = \lambda N_0 + \zeta_r N_r + \zeta_r^2 N_{rr} + \ldots
\end{aligned} \tag{3}$$

where λ—load factor, U_0, N_0—the pre-buckling (i.e., unbending) fields, U_r, N_r—the first-order nonlinear fields, U_{rr}, N_{rr}—the second-order nonlinear fields of the displacement and the sectional force, respectively. The range of indices is $[1, J]$, where J is the number of interacting modes. It is assumed that the summation is over the repeated indices.

In thin-walled structures with the initial geometric imperfections $\bar{U}$ (only the linear initial geometric imperfections corresponding to the shape of the r-th buckling mode, i.e., $\bar{U} = \zeta_r^* U_r$, are considered), the total potential energy can be described by the following formula:

$$\Pi = -\frac{1}{2}M^2\bar{a}_0 + \frac{1}{2}\sum_{r=1}^{J}\bar{a}_r\zeta_r^2\left(1 - \frac{M}{M_r}\right) + \frac{1}{3}\sum_{p}^{J}\sum_{q}^{J}\sum_{r}^{J}\bar{a}_{pqr}\zeta_p\zeta_q\zeta_r + \frac{1}{4}\sum_{r}^{J}\bar{b}_{rrrr}\zeta_r^3 - \sum_{r}^{J}\frac{M}{M_r}\bar{a}_r\zeta_r^*\zeta_r \quad (4)$$

and the equations of equilibrium corresponding to (4) are:

$$\left(1 - \frac{P}{P_r}\right)\zeta_r + a_{pqr}\zeta_p\zeta_q + b_{rrrr}\zeta_r^3 = \frac{P}{P_r}\zeta_r^* \quad r = 1, \ldots, J \quad (5)$$

where:

$$a_{pqr} = \bar{a}_{pqr}/\bar{a}_r \quad b_{rrrr} = \bar{b}_{rrrr}/\bar{a}_r \quad (6)$$

Expressions for coefficients: a_0, a_r, a_{pqr}, b_{rrrr} are calculated by known formulae [1–3,32,37]. The post-buckling coefficients a_{pqr} depend only on the buckling modes, whereas the coefficients b_{rrrr} also depend on the second-order field. The nonlinear stability problem of thin-walled structures in the first-order approximation of Koiter's theory was solved with the modified analytical-numerical method (ANM) presented in [31]. In the ANM, an exact transfer matrix method is applied as opposed to the finite strip method. The ANM should be extended by the second-order approximation of the theory. The second-order coefficients are estimated with the semi-analytical method (SAM) [32,37]. In the SAM, it is postulated to define approximated values of the coefficients b_{rrrr} based on the linear buckling problem. Such an approach allows for an exact determination of values of the coefficients a_{pqr}, according to the nonlinear Byskov–Hutchinson's theory.

The relative shortening of column ends Δ/Δ_{min} is defined as a function of P/P_{min} through differentiation of potential energy expression (4) with respect to P/P_{min} [28–32]:

$$\frac{\Delta}{\Delta_{min}} = \frac{P}{P_{min}}\left[1 + \frac{P_{min}}{P\bar{a}_0}\sum_{r=1}^{J}\frac{P_{min}}{P_r}\bar{a}_r\zeta_r(0.5\zeta_r + \zeta_r^*)\right] \quad (7)$$

where Δ_{min}—minimal shortening of the column corresponding to the minimal value of the bifurcation load P_{min}.

In the presented study, a three-modal approach was applied at most (i.e., $J = 3$ in the relationship (5)). This means that a model with a maximum of three degrees of freedom was assumed.

2.2. Finite Element Method (FEM)

To verify the SAM results, a commercial package Abaqus (Dassault Systèmes Simulia Corp., Johnston, RI, USA) [38] was applied in the finite element method (FEM) calculations. A numerical model of the thin-walled structures under analysis, with the boundary conditions adapted to the SAM, was developed. The columns under consideration were modeled with four-node shell elements with eight degrees of freedom in each node (element type: S8R). The element size equal to 2 mm was determined based on the previous experience of one of the authors and the analyses of convergence. In the FEM models of the thin-walled structures under analysis, the following was assumed: For the 300 mm-long C-channel, 12,000 elements and 96,640 degrees of freedom, whereas for the TH-channel, 13,500 elements and 108,720 degrees of freedom. In the case of longer C-channels, 20,000 elements and 160,640 degrees of freedom were assumed, while for longer TH-channel columns, 22,500 elements and 180,720 degrees of freedom were assumed. All simulations were conducted for whole thin-walled structures, because not only symmetrical modes can be discussed. In our case, symmetry boundary conditions can be applied. Both solutions are the same. Additionally, non-symmetrical buckling of the opened columns, which can follow from an interaction of anti-symmetrical modes, can be observed. They have been neglected in the present study.

Moreover, the initial conditions corresponding to the free support of the column ends under uniform compression were modified. The assumption of classic boundary conditions (Figure 2a) in which the column ends (denoted as K1 and K2) are fixed in the cross-section plane of the column, that is to say: $u_x = u_y = 0$ and $\vartheta_z = 0$, and, additionally, at the end K1, the displacements are blocked along the length, that is to say: $u_z = 0$, leads to over-stiffening of the model. The determined eigenvalues for higher global buckling modes in the columns with open cross-sections under discussion were overestimated with respect to the results attained with the analytical-numerical method. To counteract it, the boundary conditions were modified as suggested in [39]. In this case, the boundary conditions are as follows (Figure 2b,c):

- In the middle of the column, displacements along the column length are equal to 0, i.e., $u_z = 0$,
- At both ends K1 and K2, it is assumed that the intersection point of the web with the cross-section axis (denoted as point 1 in Figure 2c) cannot displace along the y-axis, i.e., $(u_y)_1 = 0$. In segment 1–2, the web cannot displace along the x-axis, i.e., $(u_x)_{1-2} = 0$. In segment 2–3, the arm displaces with respect to the y-axis identically as corner 2, i.e., $(u_y)_{2-3} = (u_y)_2$. Reinforcement 3–4 displaces along the x-axis identically as corner 3, i.e., $(u_x)_{3-4} = (u_x)_3$.

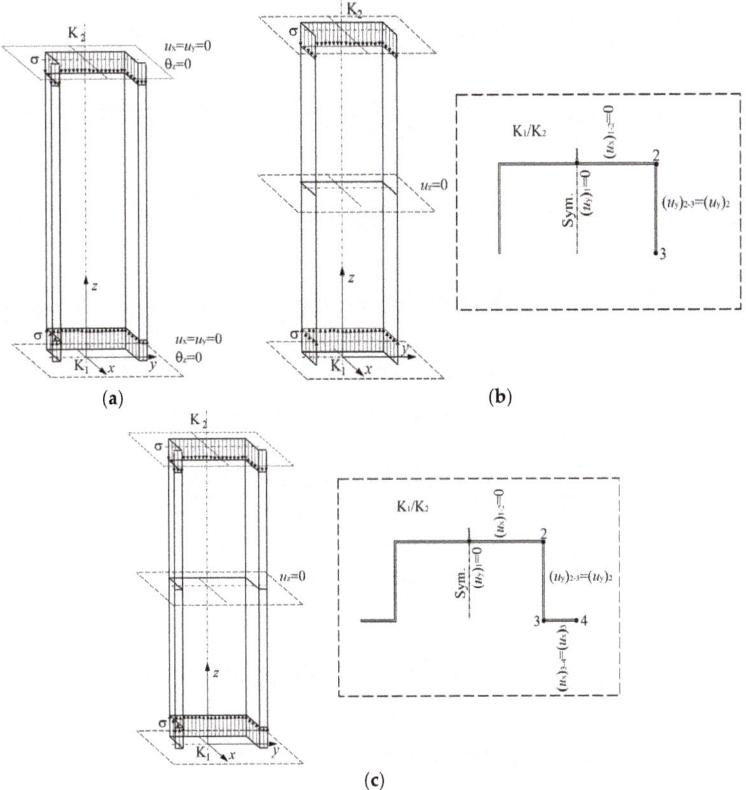

Figure 2. Boundary conditions: Classic variant of the column-free support (**a**) and modified boundary conditions for the C-channel (**b**) and the TH-channel (**c**) [39].

In Figure 2b, modified conditions for the C-column are presented, whereas those for the TH-column are to be found in Figure 2c.

Another variant of the modified boundary conditions was suggested by Szymczak and Kujawa [40–42]. Comparing the results of simulations of bifurcation loads for both the modified variants of the boundary conditions, one can state that they are consistent. The determined higher eigenvalues and post-buckling paths of equilibrium are identical in both cases of the modified boundary conditions. It is worth adding that the nonlinear stability problem was solved both with the Riks method and the Newton-Raphson method. The results did not differ; thus, in the present study, only the results for the Riks algorithm are presented. To attain numerical convergence, the computational timestep was highly diminished.

It should be mentioned that the cost/time ratio of the SAM calculations is more than two orders of magnitude lower than in the case of the FEM. Moreover, the SAM enables much easier analysis of the phenomena under investigation and their interpretation when compared to the FEM.

3. Results and Discussion

3.1. C-Channel Columns

In the first stage of the calculations, the stability of C-channel columns was investigated. The geometrical dimensions of the column cross-sections are presented in Figure 1, whereas their material constants were as follows: Young's modulus—200 GPa and Poisson's ratio—0.3. A linear-elastic model of the material and a column length equal to 300 and 500 mm were assumed. Solving the eigenproblem with both the methods (i.e., the SAM and the FEM), we determined eigenvalues and modes. The attained values of bifurcation loads are listed in Table 1. Additionally, the number of halfwaves m, which form on the column surface along the longitudinal direction, is given in brackets for the local modes. The following indices are used: 1—local buckling mode (when $m > 1$), 2—primary global buckling mode, 3—secondary global distortional-flexural mode, 4—third global distortional-flexural mode. The buckling modes where $m = 1$ are referred to as the global ones in the present paper.

Table 1. Bifurcation stresses in MPa for the C-channel (Figure 1a) and TH-channel (Figure 1b) columns under analysis; the semi-analytical method (SAM) and finite element method (FEM) results.

Length (mm)	SAM				FEM			
	Mode1	Mode2	Mode3	Mode4	Mode1	Mode2	Mode3	Mode4
	C-channel (Figure 1a)							
300	74.1 (3)	157	1168	5324	73.3	156	1155	5292
500	74.1 (5)	338	1150	5977	73.5	337	1140	6018
	TH-channel (Figure 1b)							
300	134.0 (5)	265	986	6355	137.8	268	973	6280
500	133.5 (8)	288	1234	5724	137.5	290	1223	5633

Very good compatibility of the values of bifurcation loads (the so-called eigenvalues) from the SAM and the FEM was obtained owing to the modified boundary conditions assumed in the FEM. The determined values of local bifurcation forces are lower than the primary global values, as, for the column of the length of 300 mm, we have $P_2/P_1 = 2.1$, and for the length of 500 mm, $P_2/P_1 = 4.5$. The load ratios corresponding to higher modes are as follows: $P_3/P_1 \approx 15$ and $P_4/P_1 \approx 70$, and they have not been considered in the literature devoted to interactive buckling. The authors, despite this "impracticality", deal with this problem as the main object in this study.

The FEM buckling modes are presented in Figure 3 for the 300 mm-long column and in Figure 4 for the 500 mm-long column. The buckling mode, i.e., the eigenvector, is determined with an accuracy up to a certain constant C. This remark is necessary due to the buckling modes presented in Figure 3a (local mode $i = 1$) and in Figure 3b (primary global mode $i = 2$) for 300 mm. The amplitude of global buckling (Figure 3b) can be transformed through a change in the constant C into the constant −C, i.e., into the amplitude of the local mode in practice (Figure 3a) (cf. the shape of the mode amplitude at the right-hand side (Figure 3a,b)). Thus, these modes differ only in the number of halfwaves m along the longitudinal direction. Global modes for $i = 3$ (Figure 3c) and $i = 4$ (Figure 3d) are distortional-flexural modes, where, for the first of them, maximal displacements occur for the web, and for the second one, those occur for the plates.

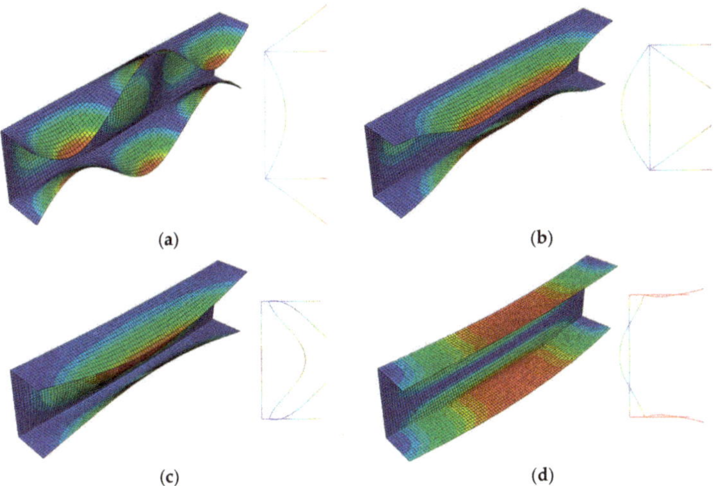

Figure 3. Buckling modes for the 300 mm-long C-section, FEM results. Subsequent modes are as follows: Local $m > 1$ (**a**), primary global (**b**), higher global distortional-flexural (**c**), and third global distortional-flexural (**d**).

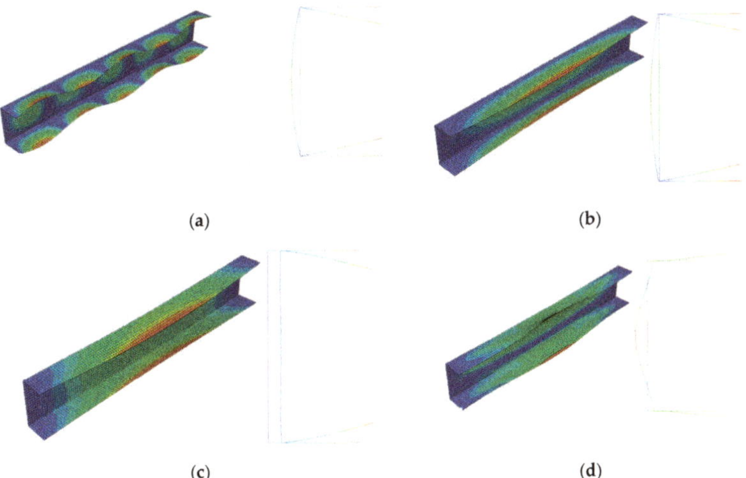

Figure 4. Buckling modes for the 500 mm-long C-section, FEM results. Subsequent modes are as follows: Local $m > 1$ (**a**), primary global (**b**), higher global distortional-flexural (**c**), and third global distortional-flexural (**d**).

The amplitudes of the local buckling mode ($i = 1$, $m = 5$—Figure 4a) and the primary global mode ($i = 2$—Figure 4b) of the C-channel, which is 500 mm long, are the same, but they differ as far the number of halfwaves along the longitudinal direction is concerned. The secondary global mode ($i = 3$—Figure 4c) is the distortional-flexural mode, where the web displaces as for the "pure" flexural mode. The third global mode ($i = 4$—Figure 4d) is also the distortional-flexural one, but 'simpler' with respect to the third mode for 300 mm (Figure 3d). A similar statement can be made when the secondary global modes for 300 (Figure 3c) and 500 mm (Figure 4c) are compared. In Figures 5 and 6, four buckling modes for 300 and 500 mm, obtained with the SAM, are shown, respectively. A very good agreement of the mode attained with the SAM and FEM for the given length (compare Figures 3 and 5 to Figures 4 and 6, correspondingly) can be seen.

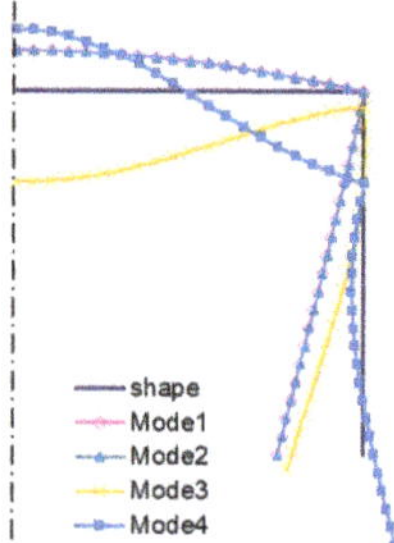

Figure 5. Buckling modes for the 300 mm-long C-section, SAM results. Subsequent modes are as follows: Local $m > 1$ (Mode1), primary global (Mode2), higher global distortional-flexural (Mode3), and third global distortional-flexural (Mode4).

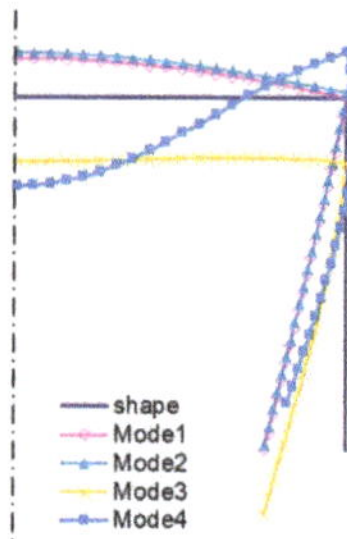

Figure 6. Buckling modes for the 500 mm-long C-section, SAM results. The same notation of modes as in Figure 5.

The next stage consisted of an analysis of post-buckling equilibrium paths for the C-columns. Two-mode ($J = 2$ in Equation (5)) and three-mode ($J = 3$) approaches were considered for the SAM to facilitate an interpretation of the results. Moreover, the initial imperfections $\zeta_r^* = |0.2|$, where the signs were selected in the most disadvantageous way [30–32,34], were assumed.

Figure 7 presents a collection of the SAM and FEM solutions showing post-buckling equilibrium paths in the system of the dimensionless load $P/P_{min} = P/P_1$ as a function of the dimensionless shortening $\Delta/\Delta_{min} = \Delta/\Delta_1$ (where P_{min} corresponds to the lowest value of the bifurcation load, and Δ_{min} denotes the minimal shortening caused by the load P_{min}) for the 300 mm-long column. The post-buckling paths for a coupled interaction of two modes ($J = 2$): The local mode ($i = 1$) with the primary global one ($i = 2$) (denoted here as SAM-1,2) and for two cases of interaction of three modes ($J = 3$): The local mode ($i = 1$) with the two lowest global modes (i.e., for $i = 2,3$) (denoted as SAM-1,2,3), and the local mode ($i = 1$) with the global primary mode ($i = 2$) and the third mode ($i = 4$) (denoted as SAM-1,2,4), are presented. For the two-mode approach, the post-buckling path grows monotonically. For the three-mode approach, the SAM-1,2,3 curve flattens, not reaching the ultimate load-carrying capacity, whereas, for the SAM-1,2,4, it attains the ultimate value of $P/P_{min} \approx 1.1$ and then falls, which corresponds to an unstable equilibrium path. It follows from these comparisons that an effect of the third global mode ($i = 4$) is crucial for interactive buckling of the 300 mm-long C-column. An application of the SAM thus enables an easy analysis of the effect of the modes under study on post-buckling paths, which is not possible virtually with the FEM. The FEM was used to verify the SAM results. In the FEM calculations, the amplitude of the initial deflection equal to 0.2 of the component plate thickness was assumed for the local mode ($i = 1$) with the global primary mode ($i = 2$) and the second mode ($i = 3$) or third global mode ($i = 4$). The results were the same. A very good agreement between the FEM curve and the SAM-1,2,3 curve was obtained. The post-buckling curve SAM-1,2 begins to differ more and more from the FEM curve for the load $P/P_{\min} \geq 1.8$. It was not possible to verify the curve SAM-1,2,4 with the FEM. Attention should be paid to the FEM curve. It breaks catastrophically for $P/P_{\min} \approx 1.95$ and $\Delta/\Delta_{min} \approx 3.4$ due to a lack of convergence of the Riks method. Thus, it is the ultimate value. None of the changes in the parameters of the Riks method allowed the calculations to continue. The authors put forward a hypothesis that when the secondary ($i = 3$) and third ($i = 4$) global modes begin to interact, numerical convergence of the FEM is lost, and the computations are catastrophically disrupted. This hypothesis requires further thorough investigations to be carried out. In Figure 8, a mode for the ultimate FEM value is shown. As can be easily seen in the central part of the C-channel, deflections of the plates and the web flattened out, with the symmetry of displacements remaining with respect to the cross-section axis.

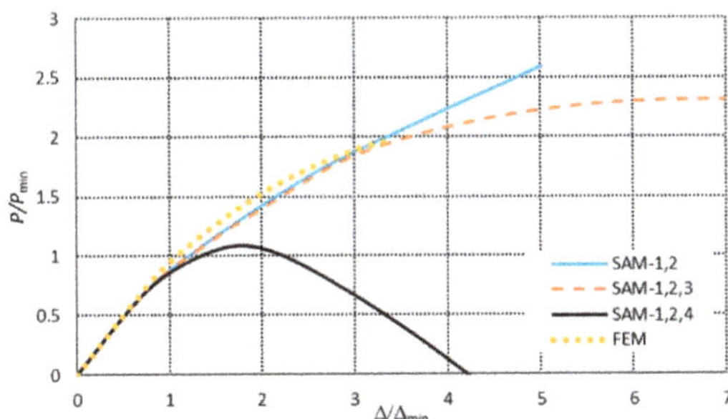

Figure 7. Post-buckling equilibrium paths in the dimensionless system for the 300 mm-long C-section.

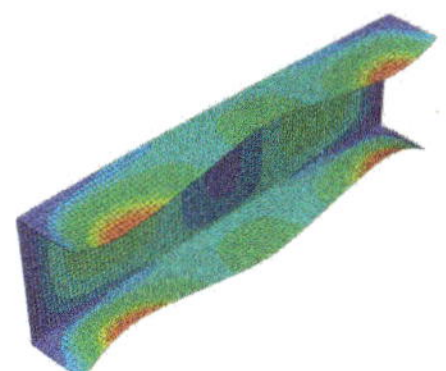

Figure 8. Mode of the 300-mm long C-section in the ultimate state, FEM results.

In the next figures (Figures 9 and 10), results for the 500 mm-long C-channel are presented. In Figure 9, the post-buckling equilibrium paths $P/P_{\min} = P/P_1$ as a function of the shortening $\Delta/\Delta_{min} = \Delta/\Delta_1$ are shown for the SAM and the FEM. The curves SAM-1,2 for the two-mode approach ($J = 2$) and SAM-1,2,3 for the three-mode approach ($J = 3$) are situated close to each other up to the load $P/P_{\min} \approx 2.5$. The post-buckling curve SAM-1,2,4 differs from the remaining curves already at $P/P_{\min} > 0.85$ and attains the ultimate load-carrying capacity for the load $P/P_{\min} \approx 2$ and the shortening $\Delta/\Delta_{min} \approx 5.1$. The FEM post-buckling equilibrium path is similar to the curves SAM-1,2 and SAM-1,2,3, but lies above them. In this case, we were not able to obtain a curve close to SAM-1,2,4 either. For example, for the 300 mm-long C-column, the FEM computations were disrupted catastrophically by a lack of numerical convergence for $P/P_{\min} \approx 1.35$ and $\Delta/\Delta_{min} \approx 1.65$, which should be treated as the ultimate value. Figure 10 presents the buckling mode of the C-section at the ultimate point.

Non-symmetrical buckling of C-section strips, which can follow from an interaction of anti-symmetrical modes, can be observed. They have been neglected in the present study.

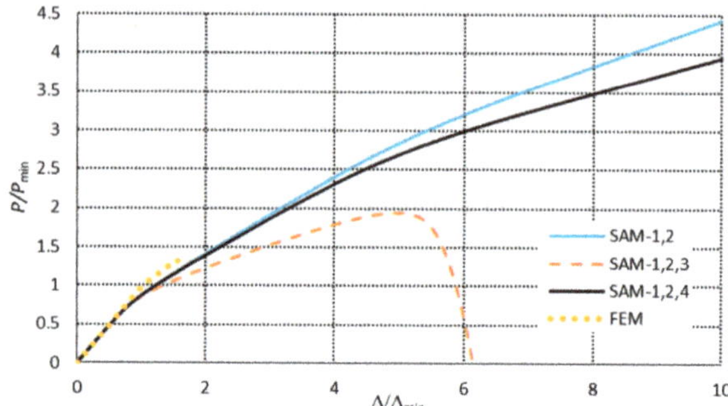

Figure 9. Post-buckling equilibrium paths in the dimensionless system for the 500 mm-long C-section.

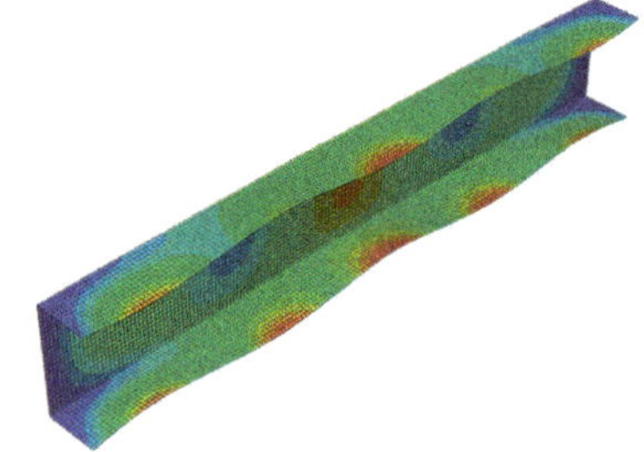

Figure 10. Mode of the 500 mm-long C-section in the ultimate state, FEM results.

For both the lengths of C-columns, a huge impact of high values of global buckling modes, which are unrealistic in the theory of interactive buckling from the practical point of view, can be seen. The present study is just an attempt to identify and to describe the phenomenon of the influence of secondary and third global distortional-flexural modes on the load-carrying capacity of medium-long C-channels.

3.2. TH-Channel Columns

In this section, the stability and post-buckling equilibrium paths of "top-hat"- section columns (Figure 1b) of the following lengths: 300 and 500 mm, will be discussed. Table 1 includes the bifurcation values of stresses obtained with the SAM and the FEM. Moreover, for the local mode ($i = 1$), the number of halfwaves m along the longitudinal direction is given in brackets. The same index notations are used as for the C-channels.

A very good agreement of the SAM and FEM results was obtained, similarly to the case of the C-channels. It should be noticed that an introduction of edge reinforcements in TH-section columns in comparison to C-section columns results is almost twice as high an increase in local bifurcational loads. What is more, the values of primary global loads for both lengths are very close. The ratio of the value of the second global mode to the local one is more than seven times higher, whereas for the third mode, it is 40 times higher, correspondingly.

Figure 11 shows bifurcation modes for the 300 mm column, whereas Figure 12 shows those for 500 mm. Local modes (Figures 11a and 12a) are the 'pure' local ones, for which the edges do not displace. They differ in the number of halfwaves along the longitudinal direction. The amplitudes of modes (i.e., the shape of the mode in the cross-section) are identical and, analogously, primary global modes ($i = 2$) (Figures 11b and 12b). In this case, the mode is distortional as the reinforced edge corner rotates together with the plate with respect to the web. Global secondary modes ($i = 3$) (Figures 11c and 12c) are the distortional-flexural ones and they are similar, provided the interpretation for the C-channel from Figure 3a,b is assumed. The global secondary mode ($i = 3$) for the 500 mm column is even "purer" than that for 300 mm, i.e., it is closer to the flexural one. Third modes ($i = 4$) are the distortional-flexural ones. In Figures 13 and 14, buckling modes of columns 300 and 500 mm long, respectively, are presented for the SAM. When Figures 11–14 are compared, it can be easily noticed that the buckling modes determined with the SAM and FEM agree very well.

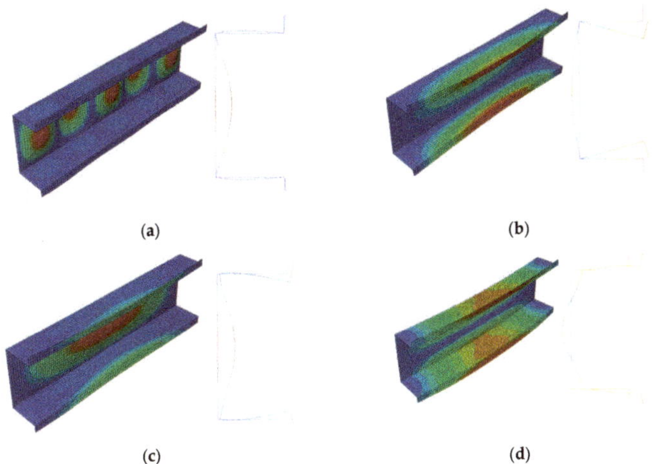

Figure 11. Buckling modes for the 300 mm-long TH-channel column, FEM results. Subsequent modes are as follows: Local $m > 1$ (**a**), primary global (**b**), higher global distortional-flexural (**c**), and third global distortional-flexural (**d**).

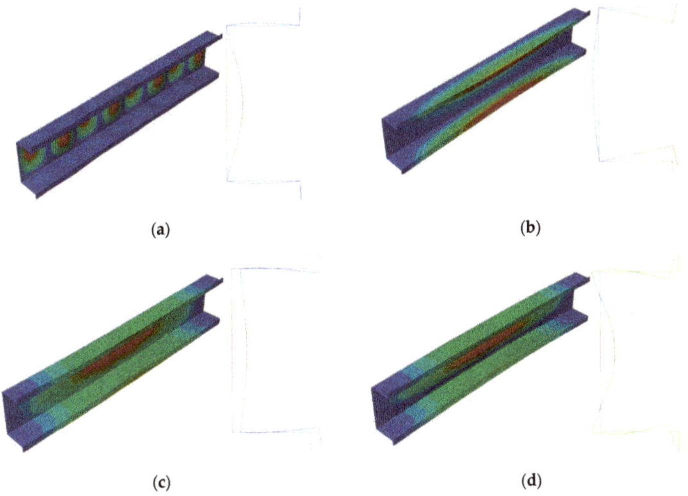

Figure 12. Buckling modes for the 500 mm-long TH-channel column, FEM results. Subsequent modes are as follows: Local $m > 1$ (**a**), primary global (**b**), higher global distortional-flexural (**c**), and third global distortional-flexural (**d**).

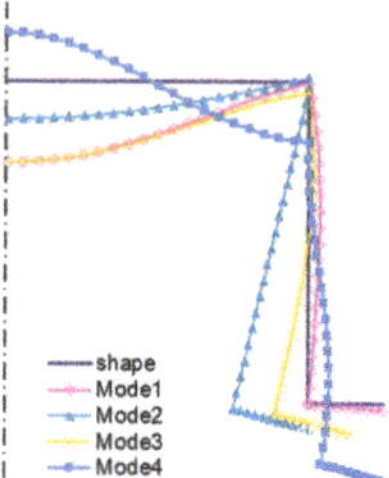

Figure 13. Buckling modes for the 300 mm-long TH-channel column, SAM results. The same notation of modes as in Figure 5.

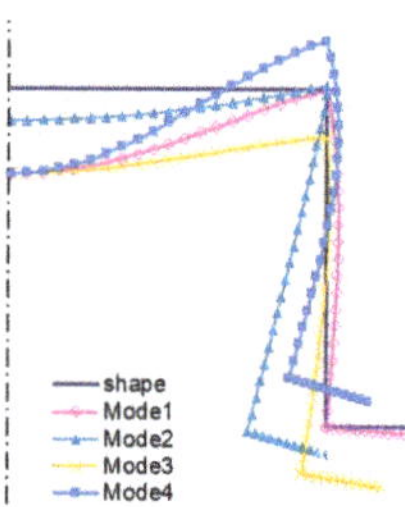

Figure 14. Buckling modes for the 500 mm-long TH-channel column, SAM results. The same notation of modes as in Figure 5.

The SAM and FEM calculations for the post-buckling states were conducted analogously as for the C-channels, that is to say, the same level of mode imperfections and two- and three-mode approaches for the SAM were assumed. The post-buckling paths in the systems $P/P_{min} = P/P_1$ versus the shortening $\Delta/\Delta_{min} = \Delta/\Delta_1$ are presented in Figure 15 for 300 mm and in Figure 16 for 500 mm. For both the lengths, the ultimate load-carrying capacity was obtained both for the two-mode ($J = 2$) and three-mode ($J = 3$) approaches. The curves SAM-1,2, SAM-1,2,3, and SAM-1,2,4 overlap up to the ultimate point, and then they fall, but at 300 mm, the SAM-1,2,4 (Figure 15) falls most dramatically. For the length of 500 mm, the curves SAM-1,2,3 and SAM-1,2,4 fall much more sharply than SAM-1,2. For 300 mm (Figure 15), the ultimate point occurs at $P/P_{min} \approx 1.7$ and $\Delta/\Delta_{min} \approx 2.25$, whereas for 500 mm (Figure 16), it occurs at $P/P_{min} \approx 2$ and $\Delta/\Delta_{min} \approx 2.6$. The FEM paths obtained for 300 (Figure 15) and 500 mm (Figure 16) are close to the ultimate load-carrying capacity for the SAM, which occurs at $P/P_{min} > 2.5$ and $\Delta/\Delta_{min} > 5.4$. Similarly, as for the C-channels, the FEM computations were catastrophically disrupted due to a lack of numerical convergence.

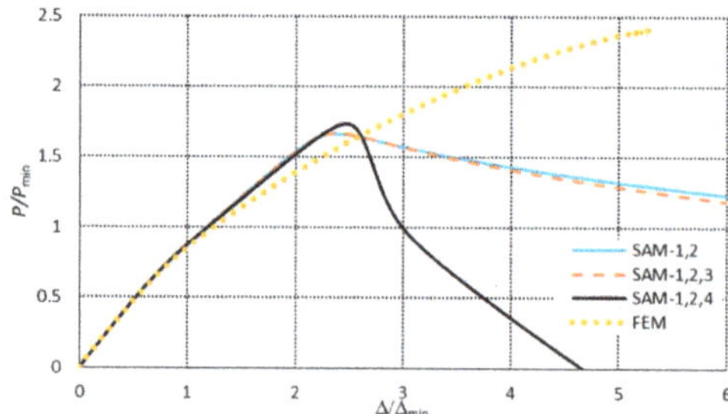

Figure 15. Post-buckling equilibrium paths in the dimensionless system for the 300 mm-long TH-section.

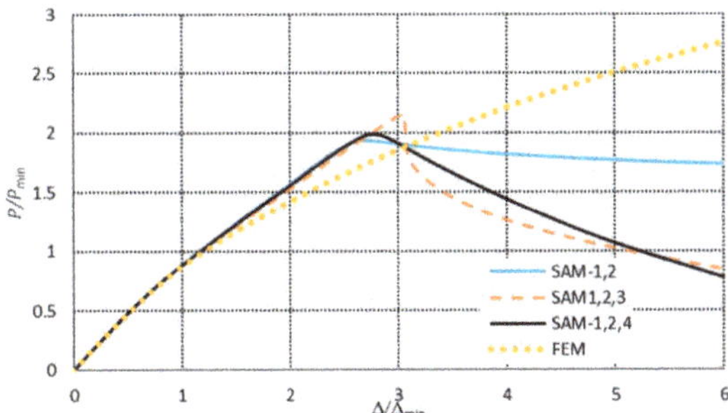

Figure 16. Post-buckling equilibrium paths in the dimensionless system for the 500 mm-long TH-section.

In Figures 17 and 18, buckling modes for the FEM ultimate points at 300 and 500 mm, respectively, are presented. The mode for 300 mm (Figure 17) is more asymmetrical than for the 500 mm C-channel (Figure 10). On the other hand, the mode for 500 mm (Figure 18) is symmetrical in practice.

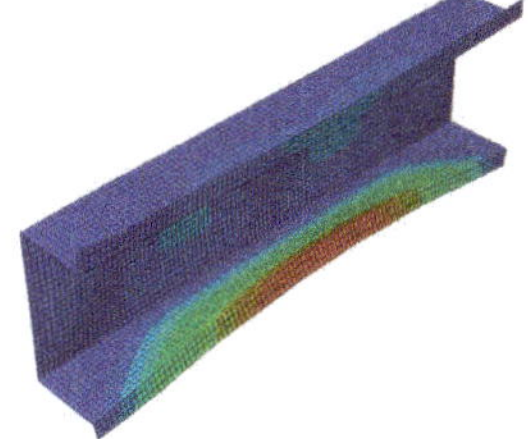

Figure 17. Mode of the 300 mm-long TH-channel column in the ultimate state, FEM results.

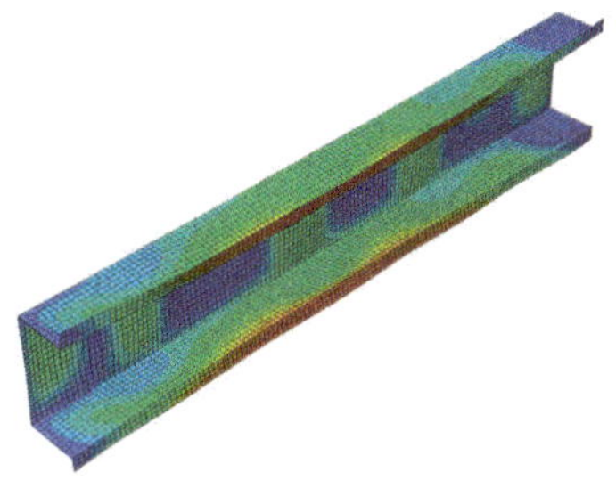

Figure 18. Mode of the 500 mm-long TH-channel column in the ultimate state, FEM results.

In the FEM, the modes considered in the analysis are not under control as opposed to the SAM. The asymmetrical mode (Figure 17) could be attained via an automatic introduction of anti-symmetrical eigenvalues (i.e., bifurcation ones) into the computations. Edge reinforcements for the top-hat channels result in higher load-carrying capacity when compared to the C-channels.

In the SAM calculations for C- and TH-channels for the SAM-1,2,4 three-mode approach, an unexpected effect of the third global distortional-flexural mode on the interactive buckling and the load-carrying capacity was found. It was not possible to verify the SAM calculations within the FEM; nevertheless, they unexpectedly cast light on a catastrophic disruption of the computations for post-buckling paths due to a lack of convergence, although Rik's procedure with various initial parameters was applied. The observed numerical phenomena for the SAM and the FEM thus require further intensive investigations with respect to modeling and the methods applied.

4. Conclusions

Multimodal buckling of the C- and TH-section (the so-called "top-hat") columns, made of isotropic material and subjected to compression, was considered. The column lengths were selected to observe strong interactions between various eigenvalues. The problem of nonlinear stability was solved with two methods, namely the semi-analytical method (SAM) using Koiter's perturbation approach and the finite element method (FEM). The results discussed prove that there is a strong effect of higher global distortional-flexural buckling modes (i.e., secondary and third) on the interactive buckling of C- and TH-columns and their load-carrying capacity. On the post-buckling equilibrium paths prepared within the SAM in the case of an interaction of the lowest local and other modes, when higher global distortional-flexural modes are neglected, the curves grow monotonically. However, if the global distortional flexural buckling mode is considered, the post-buckling equilibrium path grows up to the ultimate point, and then, it falls. This regularity repeats in all the columns under analysis. Moreover, in both the methods, attention is drawn to their conditions, particularly to the FEM catastrophic numerical behavior in all the cases of computations under discussion. Hence, the observed phenomena require further thorough and complex investigations, including the experimental ones. Finally, it should be added that all the considerations presented here refer to the elastic state. Thus, all plastic effects and plastic destruction mechanisms that occurred have been neglected.

Author Contributions: Conceptualization, Z.K.; methodology, A.T. and Z.K.; software, A.T. and Z.K.; formal analysis, A.T. and Z.K.; writing—original draft, A.T. and Z.K.; writing—review and editing, A.T. All authors have read and agree to the published version of the manuscript.

Funding: The studies have been conducted as a part of the research projects financed by the National Science Centre, Poland—decision numbers: UMO-2017/25/B/ST8/01046.

Conflicts of Interest: The authors declare no conflict of interest.

References

1. Van der Heijden, A.M.A. *Koiter's Elastic Stability of Solids and Structures*; Cambridge Universitet Delft: Delft, The Netherlands, 2008. [CrossRef]
2. Doyle, J.F. *Nonlinear Analysis of Thin-Walled Structures*; Springer: New York, NY, USA, 2001. [CrossRef]
3. Kołakowski, Z.; Teter, A. Coupled static and dynamic buckling modelling of thin-walled structures in elastic range. Review of selected problems. *Acta Mech. Et Autom.* **2016**, *10*, 141–149. [CrossRef]
4. Teter, A.; Kołakowski, Z. Interactive buckling and load-carrying capacity of thin-walled beam-columns with intermediate stiffeners. *Thin Walled Struct.* **2004**, *42*, 211–254. [CrossRef]
5. Kołakowski, Z.; Teter, A. Interactive buckling of thin-walled beam-columns with intermediate stiffeners or/and variable thickness. *Int. J. Solid Struct.* **2000**, *37*, 3323–3344. [CrossRef]
6. Adany, S. Constrained shell Finite Element Method for thin-walled members, Part 1: Constrains for a single band of finite elements. *Thin Walled Struct.* **2018**, *128*, 43–55. [CrossRef]
7. Adany, S.; Visy, D.; Nagy, R. Constrained shell Finite Element Method, Part 2: Application to linear buckling of thin-walled members. *Thin Walled Struct.* **2018**, *128*, 56–70. [CrossRef]
8. Adany, S.; Schafer, B.W. A full modal decomposition of thin-walled, single-branched open cross-section members via the constrained finite strip method. *J. Constr. Steel Res.* **2008**, *64*, 12–29. [CrossRef]
9. Schafer, B.W. Advances in the Direct Strength Method of cold-formed steel design. *Thin Walled Struct.* **2019**, *140*, 533–541. [CrossRef]
10. Becque, J.; Xilin, L.; Davison, B. Modal decomposition of coupled instabilities: The method of the equivalent nodal forces. *Thin Walled Struct.* **2019**, *143*, 106229. [CrossRef]
11. Hanock, G.J. Coupled Instabilities in Metal Structures (CIMS)–What have we learned and are we going? *Thin Walled Struct.* **2018**, *128*, 2–11. [CrossRef]
12. Niu, R.; Rasmussen, K.J.R.; Fan, F. Distortional–global interaction buckling of stainless steel C-beams: Part I–Experimental investigation. *J. Constr. Steel Res.* **2014**, *96*, 127–139. [CrossRef]
13. Niu, R.; Rasmussen, K.J.R.; Fan, F. Distortional–global interaction buckling of stainless steel C-beams: Part II–Numerical study and design. *J. Constr. Steel Res.* **2014**, *96*, 40–53. [CrossRef]

14. Abambres, M.; Camotim, D.; Silvestre, N. Modal Decomposition of Thin-Walled Member Collapse Mechanisms. *Thin Walled Struct.* **2014**, *74*, 269–291. [CrossRef]

15. Szymczak, C.; Kujawa, M. Buckling and initial post–local buckling behaviour of cold–formed channel member flange. *Thin Walled Struct.* **2019**, *137*, 177–184. [CrossRef]

16. Bebiano, R.; Camotim, D.; Gonçalves, R. GBTUL 2.0–A second-generation code for the GBT-based buckling and vibration analysis of thin-walled members. *Thin Walled Struct.* **2018**, *124*, 235–257. [CrossRef]

17. Camotim, D.; Dinis, P.B.; Martins, A.D.; Young, B. Review: Interactive behaviour, failure and DSM design of cold–formed steel members prone to distortional buckling. *Thin Walled Struct.* **2018**, *128*, 12–42. [CrossRef]

18. Garcea, G.; Leonetti, L.; Magisano, D.; Goncalves, R.; Camotim, D. Deformation modes for the post–critical analysis of thin–walled compressed members by a Koiter semi–analytic approach. *Int. J. Solids Struct.* **2017**, *110–111*, 367–384. [CrossRef]

19. Martins, A.D.; Camotim, D.; Dinis, P.B. Local-distortional interaction in cold-formed steel beams: Behaviour, strength and DSM design. *Thin Walled Struct.* **2017**, *119*, 879–901. [CrossRef]

20. Martins, A.D.; Camotim, D.; Dinis, P.B. On the distortional-global interaction in cold-formed steel columns: Relevance, post-buckling behaviour, strength and DSM design. *J. Constr. Steel Res.* **2018**, *145*, 449–470. [CrossRef]

21. Martins, A.D.; Camotim, D.; Goncalves, R.; Dinis, P.B. GBT–based assessment of the mechanics of distortional–global interaction in thin–walled lipped channel beams. *Thin Walled Struct.* **2018**, *124*, 32–47. [CrossRef]

22. Martins, A.D.; Camotim, D.; Gonçalves, R.; Dinis, P.B. On the mechanics of local–distortional interaction in thin–walled lipped channel beams. *Thin Walled Struct.* **2018**, *128*, 108–125. [CrossRef]

23. Martins, A.D.; Camotim, D.; Dinis, P.B. Distortional-global interaction in lipped channel and zed-section beams: Strength, relevance and DSM design. *Thin Walled Struct.* **2018**, *129*, 289–308. [CrossRef]

24. Martins, A.D.; Gonçalves, R.; Camotim, D. On the local and distortional post–buckling behaviour of thin–walled regular polygonal tubular columns. *Thin Walled Struct.* **2019**, *138*, 46–63. [CrossRef]

25. Martins, A.D.; Gonçalves, R.; Camotim, D. Post-buckling behaviour of thin–walled regular polygonal tubular columns undergoing local–distortional interaction. *Thin Walled Struct.* **2019**, *138*, 373–391. [CrossRef]

26. Martins, A.D.; Landesmann, A.; Camotim, D.; Dinis, P.B. Distortional failure of cold–formed steel beams under uniform bending: Behaviour, strength and DSM design. *Thin Walled Struct.* **2017**, *118*, 196–213. [CrossRef]

27. Jakubczak, P.; Gliszczynski, A.; Nienias, J.; Majerski, K.; Kubiak, T. Collapse of channel section composite profile subjected to bending, Part II: Failure analysis. *Compos. Struct.* **2017**, *179*, 1–20. [CrossRef]

28. Kolakowski, Z.; Urbaniak, M. Influence of the distortional–lateral buckling mode on the interactive buckling of short channels. *Thin Walled Struct.* **2016**, *109*, 296–303. [CrossRef]

29. Kolakowski, Z.; Jankowski, J. Interactive buckling of steel C–beams with different lengths–From short to long beams. *Thin Walled Struct.* **2018**, *125*, 203–210. [CrossRef]

30. Kolakowski, Z.; Jankowski, J. Interactive buckling of steel LC-beams under bending. *Materials* **2019**, *12*, 1440. [CrossRef]

31. Kolakowski, Z.; Krolak, M. Modal coupled instabilities of thin–walled composite plate and shell structures. *Compos. Struct.* **2006**, *76*, 303–313. [CrossRef]

32. Kolakowski, Z.; Mania, J.R. Semi–analytical method versus the FEM for analysis of the local post–buckling. *Compos. Struct.* **2013**, *97*, 99–106. [CrossRef]

33. Kolakowski, Z.; Kubiak, T.; Zaczynska, M.; Kazmierczyk, F. Global-distortional buckling mode influence on post-buckling behaviour of lip-channel beams. *Int. J. Mech. Sci.* **2020**, in press. [CrossRef]

34. Zaczynska, M.; Kolakowski, Z. The influence of the internal forces of the buckling modes on the load-carrying capacity of composite medium-length beams under bending. *Materials* **2020**, *13*, 455. [CrossRef] [PubMed]

35. Nazir, A.; Jeng, J.-Y. Buckling behavior of additively manufactured cellular columns: Experimental and simulation validation. *Mater. Des.* **2020**, *186*, 108349. [CrossRef]

36. Nazir, A.; Arshad, A.B.; Jeng, J.-Y. Buckling and Post-Buckling Behavior of Uniform and Variable-Density Lattice Columns Fabricated Using Additive Manufacturing. *Materials* **2019**, *12*, 3539. [CrossRef]

37. Kołakowski, Z. A semi-analytical method of interactive buckling of thin-walled elastic structures in the second-order approximation. *Int. J. Solids Struct.* **1996**, *33*, 3779–3790. [CrossRef]

38. Abaqus HTML Documentation. Dassault Systems. 2019. Available online: https://www.3ds.com/products-services/simulia/products/abaqus/ (accessed on 24 July 2020).

39. Abbasia, M.; Khezria, M.; Rasmussena, K.J.R.; Schaferb, B.W. Elastic buckling analysis of cold-formed steel built-up sections with discrete fasteners using the compound strip method. *Thin Walled Struct.* **2018**, *124*, 58–71. [CrossRef]

40. Szymczak, C.; Kujawa, M. On local buckling of cold-formed channel members. *Thin Walled Struct.* **2016**, *106*, 93–101. [CrossRef]

41. Szymczak, C.; Kujawa, M. Buckling of thin-walled columns accounting for initial geometrical imperfections. *Int. J. Non-Linear Mech.* **2017**, *95*, 1–9. [CrossRef]

42. Szymczak, C.; Kujawa, M. Flexural buckling and post-buckling of columns made of aluminium alloy. *Eur. J. Mech. A Solids* **2018**, *73*, 420–429. [CrossRef]

Article

Finite Beam Element for Curved Steel–Concrete Composite Box Beams Considering Time-Dependent Effect

Guang-Ming Wang, Li Zhu *, Xin-Lin Ji and Wen-Yu Ji

School of Civil Engineering, Beijing Jiaotong University, Beijing 100044, China; 14115279@bjtu.edu.cn (G.-M.W.); 19121058@bjtu.edu.cn (X.-L.J.); wyji@bjtu.edu.cn (W.-Y.J.)
* Correspondence: zhuli@bjtu.edu.cn

Received: 19 June 2020; Accepted: 20 July 2020; Published: 22 July 2020

Abstract: Curved steel–concrete composite box beams are widely used in urban overpasses and ramp bridges. In contrast to straight composite beams, curved composite box beams exhibit complex mechanical behavior with bending–torsion coupling, including constrained torsion, distortion, and interfacial biaxial slip. The shear-lag effect and curvature variation in the radial direction should be taken into account when the beam is sufficiently wide. Additionally, long-term deflection has been observed in curved composite box beams due to the shrinkage and creep effects of the concrete slab. In this paper, an equilibrium equation for a theoretical model of curved composite box beams is proposed according to the virtual work principle. The finite element method is adopted to obtain the element stiffness matrix and nodal load matrix. The age-adjusted effective modulus method is introduced to address the concrete creep effects. This 26-DOF finite beam element model is able to simulate the constrained torsion, distortion, interfacial biaxial slip, shear lag, and time-dependent effects of curved composite box beams and account for curvature variation in the radial direction. An elaborate finite element model of a typical curved composite box beam is established. The correctness and applicability of the proposed finite beam element model is verified by comparing the results from the proposed beam element model to those from the elaborate finite element model. The proposed beam element model is used to analyze the long-term behavior of curved composite box beams. The analysis shows that significant changes in the displacement, stress and shear-lag coefficient occur in the curved composite beams within the first year of loading, after which the variation tendency becomes gradual. Moreover, increases in the central angle and shear connection stiffness both reduce the change rates of displacement and stress with respect to time.

Keywords: curved steel–concrete composite box beam; two-node finite beam element with 26 DOFs; long-term behavior; age-adjusted effective modulus method

---

1. Introduction

Curved steel–concrete composite box beams have gradually become one of the main design types for urban overpasses and ramp bridges due to their light weight, high torsional rigidity, and rapid construction. However, in contrast to straight beam bridges, typical bending–torsion coupling mechanical behavior can be found in curved beam bridges [1]. Hence, when the structure is subjected to a vertical load, such as its own weight, certain bending and torsional effects are generated, leading to torsional and distortional warping. For a steel–concrete composite structure, longitudinal and transverse slips exist at the interface between the steel beam and the concrete slab of the curved composite box beam. In addition, for large-curvature curved beams with a width-to-radius ratio greater than 1/10, the change in the radius of curvature along the radial direction of the beam and the shear-lag effect caused by the wide flange still need to be considered.

For the numerical analysis of curved beams, the use of a three-dimensional finite element model can accurately simulate its mechanical behavior. However, this modeling and analysis process is complex and requires high computer performance, and the calculation efficiency with a three-dimensional model is not as good as that with a one-dimensional model. The study of one-dimensional theoretical models of curved beams first began in the 1960s. Vlasov [2] established governing differential equations for curved beams based on the three major equations of mechanics, which laid the foundation for subsequent theoretical research. Later, some Japanese scholars [3–5] conducted related research on one-dimensional theoretical models for curved box beams and proposed relevant design proposals for the behavior of shear lag and distortion. At present, the main idea of the research on one-dimensional beam models of curved beams is to introduce functions with relevant degrees of freedom (DOFs) on the basis of Vlasov's curved beam theoretical model to enable corresponding analyses.

The time-dependent (shrinkage and creep) effects of concrete have a significant influence on the mechanical behavior in steel–concrete composite beams. Due to the effects of concrete shrinkage and creep, the deformation of the structure under long-term loads is often much greater than that under short-term loads. The long-term deflection of composite beams has become a key issue for composite beam design and analysis. The authors of Reference [6] developed a two-node finite beam element with 26 DOFs for a curved composite box girder considering constrained torsion, distortion, shear lag, biaxial slip at the interface, and curvature variation along the radial direction. This paper introduces a constitutive relationship for concrete shrinkage and creep on the basis of the proposed beam finite element and further develops a new finite beam element that can consider the long-term behavior of curved composite box beams.

The shrinkage behavior of concrete is not related to its stress state; thus, shrinkage can be easily achieved by applying an initial strain to the model. The creep behavior of concrete is directly related to the stress state of concrete. Research on the long-term behavior of composite beams began in the 1970s. Bazant et al. [7] first simplified composite beams as a completely rigid interface and established a one-dimensional theoretical model of long-term behavior by using the three major equations of material mechanics. Afterward, Italian and Australian scholars became the primary researchers studying the long-term behavior of composite beams. Tarantino and Dezi used a stepwise calculation method to analyze the time-dependent effects of concrete, deduced a theoretical equation for simply supported composite beams, considering the time-dependent effects [8], and then derived theoretical equations for continuous composite beams, using the force method [9,10]. Then, Dezi et al. [11,12] adopted a single-step method (effective modulus method, average stress method, and age-adjusted effective modulus method) to consider the time-dependent effects of concrete and established a one-dimensional simplified theoretical model of the long-term behavior of simply supported and continuous composite beams. After entering the 21st century, based on the governing differential equations for composite beams that can consider the shear-lag behavior of concrete slabs, Dezi et al. [13] introduced a time-dependent constitutive model and established a one-dimensional theoretical model that can consider the shear-lag effect and the long-term behavior. Based on this model, the long-term behavior of composite beams was studied under prestressed loads [14]. With the increasing application of finite element technology in structural numerical analysis, a beam element model of composite beams was also proposed for analyzing their long-term behavior. Gara et al. [15] introduced a shape function to develop a beam element with 13 DOFs that can consider the interfacial slip in composite beams and the shear-lag effect of concrete slabs and analyzed the long-term behavior of the structure by using a stepwise calculation method. Ranzi and Bradford [16] proposed beam elements for the long-term behavior of composite beams, considering interfacial slip and concrete time-dependent effects according to the direct stiffness method and single-step method. Later, Ranzi, and Bradford [17] and Nguyen et al. [18] proposed long-term performance beam elements, considering interfacial slip and the time-dependent effects of concrete on the basis of the direct stiffness method and the stepwise calculation method. The abovementioned studies are all on beam element models that can analyze the long-term behavior of straight composite beams. Due to the complex mechanical behavior and time-dependent effect of curved composite box beams, it is challenging to develop a finite beam element for curved composite box beams that can account for constrained torsion, distortion, shear lag, interfacial biaxial slip, time-dependent effects, and curvature variation in the radial direction.

Based on the Vlasov's one-dimensional theoretical model of curved beams, this paper introduces a torsional warping function, a distortional angle, an interfacial longitudinal and an interfacial transverse sliding function, and a shear-lag intensity function for concrete slabs and a shear-lag intensity function for steel beams and adopts the age-adjusted effective modulus method to analyze the effects of concrete creep. Moreover, this study establishes a finite beam element for curved composite box beams that can consider constrained torsion, distortion, shear lag, interfacial biaxial slip, time-dependent effects, and curvature variation in the radial direction. Finally, this study analyzes the long-term mechanical behavior of curved composite box girders by use of the developed finite beam element model.

2. Two-Node Finite Beam Element with 26 DOFs for Curved Composite Box Beams

2.1. Basic Hypotheses of the Model

The basic hypotheses made in the derivation process in this paper are as follows: (1) The sections of the concrete slab are rectangular, and the slab's curvature is the same as that of the undeformed structure; (2) the vertical displacements of the concrete and the steel beam are the same; (3) the curvature radius of the composite beam does not change along the length of the beam; (4) the stiffness of the shear connections of the composite beam is uniformly distributed in the longitudinal and transverse directions; (5) the shear stress produced by the distortional effect is ignored; (6) shear deformation produced by bending and distortional warping is ignored; and (7) the shear-lag effect is taken into account under vertical bending.

2.2. Geometric Dimensions and Coordinate System of the Curved Composite Box Beam

The geometric dimensions and coordinate system of the curved composite box beam are shown in Figure 1, wherein $oxyz$ is a three-dimensional flow coordinate system that passes through the centroid line, the ox-axis is parallel to the concrete slab and points to the curvature center of the box beam, the oy-axis is straight downward, and the oz-axis coincides with the undeformed axis of the beam. Moreover, the tangential directions of every thin-walled bar in the sections of the box beam are set as the s-axis, and the normal directions of every thin-walled bar in the sections of the box beam are set as the n-axis.

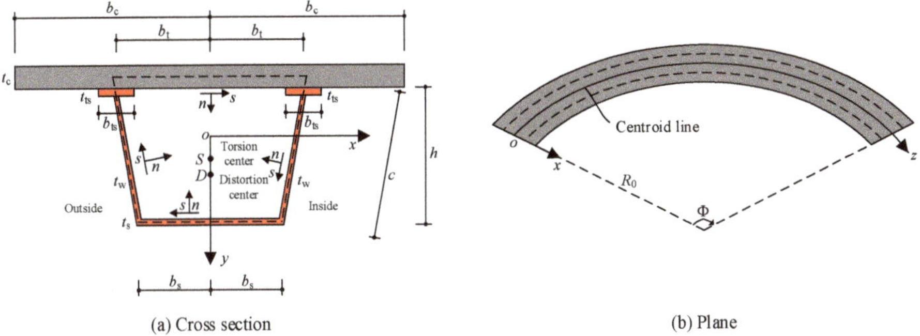

(a) Cross section   (b) Plane

Figure 1. Geometric dimensions and coordinate system of the curved composite box beam: (**a**) cross section, (**b**) plane.

2.3. Displacement Modes and Strain Components of Curved Composite Box Beams

The displacements along the ox, oy, and oz axes of the centroid of the curved composite box beam are denoted $u(z)$, $v(z)$, and $w(z)$, respectively. The torsion angle around the torsion center and the distortion angle around the distortion center are denoted $\theta(z)$ and $\theta_d(z)$, respectively. The transverse and longitudinal slip of the steel beam and concrete slab are denoted $a_k\Omega_x(z)$ and $a_k\Omega_s(z)$, respectively, where k is s or c, which represent the steel beam or concrete slab, respectively; $a_s = 1$ and $a_c = -1$; and $\Omega_x(z)$ and $\Omega_s(z)$ are the transverse and longitudinal slip functions, respectively. According to Ullmanski's second torsion theory, an independent function of torsional warping displacement, $\beta(z)$, is introduced. Additionally, f_k ($k = $ s,c) is the shear-lag warping strength functions that represent the steel beam or concrete slab. Based on the coordinate system shown in Figure 1, according to the curved shell theory, combining the introduced transverse and longitudinal slip functions and the shear-lag warping strength functions of the steel beam and concrete slab, the longitudinal displacement $W_{zk}(z,s)$, tangential displacement $W_{sk}(z,s)$, and normal displacement $W_{nk}(z,s)$ at any point of the composite box beam can be obtained as follows:

$$
\begin{aligned}
W_{zk}(z,s) &= w - (u' + wk_0 + a_k\Omega_z k_0)x - v'y + a_k\Omega_z \\
&\quad - \omega(x,y)(\beta + v'k_0) - \omega_d(x,y)\theta_d' + \psi_k(x)f_k
\end{aligned}
\tag{1}
$$

$$
W_{sk}(z,s) = -(u + a_k\Omega_x)\sin\alpha - v\cos\alpha + \theta\rho_s + \theta_d D_s
\tag{2}
$$

$$
W_{nk}(z,s) = (u + a_k\Omega_x)\cos\alpha - v\sin\alpha - \theta\rho_n - \theta_d D_n
\tag{3}
$$

where $k_0 = 1/R_0$ is the initial curvature of the centroid line of the box beam sections; R_0 is the curvature radius of the centroid line of the box beam sections; α is the angle from the x-axis to the n-axis, which is anticlockwise positive; $\omega(x,y)$ and $\omega_d(x,y)$ are the torsional warping and the distortional warping principal polar coordinates, respectively; and $\psi_k(x)$ is the shape function for the shear-lag warping of the concrete slab or steel beam, the expressions of which are as shown in Appendix A.

Transformed sections should be adopted for calculating geometric characteristics. For the calculation of the axial and bending characteristics of the composite beam, the thickness of the concrete slab is unchanged, whereas the width decreases. For the calculation of torsional and distortional effects of the composite beam, the width of the concrete slab remains unchanged, whereas the thickness decreases.

In addition, according to the literature [19,20], in Equations (1)–(3), $\rho_s = (y - y_s)\sin\alpha - (x - x_s)\cos\alpha$ is the tangential displacement distribution of the cross-sections of the composite beam under a unit torsion angle; $\rho_n = (y - y_s)\cos\alpha + (x - x_s)\sin\alpha$ is the normal displacement distribution of the cross-sections of the composite beam under a unit torsion angle; $D_s = [(y - y_d)\sin\alpha - (x - x_d)\cos\alpha]\Psi_d$ is the tangential displacement distribution of the composite beam under a unit distortion angle; $D_n = [(y - y_d)\cos\alpha + (x - x_d)\sin\alpha]\Psi_d$ is the normal displacement distribution of the composite beam under a unit distortion angle; (x_s, y_s) and (x_d, y_d) are the coordinates of the torsion center, S, and the distortion center, D, of the transformed sections of the composite beam, respectively; $\Psi_d = -1/(1+v)$ is the distortion parameter of the top and bottom plates, and $\Psi_d = v/(1+v)$ is the distortion parameter of the web, in which $v = (2h_1 b_t + 2h_3 b_s)/(h_2 c + h_4 c)$ is a constant related only to the section shape [20]; h_1 and h_3 are the vertical distances from the distortion center to the top and bottom plates; and h_2 and h_4 are the vertical distances from the distortion center to the left and right webs, respectively.

Note that a_s is 1 for the steel beam and -1 for the concrete slab. According to Equations (1) and (2), the relative slips on the interface between the steel beam and concrete slab are expressed as follows:

$$
u_{sp} = 2\Omega_x
\tag{4}
$$

$$
w_{sp} = 2\Omega_z
\tag{5}
$$

where u_{sp} and w_{sp} denote the transverse and longitudinal interfacial slip, respectively.

Based on the abovementioned displacement mode, according to the curved shell theory and taking the curvature variation in the radial direction of the beam into consideration, the normal strain component at any point, P, in the three-dimensional flow coordinate system is expressed as follows:

$$
\begin{aligned}
\varepsilon_{pk} &= \frac{\partial W_{zk}}{\rho(x)\partial\phi} + \frac{W_s \sin\alpha}{\rho(x)} - \frac{W_n \cos\alpha}{\rho(x)} \\
&= \frac{R_0}{\rho(x)}\left(\frac{\partial W_{zk}}{R_0\partial\phi} + \frac{W_s \sin\alpha}{R_0} - \frac{W_n \cos\alpha}{R_0}\right) \\
&= \frac{R_0}{\rho(x)}[w' - xu'' - yv'' + a_k(1-k_0 x)\Omega_z' - k_0 u - k_0 x w' - a_k k_0 \Omega_x + \\
&\quad k_0(y-y_s)\theta - \omega(\beta' + k_0 v'') + k_0(y-y_d)\Psi_d\theta_d - \omega_d\theta_d'' + \psi_k f_k']
\end{aligned}
\tag{6}
$$

According to Hypothesis (5), based on the curved shell theory, the shear strain component at any point, P, of the cross-section is expressed as follows:

$$
\gamma_{pk} = \frac{R_0}{\rho(x)}\left[(\theta' + v'k_0)r^* - (\beta + v'k_0)\frac{\partial\omega}{\partial s} + \psi_{k,x}f_k\right]
\tag{7}
$$

where r^* denotes the vertical distance between the torsion center and the tangent line of point, P.

$$
\frac{\partial\omega}{\partial s} = r^* - \frac{\Omega}{\tilde{t}\oint(ds/\tilde{t})}
\tag{8}
$$

where $\Omega = \oint \tilde{t}ds$ is twice the surrounding area, and $\tilde{t}$ is the thickness of each thin-walled plate of the transformed section.

2.4. Equilibrium Equation for Curved Composite Box Beams

The equilibrium equation for a curved steel–concrete composite beam can be derived according to the virtual work principle as follows:

$$
\begin{aligned}
\delta\Pi = &\int_\Phi \iint_{A_s} \delta\varepsilon_s^T \sigma_s da\rho(x)d\phi + \int_\Phi \iint_{A_c} \delta\varepsilon_c^T \sigma_c da\rho(x)d\phi + \int_\Phi \int_{2b_{ts}} \delta d_{slip}^T q_{sh} dx\rho(x)d\phi + \\
&\int_\Phi \delta\theta_d K_R \theta_d \rho(x)d\phi - \Sigma\, \delta W^T Q - \int_\Phi \delta W^T qp(x)d\phi = 0 \\
&\forall \delta\varepsilon_s, \delta\varepsilon_c, \delta d_{slip}, \delta\theta_d, \delta W
\end{aligned}
\tag{9}
$$

where $\int_\Phi \iint_{A_s} \delta\varepsilon_s^T \sigma_s da\rho(x)d\phi$ and $\int_\Phi \iint_{A_c} \delta\varepsilon_c^T \sigma_c da\rho(x)d\phi$ are the internal virtual work produced by the deformation of the steel beam and concrete slab, respectively; As and Ac denote the cross-sectional areas of the steel beam and concrete slab, respectively; and Φ is the central angle of the curved composite box beam. In terms of Equations (6) and (7), the strain variables of the steel beam and concrete slab can be written as follows:

$$
\varepsilon_k = S B_k d
\tag{10}
$$

where the dimensions of the matrix B_k are 11×23, and its exact value is given in Appendix B. The matrix S and displacement vector d are expressed as follows:

$$
S = \begin{bmatrix}
\frac{R_0}{\rho(x)} & \frac{R_0}{\rho(x)}x & \frac{R_0}{\rho(x)}y & \frac{R_0}{\rho(x)}\omega & \frac{R_0}{\rho(x)}\omega_d & \frac{R_0}{\rho(x)}\psi_c & \frac{R_0}{\rho(x)}\psi_s & 0 & 0 & 0 & 0 \\
0 & 0 & 0 & 0 & 0 & 0 & 0 & \frac{R_0}{\rho(x)}r^* & \frac{\partial\omega}{\partial s} & \psi_{c,x} & \psi_{s,x}
\end{bmatrix}
\tag{11}
$$

$$
d = \left\{\ [u]\quad [v]\quad [w]\quad [\theta]\quad [\beta]\quad [\theta_d]\quad [\Omega]\quad [f]\ \right\}^T
\tag{12}
$$

where

$$
[u] = \begin{pmatrix} u & u' & u'' \end{pmatrix}
\tag{13}
$$

$$
[v] = \begin{pmatrix} v & v' & v'' \end{pmatrix}
\tag{14}
$$

$$[w] = \begin{pmatrix} w & w' \end{pmatrix} \tag{15}$$

$$[\theta] = \begin{pmatrix} \theta & \theta' \end{pmatrix} \tag{16}$$

$$[\beta] = \begin{pmatrix} \beta & \beta' \end{pmatrix} \tag{17}$$

$$[\theta_d] = \begin{pmatrix} \theta_d & \theta_d' & \theta_d'' \end{pmatrix} \tag{18}$$

$$[\Omega] = \begin{pmatrix} 2\Omega_x & 2\Omega_x' & 2\Omega_s & 2\Omega_s' \end{pmatrix} \tag{19}$$

$$[f] = \begin{pmatrix} f_c & f_c' & f_s & f_s' \end{pmatrix} \tag{20}$$

When the bridge is in use, the steel beam is in the elastic phase, and the stress–strain relationship of the steel beam is expressed as follows:

$$\sigma_s = \left\{ \begin{array}{c} \sigma_s \\ \tau_s \end{array} \right\} = E_s \begin{bmatrix} 1 & 0 \\ 0 & 1/2(1+\mu_s) \end{bmatrix} \left\{ \begin{array}{c} \varepsilon_s \\ \gamma_s \end{array} \right\} \tag{21}$$

where σ_s and τ_s are the normal stress and shear stress of the steel beam, respectively; E_s is the elastic modulus of steel; and μ_s is the Poisson's ratio of steel.

To take the effects of concrete shrinkage and creep into consideration, the stress–strain relationship of the concrete slab can be expressed as follows:

$$\left\{ \begin{array}{c} \varepsilon_c(t) - \varepsilon_c^{sh}(t) \\ \gamma_c(t) \end{array} \right\} = J(t,t_0) \begin{bmatrix} 1 & 0 \\ 0 & 2(1+\mu_c) \end{bmatrix} \left\{ \begin{array}{c} \sigma_c(t_0) \\ \tau_c(t_0) \end{array} \right\} + \int_{t_0}^{t} J(t,\tau) \begin{bmatrix} 1 & 0 \\ 0 & 2(1+\mu_c) \end{bmatrix} \left\{ \begin{array}{c} d\sigma_c(\tau) \\ d\tau_c(\tau) \end{array} \right\} \tag{22}$$

To solve the integral problem in Equation (22), Bazant [21] proposed the age-adjusted effective modulus method to calculate the stress–strain relationship of concrete, which is expressed as follows:

$$\sigma_c(t) = \left\{ \begin{array}{c} \sigma_c(t) \\ \tau_c(t) \end{array} \right\} = E_{ce}(t,t_0) \begin{bmatrix} 1 & 0 \\ 0 & 1/(2+2\mu_c) \end{bmatrix} \left\{ \begin{array}{c} \varepsilon_c(t) - \varepsilon_c^{sh}(t) \\ \gamma_c(t) \end{array} \right\} + \xi(t,t_0) \left\{ \begin{array}{c} \sigma_c(t_0) \\ \tau_c(t_0) \end{array} \right\} \tag{23}$$

where t_0 is the initial loading age; ε_c^{sh} is the concrete shrinkage strain; μ_c is the Poisson's ratio of concrete; $J(t,t_0)$ is the concrete creep function at time t; σ_c and τ_c are the normal stress and shear stress of the concrete slab, respectively; $E_{ce}(t,t_0)$ is the concrete age-adjusted effective elastic modulus; and $\xi(t,t_0)$ is the influence factor of the creep effects. $E_{ce}(t,t_0)$ and $\xi(t,t_0)$ can be calculated by Equations (24) and (25), respectively.

$$E_{ce}(t,t_0) = \frac{E_c(t_0)}{1 + \chi(t,t_0)\varphi(t,t_0)} \tag{24}$$

$$\xi(t,t_0) = \frac{\varphi(t,t_0)(\chi(t,t_0) - 1)}{1 + \chi(t,t_0)\varphi(t,t_0)} \tag{25}$$

where $E_c(t_0)$ is the concrete elastic modulus at time t_0, $\varphi(t,t_0)$ is the concrete creep coefficient, and $\chi(t,t_0)$ is the concrete aging coefficient, which can be derived from Equation (26).

$$\chi(t,t_0) = \frac{E_c(t_0)}{E_c(t_0) - R(t,t_0)} - \frac{1}{\varphi(t,t_0)} \tag{26}$$

where $R(t,t_0)$ is the concrete relaxation function, which can be calculated according to the literature [21].

Note that $\int_{\Phi} \int_{2b_{ts}} \delta d_{slip}^{T} q_{sh} dx \rho(x) d\phi$ in Equation (9) is the internal virtual work produced by interfacial slip between the steel beam and concrete slab. According to Equations (4) and (5), the interfacial slip between the steel beam and concrete slab can be written as follows:

$$d_{slip} = \left\{ \begin{array}{cc} u_{sp} & w_{sp} \end{array} \right\}^{T} = S_\Omega B_\Omega d \tag{27}$$

where the dimensions of matrix B_Ω are 2×23, as shown in Appendix B, and S_Ω is an element matrix with dimensions of 2×2.

During use, the shear stress flow of a unit area at the interface between the steel beam and concrete slab can be expressed as follows:

$$q_{sh} = \left\{ \begin{array}{cc} q_{us} & q_{ws} \end{array} \right\}^{T} = \left[\begin{array}{cc} \rho_u & 0 \\ 0 & \rho_w \end{array} \right] \left\{ \begin{array}{c} u_{sp} \\ w_{sp} \end{array} \right\} = \rho_{slip} d_{slip} \tag{28}$$

where q_{us} and q_{ws} are transverse and longitudinal shear stress flow of a unit area in the composite beam interface, respectively; ρ_{slip} is the shear connection stiffness matrix of a unit area in the composite beam interface; and ρ_u and ρ_w are the transverse and longitudinal shear connection stiffness of a unit area in the composite beam, respectively.

Note that $\int_{\Phi} \delta\theta_d K_R \theta_d \rho(x) d\phi$ in Equation (9) is the internal virtual work produced when the frame resists distortion. K_R is the frame distortion-resisting stiffness, which can be calculated according to the literature [22].

Note that $\sum \delta W^T Q$ and $\int_{\Phi} \delta W^T q \rho(x) d\phi$ in Equation (9) are the external virtual work produced by external loads, wherein $Q = \left(\begin{array}{ccc} Q_n & Q_s & Q_z \end{array} \right)^{T}$ and $q = \left(\begin{array}{ccc} q_n & q_s & q_z \end{array} \right)^{T}$ are the concentrated load vector and the uniformly distributed load vector, respectively; Q_n, Q_s, and Q_z are concentrated forces in the directions of the n-axis, s-axis, and z-axis, respectively; q_n, q_s, and q_z are uniformly distributed forces in the directions of the n-axis, s-axis, and z-axis, respectively; and W is the displacement vector under external loads. The equation for this displacement vector is given in Equation (29), which is based on Equations (1)–(3).

$$W = \left\{ \begin{array}{ccc} W_{nk} & W_{sk} & W_{zk} \end{array} \right\}^{T} = H_1 A + H_2 A' \tag{29}$$

where

$$[H_1]_{3\times10} = \left\{ \begin{array}{ccc} [B_1]^{T} & [B_2]^{T} & [B_3]^{T} \end{array} \right\}^{T} \tag{30}$$

$$[H_2]_{3\times10} = \left\{ \begin{array}{ccc} [0]_{1\times10}^{T} & [0]_{1\times10}^{T} & [B_4]^{T} \end{array} \right\}^{T} \tag{31}$$

$$B_1 = \left(\begin{array}{cccccccccc} \cos\alpha & -\sin\alpha & 0 & -\rho_n & 0 & -D_n & \frac{a_k \cos\alpha}{2} & 0 & 0 & 0 \end{array} \right) \tag{32}$$

$$B_2 = \left(\begin{array}{cccccccccc} -\sin\alpha & -\cos\alpha & 0 & \rho_s & 0 & D_s & -\frac{a_k \sin\alpha}{2} & 0 & 0 & 0 \end{array} \right) \tag{33}$$

$$B_3 = \left(\begin{array}{cccccccccc} 0 & 0 & 1-xk_0 & 0 & -\omega & 0 & 0 & \frac{a_k(1-xk_0)}{2} & \psi_c & \psi_s \end{array} \right) \tag{34}$$

$$B_4 = \left(\begin{array}{cccccccccc} -x & -(y+\omega k_0) & 0 & 0 & 0 & -\omega_d & 0 & 0 & 0 & 0 \end{array} \right) \tag{35}$$

$$A = \left(\begin{array}{ccccccccc} u & v & w & \theta & \beta & \theta_d & 2\Omega_x & 2\Omega_z f_c f_s \end{array} \right)^{T} \tag{36}$$

2.5. Finite Beam Element for Curved Composite Box Beams

According to the methodology of finite element discretization, Equation (9) can be solved by adopting a two-node finite beam element, wherein thirteen DOFs exist at each of the nodes. The nodal displacement vector of the two-node finite beam element with 26 DOFs is expressed as follows:

$$d^e = \left\{ \begin{array}{cc} d_1^e & d_2^e \end{array} \right\}^{T} \tag{37}$$

$$d_i^e = \left(\begin{array}{ccccccccccccc} u_i & u_i' & v_i & v_i' & w_i & \theta_i & \beta_i & \theta_{di} & \theta_{di}' & 2\Omega_{xi} & 2\Omega_{zi} & f_{ci} & f_{si} \end{array} \right) \text{ for } i = 1,2 \tag{38}$$

The shape function matrix $[N]_{23\times26}$ is introduced into Equation (12), and the shape function matrix $[N_F]_{10\times26}$ is introduced into Equation (36); both of the shape function matrices are shown in Appendix C. Substituting Equations (10)–(36) and the shape function matrices N and N_F into Equation (9), the finite beam element equation for the curved composite box beam can be derived as follows:

$$Kd^e = F \tag{39}$$

$[K]_{26 \times 26}$ in Equation (39) is the element stiffness matrix, which can be calculated as follows:

$$K = \int_{l_e} N^T \left[\iint_{A_s} B_s^T S^T E_s S B_s da + \iint_{A_c} B_c^T S^T E_c S B_c da + \int_{2b_{ts}} B_\Omega^T S_\Omega^T \rho_{slip} S_\Omega B_\Omega dx + T_R \right] N dz \tag{40}$$

where $[T_R]_{23 \times 23}$ is the distortion-resisting stiffness matrix of the frame, in which $T_R(13,13) = K_R$ and the other elements are 0, and l_e is the length of the finite beam elements.

$[F]_{26 \times 1}$ in Equation (39) is the equivalent load vector of the node, which can be calculated as follows:

$$F = F_q + F_\xi + F_{sh} \tag{41}$$

where F_q is the equivalent load vector of the node caused by external loads, which can be calculated by Equation (42); F_ξ is the equivalent load vector of the node caused by the concrete creep effect, which can be calculated by Equation (43); and F_{sh} is the equivalent load vector of the node caused by the concrete shrinkage effect, which can be calculated by Equation (44).

$$F_q = \int_{l_e} \left(N_F^T H_1^T + N_F'^T H_2^T \right) q dz + \left(N_F^T H_1^T + N_F'^T H_2^T \right) Q \tag{42}$$

$$F_\xi = -\int_{l_e} N^T B_c{}^T \left(\iint_{A_c} S^T \xi(t, t_0) \sigma_c(t_0) da \right) dz \tag{43}$$

$$F_{sh} = \int_{l_e} N^T B_c{}^T \left(\iint_{A_c} S^T E_{ce}(t, t_0) \varepsilon_c^{sh}(t) da \right) dz \tag{44}$$

where

$$\sigma_c(t_0) = \left\{ \begin{array}{cc} \sigma_c(t_0) & \tau_c(t_0) \end{array} \right\}^T \tag{45}$$

$$\varepsilon_c^{sh}(t) = \left\{ \begin{array}{cc} \varepsilon_c^{sh}(t) & 0 \end{array} \right\}^T \tag{46}$$

Notably, the calculation methods for the characteristic geometrical parameters of the section under torsion and distortion in the finite beam element model can be found in the literature [6,23].

3. Numerical Validation of the Beam Element Model

The proposed finite beam element model is applied to calculate the long-term behavior of a typical curved composite box beam. The cross-section and load layout of the curved composite box beam are shown in Figure 2: The arc length of the longitudinal section centroid line is 6200 mm, the central angle is 45°, and seven transverse diaphragms are uniformly distributed along the longitudinal direction; the slab is composed of C30 concrete and is 750 mm wide and 50 mm thick; the steel beam has a total height of 300 mm; the web has a thickness of 12 mm; the bottom flange is 350 mm wide and 12 mm thick; the interfacial shear connection stiffness along the transverse and longitudinal directions, ρ_u and ρ_w, are both equal to 100 N/mm³; and a vertical load of 10 kN/mm acts at the section centroid line.

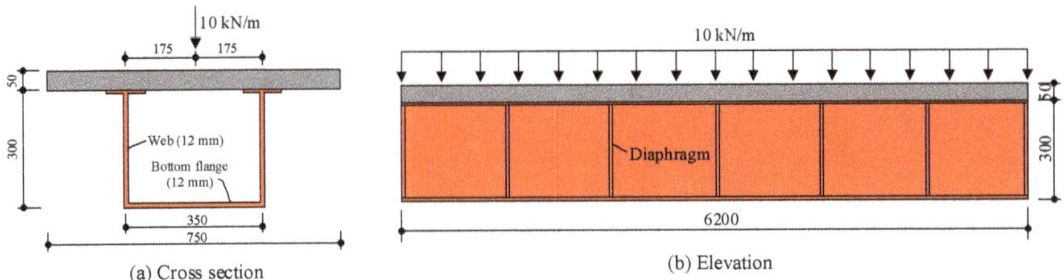

Figure 2. Geometric dimensions and load layout of the curved composite box beam used for validation (unit: mm): (**a**) cross section, (**b**) elevation.

The boundary conditions of the beam element model are as follows: At the mid-span, the longitudinal displacement and the longitudinal slip between the concrete slab and steel beam are all restrained to ensure symmetrical structural boundary constraints; at both ends of the beam, the transverse displacement, transverse slip, deflection, and torsion angle are restrained; and at the position of every diaphragm, the distortion angle is restrained.

The elaborate finite element model of a curved steel–concrete composite box beam is established in ANSYS. The concrete slabs, steel beams, and transverse diaphragms are modeled with Shell181 elements, whereas the shear connectors are modeled with Combin14 elements. The finite element model is shown in Figure 3.

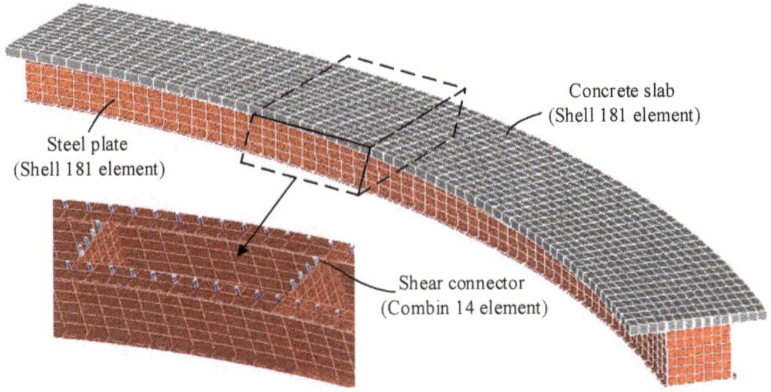

Figure 3. Elaborate finite element model of a curved composite box beam established in ANSYS.

The parameters of the material and environment of the proposed beam element model and the elaborate finite element model established in ANSYS are as follows: The elastic modulus of steel E_s and Poisson's ratio of steel μ_s are 2.06×10^5 MPa and 0.3, respectively; the calculation of the creep behavior of concrete in the two models use the age-adjusted effective modulus method; and the Poisson's ratio of concrete μ_c is 0.2. The elastic modulus formula and the shrinkage creep model are obtained within CEB-FIP90 [24]; the drying age of concrete is three days, the relative humidity in the environment is 75%, and the initial loading age t_0 is 28 days. Using the finite beam element model and the elaborate finite element model, time-dependent changes in the deflection, transverse displacement, and stress of the curved composite box beam were calculated from the loading age to 3000 days, as shown in Figures 4–6. From these three figures, the accuracy and applicability of the proposed two-node finite element with 26 DOFs are verified due to the good correlation between the results of the elaborate finite element model and those of the finite beam element model. Notably, the proposed beam element model and the elaborate FE model took 7 and 53 s to obtain the long-term behavior of the curved composite beam, respectively. It can be known that the calculation efficiency of the proposed beam element model is much higher than that of the elaborate FE model.

According to the three figures, the deflection and transverse displacement of the curved composite box beam increase over time due to concrete shrinkage and creep, and the compressive stress in the concrete decreases over time, whereas the tensile stress in the steel beam increases over time. Within 250 days after loading, the displacement and the stress of the curved composite box beam change significantly with respect to time, and then the growth rate slows and becomes gradual after the first year. The results show that the influence of concrete shrinkage and creep is significant in the early loading stage, whereas this influence is very small after the first year.

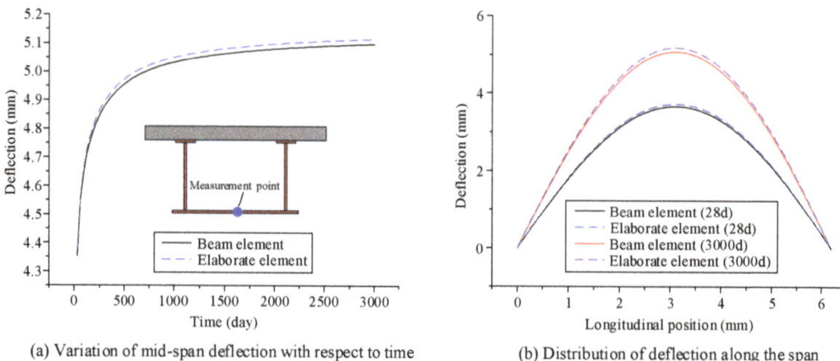

(a) Variation of mid-span deflection with respect to time

(b) Distribution of deflection along the span

Figure 4. Deflection variation in curved composite box girder, with respect to time: (**a**) variation of mid-span deflection with respect to time, (**b**) distribution of deflection along the span.

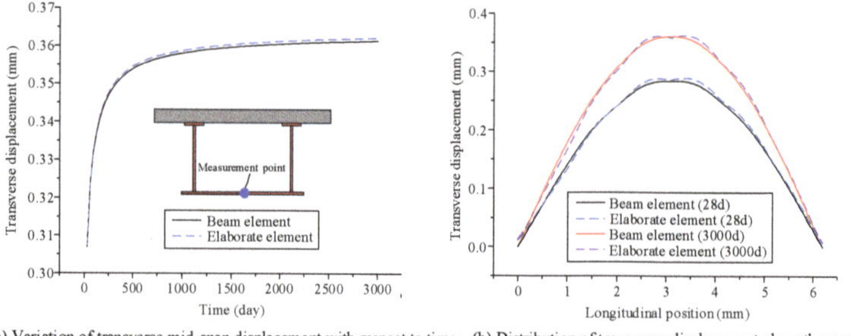

(a) Variation of transverse mid-span displacement with respect to time   (b) Distribution of transverse displacement along the span

Figure 5. Transverse displacement variation in the curved composite box girder, with respect to time: (**a**) variation of transverse mid-span displacement with respect to time, (**b**) distribution of transverse displacement along the span.

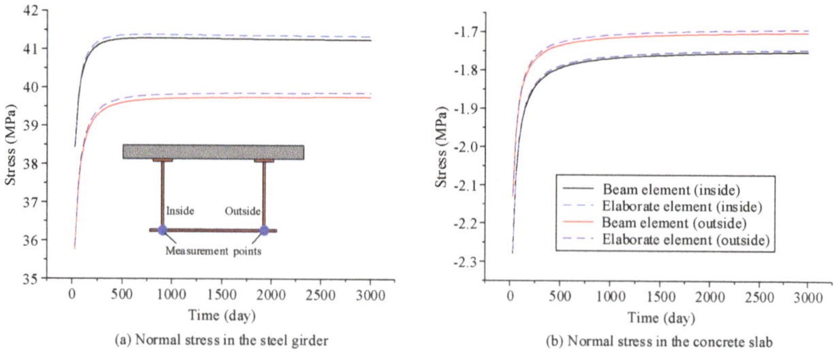

(a) Normal stress in the steel girder

(b) Normal stress in the concrete slab

Figure 6. Stress variation in the steel girder and concrete slab of the curved composite box beam, with respect to time: (**a**) normal stress in the steel girder, (**b**) normal stress in the concrete slab.

4. Parametric Analysis

4.1. Influence of the Time-Dependent Effects on Shear Lag

The shear-lag coefficient, λ, is introduced to reflect the influence of the shear-lag effect of curved composite beams. The shear-lag coefficient, λ, is defined as ratio of the stress at the intersection of the web and flange to that at the same location when the shear-lag effect is ignored. Based on the curved composite box girder in Section 3, which is used to verify the accuracy of the proposed model, the changes in the shear-lag coefficient of the concrete slab and steel bottom flange with respect to time are calculated by using the proposed finite beam element model, as shown in Figures 7 and 8, respectively. The results show that the shear-lag coefficient at the inside and the outside of the concrete slab gradually diminishes over time. A small change in the shear-lag coefficient at the inside and outside of the steel bottom flange occurs in the curved composite beams over time, and it tends to stabilize in the late stage of development. The effects of concrete shrinkage and creep can weaken the shear-lag effects in curved composite box girders.

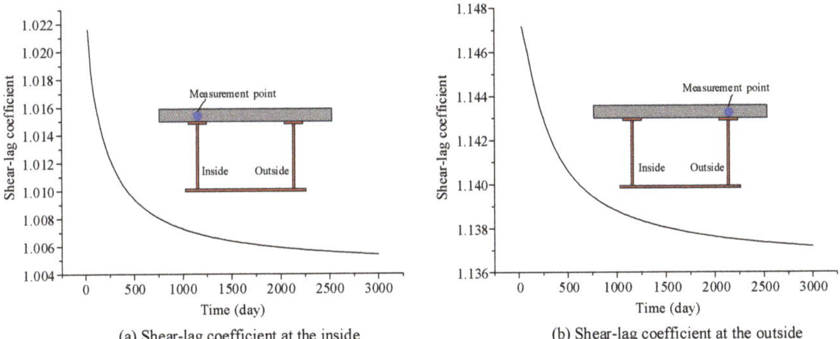

Figure 7. Change in the shear-lag warping strength function of the concrete slab, with respect to time: (**a**) shear-lag coefficient at the inside; (**b**) shear-lag coefficient at the outside.

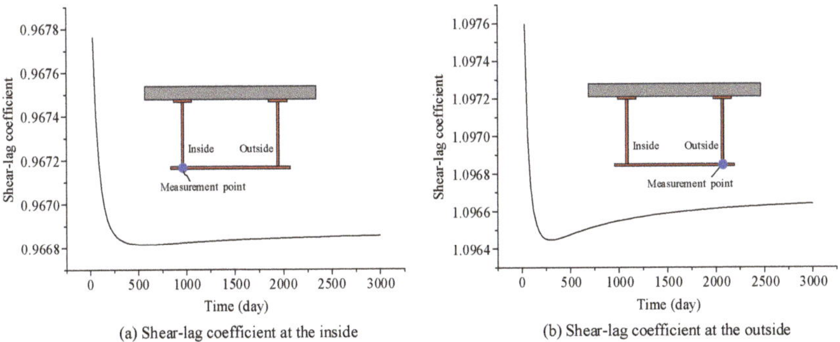

Figure 8. Change in the shear-lag warping strength function of the steel bottom flange, with respect to time: (**a**) shear-lag coefficient at the inside, (**b**) shear-lag coefficient at the outside.

4.2. Influence of Time-Dependent Effects on Mechanical Behavior for Different Initial Curvatures

The initial central angles are set as 0°, 30°, and 60°, based on the curved composite box girder in Section 3, which is applied to verify the accuracy of the model. According to the proposed finite beam element model, the deflection, transverse displacement, and longitudinal displacement distributions along the span are calculated for a curved composite box girder at the initial and ultimate loading ages, as shown in Figure 9. The normal stress distributions along the transverse direction are calculated for the concrete slab and steel bottom flange at the initial and ultimate loading ages, as shown in Figure 10. Figure 9 shows that the maximum growth rates of deflection for central angles of 0°, 30°, and 60° are 30.5%, 29.2%, and 25.8%, respectively. The maximum growth rates of longitudinal displacement for central angles of 0°, 30°, and 60° are 20.3%, 20.6%, and 21.6%, respectively. The maximum growth rates of transverse displacement of the curved composite box girder with central angles of 30° and 60° are 47.2% and 45.1%, respectively. Figure 10 illustrates the stress distributions of the curved composite box girder with central angles of 0°, 30°, and 60° due to the influence of concrete shrinkage and creep. The tensile stress growth rates at the inside of the steel bottom flange for central angles of 0°, 30°, and 60° are 50.6%, 39.0%, and 18.6%, respectively. The compressive stress reduction rates at the inside of the concrete slab for central angles of 0°, 30°, and 60° are 58.8%, 57.8%, and 53.0%, respectively.

It can be concluded that the growth rate of displacement, caused by concrete shrinkage and creep effect, becomes small with the increase in the initial curvature of the curved box girder. The tensile stress growth rate of the steel girder decreases gradually with increasing initial curvature. The compressive stress reduction rate of the concrete slab decreases gradually with increasing initial curvature.

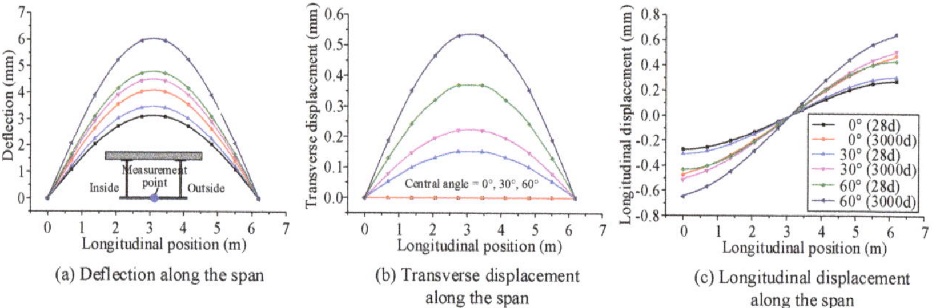

Figure 9. Displacement distribution along the span of the curved composite box girder, with different central angles at the initial loading age and ultimate loading age: (**a**) deflection along the span, (**b**)transverse displacement along the span, (**c**) longitudinal displacement along the span.

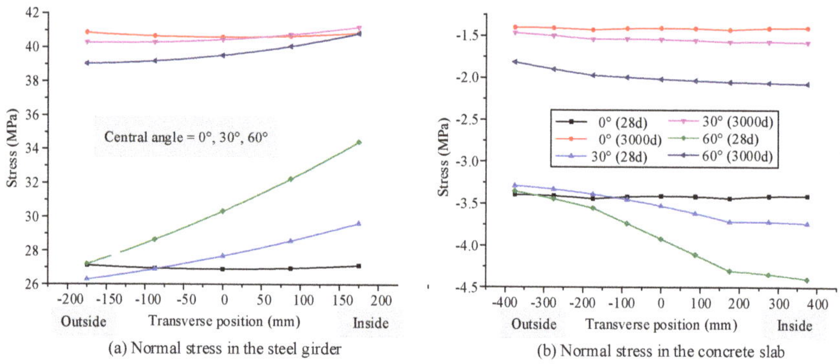

Figure 10. Normal stress distribution along the transverse direction of the curved composite box girder, with different central angles at the initial loading age and ultimate loading age: (**a**) normal stress in the steel girder, (**b**) normal stress in the concrete slab.

4.3. Influence of the Time-Dependent Effects on Mechanical Behavior for Different Interfacial Shear Connection Stiffness Values

The curved composite box girder in Section 3 was selected as the research object. The interfacial shear connection stiffness values are set as 0.1, 1, and 10 N/mm^3. According to the proposed finite beam element model, the deflection and transverse and longitudinal displacement distributions along the span of the curved composite box girder at the initial and ultimate loading ages are calculated, as shown in Figure 11. The normal stress distributions along the transverse direction are calculated for the concrete slab and steel bottom flange at the initial loading age and ultimate loading age, as shown in Figure 12. Figure 11 shows that the maximum deflection growth rates are 54.4%, 32.3%, and 28.9% for curved composite box girders with shear connection stiffness values of 0.1, 1, and 10 N/mm^3, respectively, due to the influence of concrete shrinkage and creep. Moreover, for shear connection stiffness values of 0.1, 1, and 10 N/mm^3, the maximum growth rates of transverse displacement are 116.1%, 67.2%, and 49.2%, respectively, whereas the maximum growth rates of longitudinal displacement are 86.9%, 75.8%, and 71.3%, respectively. Figure 12 shows that the tensile stress growth rates at the inside of the steel bottom flange are 47.1%, 35.2%, and 37.8% for curved composite box girders with shear connection stiffness values of 0.1, 1, and 10 N/mm^3, respectively, due to the influence of concrete shrinkage and creep. Under these same stiffness values, the compressive stress reduction rates at the inside of the concrete slab are 44.6%, 59.4%, and 57.1%, respectively.

It can be concluded that the growth rate of displacement, which is caused by concrete shrinkage and creep, becomes small as the interfacial shear connection stiffness of the curved composite box girder increases. The tensile stress growth rate of the steel girder decreases with increasing interfacial shear connection stiffness. The compressive stress reduction rate of the concrete slab becomes large with increasing interfacial shear connection stiffness.

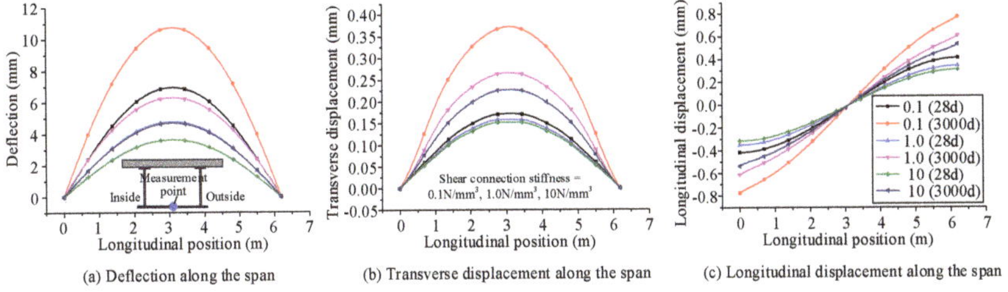

(a) Deflection along the span   (b) Transverse displacement along the span   (c) Longitudinal displacement along the span

Figure 11. Displacement distribution along the span of curved composite box beam at the initial and ultimate loading ages for different interfacial shear connection stiffness: (**a**) deflection along the span, (**b**) Transverse displacement along the span, (**c**) longitudinal displacement along the span.

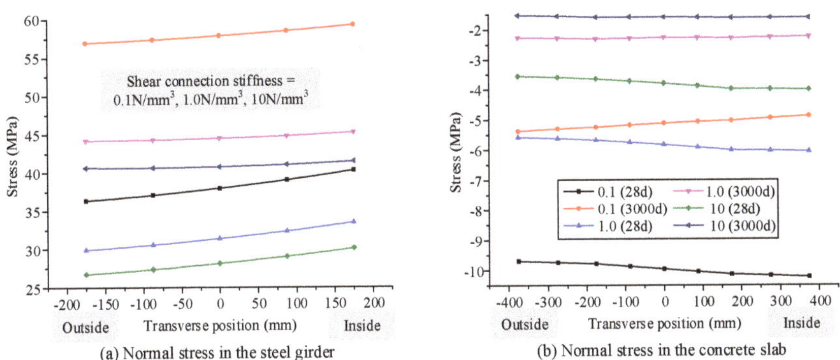

(a) Normal stress in the steel girder   (b) Normal stress in the concrete slab

Figure 12. Normal stress distribution along the transverse direction of the curved composite box girder at the initial and ultimate loading ages for different interfacial shear connection stiffness: (**a**) normal stress in the steel girder, (**b**) normal stress in the concrete slab.

5. Conclusions

(1) Based on the Vlasov's one-dimensional theoretical model of curved beams, a torsional warping function, a distortional angle, transverse and longitudinal slip functions, a shear-lag intensity function

for the concrete slab, a shear-lag intensity function for the steel beam, and constitutive relationships for concrete shrinkage and creep were introduced. The variation of the curvature in the radial direction was also taken into account. The equilibrium equation for the curved composite box beam was proposed according to the virtual work principle. Then, the stiffness matrix and equivalent nodal load vector of the two-node finite beam element with 26 DOFs, which analyzes the long-term behavior of curved composite box beams, were obtained by adopting the finite element method and the age-adjusted effective modulus method.

(2) An elaborate finite element model of a typical curved composite box beam was established. The correctness and applicability of the two-node finite beam element with 26 DOFs were verified through a comparison of the results from the finite beam element model and those from the elaborate finite element model.

(3) The proposed finite beam element model was adopted to analyze the influence of the concrete time-dependent effects on the long-term behavior of curved composite box beams. The analysis shows that significant changes of the displacement, stress, and shear-lag coefficient occur in the curved composite beams within the first year of loading, due to the time-dependent effects of concrete, after which the variation tendency becomes gradual.

(4) The proposed finite beam element model was adopted to analyze the influence of the concrete shrinkage and creep effects on the long-term behavior of curved composite box beams under different central angles and shear connection stiffness. The results show that increases of the central angle and the shear connection stiffness reduce the change rates of both displacement and stress, with respect to time.

Author Contributions: Data curation, X.-L.J.; Formal analysis, G.-M.W. and L.Z.; Funding acquisition, L.Z.; Investigation, L.Z.; Resources, W.-Y.J. All authors have read and agreed to the published version of the manuscript.

Funding: This research was funded by the Transportation Science and Technology Program of Hebei Province, China (grant number QG2018-8), the Science and Technology Program of Qinghai Province, China (grant number 2017-SF-139), and the National Natural Science Foundation of China (grant number 51608031).

Acknowledgments: The authors express their thanks to the people helping with this work, and acknowledge the valuable suggestions from the peer reviewers.

Appendix A

A quadratic polynomial is used to characterize the shear-lag warping shape function, $\psi_c(x)$, of the concrete slab and the shear-lag warping shape function, $\psi_s(x)$, of the steel beam, which are expressed in Equations (A1) and (A2), respectively. The physical meaning of the polynomial is the shear-lag warping displacement at any position on the section of the concrete slab or steel beam when a unit shear-lag warping intensity occurs.

$$\psi_c(x) = \begin{cases} \left[1 - \left(\frac{x+b_c}{b_c-b_t}\right)^2\right]\left(\frac{b_c-b_t}{b_t}\right)^2 & x > b_t \\ 1 - \left(\frac{x}{b_t}\right)^2 & -b_t \leq x \leq b_t \\ \left[1 - \left(\frac{x-b_c}{b_c-b_t}\right)^2\right]\left(\frac{b_c-b_t}{b_t}\right)^2 & x < -b_t \end{cases} \tag{A1}$$

$$\psi_s(x) = \begin{cases} 0 & \text{top flange and web of the steel girder} \\ 1 - \left(\frac{x}{b_s}\right)^2 & \text{bottom flange of the steel girder} \end{cases} \tag{A2}$$

Appendix B

(1) The nonzero elements of matrix B_k are shown hereafter:

$B_k(1,1) = -k_0$; $B_k(1,8) = 1$; $B_k(1,9) = -y_s k_0$; $B_k(1,13) = -y_d k_0 \Psi_d$; $B_k(1,16) = -a_k k_0/2$; $B_k(1,19) = a_k/2$;
$B_k(2,3) = -1$; $B_k(2,8) = -k_0$; $B_k(2,19) = -a_k k_0/2$;

$B_k(3,6) = -1; B_k(3,9) = k_0; B_k(3,13) = k_0\Psi_d;$
$B_k(4,6) = -k_0; B_k(4,12) = -1;$
$B_k(5,15) = -1;$
$B_k(6,21) = 1;$
$B_k(7,23) = 1;$
$B_k(8,5) = k_0; B_k(8,10) = 1;$
$B_k(9,5) = -k_0; B_k(9,11) = -1;$
$B_k(10,20) = 1;$ and
$B_k(11,22) = 1.$
In these elements, $k = s,c$ represents the steel girder and concrete slab, respectively.
(2) The nonzero elements of matrix B_Ω are given hereafter:
$B_\Omega(1,16) = 1$ and $B_\Omega(2,18) = 1.$

Appendix C

The nonzero elements of the shape function matrix N are presented hereafter:
$N(1,1) = n_1; N(1,2) = n_2; N(1,14) = n_3; N(1,15) = n_4;$
$N(2,1) = n_1'; N(2,2) = n_2'; N(2,14) = n_3'; N(2,15) = n_4';$
$N(3,1) = n_1''; N(3,2) = n_2''; N(3,14) = n_3''; N(3,15) = n_4'';$
$N(4,3) = n_1; N(4,4) = n_2; N(4,16) = n_3; N(4,17) = n_4;$
$N(5,3) = n_1'; N(5,4) = n_2'; N(5,16) = n_3'; N(5,17) = n_4';$
$N(6,3) = n_1''; N(6,4) = n_2''; N(6,16) = n_3''; N(6,17) = n_4'';$
$N(7,5) = m_1; N(7,18) = m_2;$
$N(8,5) = m_1'; N(8,18) = m_2';$
$N(9,6) = m_1; N(9,19) = m_2;$
$N(10,6) = m_1'; N(10,19) = m_2';$
$N(11,7) = m_1; N(11,20) = m_2;$
$N(12,7) = m_1'; N(12,20) = m_2';$
$N(13,8) = n_1; N(13,9) = n_2; N(13,21) = n_3; N(13,22) = n_4;$
$N(14,8) = n_1'; N(14,9) = n_2'; N(14,21) = n_3'; N(14,22) = n_4';$
$N(15,8) = n_1''; N(15,9) = n_2''; N(15,21) = n_3''; N(15,22) = n_4'';$
$N(16,10) = m_1; N(16,23) = m_2;$
$N(17,10) = m_1'; N(17,23) = m_2';$
$N(18,11) = m_1; N(18,24) = m_2;$
$N(19,11) = m_1'; N(19,24) = m_2';$
$N(20,12) = m_1; N(20,25) = m_2;$
$N(21,12) = m_1'; N(21,25) = m_2';$
$N(22,13) = m_1; N(22,26) = m_2;$
$N(23,13) = m_1';$ and $N(23,26) = m_2'.$
The nonzero elements of the shape function matrix N_F are given hereafter:
$N_F(1,1) = n_1; N_F(1,2) = n_2; N_F(1,14) = n_3; N_F(1,15) = n_4;$
$N_F(2,3) = n_1; N_F(2,4) = n_2; N_F(2,16) = n_3; N_F(2,17) = n_4;$
$N_F(3,5) = m_1; N_F(3,18) = m_2;$
$N_F(4,6) = m_1; N_F(4,19) = m_2;$
$N_F(5,7) = m_1; N_F(5,20) = m_2;$
$N_F(6,8) = n_1; N_F(6,9) = n_2; N_F(6,21) = n_3; N_F(6,22) = n_4;$
$N_F(7,10) = m_1; N_F(7,23) = m_2;$
$N_F(8,11) = m_1; N_F(8,24) = m_2;$
$N_F(9,12) = m_1; N_F(9,25) = m_2;$
$N_F(10,13) = m_1;$ and $N_F(10,26) = m_2.$

In these elements, $n_1 = \left(1 + \frac{2z}{l_e}\right)\left(\frac{z-l_e}{l_e}\right)^2$; $n_2 = z\left(\frac{z-l_e}{l_e}\right)^2$; $n_3 = \left(1 - \frac{2(z-l_e)}{l_e}\right)\left(\frac{z}{l_e}\right)^2$; $n_4 = (z-l_e)\left(\frac{z}{l_e}\right)^2$; $m_1 = 1 - \frac{z}{l_e}$; and $m_2 = \frac{z}{l_e}$.

References

1. Pnevmatikos, N.; Sentzas, V. Preliminary estimation of response of curved bridges subjected to earthquake loading. *J. Civ. Eng. and Archit.* **2012**, *6*, 1530–1535.
2. Vlasov, V.Z. *Thin-Walled Elastic Beams*, 2nd ed.; Israel Program for Scientific Translation: Jerusalem, Israel, 1961.
3. Nakai, H.; Yoo, C.H. *Analysis and Design of Curved Steel Bridges*; McGraw-Hill Co.: New York, NY, USA, 1988.
4. Nakai, H.; Murayama, Y. Distortional stress analysis and design aid for horizontally curved box girder bridges with diaphragms in steel box girder bridges. *Proc. Jpn. Soc. Civ. Eng.* **1981**, *309*, 25–39. [CrossRef]
5. Arizumi, Y.; Hamada, S.; Oshiro, T. *Experimental and Analytical Studies on Behavior of Curved Composite Box Girders*; Bulletin of the Faculty of Engineering, University of the Ryukyus: Okinawa, Japan, 9 May 1987; pp. 175–195.
6. Zhu, L.; Wang, J.J.; Su, R.K.L. Finite beam element with 26 DOFs for curved composite box girders considering constrained torsion, distortion, shear lag and biaxial slip. *Eng. Struct.* **2020**, under review.
7. Bazant, Z.P. Numerical analysis of creep of an indeterminate composite beam. *J. Appl. Mech.* **1970**, *37*, 1161–1164. [CrossRef]
8. Tarantino, A.M.; Dezi, L. Creep effects in composite beams with flexible shear connectors. *J. Struct. Eng.* **1992**, *118*, 2063–2081. [CrossRef]
9. Dezi, L.; Tarantino, A.M. Creep in composite continuous beams. I: Theoretical treatment. *J. Struct. Eng.* **1993**, *119*, 2095–2111. [CrossRef]
10. Dezi, L.; Tarantino, A.M. Creep in composite continuous beams. II: Parametric Study. *J. Struct. Eng.* **1993**, *119*, 2112–2133. [CrossRef]
11. Dezi, L.; Ianni, C.; Tarantino, A.M. Simplified creep analysis of composite beams with flexible connectors. *J. Struct. Eng.* **1993**, *119*, 1484–1497. [CrossRef]
12. Dezi, L.; Leoni, G. Tarantino AM. Algebraic methods for creep analysis of continuous composite beams. *J. Struct. Eng.* **1996**, *122*, 423–430. [CrossRef]
13. Dezi, L.; Gara, F.; Leoni, G.; Tarantino, A.M. Time-dependent analysis of shear-lag effect in composite beams. *J. Eng. Mech.* **2001**, *127*, 71–79. [CrossRef]
14. Dezi, L.; Leoni, G. Effective slab width in prestressed twin-girder composite decks. *J. Struct. Eng.* **2006**, *132*, 1358–1370. [CrossRef]
15. Gara, F.; Leoni, G.; Dezi, L. A beam finite element including shear lag effect for the time-dependent analysis of steel-concrete composite decks. *Eng. Struct.* **2009**, *31*, 1888–1902. [CrossRef]
16. Ranzi, G.; Bradford, M.A. Analytical solutions for the time-dependent behavior of composite beams with partial interaction. *Int. J. Solids Struct.* **2006**, *43*, 3770–3793. [CrossRef]
17. Ranzi, G.; Bradford, M.A. Analysis of composite beams with partial interaction using the direct stiffness approach accounting for time effects. *Int. J. Numer. Methods Eng.* **2009**, *78*, 564–586. [CrossRef]
18. Nguyen, Q.H.; Hjiaj, M.; Uy, B. Time-dependent analysis of composite beams with continuous shear connection based on a space-exact stiffness matrix. *Eng. Struct.* **2010**, *32*, 2902–2911. [CrossRef]
19. Li, G.H. Torsion and bending of thin-walled box girder with great initial curvature. *Chin. Civ. Eng. J.* **1987**, *20*, 65–75. (In Chinese)
20. Lu, P.Z. Triaxial Theoretic Analysis and Application Research of Steel-Concrete Composite Box Beams. Ph.D. Dissertation, Southwest Jiaotong University, Chengdu, China, 2010. (In Chinese)
21. Bazant, Z.P. Prediction of concrete creep effects using age-adjusted effective modulus method. *J. Am. Concr. Inst.* **1972**, *69*, 212–217.
22. Guo, J.Q.; Fang, Z.Z.; Zheng, Z. *Design Theory of Box Girder*; China Communication Press: Beijing, China, 2008. (In Chinese)

23. Li, M.J. Finite Beam Element Considering Multi-Mechanical, Geometrical and Time-Dependent Effects of Curved Composite Box-Shape Beams. Master's Dissertation, Beijing Jiaotong University, Beijing, China, 2019. (In Chinese)
24. Comite Euro-International du Beton-Federation International de la Precontrainte (CEB-FIP). *CEB-FIP Model Code 1990: Design Code*; Thomas Telford Ltd.: London, UK, 1993.

Article

Axial Compressive Behaviour of Square Through-Beam Joints between CFST Columns and RC Beams with Multi-Layers of Steel Meshes

Weining Duan [1], Jian Cai [1,2], Xu-Lin Tang [3], Qing-Jun Chen [1,2], Chun Yang [1,*] and An He [4]

1   School of Civil Engineering and Transportation, South China University of Technology, Guangzhou 510640, China; duan.wn@mail.scut.edu.cn (W.D.); cvjcai@scut.edu.cn (J.C.); qjchen@scut.edu.cn (Q.-J.C.)
2   State Key Laboratory of Subtropical Building Science, South China University of Technology, Guangzhou 510640, China
3   Guangzhou Jishi Construction Group Co., Ltd., Guangzhou 510115, China; xulintang@foxmail.com
4   School of Civil and Environmental Engineering, Nanyang Technological University, 50 Nanyang Avenue, Singapore 639798, Singapore; he.an@ntu.edu.sg
*   Correspondence: chyang@scut.edu.cn

Received: 11 May 2020; Accepted: 27 May 2020; Published: 29 May 2020

Abstract: The axial compressive behaviour of an innovative type of square concrete filled steel tube (CFST) column to reinforced concrete (RC) beam joint was experimentally investigated in this paper. The innovative joint was designed such that (i) the steel tubes of the CFST columns were completely interrupted in the joint region, (ii) the longitudinal reinforcements from the RC beams could easily pass through the joint area and (iii) a reinforcement cage, including a series of reinforcement meshes and radial stirrups, was arranged in the joint area to strengthen the mechanical performance of the joint. A two-stage experimental study was conducted to investigate the behaviour of the innovative joint under axial compression loads, where the first stage of the tests included three full-scale innovative joint specimens subjected to axial compression to assess the feasibility of the joint detailing and propose measures to further improve its axial compressive behaviour, and the second stage of the tests involved 14 innovative joint specimens with the improved detailing to study the effect of the geometric size of the joint, concrete strength and volume ratio of the steel meshes on the bearing strengths of the joints. It was generally found from the experiments that (i) the innovative joint is capable of achieving the design criterion of the 'strong joint-weak member' with appropriate designs, and (ii) by decreasing the height factor and increasing the volume ratio of the steel meshes, the axial compressive strengths of the joints significantly increased, while the increase of the length factor is advantageous but limited to the resistances of the joint specimens. Because of the lack of existing design methods for the innovative joints, new design expressions were proposed to calculate the axial compression resistances of the innovative joints subjected to bearing loads, with the local compression effect, the confinement effect provided by the multi-layers of steel meshes and the height effect of concrete considered. It was found that the proposed design methods were capable of providing accurate and safe resistance predictions for the innovative joints.

Keywords: through-beam joint; concrete filled steel tube (CFST) columns; reinforced concrete (RC); axial compressive behaviour; steel mesh; local compression; confined concrete; height factor

---

1. Introduction

The concrete-filled steel tube (CFST) structure is a high-efficiency solution for high-rise buildings and bridges due to its high bearing capacity, excellent ductility and great energy dissipation capacity [1].

In CFST structures, the joints connecting the CFST columns and reinforced concrete (RC) beams play an important role in achieving the structural integrity of the CFST frame [2,3], and thus researches related to the detailing of the joints have become a hot spot during the past few years. One of the typical joints connecting the CFST columns and RC beams is the through-column joint, in which steel cleats are attached to the outer surface of the steel tube of the CFST column at the joint area and the longitudinal reinforcements in the RC beams were welded on these steel cleats. The seismic behaviour and the failure modes of the through-column joints have been widely investigated [4–7], indicating that the through-column joints possess excellent seismic behaviour but require a significant amount of site welding.

Recently, through-beam joints connecting the CFST columns and RC beams have attracted the interest of researchers because of their simple detailing in the joint area, with a brief review of the recently developed through-beam joints summarised herein. Nie et al. [8] and Bai et al. [9] developed a representative through-beam joint connecting the concrete encased CFST columns and RC beams, where the steel tubes of the CFST columns were completely interrupted in the joint area, and the steel-reinforcement bars in the RC beams were continuous in the floor. Zhang et al. [10] developed a through-beam joint for concrete-filled double-skin steel tubes structural members, where the outer tube was interrupted in the joint area and an octagonal RC ring beam was arranged outside the joint to connect the reinforcements from the RC beams. Chen et al. [11,12] reported a new type of through-beam connection for CFST columns and RC beams, where the steel tube of the CFST column is completely or partially interrupted, and a ring beam was used to strengthen the load-carrying capacity of the joint. Zhou et al. [13,14] proposed a tubed-reinforced-concrete column to RC beam joint, where the steel tubes are completely interrupted in the joint area, with different manners to strengthen the joint, including the strengthening stirrup, horizontal hunches, and internal diaphragm. The seismic behaviour of the aforementioned through-beam joints was examined through cyclic loading tests. It was generally found that the through-beam joints are capable of achieving the design criterion of 'strong joint-weak member' with excellent seismic performance. Provided that the steel tubes of the CFST columns were generally interrupted at the joint area, axial compressive tests on the through-beam joints were also conducted by Nie et al. [8], Bai et al. [9], Chen et al. [11] and Zhou et al. [13], to investigate the axial compressive strengths of the joints; it was found that the strengths of the joints were higher than those of the corresponding CFST columns, and thus the design criterion of the 'strong joint-weak member' was deemed to be achieved. Recently, an innovative joint connecting the square CFST columns and RC beams has been developed by the authors [15]. The configuration and detailing of the innovative joint are displayed in Figure 1, where the outer steel tube was completely interrupted in the joint zone, and the longitudinal reinforcements from the RC beams could easily pass through the joint area. A reinforcement cage, including a series of reinforcement meshes and radial stirrups, was arranged in the joint area to strengthen the mechanical performance of the joint because of the interruption of the steel tube. The benefits of the proposed configuration of the innovative joints over the existing through-column and through-column beam joints include (i) the reinforcements from both structural columns and beams could easily pass through the joint area, ensuring both robustness and integrity of the joints, and (ii) the reinforcement cage in the joint can be precast and on-site welding is not required, leading to a simplified construction process. Cyclic loading tests on the innovative joints have been conducted by the authors [15] to study their seismic behaviour, while the axial compression behaviour of the innovative joints subjected to bearing loads remained unexplored and will be studied in the present paper.

Because of the lack of existing experimental investigation on the compressive behaviour of innovative joints, a two-stage experimental study was first conducted to investigate the compressive behaviour of the innovative joint subjected to bearing loads, where the first stage of the tests included three full-scale innovative joint specimens to assess the feasibility of the joint detailing and propose measures to further improve the axial compressive behaviour of the joint, and the second stage of the tests involved 14 innovative joint specimens with the improved detailing to further study the effects of the geometric size of the joint, concrete strength and volume ratio of the steel meshes on the bearing strengths of the joints. Because of the lack of existing design methods for innovative joint, new design expressions for the bearing strengths of the innovative joints were developed and validated by the experimental results.

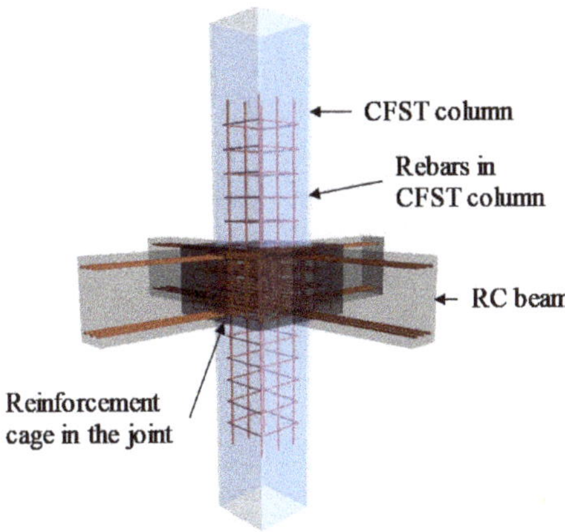

Figure 1. Square through-beam joint system.

2. Tests of Joint with CFST Columns (Series I Tests)

2.1. General

The Series I axial compression tests included three geometrically identical full-scale specimens, namely SC1, SC2, and SC3, and were designed with various rebar diameters of the steel meshes in the joints. Each specimen included the upper and lower CFST columns and the joint, as shown in Figure 2, while the slab and the RC beams were not fabricated because of the size limitation of the testing machine. Note that the longitudinal reinforcements of the RC frame beams were still arranged in the joint. Figure 3 and Table 1 report the geometric dimensions and reinforcements of each part of the specimens. Specifically, the longitudinal reinforcements of the column were inserted through the joint area with the anchorage lengths of 1000 mm in the CFST columns, as determined based on the Chinese building code [16]. To avoid the compressive stress concentrating at the end of the steel tube, a steel ring whose width and thickness were 18 mm and 6 mm, respectively, was welded to each end of the steel tubes. Five layers of steel meshes with the adjacent distance of 75 mm were arranged in the joint area. The diameters of the rebars in the steel meshes were 11 mm, 8.5 mm and 5.5 mm for specimens SC1, SC2 and SC3, respectively, as shown in Figure 3.

Table 1. Specimen parameters and the experimental and calculated loads for Series I tests.

Specimen	$D \times t$ (mm × mm)	$A \times A \times H$ (mm × mm × mm)	$f_{cu,u}$ (MPa)	f_{cu} (MPa)	$f_{cu,l}$ (MPa)	d_{bar} (mm)	ρ_v (%)	N_{cr} (kN)	N_u (kN)	N_{cr}/N_u	N_{cal} (kN)	N_{cal}/N_u
SC1	500 × 6	1000 × 1000 × 600	46.3	40.8	44.7	11	1.65	6000	10,534	0.57	20,112	1.91
SC2	500 × 6	1000 × 1000 × 600	46.3	40.8	44.7	8.5	0.99	5300	12,004	0.44	18,395	1.53
SC3	500 × 6	1000 × 1000 × 600	46.3	40.8	44.7	5.5	0.41	5100	11,404	0.45	15,954	1.40

Notes: D and t are diameter and thickness of steel tube; A and H are width and height of the joint, respectively; $f_{cu,u}$, f_{cu} and $f_{cu,l}$ are the cubic compressive strengths of the concrete for upper column, joint area and lower column, respectively; d_{bar} is the diameter of rebar in the steel meshes; ρ_v is the volume ratio of steel mesh in the joint, which is calculated as $V_{sm}/(A \times A \times H)$; V_{sm} is the total volume of steel meshes in the joint; N_{cr} is the load when the first crack appeared; N_u is the peak load of the specimen; N_{cal} is the calculated strength obtained by Equations (6)–(8).

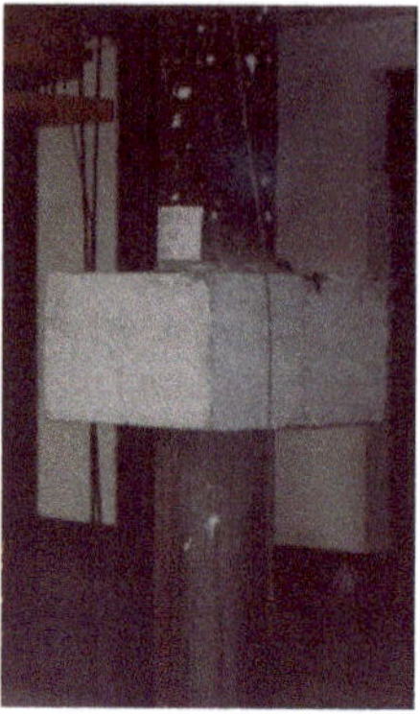

Figure 2. Overview of the specimens.

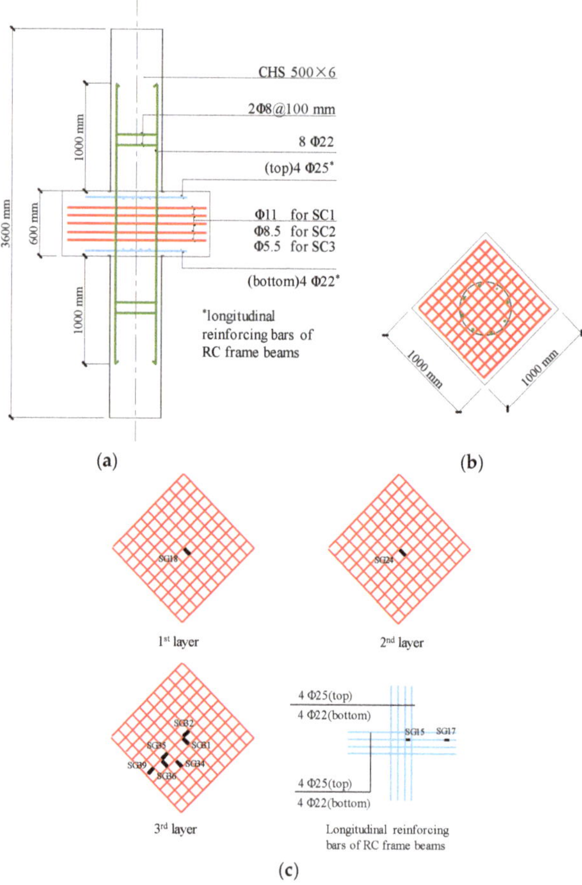

Figure 3. Details of the specimens in Series I (**a**) Elevation view; (**b**) Top view; (**c**) Steel meshes and beam rebars in the joints (SG: strain gauge).

During the fabrication of the specimens, four concrete cube specimens were respectively reserved when casting the concrete of the lower CFST column, joint area and upper CFST column. Compression tests were conducted on the concrete cubes to derive their cubic strengths at the day of the column tests, with the average cubic compressive strength of each part of the specimen reported in Table 1. Tensile tests were conducted to derive the material properties of the reinforcements and the steel tube, with the key results from the tensile tests reported in Table 2. The Young's modulus of the steels may be taken as 200,000 MPa, according to the Chinese building code [16].

Table 2. Measured material properties of reinforcing bars and steel tube for Series I test.

Member	Bar Φ5.5	Bar Φ8	Bar Φ8.5	Bar Φ11	Bar Φ22	Bar Φ25	Steel Tube
Yield strength f_y (MPa)	517	318	510	448	386	368	330
Ultimate strength f_u (MPa)	578	460	608	540	575	575	460

2.2. Experimental Test Setup

The Changchun CSS-254 15 MN universal testing machine was employed to apply axial compression force to the specimens, with the test setup shown in Figure 4. During the tests, four LVDTs were arranged in the joint area and the CFST columns to measured axial shortening of the specimen (see Figure 4), and a series of strain gauges were attached to the steel meshes in the joint to measure their deformations during loading (See Figure 3). Both load and displacement control modes were employed to drive the testing machine. Specifically, the initial loading rate was set to be equal to 5 kN/s until the material yielding was observed, according to the readings from the attached strain gauges, after which a displacement rate of 2 mm/min was used to complete the post-yield loading stage. The tests were terminated when the resistance of the specimen had deteriorated below 85% of its maximum strength.

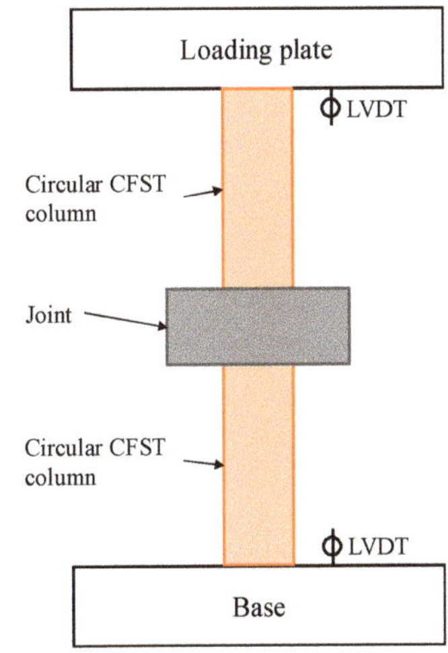

Figure 4. Schematic diagram of the test setup.

2.3. Experimental Results

2.3.1. Failure Modes

All the specimens exhibited similar overall behaviour under axial compression during the tests, as summarised below. The first vertical crack was found on the side surface of the joint when the upper load reached 44~57% of the peak loads. With the load increased, the horizontal cracks occurred and gradually crossed the first vertical crack. The dilation and local buckling of the steel tubes were observed when the specimens reached their peak loads, while the concrete in the joint area remained intact without series spalling, which indicates that the specimens were failed by the local buckling of outer steel tubes of the CFST columns and thus satisfied the design criterion of the 'strong joint-weak member'. The crack patterns on each side surface of the joint at failure are shown in Figure 5.

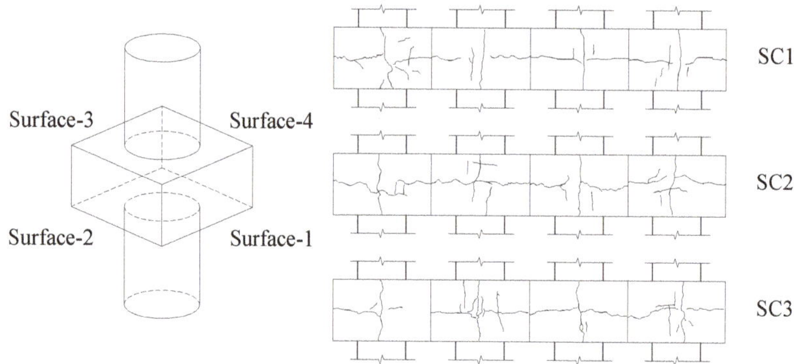

Figure 5. Cracks on the side surfaces of the specimens.

Taken specimen SC2 as an example to illustrate the experimental observation in detail, when the load reached about 45% of N_u (N_u is the peak load of the specimen, for SC2 N_u = 12,004 kN), the first crack was found at the middle of the side surface of the joint in the vertical direction. As the load increased to about 50% of N_u, the vertical crack developed with a maximum width of about 0.1 mm, and a short horizontal crack occurred on the side surface of the concrete at the same time. At about 70% of N_u, the maximum widths of the vertical and horizontal cracks on the side surface of the joint reached 0.25 mm and 0.15 mm, respectively. When the specimen reached its maximum load N_u, dilation and local buckling of the steel tubes of the CFST columns were observed. At the post-peak loading stage, the steel tubes of the CFST columns continued to dilate with the load capacity of the specimen gradually dropped to about 85% N_u.

2.3.2. Load-Deformation and Load-Strain Curves

The axial load-deformation curves of the three specimens are shown in Figure 6, where the vertical axis is the force applied at two ends of the columns, and the horizontal axis is the displacement of the whole specimen consisted of the upper and lower CFST columns and the joint zone. As shown in Figure 6, the initial elastic stiffnesses and the shapes of the curves for the three specimens are quite close to each other at the elastic stages. After reaching the peak loads of the specimens, the load-deformation curves descend gradually and show ductile post-peak behaviour. The loads corresponding to the first cracks N_{cr} and the peak loads N_u for all specimens are listed in Table 1. It can be seen that the value of N_{cr} for SC1 is 1.13 and 1.18 times than those of SC2 and SC3 respectively, which indicates that the crack resistance of the joint is enhanced by improving the volume ratio of steel meshes in the joint (ρ_v). However, the ultimate axial compressive strength of SC1 is smaller than those of SC2 and SC3. It may be attributed to the discreteness of concrete material in the CFST columns, as all specimens eventually failed in the CFST columns.

The load-strain relationships of the steel meshes in all specimens are similar, as Figure 7 shows. It can be found that the strains of the steel meshes grew slowly in the early loading phase. As the upper load increased to about 50% of N_u when the first crack appeared, the strains developed dramatically and subsequently reached the yield strains at the peak loads. It was also found that as the volume ratio of the steel meshes in the joint decreases, the strain values corresponding to the peak loads increase, which exhibits more adverse elongation deformation of the steel meshes.

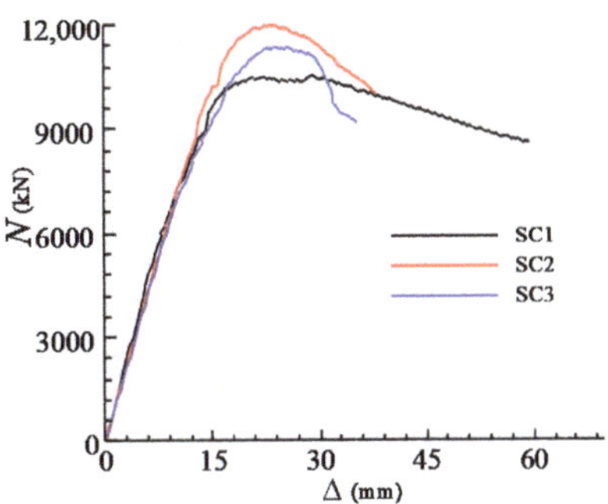

Figure 6. Axial load and displacement relationship.

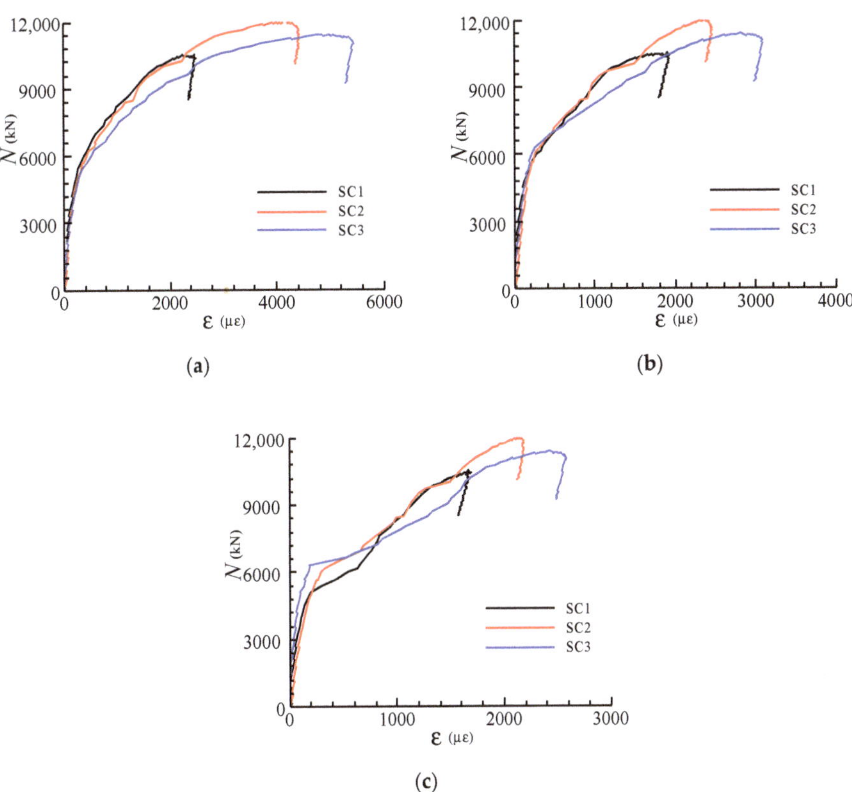

Figure 7. Load (*N*)-strain (ε) curves for steel meshes (**a**) SG24; (**b**) SG34; (**c**) SG36.

Figure 8 shows the load-strain curves for the steel meshes at different layers in specimen SC1. The strain gauges were arranged at the centre of the cross-section of the joints (see Figure 3). It can be seen that the strains in the steel meshes at the mid-height of the joint are larger than those in the upper and lower layers. Besides, the steel meshes reach their tensile yield strains at the peak load except the one located in the first layer (SG18).

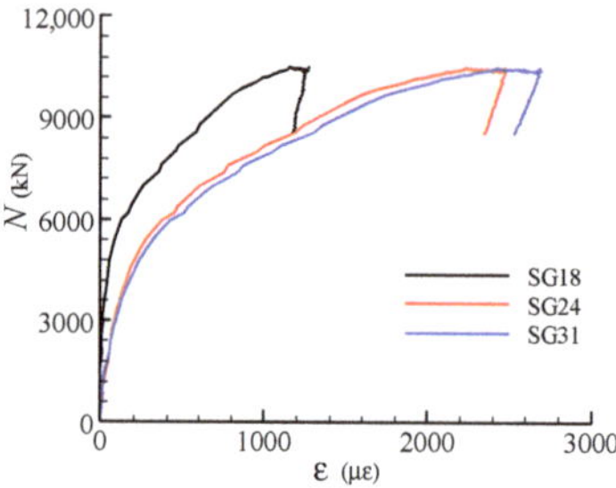

Figure 8. Load-strain curves for steel meshes at different layers in specimen SC1.

Figure 9 shows the load-strain curves for the steel mesh at the third layer with different plan positions in specimen SC1. It can be observed that the strains which are closer to the central section exhibit more serious tensile deformation. In addition, the steel meshes suffered from small compressive strain at the outermost positions (SG32) in the early loading stage.

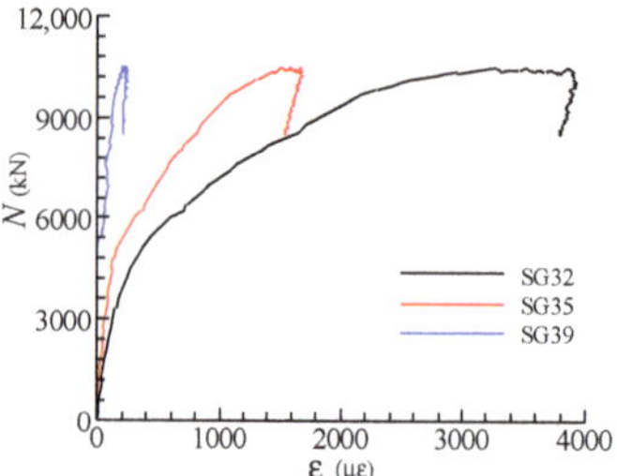

Figure 9. Load-strain curves for steel meshes at different plan positions in specimen SC1.

Figure 10 illustrated the load-strain histories for the longitudinal bars in RC frame beam for specimen SC2. It can be found that the tensile strain of the longitudinal bars is 900 $\mu\varepsilon$ at the peak loads and did not reach its tensile yield strain, which indicates that the longitudinal bars could provide favourable but limited confinement to the concrete of the joint.

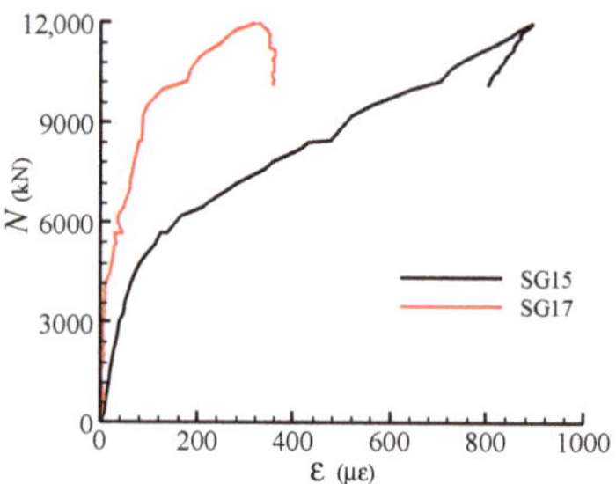

Figure 10. Load-strain curves for longitudinal reinforcing bars of the reinforced concrete (RC) beams.

2.4. Summary for the Series I Tests

According to the experimental results and analysis of the Series I tests, it can be concluded that the axial load carrying capacity of the joint can be higher than those of the CFST columns by proper design, which verifies the feasibility of the innovative joint. However, the multi-layers of steel meshes in the joint is difficult to rig up because of the lack of efficient supports. Moreover, square CFST columns are extensively used as structural members due to aesthetic consideration. In order to improve the construction operability of the joints, an improved joint detailing specifically for square CFST columns is developed. In the improved joint detailing, the stirrups were added in and arranged radially around the multi-layers of steel meshes to assemble into a steel cage, as shown in Figure 11. Therefore, the axial compression behaviour of the joint with improved detailing which comprises of the multi-layers of steel meshes and radial stirrups will be further studied in the following section.

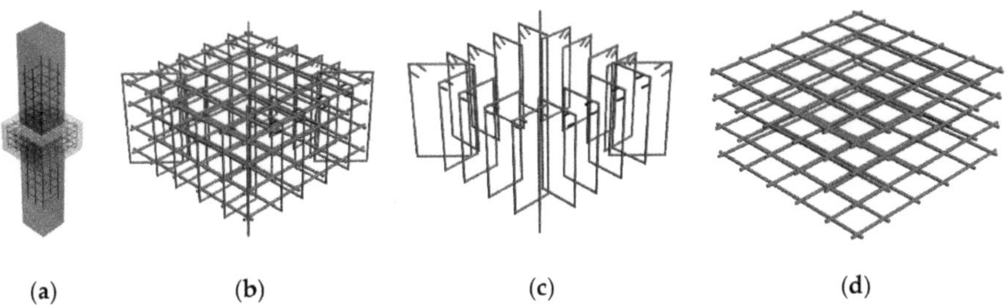

(a)   **(b)**   **(c)**   **(d)**

Figure 11. New type of joint between square concrete filled steel tube (CFST) columns and RC beams (a) Joints; (b) Steel cage of joint; (c) Radial stirrups; (d) Multi layers of steel meshes.

3. Axial Compression Tests on Innovative Joints (Series II Tests)

3.1. General

In order to study the axial compressive behaviour of the innovative joints with the improved detailing of the steel meshes, the tests of Series II were conducted subsequently. A total of 14 full-scale specimens were designed and fabricated. The details of the joint specimens are shown in Figure 12 and Table 3. The influence factors of Series II tests include the volume ratio of the steel meshes in the joint (ρ_v), the dimension of the joint and the concrete strength. The test specimens were named based on the following parameters, where C20 and C30 represent the cubic compressive strengths of the concretes are 15.34 MPa and 32.96 MPa respectively, L1, L2 and L3 refer to α = 1.6, 1.8 and 2.0 respectively; H1, H2 and H3 indicate β = 0.6, 0.8 and 1.0 respectively; and S1, S2 and S3 represent ρ_v = 1.0%, 1.5% and 2.0% respectively.

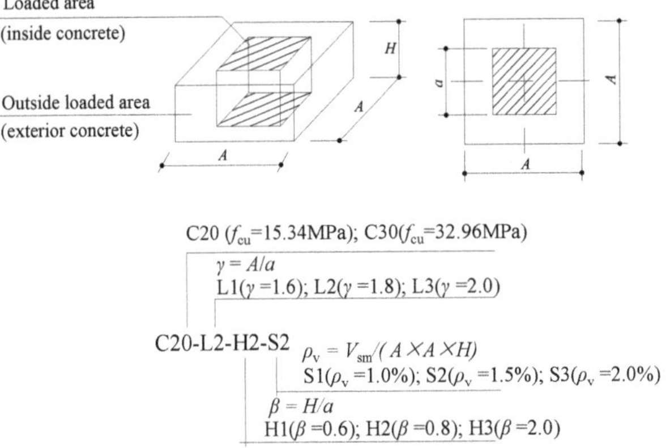

Figure 12. Specimens for Series II tests.

Table 3. Specimen parameters for Series II tests.

Specimen	$A \times A \times H$ (mm × mm × mm)	$a \times a$ (mm)	f_{cu} (MPa)	γ	β	ρ_v (%)	Layers of Steel Meshes	1st Layer Steel Mesh	2nd Layer Steel Mesh	3rd Layer Steel Mesh	4st Layer Steel Mesh
C20-L1-H3-S2	480 × 480 × 300	300 × 300	15.35	1.6	1.0	1.5	4	6Φ8	6Φ8	6Φ8	6Φ8
C20-L2-H1-S2	540 × 540 × 180	300 × 300	15.35	1.8	0.6	1.5	3	6Φ8	4Φ6 + 2Φ8	6Φ8	-
C20-L2-H2-S2	540 × 540 × 240	300 × 300	15.35	1.8	0.8	1.5	4	6Φ8	3Φ6 + 3Φ8	3Φ6 + 3Φ8	6Φ8
C20-L2-H3-S1	540 × 540 × 300	300 × 300	15.35	1.8	1.0	1.0	4	6Φ6	5Φ8 + 1Φ6	5Φ8 + 1Φ6	6Φ6
C20-L2-H3-S2	540 × 540 × 300	300 × 300	15.35	1.8	1.0	1.5	4	6Φ6	3Φ8 + 3Φ10	3Φ8 + 3Φ10	6Φ6
C20-L2-H3-S3	540 × 540 × 300	300 × 300	15.35	1.8	1.0	2.0	4	6Φ10	1Φ8 + 5Φ10	1Φ8 + 5Φ10	6Φ10
C20-L3-H3-S2	600 × 600 × 300	300 × 300	15.35	2.0	1.0	1.5	4	6Φ8	5Φ10 + 1Φ8	5Φ10 + 1Φ8	6Φ8
C30-L2-H1-S2	540 × 540 × 180	300 × 300	32.96	1.8	0.6	1.5	3	6Φ8	4Φ6 + 2Φ8	6Φ8	-
C30-L2-H2-S1	540 × 540 × 240	300 × 300	32.96	1.8	0.8	1.0	4	6Φ6	5Φ6 + 1Φ8	5Φ6 + 1Φ8	6Φ6
C30-L2-H2-S2	540 × 540 × 240	300 × 300	32.96	1.8	0.8	1.5	4	6Φ8	3Φ6 + 3Φ8	3Φ6 + 3Φ8	6Φ8
C30-L2-H2-S3	540 × 540 × 240	300 × 300	32.96	1.8	0.8	2.0	4	6Φ10	4Φ8 + 2Φ6	4Φ8 + 2Φ6	6Φ10
C30-L2-H3-S1	540 × 540 × 300	300 × 300	32.96	1.8	1.0	1.0	4	6Φ6	5Φ8 + 1Φ6	5Φ8 + 1Φ6	6Φ6
C30-L2-H3-S2	540 × 540 × 300	300 × 300	32.96	1.8	1.0	1.5	4	6Φ8	3Φ8 + 3Φ10	3Φ8 + 3Φ10	6Φ8
C30-L2-H3-S3	540 × 540 × 300	300 × 300	32.96	1.8	1.0	2.0	4	6Φ10	1Φ8 + 5Φ10	1Φ8 + 5Φ10	6Φ10

Notes: A and H are width and height of the joint, respectively; f_{cu} is the cubic compressive strength of the concrete; γ is the length factor, which is calculated as A/a; β is the height factor and is calculated as $\beta = H/a$, ρ_v is the volume ratio of steel mesh in the joint and is calculated as $V_{sm}/(A \times A \times H)$, where V_{sm} is the total volume of steel meshes in the joint.

The specimens were all loaded concentrically. The compression zones on surfaces of the joints were square with the side length *a* of 300 mm. In order to examine the actual bearing strengths of the joints, the upper and bottom CFST columns adopted in the Series I tests were removed and replaced by a pair of thick steel plates, with their cross-section dimensions identical with the local compression area. A total of 3 (β = 0.6) or 4 (β = 0.8 and 1.0) layers of steel meshes were horizontally and symmetrically arranged in the joint area. The diameters of the rebars in each steel mesh were determined according to the volume ratio of the steel meshes in the joint (ρ_v) and the joint volume, with the detailed steel bars adopted in each layer of the steel mesh listed in Table 3. It should be noted that during the fabrication of the steel meshes, the steel bars with larger diameters were prior to being arranged in the central of the joint area. The concrete protective cover was 20 mm. The diameter of the stirrup is 6 mm. The mechanical properties of the steel bars are summarized in Table 4, while the concrete cubic strengths for C20 and C30 grades were 15.34 MPa and 32.96 MPa, respectively.

Table 4. Measured material properties of reinforcing bars for Series II tests.

Member	Series C20			Series C30		
	BarΦ6	Bar Φ8	Bar Φ10	Bar Φ6	Bar Φ8	Bar Φ10
Yielding strength f_y (MPa)	259.9	472	422	260	300	340
Ultimate strength f_u (MPa)	430	536.1	480.4	440	340	430

3.2. Experimental Test Setup

The test setup for the Series II tests is shown in Figure 13, where a pair of thick steel plates with their cross-section sizes corresponding to the local compression zone were respectively attached to the top and bottom surfaces of the specimen to simulate the actions of the CFST columns, and four LVDTs were adopted to record the displacements between the steel plates. Additionally, strain gauges were attached to the steel meshes and radial stirrups, to measure the deformations of the rebars in the joint. The loading rate utilised in the Series II tests was the same as that used for the Series II tests, as described in Section 2.2.

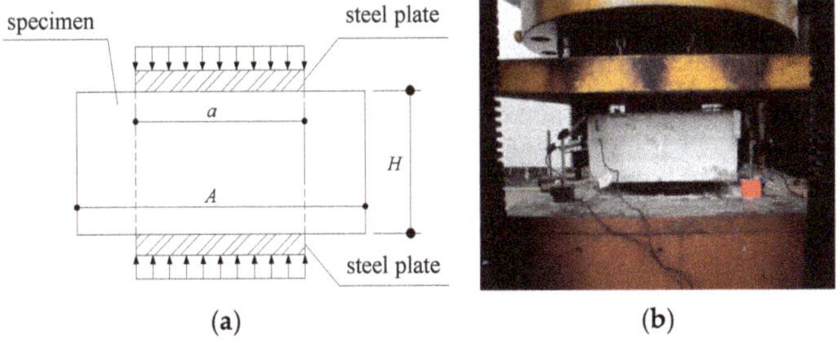

Figure 13. Test setup for the Series II tests (a) Elevation View; (b) Photo of the setup.

3.3. Failure Modes

Table 5 summarises the key experimental observations, including the loads corresponding to the appearance of the first crack N_{cr}, the loads when the maximum crack width developed to 0.2 mm $N_{0.2}$ and the loads when the maximum crack width developed to 0.3 mm $N_{0.3}$. It can be concluded that all the specimens had similar performance during the tests, with cracking of the concrete, yielding of the steel meshes and eventual spalling and crushing of the concrete. At the ultimate loads, several vertical penetrating cracks occurred on the surface of the joint. However, the distributions of the horizontal cracks were different with various height factor β. Accordingly, three different crack patterns (i.e., Mode I, Mode II and Mode III) on the surface of the specimens can be classified according to the distributions of the horizontal cracks, as shown in Figure 14. As the height factor β increases from 0.6 to

1.0, the horizontal cracks begin to generate, and the number of horizontal penetrating cracks gradually rises. The crack pattern modes of all the 14 specimens are summarised in Table 5.

Table 5. Ultimate strength, characteristic load and failure modes for Series II tests.

Specimens	N_{cr} (kN)	N_{cr}/N_u	$N_{0.2}$ (kN)	$N_{0.2}/N_u$	$N_{0.3}$ (kN)	$N_{0.3}/N_u$	N_u (kN)	Crack Mode	N_{cal} (kN)	N_{cal}/N_u
C20-L1-H3-S2	300	0.122	800	0.325	1000	0.406	2465	III	3097	1.26
C20-L2-H1-S2	400	0.116	1100	0.320	1500	0.437	3435	I	3763	1.10
C20-L2-H2-S2	600	0.178	1400	0.416	1500	0.445	3369	II	3592	1.07
C20-L2-H3-S1	600	0.280	1200	0.559	1400	0.652	2146	III	3103	1.45
C20-L2-H3-S2	600	0.240	1000	0.400	1400	0.560	2498	III	3484	1.39
C20-L2-H3-S3	600	0.221	750	0.277	1000	0.369	2712	III	3813	1.41
C20-L3-H3-S2	600	0.235	1050	0.412	1400	0.549	2550	III	3871	1.52
C30-L2-H1-S2	800	0.103	1100	0.141	1800	0.231	7801	I	6845	0.88
C30-L2-H2-S1	900	0.156	2100	0.363	2700	0.467	5782	II	5938	1.03
C30-L2-H2-S2	1200	0.184	2400	0.369	3300	0.507	6505	II	6380	0.98
C30-L2-H2-S3	1200	0.180	2400	0.360	2700	0.405	6659	II	6789	1.02
C30-L2-H3-S1	900	0.193	1500	0.321	2400	0.514	4673	III	5612	1.20
C30-L2-H3-S2	900	0.162	1500	0.270	1800	0.324	5550	III	6081	1.10
C30-L2-H3-S3	900	0.147	1500	0.245	1800	0.294	6122	III	6512	1.06

Notes: N_{cr} is the load when the first crack appeared, N_u is the peak load of specimen, $N_{0.2}$ and $N_{0.3}$ are the loads when the maximum crack width developed to 0.2 mm and 0.3 mm respectively, N_{cal} is the calculated strength obtained by Equations (6)–(8).

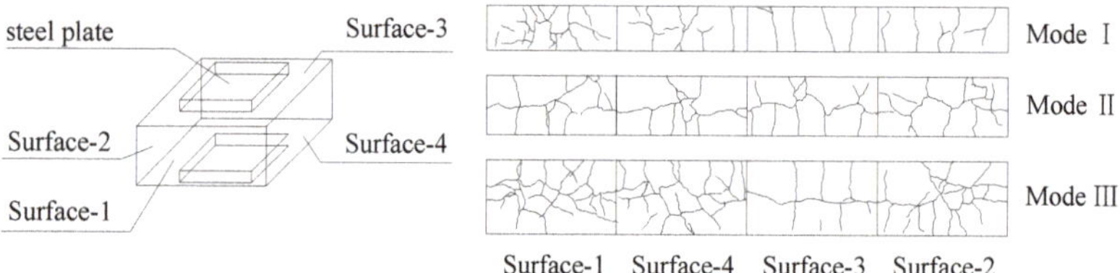

Figure 14. Three crack patterns on the side surfaces for the Series II specimens.

Regarding the crack pattern of Mode I, several vertical penetrating cracks can be found on the surfaces of the joint. The horizontal cracks may appear on the surfaces of the joint, but they were not penetrating (Figure 14). The specimens with the height factor β of 0.6 were categorised as the crack pattern of Mode I. Taking specimen C30-L2-H1-S2 as an example to describe the experimental observation in detail, the first vertical crack appeared when the upper load reached 800 kN. The crack extended gradually and became a penetrating crack when the load reached 1200 kN. The steel meshes started to reach the yield strains as the load increased to about 4000 kN. After that, the specimen reached its maximum load-carrying capacity of 7800 kN. During the post-peak loading stage, the concretes began to spall and crush. The test was terminated when the bearing capacity of the specimen declined to 85% of its peak load.

Regarding the crack pattern of Mode II, several vertical penetrating cracks and a main horizontal penetrating crack can be observed on each side surface of the joint (see Figure 14). The specimens with the height factor β of 0.8 were categorised as the crack pattern of Mode II. Taking specimen C30-L2-H2-S2 as an example to describe the experimental observation in detail, the first vertical crack occurred at the load of 1200 kN and developed into a penetrating crack at the load of 1500 kN. When the upper load reached 3300 kN, several horizontal cracks began to appear and gradually developed into a penetrating crack at the load of 5000 kN. At the load of about 6500 kN, the specimen reached its maximum load-carrying capacity. The test was ended when the upper load dropped to 85% of its peak load.

Regarding the crack pattern of Mode III, several vertical penetrating cracks and more than one horizontal penetrating cracks could be found on each side surface of the specimen (see Figure 14).

The specimens with the height factor β of 1.0 were categorised as the crack pattern of Mode III. Taking specimen C30-L2-H3-S2 as an example to describe the experimental observation in detail, the first vertical crack appeared at the load of 900 kN. When the upper load reached 2700 kN, the first horizontal crack occurred. After that, several horizontal cracks could be observed on each side surface of the specimen and eventually generated two main horizontal penetrating cracks when the load increased to 5000 kN. The specimen reached its load-carrying capacity of 5550 kN. The test was terminated when the load dropped to 85% of its peak load.

The comparison of failure modes between different concrete strengths is displayed in Figure 15. For the specimen C20-L1-H3-S2 with lower concrete strength (see Figure 15a), serious spalling and crushing of the concretes could be observed in the experiment when the specimen reached its ultimate strength, while for the specimen C30-L2-H3-S3 with higher concrete strength, the concretes of the joint remained intact at the post-peak loading stage even though serious vertical and horizontal cracks had been developed (Figure 15b).

(a)   (b)

Figure 15. Comparison of failure modes for specimens with different concrete strengths (a) C20 L1 H3 S2; (b) C30 L2 H3 S2.

3.4. Load-Deformation Curves and Load-Strain Curves

The load (N) versus longitudinal displacement (Δ) curves for typical specimen series are plotted in Figure 16. It can be seen that the strengths of the joints were sustained or declined slowly after the peak loads, which indicates excellent ductility of the joints. Figure 16a shows the load-deformation curves for specimen series C20-H3-S2, where all the parameters are the same except for the length factor α. It can be observed that the curves are quite close to each other. As shown in Figure 16a and Table 5, the ultimate strengths for specimens C20-L2-H3-S2 and C20-L3-H3-S2 are 1.013 and 1.034 times higher than that of specimen C20-L1-H3-S2. It can be concluded that the increase of α is advantageous but limited to the peak loads of the specimens in this experiment. Figure 16b shows the load-deformation curves for specimen series C30-L2-S2. In these three specimens, the height factors β are 0.6, 0.8 and 1.0 for specimens C30-L2-H1-S2, C30-L2-H2-S2 and C30-L2-H3-S2, respectively. As shown in Figure 16b and Table 5, the height factor β is a significant parameter affecting the behaviour of the specimen. The increase of β could evidently reduce the initial stiffnesses and the peak loads of the specimens. Figure 16c shows the load-deformation curves for specimen series C30-L2-H2. The volume ratios of steel meshes in the joint ρ_v for specimens C30-L2-H2-S1, C30-L2-H2-S2 and C30-L2-H2-S3 are 1.0%, 1.5% and 2.0%, respectively. As shown in Figure 16c and Table 5, the peak load of the specimens C30-L2-H2-S2 and C30-L2-H2-S3 are 1.125 and 1.152 times higher than that of the specimen C30-L2-H2-S1, which states that the increase of ρ_v could significantly increase the peak strengths of the specimens. However, the strengthen efforts could be weakening once ρ_v is larger than 1.5%. In addition, the effect of ρ_v on initial stiffness of the specimen is limited.

The load-strain curves of the radial stirrups for a typical specimen C30-L2-H3-S1 are present in Figure 17, where the outer and upper parts of the radial stirrups were in tension, while the internal part of the stirrup was in compression. The internal part of the stirrup approximately reached compressive yield strain at the ultimate load stage, which states that the stirrup could provide certain axial compressive strength for the joint.

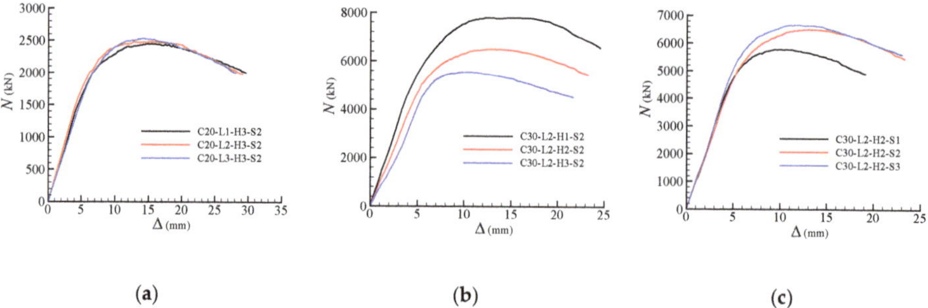

Figure 16. Axial load (*N*) verse displacement (Δ) relationships (**a**) Series C20 H3 S2; (**b**) Series C30 L2 S2; (**c**) Series C30 L2 H2.

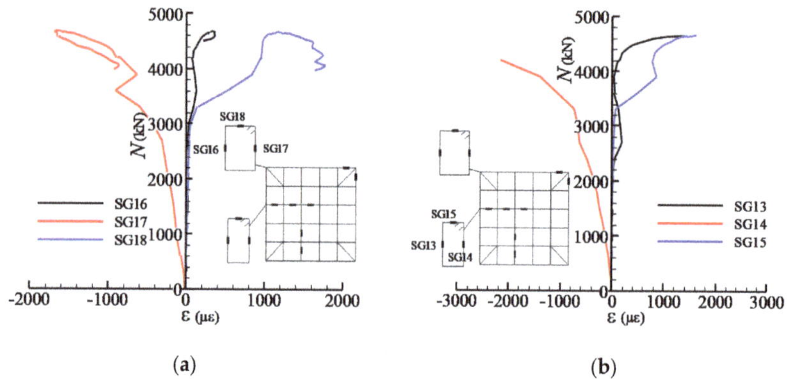

Figure 17. Load-strain curves for radial stirrups in specimen C30-L2-H3-S1 (**a**) SG16, SG17 and SG18; (**b**) SG13, SG15 and SG15.

3.5. Summary for the Series II Tests

According to the experimental results and analysis of the Series II, it can be concluded that (i) the resistances of the joints were sustained or declined slowly after the peak loads, indicating excellent ductility of the joints when subjected to bearing loads, and (ii) with the decrease of height factor β and the increase of volume ratio of steel meshes in the joint ρ_v, the axial compressive strengths of the joints significantly increased, while the increase of the length factor α is advantageous but limited to the peak loads of the specimens.

4. Calculation of Axial Compression Capacity of the Joints

4.1. General

Because of the lack of existing guidelines for the design of the innovative joints, a theoretical investigation was conducted in this section to determine the axial compression resistances of the innovative joints subjected to bearing loads, based on the test observations and analysis in Sections 2 and 3. In the following subsections, the local compression effect, the confinement effect provided by the multi-layers of steel meshes and the height effect of concrete will be first discussed. New proposal for calculating the axial compression resistances of the innovative joints subjected to bearing loads was then proposed and validated with the experimental results.

4.2. Effect of Local Compression

Several experimental studies have been conducted to investigate the resistances of plain concretes under local compression [17–22], and demonstrated that local compressive strengths of the plain concretes increased because the concrete outside the local compression area provided confined stress to the concretes within the local compression area, in which the degree of confinement can be quantified by using the bearing ratio (i.e., the ratio of the total surface area to the bearing area). A relationship representing the degree of confinement has been proposed by Komendant [23] and expressed as the square root function between the bearing area and the total surface area, which has been widely used in the existing international design codes, including the American Concrete Institute (ACI) design code ACI 318-08 [24] and the Chinese building codes GB 50010 [16]. In the present study, the expression specified in ACI 318-08 [24] was adopted to determine the nominal bearing strength of plain concrete, as given by Equation (1), where N_{plain} is the design bearing compression of unconfined concrete; f_c is the characteristic compressive cylinder strength of unconfined concrete; A_1 is the loaded area; A_2 is the area of the lower base of the largest frustum of a pyramid, cone or tapered wedge contained wholly within the support and having its upper base equal to the loaded area.

$$N_{plain} = 0.85 f_c A_1 \sqrt{\frac{A_2}{A_1}} \quad \text{with} \quad \sqrt{\frac{A_2}{A_1}} \le 2.0 \tag{1}$$

4.3. Confinement Effect Provided by the Multi-Layers of Steel Meshes

It is certified from the experimental results that the ductility and the bearing strength of concrete improved due to the confinement from the multi-layers of steel meshes. However, the advantage contribution of the confinement effect on the concrete strength is not included in Equation (1), as set out in ACI 318-08 [24], which would underestimate the effective compressive strength of the concrete. In order to take into account the beneficial confinement effects provided by the multi-layers of steel meshes, the confined concrete model, as proposed by Mander et al. [25], was used to quantify the degree of confinement provided by the multi-layers of steel meshes; this confined concrete model has also been used in the previous experimental and theoretical studies on the axial compressive strength of a circular through-beam joint between the CFST columns and RC beams [11], indicating that the confined concrete model proposed by Mander et al. [25] was capable of precisely predicting the compressive strength of the circular through-beam joint. The expression for calculating the confined concrete strength is given in Equation (2), where f'_{cc} is the peak compressive strength of the confined concrete; f'_{co} is the peak compressive strength of the unconfined concrete; f'_{ℓ} is the effective lateral confining pressure applied by reinforcing bars.

$$f'_{cc} = \left[-1.254 + 2.254\left(1 + 7.94\frac{f'_{\ell}}{f'_{co}}\right)^{0.5} - 2.0\frac{f'_{\ell}}{f'_{co}} \right] f'_{co} \tag{2}$$

Given that the amount of the reinforcements in both directions of the multi-layers of steel meshes are approximately the same, the effective lateral confining pressure f'_{ℓ} is then determined by Equation (3) [25], where f_l is the lateral pressure provided by the transverse reinforcements and assumed to be uniformly distributed in the concrete core; f_y is the yield strength of the reinforcing bars; k_e is the confinement effectiveness coefficient, which takes into account the reduction in the confinement effect due to the spalling off of the cover concrete. Although vertical and horizontal cracks were observed on the side surfaces of the joint specimens, the concrete in the loaded area remained intact in the experiments. Therefore, the reduction in the confinement effect due to the spalling of the cover concrete is neglected, with the confinement effectiveness coefficient $k_e = 1$, leading to the expression to calculate the lateral pressure given by Equation (4).

$$f_l' = k_e f_l = \frac{k_e \rho_v f_y}{2} \tag{3}$$

$$f_l' = k_e f_l = \frac{\rho_v f_y}{2} \tag{4}$$

4.4. Height Effect of Concrete

Existing experimental results have demonstrated that the height factor β is a critical factor affecting the ultimate strength of the joint [11,20]. With the height of the specimen increasing, the ultimate strength of the specimen declined rapidly, as evident in Section 3.4. The reason for the height effect of concrete can be explained by the friction generated at the surface between the steel plates and the concretes. The friction gradually transferred from the contact surface to the mid-height cross-section of the specimen, which plays a role of confinement and thus strengthens the compressive strength of the concrete. In order to quantify the degree of confinement at the mid-height cross section of the joint due to the friction effect, a regression analysis was conducted, based on the existing experimental results on the height effect of plain concrete. First, a total of 99 experimental data on the height effect of concrete were collected from the existing experiments carried out by Dehestani et al. [26] and Yi et al. [27]. The lateral pressure $f'_{\ell h}$ for the height effect of each data can then be derived by substituting the corresponding unconfined concrete strengths and the confine concrete strengths into Equation (2). The ratio of $f'_{\ell h}/f'_{co}$ for each data was plotted against the corresponding h/d ratio and displayed in Figure 18, in which d is the width of the local compression zone. As shown in Figure 18, the increase of h/d ratio results in a downtrend of $f'_{\ell h}/f'_{co}$ and the regression equation of lateral pressure for the height effect of concrete can be obtained and given by Equation (5).

$$\frac{f'_{\ell h}}{f'_{co}} = \frac{1}{12.820h/d} - 0.039, \quad 0.5 < h/d \le 2 \tag{5}$$

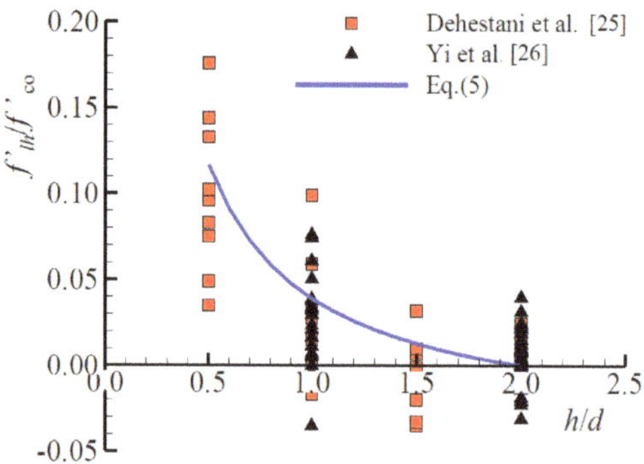

Figure 18. Variation in f'_{lh}/f'_{co} with h/d ratio.

In Equation (5), it is discovered that $f'_{\ell h}/f'_{co} = 0$ when $h/d = 2$, which represents the frictional force between the loading plates and specimens is neglected if the height-to-width ratios of the concrete specimens are greater than 2. According to Equations (2) and (5), when $h/d = 1$, the value of $f'_{\ell h}/f'_{co}$ is equal to 0.8, which were consistent with the conversion relationship of BSI for the concrete compressive strength between the cylindrical specimen and the cubic specimen [28].

4.5. Prediction and Discussion

Through considering the local compression effect, the confinement effect provided by the multi-layers of steel meshes and the height effect of concrete, the formulas for calculating the compressive resistances of the joints are summarised in Equations (6)–(8).

$$N_{\text{cal}} = 0.85 f'_{\text{cc}} A_1 \sqrt{\frac{A_2}{A_1}} \tag{6}$$

$$f'_{\text{cc}} = \left[-1.254 + 2.254 \left(1 + 7.94 \frac{f'_\ell}{f'_{\text{co}}} \right)^{0.5} - 2.0 \frac{f'_\ell}{f'_{\text{co}}} \right] f'_{\text{co}} \tag{7}$$

$$f_l' = \frac{\rho_v f_y}{2} + \left(\frac{1}{12.820 H/a} - 0.039 \right) f'_{\text{co}} \tag{8}$$

Assessment of the accuracy of the new proposal was then conducted, through the comparisons between the experimental ultimate strengths N_u and the predicted results N_{cal}, as listed in Tables 1 and 5. It can be found that the resistance predictions N_{cal} for the joint specimens in Series I are higher than the corresponding experimental counterparts, which could be attributed to the fact that all the specimens in Series I failed by the CFST columns, and the ultimate strengths of the joints were higher than those of the CFST columns. For Series II, the resistance predictions derived by Equations (6)–(8) are in good agreement with the experimental ones for normal strength concrete (C30), with a mean value and standard deviation of N_{cal}/N_u equal to 1.038 and 0.100, respectively. However, the ultimate strengths were overestimated for the specimen series with low strength concrete (C20). It may be because the outer concrete of the specimens with lower concrete strength was prematurely spalled and crushed when the specimens were reaching their ultimate strengths, which would weaken the favourable effect of local compression.

Therefore, the calculation model for the ultimate strengths of the joints subjected to bearing loads for practical engineering design is recommended herein. The confinement effect provided by the multi-layers of steel meshes and the height effect of concrete are considered in the model, while the effect of local compression is neglected because of its uncertainty, leading to the final formulations for the design of the ultimate strengths of the square joints given by Equations (9) and (10).

$$N_{\text{design}} = \left[-1.254 + 2.254 \left(1 + 7.94 \frac{f'_\ell}{f'_{\text{co}}} \right)^{0.5} - 2.0 \frac{f'_\ell}{f'_{\text{co}}} \right] f'_{\text{co}} A_1 \tag{9}$$

$$f_l' = \frac{\rho_v f_y}{2} + \left(\frac{1}{12.820 H/a} - 0.039 \right) f'_{\text{co}} \tag{10}$$

Figure 19 shows the comparison between the calculated results obtained by Equations (9) and (10) and experimental results of Series II tests, where the calculated strengths are relatively accurate and conservative in contrast to the experimental strengths.

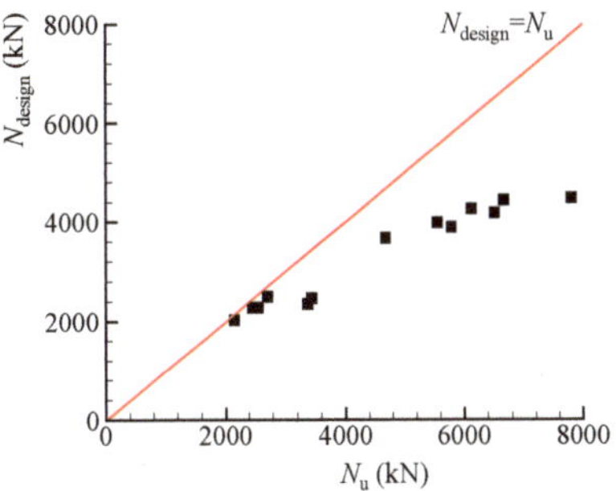

Figure 19. Comparison of the calculated load for design (N_{design}) and experimental load (N_u).

5. Conclusions

The axial compressive behaviour of square through-beam joint between CFST column and RC beams with multi-layers of steel meshes was experimentally and theoretically investigated in this paper. The following conclusions can be drawn based on the results of this study.

The first stage of the tests included three full-scale innovative joint specimens, indicating that the innovative joint is capable of achieving the design criterion of the 'strong joint-weak member' with appropriate designs.

The second stage of the tests involved 14 innovative joint specimens with the improved detailing to study the effects of the geometric size of the joint, concrete strength and volume ratio of the steel meshes on the bearing strengths of the joints. It is shown that by increasing the height factor β of the joint specimens from 0.6 to 1.0, the axial compression strengths of the joints significantly decreased with a maximum reduction ratio in strength equal to 28.9%. The increase of the volume ratio of the steel meshes in the joint ρ_v is advantageous for the axial compression strengths of the joints. As ρ_v increased from 1.0 to 2.0, the axial compression strengths of the joints increased by 26.4% and 15.2% for the specimens with concrete grades of C20 and C30, respectively. The increase of the length factor α is advantageous but limited to the peak loads of the specimens.

Because of the lack of existing design methods for the innovative joints, new design expressions were proposed to calculate the axial compression resistances of the innovative joints subjected to bearing loads, with the local compression effect, the confinement effect provided by the multi-layers of steel meshes and the height effect of the concrete considered. The strength predictions of the joints are in good agreement with the experimental ones for normal strength concrete. Through the discussion of the calculated results, recommended formulations for the design of the joints were proposed, which is capable of providing accurate and safe resistance predictions for the innovative joints. However, because of the limited range of variables chosen for the specimens in the tests, extensive applications of the proposed calculation method should be verified by further experimental researches.

Author Contributions: Conceptualization, W.D., J.C.; methodology, W.D., X.-L.T., Q.-J.C.; validation, Q.-J.C., C.Y., A.H.; formal analysis, W.D., X.-L.T.; investigation, W.D., X.-L.T.; writing—original draft preparation, W.D., X.-L.T.; writing—review and editing, Q.-J.C., C.Y., A.H.; visualization, W.D., X.-L.T.; supervision, J.C.; project administration, J.C.; funding acquisition, J.C., X.-L.T. All authors have read and agreed to the published version of the manuscript.

Funding: This research was funded by Science and Technology Program of Guangzhou (201902010073), Pearl River S and T Nova Program of Guangzhou (201806010137) and Science and Technology Planning Project of Guangzhou Municipal Construction ([2019]-KJ027).

Conflicts of Interest: The authors declare no conflict of interest.

References

1. Roeder, C.W. Overview of hybrid and composite systems for seismic design in the United States. *Eng. Struct.* **1998**, *20*, 355–363. [CrossRef]
2. Miller, D.K. Lessons learned from the Northridge earthquake. *Eng. Struct.* **1998**, *20*, 249–260. [CrossRef]
3. Popov, E.P.; Yang, T.-S.; Chang, S.-P. Design of steel MRF connections before and after 1994 Northridge earthquake. *Eng. Struct.* **1998**, *20*, 1030–1038. [CrossRef]
4. Fang, X.; Li, S.; Qian, J.; Yang, R. Experimental Research on Seismic Behavior of Concrete Filled Steel Tubular Column Ring Beam Joint Under Cyclic Loading. *J. Build. Struct.* **2002**, *23*, 10–18.
5. Han, L.-H.; Qu., H.; Tao, Z.; Wang, Z.-F. Experimental behaviour of thin-walled steel tube confined concrete column to RC beam joints under cyclic loading. *Thin-Walled Struct.* **2009**, *47*, 847–857. [CrossRef]
6. Pan, P.; Lam, A.; Lin, X.; Li, Y.; Ye, L. Cyclic loading tests and finite element analyses on performance of ring beam connections. *Eng. Struct.* **2013**, *56*, 682–690. [CrossRef]
7. Zha, X.; Wan, C.; Yu, H.; Dassekpo, J.-B.M. Seismic behavior study on RC-beam to CFST-column non-welding joints in field construction. *J. Constr. Steel Res.* **2016**, *116*, 204–217. [CrossRef]
8. Nie, J.; Bai, Y.; Cai, C. New connection system for confined concrete columns and beams. I: Experimental study. *J. Struct. Eng.* **2008**, *134*, 1787–1799. [CrossRef]

9. Bai, Y.; Nie, J.; Cai, C. New connection system for confined concrete columns and beams. II: Theoretical modeling. *J. Struct. Eng.* **2008**, *134*, 1800–1809. [CrossRef]

10. Zhang, Y.; Zhao, J.; Cai, C. Seismic behavior of ring beam joints between concrete-filled twin steel tubes columns and reinforced concrete beams. *Eng. Struct.* **2012**, *39*, 1–10. [CrossRef]

11. Chen, Q.; Cai, J.; Bradford, M.A.; Liu, X.; Wu, Y. Axial compressive behavior of through-beam connections between concrete-filled steel tubular columns and reinforced concrete beams. *J. Struct. Eng.* **2015**, *141*, 04015016. [CrossRef]

12. Chen, Q.-J.; Cai, J.; Bradford, M.A.; Liu, X.; Zuo, Z.-L. Seismic behaviour of a through-beam connection between concrete-filled steel tubular columns and reinforced concrete beams. *Eng. Struct.* **2014**, *80*, 24–39. [CrossRef]

13. Zhou, X.; Cheng, G.; Liu, J.; Gan, D.; Frank Chen, Y. Behavior of circular tubed-RC column to RC beam connections under axial compression. *J. Constr. Steel Res.* **2017**, *130*, 96–108. [CrossRef]

14. Zhou, X.; Zhou, Z.; Gan, D. Cyclic testing of square tubed-reinforced-concrete column to RC beam joints. *Eng. Struct.* **2018**, *176*, 439–454. [CrossRef]

15. Tang, X.-L.; Cai, J.; Chen, Q.-J.; Liu, X.; He, A. Seismic behaviour of through-beam connection between square CFST columns and RC beams. *J. Constr. Steel Res.* **2016**, *122*, 151–166. [CrossRef]

16. *Code for Design of Concrete Structures, National Standard of PRC GB 50010*; National Standards of People's Republic of China: Beijing, China, 2010.

17. Hawkins, N.M. The bearing strength of concrete loaded through rigid plates. *Mag. Concr. Res.* **1968**, *20*, 31–40. [CrossRef]

18. Niyogi, S.K. Bearing strength of reinforced concrete blocks. *J. Struct. Div.* **1975**, *101*, 1125–1137.

19. Ahmed, T.; Burley, E.; Rigden, S. Bearing capacity of plain and reinforced concrete loaded over a limited area. *Struct. J.* **1998**, *95*, 330–342.

20. Ince, R.; Arici, E. Size effect in bearing strength of concrete cubes. *Constr. Build. Mater.* **2004**, *18*, 603–609. [CrossRef]

21. Au, T.; Baird, D.L. Bearing capacity of concrete blocks. *ACI Struct. J.* **1960**, *56*, 869–880.

22. Meyerhof, G.G. The bearing capacity of concrete and rock. *Mag. Concr. Res.* **1953**, *4*, 107–116. [CrossRef]

23. Komendant, A.E. *Prestressed Concrete Structures*, 1st ed.; McGraw-Hill: New York, NY, USA, 1952.

24. ACI Committee, American Concrete Institute; International Organization for Standardization. *Building Code Requirements for Structural Concrete (ACI 318-08) and Commentary*; American Concrete Institute: Farmington Hills, MI, USA, 2008.

25. Mander, J.B.; Priestley, M.J.; Park, R. Theoretical stress-strain model for confined concrete. *J. Struct. Eng.* **1988**, *114*, 1804–1826. [CrossRef]

26. Dehestani, M.; Nikbin, I.; Asadollahi, S. Effects of specimen shape and size on the compressive strength of self-consolidating concrete (SCC). *Constr. Build. Mater.* **2014**, *66*, 685–691. [CrossRef]

27. Yi, S.-T.; Yang, E.-I.; Choi, J.-C. Effect of specimen sizes, specimen shapes, and placement directions on compressive strength of concrete. *Nucl. Eng. Des.* **2006**, *236*, 115–127. [CrossRef]

28. British Standards Institution. *Eurocode 2: Design of Concrete Structures: Part 1-1: General Rules and Rules for Buildings*; British Standards Institution: London, UK, 2004.

 *materials*

Article

Experimental Research of the Time-Dependent Effects of Steel–Concrete Composite Girder Bridges during Construction and Operation Periods

Guang-Ming Wang [1], Li Zhu [1,*], Guang-Pan Zhou [2], Bing Han [1] and Wen-Yu Ji [1]

[1] School of Civil Engineering, Beijing Jiaotong University, Beijing 100044, China;
    14115279@bjtu.edu.cn (G.-M.W.); skyiflv@163.com (B.H.); wyji@bjtu.edu.cn (W.-Y.J.)
[2] School of Science, Nanjing University of Science and Technology, Nanjing 210094, China;
    newsypaper@163.com
*   Correspondence: zhuli@bjtu.edu.cn

Received: 8 April 2020; Accepted: 27 April 2020; Published: 3 May 2020

Abstract: The present work aimed to study the effects of temperature changes and concrete creep on I-shaped steel–concrete composite continuous girder bridges during construction and operation processes. This study combined structural health monitoring data, an ANSYS finite element simulation, and the age-adjusted effective modulus method to obtain the variation laws of temperature and internal force in composite girders. Moreover, a temperature gradient model was proposed that is suitable for bridges in Hebei, China. In addition, a concrete creep experiment under unidirectional axial compression was performed using concrete specimens prepared from the concrete batch used to create the composite girder. The long-term evolution laws of the deflection and internal force of the composite girder were obtained by predicting the concrete creep effect. The measured data showed that the temperature variation trends of the steel beam and concrete slab were characterized by a sinusoidal curve without a temperature lag. The heating rate of the concrete slab was higher than the cooling rate. The prediction results showed that the internal force changes in the composite girder were characterized by three stages. The stress changes in the composite girder during the first 10 days were significant and the stress charge rate of the concrete slab, the steel girder and the shear stud can reach 5%–28%. The stress change rate decreased continuously during 10–90 days. The stress changed slowly and smoothly after 90 days. This research can provide feedback and reference for structural health monitoring and service safety control of similar I-shaped steel–concrete composite bridges.

Keywords: steel–concrete composite bridge; I-shaped beam; concrete creep; temperature; prediction; experiment

1. Introduction

Steel–concrete composite girder bridges are becoming increasingly popular with increasing bridge spans and advantages in terms of reliable performance, reasonable cost, and structural rationality. However, some shortcomings of these structures need to be solved, including the stress redistribution caused by ambient temperature changes and concrete shrinkage and creep. Therefore, the time-dependent effects and the evolution law of structural performance during bridge construction and operation periods should be studied, which is beneficial to the further promotion and application of composite girder bridges [1].

First, the maximum stress caused by temperature changes can reach 20%–30% of the allowable stress, which cannot be ignored [2]. Moreover, thermal residual stresses and relative slip between steel and concrete will be caused by the temperature difference between the steel beam and concrete slab [3–5]. Chen [6], Gu [7], Zhou [8], and Liu [9] deduced formulas for the internal force and relative slip in steel–concrete composite girders under the influence of temperature. Chen [10], Su [11], Wang [12], and Xiao [13] verified the existence of a large temperature difference and gradient in a steel–concrete composite box girder based on monitoring data of the temperature field, and the temperature distribution was different from that in the existing Chinese specifications. Moreover, the daily temperature differences, convection, and solar radiation were the main factors affecting the temperature field of the composite girder. However, the above studies were mainly directed to the single-piece I-beam girders or steel box girders, whereas few studies involved multiple I-beam girders. Furthermore, only the calculation method for the temperature difference between the steel and concrete is regulated by the Chinese regulations. The influences of the variation of the transverse temperature on the mechanical behavior of the bridges remain to be further studied.

The theories describing the stress–strain constitutive relationship caused by concrete creep include the aging theory (creep rate method), elastic creep theory (superposition method), elastic aging theory (flow rate method), secondary effect flow theory, and age adjustment effective modulus method [14–18]. The concrete creep effects of composite bridges were analyzed by several researchers considering the shear connections, the shear lag effect in box girders, the randomness of shrinkage and creep effects, and the changes in temperature and humidity [19–26]. In terms of experimental research, Gara, Ranzi, and Sullivan obtained an analysis model of concrete creep by conducting long-term full-scale tests with simply supported composite girder bridges [27–30]. Fan and Nie studied the concrete shrinkage and creep effects and cracking of composite girder bridges based on long-term loading tests and proposed an improved composite deck system for cable-stayed bridges [31,32]. However, the above studies were mainly aimed at analyzing the concrete shrinkage and creep effects of composite girders under specific conditions. The establishment of concrete constitutive relations and the realization of corresponding numerical models still needs to be analyzed. Numerical simulation and scale model test methods are mainly used in research. However, there are few studies aimed at long-term monitoring of concrete shrinkage and creep effects for real bridges because of the difficulty in long-term field monitoring.

In addition, the significances of the long-term monitoring of bridges are as follows: (1) the difference between the monitored experimental results and the numerical model for prediction analysis of bridges can be identified to correct the model. The corrected model can be further applied in the subsequent analysis of the mechanical behavior of the bridges. (2) The bridges are monitored for the long term to ensure their safety. This can provide a steady foundation for superior behavior of bridges in their life cycle and meet the requirement of the long-cycle design idea. (3) The monitored experimental results of the mechanical behavior of the bridges in a complicated and variable environment may create a new research issue. Therefore, long-term monitoring of the bridges is important for the development of investigation of the bridges.

In this paper, the effects of temperature changes and concrete creep on I-shaped steel–concrete composite continuous girder bridges were studied. This study combined field monitoring data, an ANSYS finite element (FE) simulation, and the age-adjusted effective modulus method to obtain the variation laws of temperature and internal force of the composite girders. Moreover, a temperature gradient model suitable for bridges in the Hebei region of China was proposed. In addition, a concrete creep experiment under unidirectional axial compression was carried out using concrete material from the batch used to create the composite girder. Then, concrete creep curves under 28-day loading were obtained. The long-term evolution laws of the deflection and internal force of the composite girder in response to concrete creep were predicted.

2. Background Project Overview

2.1. Bridge Overview

The background project is located in Baoding city, Hebei Province in China, and the first west segment of the bridge is shown in Figure 1. Sections S1, S2, S3, S4, S5, and S6 in Figure 1 represent the control sections located at the side support, 1/2 and 3/4 sections of the 4th span, the 2nd pier top, 1/4 and 1/2 sections of the 3rd span, respectively.

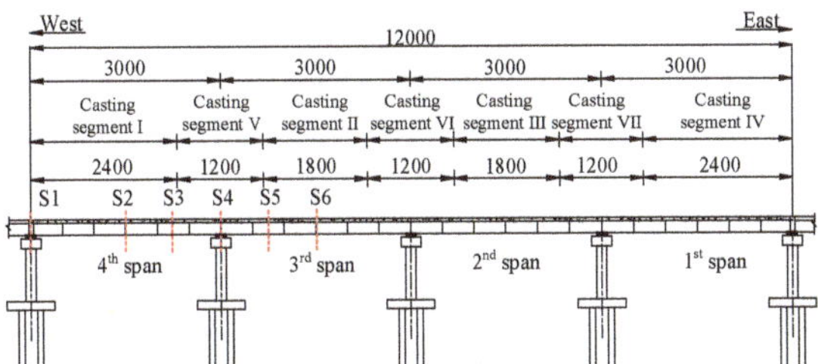

Figure 1. Layout of the first west standard segment of the bridge (units: cm).

The I-shaped steel–concrete composite continuous girder was used in the superstructure of the bridge. The north and south side girders are symmetrical relative to the road centerline. The cross-section of the northern half of the composite girder is shown in Figure 2. The steel beams are labeled A, B, C, and D from the road centerline to the outside. The thickness of the web of the steel beam was 12 mm.

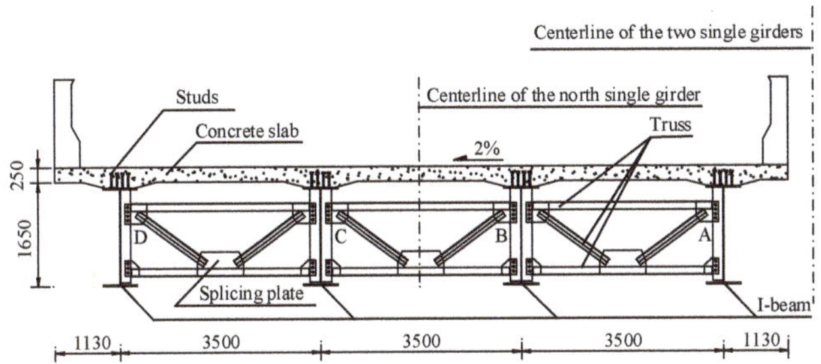

Figure 2. Cross-section of the north half composite girder (units: mm).

2.2. Structural Health Monitoring (SHM) System

Figure 3 shows the arrangement of the vibrating string strain gauges attached to the steel beam web and the ones embedded in the concrete slab in Sections S1–S6. The arrangement of the fiber Bragg grating sensors attached to the top flange surface of the steel beam is shown in Figure 4. In addition, 11 studs symmetrical to S4 were selected, as shown in Figure 5. For the girder deflection, four measuring points were uniformly arranged in the transverse direction for each span, as shown in Figure 6.

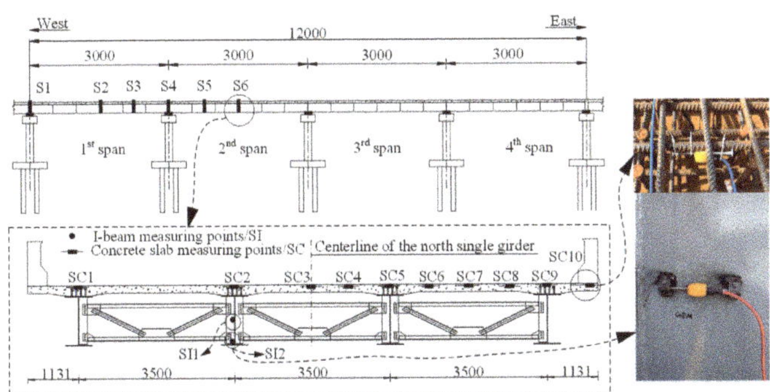

Figure 3. Layout of the strain gauges for the steel beam web and concrete slab (units: cm).

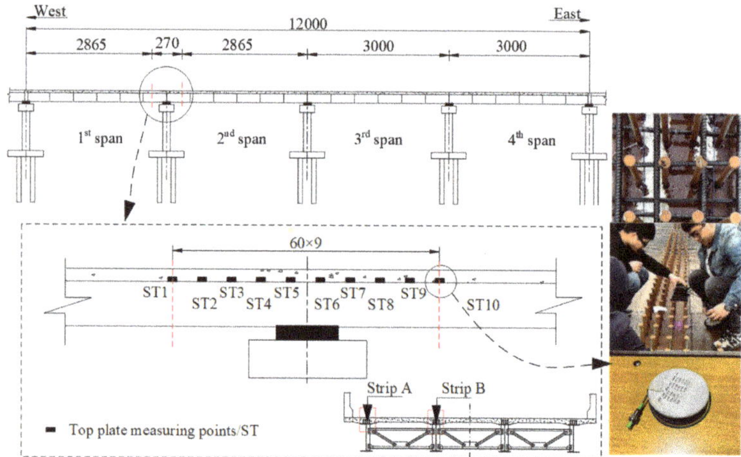

Figure 4. Layout of the fiber Bragg grating sensors for the top plate of the steel beam (units: cm).

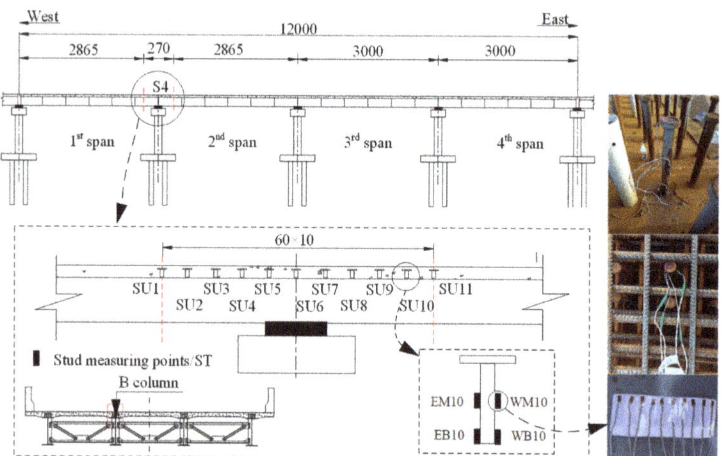

Figure 5. Layout of the strain gauges for the studs (units: cm).

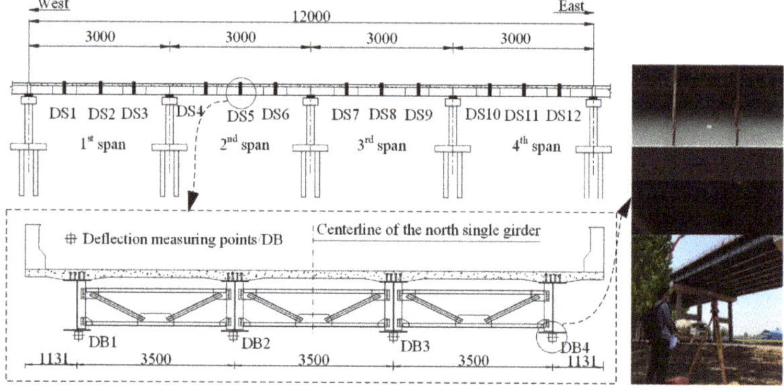

Figure 6. Layout of the measuring points for girder deflection (units: cm).

3. Thermal Effect during the Construction Process

The regulations of the temperature gradient model in the codes of every country are different because the selection of the temperature gradient model is related to the bridge type, cross-section form, sunshine conditions, and geographical location. In this paper, the temperature distribution law of a multiple I-steel composite girder was obtained based on field-measured data, and the most unfavorable temperature gradient mode was determined. Moreover, the transverse gradient model of temperature was proposed, which can guide structural designs.

3.1. Variation Law of Girder Temperature

The measured temperature changes in the composite girder in Sections S4 and S6 from May 1 to May 8 of 2018 are shown in Figures 7 and 8, respectively. The temperature variation at each measuring point was characterized by an obvious sinusoidal performance with peak and valley values occurring on each day. Moreover, the temperature change trends were consistent without a lag phenomenon. In addition, the heating rate of the concrete slab was greater than the cooling rate. The fastest heating and cooling rates were 1.5 °C/h at 13:00 and −0.9 °C/h at 21:00, respectively, and the trends of both increased first and then decreased. Moreover, there was a gradient change along the vertical direction of the composite girder. The temperature of the concrete slab was higher than the temperature of the steel beam. The temperature at the middle of the steel beam was higher than the temperature at the bottom flange of the steel beam.

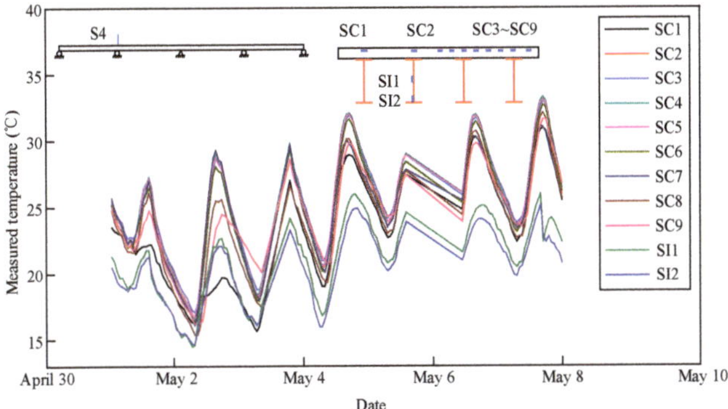

Figure 7. Measured temperature changes in Section S4.

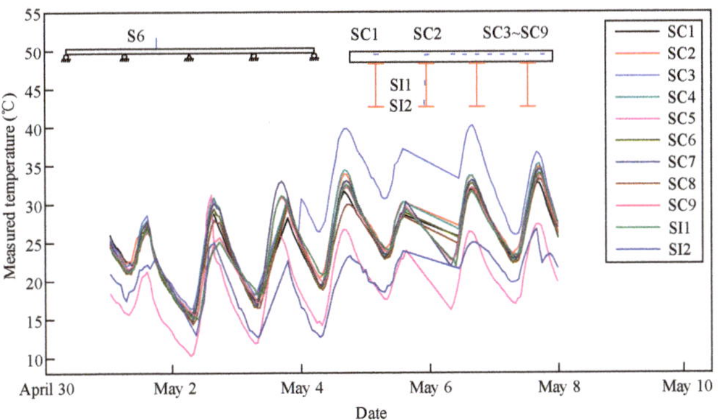

Figure 8. Measured temperature changes in Section S6.

3.2. Variation Law of Girder Strain

The measured strain changes in the composite girder in Sections S4 and S6 in one day are shown in Figures 9 and 10, respectively. Figure 9 shows that the most prominent daily strain variation in the middle of the steel beam was 30.9 µε. The daily strain variation in the concrete slabs was smaller: 8 µε. Figure 10 shows that the most prominent daily strain variation in the middle of the steel beam was 83.4 µε, and that of the concrete slabs was 23.9 µε. The daily strain change in the composite girder in Section S4 was smaller than that in Section S6. In addition, the changes in ambient temperature could not be reflected by the embedded vibrating wire sensor in the concrete, and the strain changes in the concrete had a lag of 4.5 h with respect to one of the steel beams.

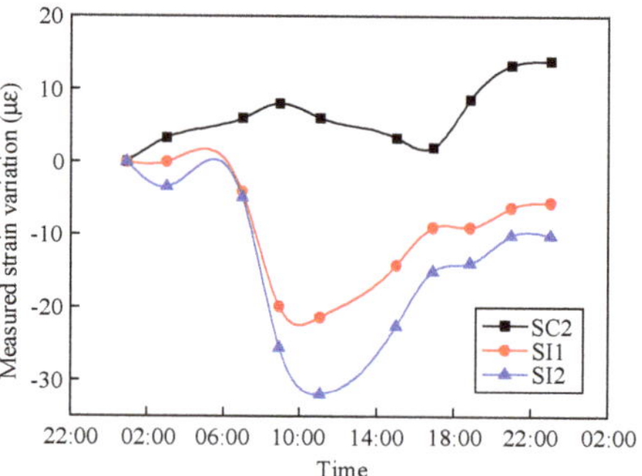

Figure 9. Measured girder strain in Section S4.

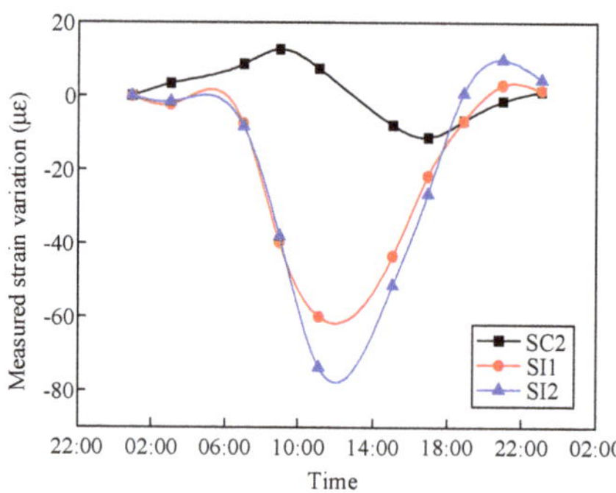

Figure 10. Measured girder strain in Section S6.

The variation laws of the internal forces of the concrete slab and steel beam in Section S6 along with temperature changes in one day were analyzed, as shown in Figures 11 and 12, respectively. The compressive strains in the concrete and steel beams were caused by the expansion of material with increasing temperature. The tensile strains were caused by material shrinkage along with decreasing temperature. The strain variation rates in the concrete and steel beams were 3.4 µε/°C and 23.9 µε/°C, respectively.

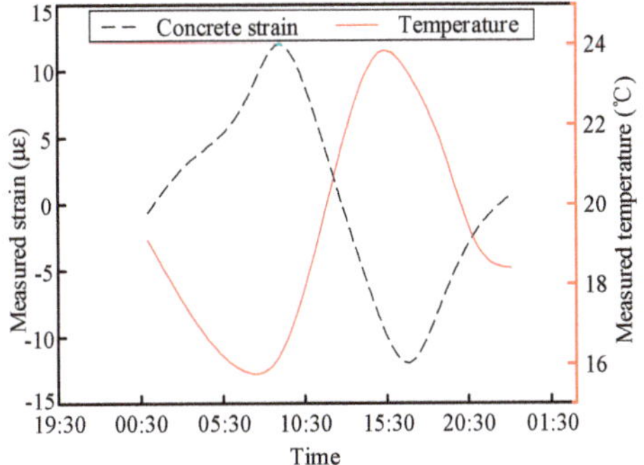

Figure 11. Measured variation in the concrete strain in Section S6 along with temperature changes.

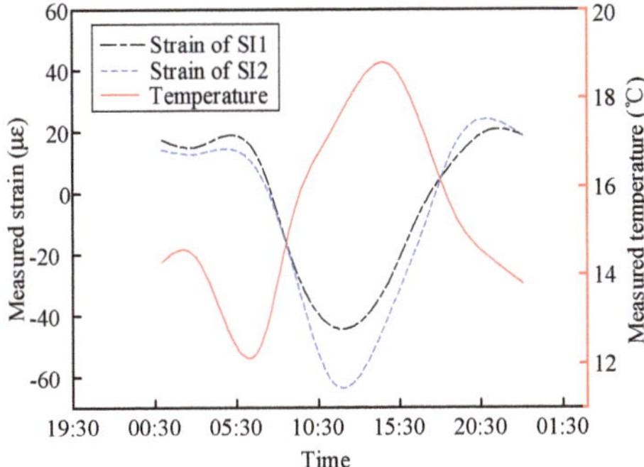

Figure 12. Measured variation in the steel beam strain in Section S6 along with temperature changes.

The variation laws of the internal forces of the concrete slab and steel beam in Section S4 along with temperature changes in one day are shown in Figures 13 and 14, respectively. The strain variation rates in the concrete and steel beams were 5 $\mu\varepsilon/°C$ and 23.8 $\mu\varepsilon/°C$, respectively.

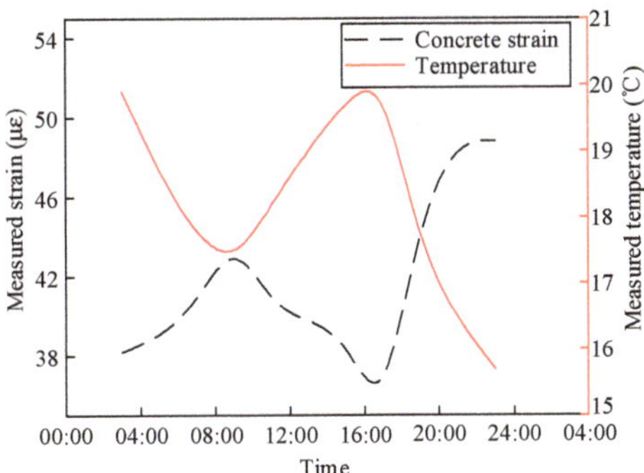

Figure 13. Measured variation in the concrete strain in Section S4 along with temperature changes.

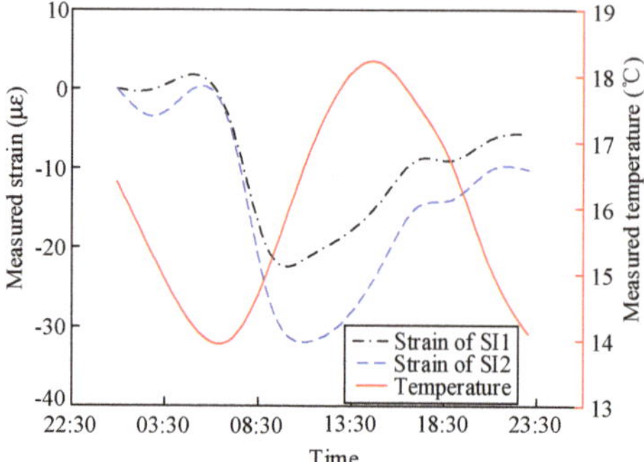

Figure 14. Measured variation in the steel beam strain in Section S4 along with temperature changes.

3.3. Proposed Gradient Model of Temperature

The temperature of the concrete slab is characterized by a large transverse gradient because of the bridge direction and environmental shielding. However, the Chinese code specifies only the transverse temperature effect of the wide box girder without cantilevers. Therefore, the suggested temperature gradient model for composite girder bridges in the Hebei Province of China was proposed by processing the temperature monitoring data with a guaranteed rate of 95%.

The transverse distribution of the concrete temperature in Section S4 is shown in Figure 15. The temperature difference along the transverse direction under normal climatic conditions between 1:00 and 11:00 was less than 1.5 °C. The transverse distribution of the concrete temperature from 11:00 to 16:00 was characterized by an isosceles trapezoidal distribution. Moreover, the temperature in the middle region of the concrete slab was higher than that in the outside region of the concrete slab. The transverse distribution of temperature in the south concrete slab gradually became a trapezoidal distribution with increasing temperature due to the long exposure to sunshine. The transverse difference in temperature reached a maximum of 10.1 °C at 17:00. The north measuring points of the concrete were shielded by the steel beams and could not be illuminated by the sun. Therefore, the temperature at these points was the lowest among the measuring points, and the temperatures at these points changed slightly.

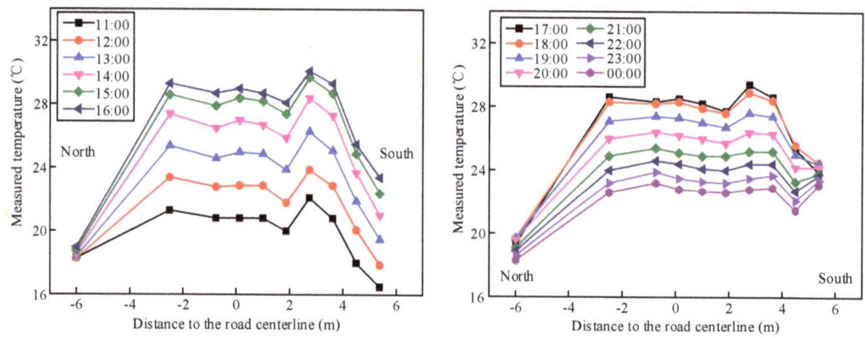

Figure 15. Transverse distribution of concrete temperature in Section S4 for one day.

The maximum differences in the girder temperatures during each day were extracted as shown in Table 1, and the mean value at each measured point was calculated.

The temperature gradient model for the composite girder bridges in Hebei, China, was proposed with a guaranteed rate of 95%. A temperature gradient of 4 °C/m was adopted for the height range from the bottom of the steel beam to the junction of the flange and web. A temperature gradient of

20 °C/m was adopted for the height range from the junction of the flange and web to the top of the concrete slab. Moreover, the temperature gradient in the pavement could be specified according to the Chinese code. In addition, the heat diffusion and conduction in the girder changed along the transverse direction because of the different boundary conditions, which resulted in the transverse gradient of the temperature. A transverse gradient of 3 °C/m could be adopted. The proposed temperature gradient model can be referenced during the design process.

Table 1. Maximum temperature difference in the girder during each day.

Section	Day	Temperature of Each Measuring Point (°C)											
		SC1	SC2	SC3	SC4	SC5	SC6	SC7	SC8	SC9	SC10	SI1	SI2
S4	1	4.90	5.10	4.60	5.00	5.40	4.70	5.00	4.70	4.50	4.10	2.90	2.60
	2	11.40	12.60	11.60	12.40	11.50	11.90	11.60	12.20	10.20	8.30	8.20	7.40
	3	11.40	11.20	11.00	11.40	11.30	11.50	11.7	11.70	9.20	8.50	8.10	7.20
	4	9.90	11.60	10.90	11.50	11.40	11.50	12.30	9.90	10.80	9.00	9.20	8.90
	5	7.60	8.30	8.50	8.10	8.50	8.30	8.30	7.50	7.70	6.50	4.30	3.90
	6	7.60	8.40	8.10	8.40	8.30	8.60	8.40	7.40	7.60	6.30	4.40	4.00
	7	8.50	9.90	9.20	9.80	9.70	9.60	9.90	8.30	9.40	8.10	5.50	5.40
S6	1	4.90	5.10	6.60	–	6.70	6.20	6.50	5.50	5.00	5.60	4.80	4.60
	2	13.00	14.20	14.20	–	14.80	14.70	14.90	13.50	10.10	10.10	9.00	8.50
	3	10.70	12.40	12.40	–	12.60	13.70	14.20	13.40	13.70	12.90	10.00	9.40
	4	12.20	13.40	13.40	–	14.10	14.00	13.70	11.10	12.20	11.70	10.60	10.20
	5	8.70	9.50	9.50	–	9.60	9.90	9.30	8.80	8.30	7.50	6.50	6.10
	6	8.90	9.60	9.60	–	9.60	9.80	9.40	8.80	8.30	7.50	6.50	6.10
	7	9.90	11.40	10.70	–	11.50	11.70	11.60	10.60	10.40	10.80	8.90	8.70
Average		9.26	10.19	10.02	9.51	10.36	10.44	10.39	9.53	9.10	8.35	7.06	6.64

4. Concrete Creep Effect during the Construction and Operation Periods

The stress redistribution in composite girders is difficult to solve by theoretical analysis alone because of the complexity of concrete creep and the property differences between steel and concrete. The variation laws of girder deflection and stress caused by concrete creep were studied based on field monitoring data, concrete creep experiments, and numerical models.

The expander was added into the concrete of the composite bridge to eliminate the shrinkage effect. Little concrete shrinkage strain was observed in the shrinkage tests of concrete batches. Therefore, only the concrete creep effect was taken into account in the below-mentioned discussion.

4.1. Concrete Creep Experiment

4.1.1. Concrete Specimen

A concrete creep experiment under unidirectional axial compression, including two ordinary concrete specimens and two steel fiber reinforced concrete specimens, was carried out using specimens made from the same batch of materials as the actual bridge. The dimension of the concrete specimens was 100 mm × 100 mm × 300 mm. The average value was taken from the 28-day creep data of the two concrete specimens. Moreover, two compensation specimens were designed to offset the deformations caused by the hydration of cement, external temperature, and humidity, including concrete condensation, hardening shrinkage, drying shrinkage, and temperature expansion.

The load on the composite girder during the normal use stage was approximately 40% of the ultimate load, which was selected as the load in the test. The 28-day compressive strength of the ordinary concrete specimen was 55.60 MPa and the 28-day compressive strength of the steel fiber reinforced concrete was 48.40 MPa.

4.1.2. Creep Experiment Process

A picture of the concrete creep experiment is shown in Figure 16. The experimental equipment included a creeper, an oil pressure jack, a pressure sensor, and a static pressure collector. The maximum pressure of the jack was 1000 kN. The sensor used as the load measuring device had a range of 700 kN and an accuracy of 0.01 kN. A paper-based strain gauge with a range of 10 cm and an accuracy of ± 1 $\mu\varepsilon$ was used, and the range was greater than the ultimate strain of the concrete. In addition, the load loss and stress decrease would be caused by concrete shrinkage and creep. Therefore, the accuracy of the experimental data was guaranteed by checking the load regularly, and the jack was used to adjust the load to ensure that the load change was always within $\pm 2\%$. The loading age of all concrete specimens t_0 was 28 days.

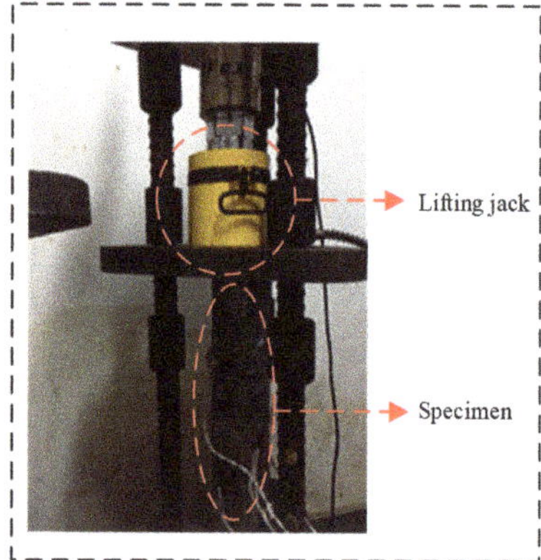

Figure 16. Photograph of a concrete creep experiment.

4.1.3. Experimental Results

According to the definition of creep degree function $C(t,t_0)$, as shown in Equation (1):

$$C(t,t_0) = \frac{\varepsilon_c(t)}{\sigma(t_0)} \tag{1}$$

where $\varepsilon_c(t)$ is concrete creep strain at time t, $\sigma(t)$ is concrete stress at time t.

The curves of the creep degree function C of the ordinary concrete and the fiber-reinforced concrete were obtained and shown in Figure 17.

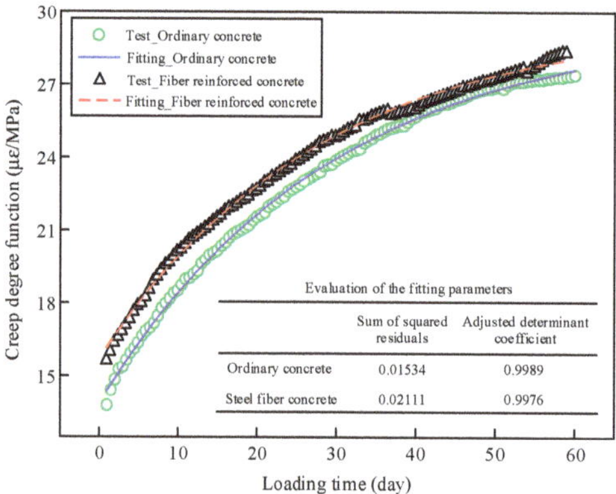

Figure 17. Comparison between the fitting results and the experimental results of the creep degree function.

A parameter regression was performed by utilizing a large number of experimental data with high prediction accuracy. A simplified fitting formula of creep degree function was proposed based on the measured data:

$$C(t, t_0) = A_0 + A_1 e^{-(t-t_0)/m} \tag{2}$$

where A_0, A_1, and m are the parameters that can be regressed by experimental results, which are shown in Table 2. t_0 is the initial loading age of concrete, which is 28 days for the concrete of the bridge.

Table 2. Parameters of the creep degree function.

Specimen	A_0	A_1	m
Ordinary concrete	29.713	−15.847	29.491
Steel fiber concrete	29.025	−12.717	28.149

A comparison between the fitting results of the simplified formula and the experimental data for the creep degree function is shown in Figure 17. The statistical results of the fitting parameters are also given in the figure. The dispersion of the parameters was small. The fitting results are credible.

4.2. Prediction Method of the Concrete Creep Effect

The aging coefficient was used to consider the effect of concrete aging on the final creep value based on the age-adjusted effective modulus method [33–36]. The creep effect of concrete can be accurately calculated by combining the age-adjusted effective modulus method and the FE method. The age-adjusted effective modulus method is expressed as follows:

The sum of elastic strain and creep strain of concrete $\varepsilon(t, t_0)$ is as follows:

$$\begin{aligned}
\varepsilon(t, t_0) &= \frac{\sigma(t)}{E_{t_0}} + \frac{\sigma(t_0)}{E_{t_0}}\varphi(t, t_0) + \frac{1}{E_{t_0}}\int_{t_0}^{t}\frac{d\sigma(\tau)}{d\tau}\varphi(t, \tau)d\tau \\
&= \frac{\sigma(t)}{E_{t_0}} + \frac{\sigma(t_0)}{E_{t_0}}\varphi(t, t_0) + \frac{1}{E_{t_0}}\int_{\sigma(t_0)}^{\sigma(t)}\varphi(t, \tau)d\sigma(\tau)
\end{aligned} \tag{3}$$

where E_{t_0} is Young's Modulus of concrete at loading time t_0; $\varphi(t, t_0)$ is creep coefficient at time t with loading time t_0, which can be calculated by Equation (4):

$$\varphi(t, t_0) = C(t, t_0)E_{t_0} \tag{4}$$

According to the integral mean value theorem, Equation (3) can be converted into the following form:

$$\begin{aligned}
\varepsilon(t, t_0) &= \frac{\sigma(t)}{E_{t_0}} + \frac{\sigma(t_0)}{E_{t_0}}\varphi(t, t_0) + \frac{\sigma(t)}{E_{t_0}}\varphi(t, t_0)\chi(t, t_0) \\
&= \frac{\sigma(t_0)}{E_{t_0}}\varphi(t, t_0) + \frac{\sigma(t)}{E_{t_0}}(1 + \varphi(t, t_0)\chi(t, t_0)) \\
&= \frac{\sigma(t_0)}{E_{t_0}}\varphi(t, t_0) + \frac{\sigma(t)}{E_{\varphi}}
\end{aligned} \tag{5}$$

where E_{φ} is the age-adjusted effective modulus. The age-adjusted effective modulus E_{φ} can be expressed as follows:

$$E_{\varphi} = \frac{E_{t_0}}{1 + \varphi(t, t_0)\chi(t, t_0)} \tag{6}$$

where $\chi(t, t_0)$ is the aging coefficient, which ranges from 0.5 to 1.0 and is generally chosen as 0.8.

According to the proposed creep degree function and the Dischinger method, the aging coefficient can be expressed as follows:

$$\chi(t, t_0) = \frac{1}{1 - e^{-\varphi(t, t_0)}} - \frac{1}{\varphi(t, t_0)} \tag{7}$$

4.3. Elaborate FE Model

The elaborate FE model of the bridge established in ANSYS is shown in Figure 18. The basic assumptions during model establishment were as follows: the concrete slabs were connected with the steel beams through the studs and only the self-weight of the newly poured concrete slab was considered (i.e., the stiffness was not considered). SHELL181 elements were used to simulate the steel beam and concrete slab. BEAM188 elements were used to simulate the stud and steel truss. The full bridge model included 24,764 nodes and 27,750 elements.

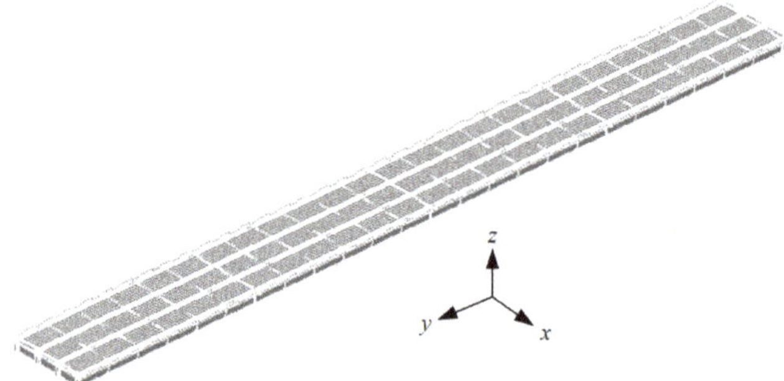

Figure 18. Elaborate FE model of the bridge.

The Young's modulus of steel is 2.06×10^5 MPa, the Poisson's ratio of steel is 0.3, and the density of steel is 7850 kg/m^3. The Poisson's ratio of concrete is 0.2, and the density of concrete is 2500 kg/m^3. The age-adjusted effective modulus method was used to simulate the concrete creep effect of the composite girder bridge. The 28-day Young's modulus of the concrete (3.65×10^4 MPa) was measured by testing specimens made from the same batch of concrete material.

The FE model was verified based on the monitoring data of structural deformation and stress during the concrete pouring process. The pouring dates of each segment are shown in Table 3 and the positions of each segment are shown in Figure 1.

The concrete construction process was simulated based on the FE model by locking and activating the concrete slab elements of each pouring segment in sequence using the life-and-death element method.

Table 3. Dates of concrete pouring.

Casting Segment	Casting Area	Date
I	Positive bending moment zone of the fourth span	9 a.m., 23 April 2018
II	Positive bending moment zone of the third span	9 a.m., 24 April 2018
III	Positive bending moment zone of the second span	2 p.m., 24 April 2018
IV	Positive bending moment zone of the first span	9 a.m., 25 April 2018
V	Negative moment zones of the third and fourth spans	2 p.m., 25 April 2018
VI	Negative moment zones of the second and third spans	9 a.m., 27 April 2018
VII	Negative moment zones of the second and first spans	12 a.m., April 27, 2018

A comparison between the calculated and measured deflections of the steel beams at the middle of each span is shown in Figure 19. The calculated deflections were in good agreement with the measured data with an error of 15%–21%. Moreover, the calculated deflections were slightly smaller than the measured data because that part of the stiffness of the concrete slab was taken into account in the FE model.

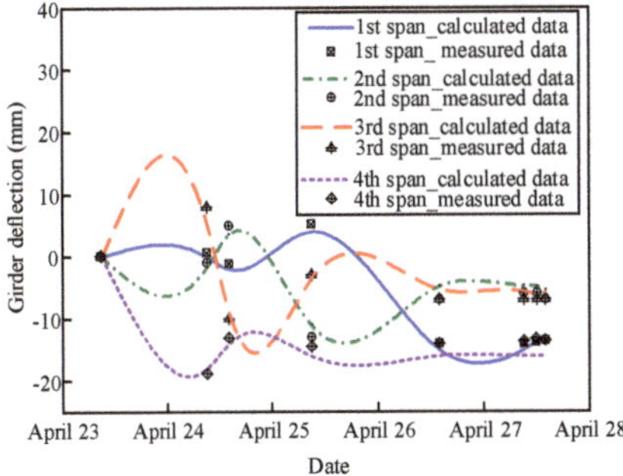

Figure 19. Comparison between the calculated and measured deflections at the middle of each span.

Figure 20 shows a comparison of the deflection variation along the transverse direction in Sections S2 and S6 when pouring the concrete in the middle positive moment zone. Due to the structural symmetry, the deflection changes in the inner two steel beams (B and C) were the same, and those of the outer two steel beams (A and D) were the same. Moreover, the deflection changes in the inner and outer beams were different with a difference of 3–4 mm because of the different constraints of the steel beams.

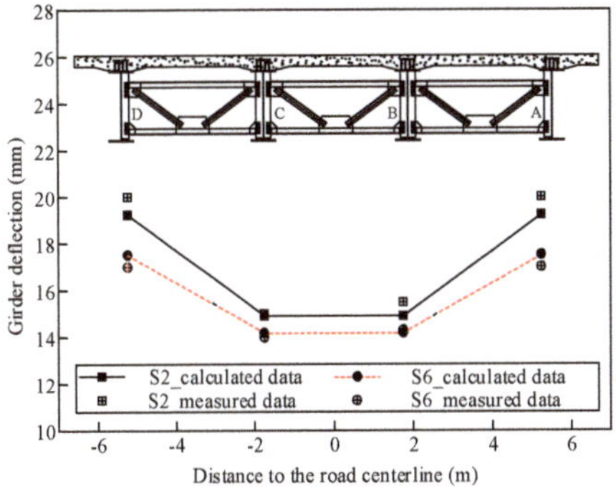

Figure 20. Comparison between the calculated and measured deflection variations along the transverse direction.

Comparisons between the calculated and measured girder strains in Sections S4 and S6 are shown in Tables 4 and 5, respectively. The deviation between the calculated and measured strains was small during the construction stage characterized by large deformation. However, the strain errors during the construction stage characterized by smaller deformation were relatively large because of environmental interference.

Table 4. Comparison between the calculated and measured girder strains in Section S4.

Casting Segment	Calculated Strain (με)		Measured Strain (με)		Error Percentage (%)	
	Bottom Plate	**Top Plate**	**Bottom Plate**	**Top Plate**	**Bottom Plate**	**Top Plate**
I	−5.1	6.4	−3.5	4.4	45.71	45.45
II	16.9	−12.5	12.7	−14.6	33.07	−14.38
III	−62.7	57.5	−53.2	46.2	17.86	24.46
IV	−130.4	124.9	−110.4	100.0	18.12	24.90
V	−87.3	124.5	−101.4	93.2	−13.91	33.58
VI	−119.9	124.9	−130.1	109.5	−7.84	14.06
VII	−131.1	118.2	−161.3	130.5	−18.72	−9.43

Table 5. Comparison between the calculated and measured girder strains in Section S6.

Casting Segment	Calculated Strain (με)			Measured Strain (με)			Error Percentage (%)		
	Bottom Plate	Top Plate	Concrete Slab	Bottom Plate	Top Plate	Concrete Slab	Bottom Plate	Top Plate	Concrete Slab
I	−5.3	5.3	0.0	−8.4	6.5	0.0	36.90	18.46	–
II	11.4	−9.5	0.0	13.6	-9.5	0.0	16.18	0.00	–
III	−47.3	28.7	0.0	−58.5	31.8	0.0	19.15	9.75	–
IV	202.0	−50.9	−120.5	239.1	−73.9	−120.4	15.52	31.12	0.08
V	202.4	−51.0	−120.5	245.2	−62.0	−126.8	17.46	17.74	−4.97
VI	202.0	−50.8	−120.5	238.8	−61.8	−132.4	15.41	17.80	−8.99
VII	199.1	−45.9	−120.8	240.1	−75.9	−136.4	17.08	39.53	−11.44

The deflection monitoring data within two months after bridge completion were used to verify the FE model. Comparisons between the calculated and measured deflections in Sections S2 and S6 are shown in Figure 21. The calculated variation trend of deflection was basically consistent with that of the monitoring results, which reflects the accuracy of the numerical simulation method. In addition, the simulation and monitoring results show that the girder deflection increased continuously in response to the concrete creep effect. Moreover, the deflection variation in Section S2 was larger than that in Section S6.

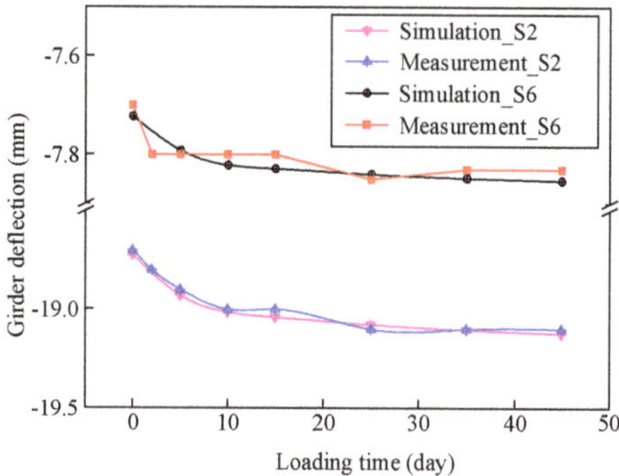

Figure 21. Comparison between the calculated and measured girder deflections within two months after bridge completion.

4.4. Predicted Long-Term Effects of Concrete Creep

The abovementioned FE model was used to predict the long-term evolution of the structural performance of the composite girder bridge. The prediction results of girder deflection under self-weight are shown in Figure 22, which were characterized by two development stages. The change in girder deflection was fast during the first 15 days and then slow afterwards. Moreover, the change trends and amplitudes of the deflection of the inner and outer steel beams were basically the same. In addition, the deflection change in the second span under the influence of concrete creep was smaller than that in the first span affected by the self-weight of the adjacent span.

The concrete creep significantly affected the stress redistribution between the steel beam and concrete slab. The girder stress changes in Sections S4 and S6 were predicted as shown in Figures 23 and 24, respectively. The compressive stress in the concrete slab in Section S4 decreased over time. The tensile stress in the upper flange of the steel beam and the compressive stress in the lower flange of the steel beam increased. The compressive stress in the concrete slab in Section S6 increased gradually over time. The compressive stress in the upper flange of the steel beam decreased, and the tensile stress in the lower flange of the steel beam increased.

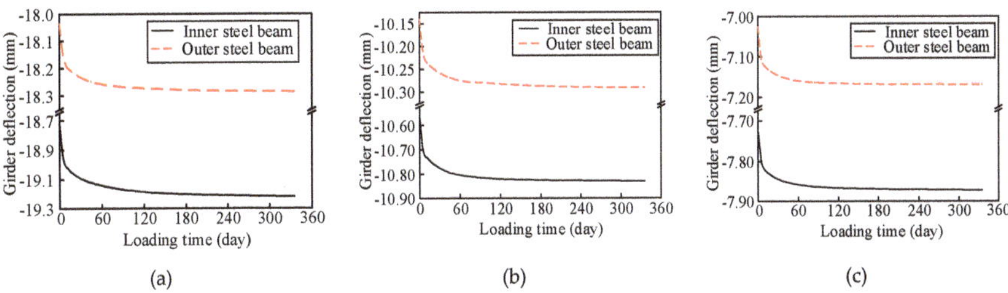

Figure 22. Girder deflection in each section caused by concrete creep (**a**) Section S2; (**b**) Section S3; (**c**) Section S6.

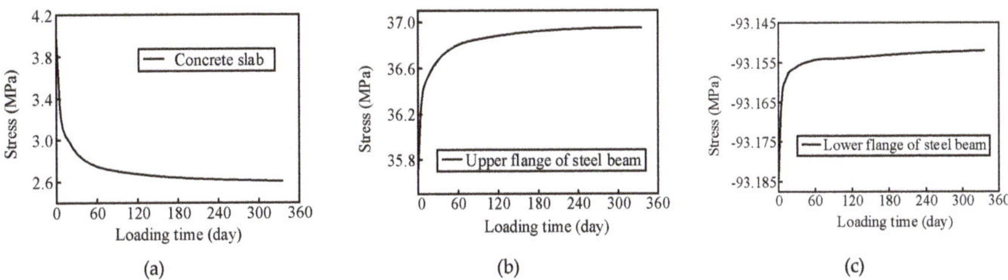

Figure 23. Girder stress changes in Section S4 (**a**) concrete slab; (**b**) upper flange of the steel beam; (**c**) lower flange of the steel beam.

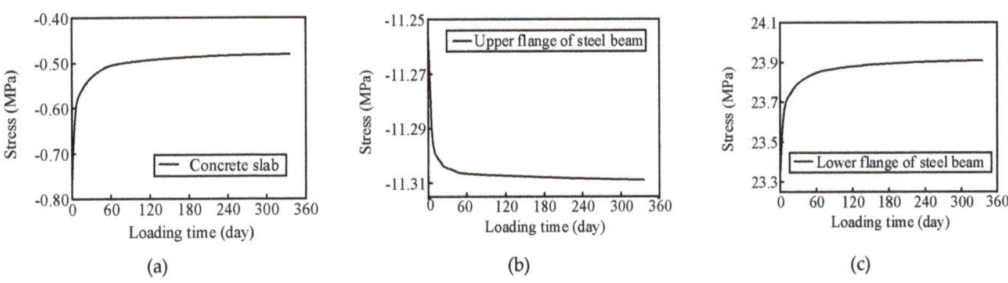

Figure 24. Girder stress changes in Section S6 (**a**) concrete slab; (**b**) upper flange of the steel beam; (**c**) lower flange of the steel beam.

The stress in the concrete slab in the negative moment zone of the composite girder decreased under the influence of concrete creep. Over time, the steel beams participated more in bearing the load. The concrete slab in the positive bending moment zone was continuously involved in load-bearing, and the compression part of the steel beam decreased gradually. The internal force in the lower flange of the steel beam in the positive moment region still increased due to the stiffness reduction in the negative moment region. In addition, the stress variation in the composite girder was characterized by three stages. The stress variations during the first 10 days were the most significant, the stress change rate decreased continuously during days 10–90, and the stress changed slowly and smoothly after 90 days.

Beam elements were used to simulate the stud to consider the interfacial slip. The internal force changes in the studs of the same column at the first and second spans were predicted under concrete creep. The bending moment of the lower end of the stud after 28 days of concrete pouring under self-weight is shown in Figure 25. The bending moment change in the stud along the longitudinal direction was characterized by a sinusoidal distribution. Moreover, the stud bending moment was the largest in Section S3. The long-term variation in the stud bending moment at each section is shown in Figure 26. The stud bending moment near the 3rd pier section increased gradually as a result of concrete creep, whereas the stud bending moments in the other sections decreased gradually.

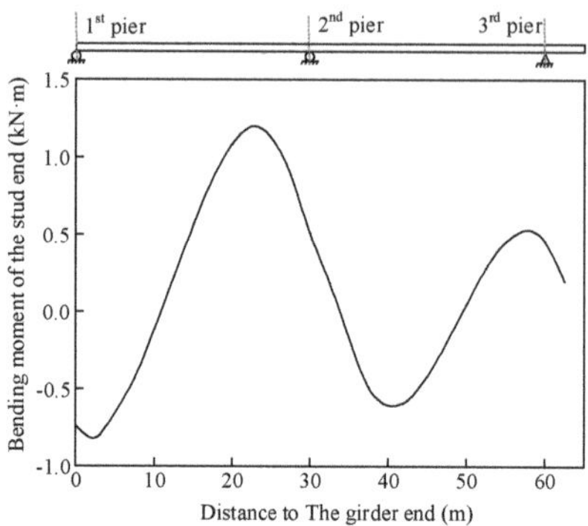

Figure 25. Bending moment of the stud end under self-weight.

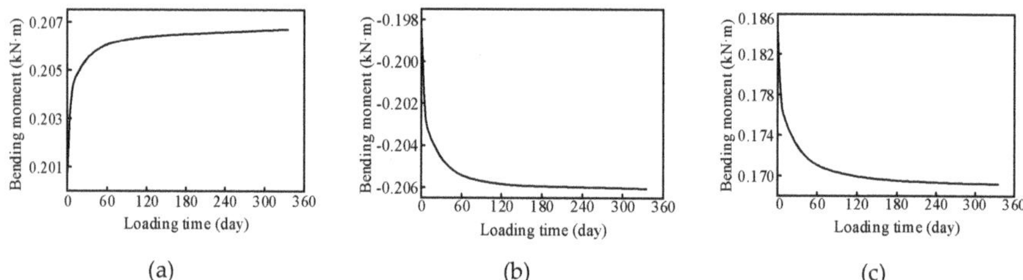

(a) (b) (c)

Figure 26. Long-term changes in the stud stress in each section (**a**) stud in the 3rd pier section; (**b**) stud in Section S6; (**c**) stud in Section S4.

5. Conclusions

In this study, we focused on the influence of time-dependent effects on an I-shaped steel–concrete composite continuous girder bridge, including temperature changes and concrete creep effect. The field monitoring data and FE simulation were combined. The main contributions of this paper are as follows:

(1) The daily temperature variations in the composite girder and the corresponding influence law were studied. The temperature variation trends of the steel beam and concrete slab were characterized by sinusoidal curves without temperature lag. In addition, the heating rate of the concrete slab was higher than the cooling rate. The rates of heating and cooling of the concrete slab first increased and then decreased. Moreover, the suggested temperature gradient model for composite girder bridges in the Hebei Province of China was proposed with a guaranteed rate of 95% by processing the maximum temperature difference during a single day.

(2) A concrete creep experiment under uniaxial compression was carried out using the same batch of concrete material as that used for the bridge. A creep degree function of concrete was obtained. The concrete creep effect of the composite girder bridge was predicted by integrating the age-adjusted effective modulus method and FE model. The accuracy of the numerical analysis was verified by comparing the calculated and monitoring results of girder deflections and strains. The prediction results show that the internal force of the concrete slab in the negative bending moment zone decreased, and the stress of the steel beam increased gradually. The concrete slab in the positive bending moment zone was continuously involved in load bearing. The compression part of the steel beam decreased gradually. However, the internal force of the lower flange of the steel beam in the positive bending moment zone still increased because of the stiffness reduction in the negative moment zone.

(3) The internal force change in the composite girder was characterized by three stages. The stress changes during the first 10 days were significant, the stress change rate decreased continuously during days 10–90, and the stress changed slowly and smoothly after 90 days. The bending

moment change in the stud along the longitudinal direction was characterized by a sinusoidal distribution. The stud bending moment was the largest in Section S3. The stud bending moment near the 3rd pier section increased gradually as a result of concrete creep, and the stud bending moments in the other sections decreased gradually.

Author Contributions: Conducting the experiment, G.-M.W.; Analyzing the experimental results and writing the paper, L.Z.; Building the FE model, G.-P.Z.; Managing the program (QG2018-8) related with the research presented in this paper, B.H.; Helping the first author to conduct the experiment, W.-Y.J. All authors have read and agreed to the published version of the manuscript.

Funding: This study was funded by the Science and Technology Program of Qinghai Province, China (grant number 2017-SF-139), the Transportation Science and Technology Program of Hebei Province, China (grant number QG2018-8), the National Natural Science Foundation of China (grant number 51608031) and the Fundamental Research Funds for the Central Universities of China (grant number 30919011246).

Acknowledgments: The authors express their thanks to the people helping with this work, and acknowledge the valuable suggestions from the peer reviewers.

Conflicts of Interest: The authors declare no conflict of interest.

References

1. Editorial Department of China Journal of Highways. Summary of Academic Research on Bridge Engineering in China 2014. *China J. Highw.* **2014**, *27*, 1–96.
2. Wang, J.Q.; Lv, Z.T.; Liu, Q.W.; Wang, F. Monitoring and numerical analysis of steel-concrete continuous composite girder bridge construction. In Proceedings of the China Bridge Society Bridge and Structural Engineering Branch 2005 National Bridge Conference, Hangzhou, China, 24–25 October 2005.
3. Li, X. Construction Control Technology of Steel-Concrete Composite Continuous Box Girder Bridge. Master's Thesis, Chongqing Jiaotong University, Chongqing, China, 2015.
4. Sun, T. Deformation analysis of steel-concrete composite beams during construction. *J. Lanzhou Univ. Technol.* **2011**, *37*, 107–109.
5. Tan, W.H.; Li, D. Monitoring and analysis of steel-composite composite beam deformation. In Proceedings of the 16th National Conference on Structural Engineering, Taiyuan, Shaanxi, China, 17–19 August 2007.
6. Chen, Y.J.; Ye, M.X. Response analysis of steel-concrete composite beam under temperature. *China Railw. Sci.* **2001**, *22*, 48–53.
7. Gu, P.; Shi, Q.C. Analysis of temperature effect of external prestressed steel-concrete composite girder bridge. *J. Highw. Transp.* **2008**, *25*, 55–59.
8. Zhou, L.; Lu, Y.C.; Li, X.F. The calculation of temperature stress of steel-concrete composite beam. *J. Highw. Transp.* **2012**, *29*, 83–88.
9. Liu, Y.J.; Liu, J.; Zhang, N.; Feng, B.W.; Xu, L. Analytical solution of temperature effect of steel-concrete composite beams. *J. Traffic Transp. Eng.* **2017**, *17*, 9–19.
10. Chen, Y.J.; Wang, L.B.; Li, Y. Study on temperature field and temperature effect of steel-concrete composite girder bridge. *J. Highw. Transp.* **2014**, *31*, 85–91.
11. Su, J.H.; Duan, S.J. Study on sunlight temperature effect of steel-concrete double-sided composite box girder. *J. Shijiazhuang Railw. Univ.* **2013**, *26*, 11–14.
12. Wang, Z.X. Analysis of Temperature Effect during Construction of Steel-Concrete Composite Girder Cable-Stayed Bridge in Alpine Region. Master's Thesis, Chang'an University, Xi'an, Shanxi, China, 2017.
13. Xiao, L.; Liu, L.F.; Wei, X. Temperature field analysis of steel-concrete composite girder bridge. In Proceedings of the China Steel Structure Association Structural Stability and Fatigue Branch 16th (ISSF-2018) Academic Exchange Conference and Teaching Seminar, Qingdao, Shandong, China, 2–14 September 2018.
14. Hui, R.Y.; Huang, G.X.; Yi, B.R.; Shen, J.H. Experimental study on concrete triaxial creep. *China J. Hydraul. Eng.* **1993**, *7*, 75–80.
15. Zhou, L. Internal force redistribution of steel-concrete composite beams caused by shrinkage and creep of concrete. *Bridge Constr.* **2001**, *2*, 1–4.
16. Liu, X.; Erkmen, R.E.; Bradford, M.A. Creep and shrinkage analysis of curved composite beams with partial interaction. *Int. J. Mech. Sci.* **2012**, *58*, 57–68. [CrossRef]

17. Chen, D.K.; Hu, B. Time-dependent analysis and fracture of steel-concrete composite structure. *J. Harbin. Technol.* **2003**, *35*, 123–128.

18. Zhao, M.; Wang, X.M.; Gao, J. Finite element numerical analysis of prestressed concrete structures. *J. Shijiazhuang Railw. Univ.* **2004**, *17*, 84–88.

19. Macorini, L.; Fragiacomo, M.; Amadio, C.; Izzuddin, B.A. Long-term analysis of steel–concrete composite beams: FE modelling for effective width evaluation. *Eng. Struct.* **2005**, *28*, 1110–1121. [CrossRef]

20. Gara, F.; Ranzi, G.; Leoni, G. Time analysis of composite beams with partial interaction using available modelling techniques: A comparative study. *J. Constr. Steel Res.* **2006**, *62*, 917–930. [CrossRef]

21. Miao, L.; Chen, D.W. Calculation of long-term performance of beams considering shear lag effect. *Eng Mech.* **2012**, *29*, 252–258.

22. Zhao, G.Y.; Xiang, T.Y.; Xu, T.F.; Zhan, Y.L. Stochastic analysis of shrinkage and creep effects of steel-concrete composite beams. *China J. Comput. Mech.* **2014**, *31*, 67–71.

23. Lu, Z.F.; Liu, M.Y.; Li, Q. Analysis of creep effect of steel-composite composite continuous beam bridge considering temperature and humidity changes. *J. Cent South Univ. Sci. Technol.* **2015**, *46*, 2650–2657.

24. Zhou, G.P.; Li, A.Q.; Li, J.H.; Duan, M.J. Structural health monitoring and time-dependent effects analysis of self-anchored suspension bridge with extra-wide concrete girder. *Appl. Sci.* **2018**, *8*, 115. [CrossRef]

25. Zhou, G.P.; Li, A.Q.; Li, J.H.; Duan, M.J.; Spencer, B.F.; Zhu, L. Beam finite element including shear lag effect of extra-wide concrete box girders. *J. Bridge Eng.* **2018**, *23*, 04018083. [CrossRef]

26. Zhou, G.P.; Li, A.Q.; Li, J.H.; Duan, M.J. Health monitoring and comparative analysis of time-dependent effect using different prediction models for self-anchored suspension bridge with extra-wide concrete girder. *J. Cent South Univ.* **2018**, *25*, 2025–2039. [CrossRef]

27. Gara, F.; Ranzi, G.; Ansourian, P. General method of analysis for composite beams with longitudinal and transverse partial interaction. *Comput Struct.* **2006**, *84*, 2373–2384.

28. Ranzi, G. Short—And long-term analyses of composite beams with partial interaction stiffened by a longitudinal plate. *Steel Compos. Struct.* **2006**, *6*, 237–255. [CrossRef]

29. Ranzi, G.; Bradford, M.A. Analytical solutions for the time-dependent behaviour of composite beams with partial interaction. *Int. J. Solids Struct.* **2005**, *43*, 3770–3793. [CrossRef]

30. Sullivan, R.W. Using a Spectrum Function Approach to Model Flexure Creep in Viscoelastic Composite Beams. *Mech. Adv. Mater. Struct.* **2012**, *19*, 39–47. [CrossRef]

31. Fan, J.S.; Nie, J.G.; Wang, H. Study on long-term mechanical behavior of composite beams considering shrinkage, creep and cracking (I): Experiment and calculation. *J. Civil Eng.* **2009**, *42*, 8–15.

32. Fan, J.S.; Nie, X.; Li, Q.W. Study on long-term mechanical behavior of composite beams considering shrinkage, creep and cracking (II): Theoretical analysis. *J. Civil Eng.* **2009**, *42*, 16–22.

33. Bazant, M.Z.; Kaxiras, E.; Justo, J.F. The environment-dependent interatomic potential applied to silicon disordered structures and phase transitions. *MRS Proc.* **1997**, *491*, 339–345. [CrossRef]

34. Justo, J.F.; Bazant, M.Z.; Kaxiras, E.; Bulatov, V.V.; Yip, S. Interatomic potential for condensed phases and bulk defects in silicon. *MRS Proc.* **1997**, *469*, 217–227. [CrossRef]

35. Bazant, Z.P.; Hauggaard, A.B.; Baweja, S.; Ulm, F.J. Microprestress-solidification theory for concrete creep. (I): Aging and drying effects. *J. Eng. Mech.* **1997**, *123*, 1189–1194.

36. Bazant, Z.P.; Hauggaard, A.B.; Baweja, S. Microprestress-solidification theory for concrete creep. (II): Algorithm and verification. *J Eng. Mech.* **1997**, *123*, 1195–1201. [CrossRef]

MDPI

St. Alban-Anlage 66

4052 Basel

Switzerland

Tel. +41 61 683 77 34

Fax +41 61 302 89 18

www.mdpi.com

Materials Editorial Office

E-mail: materials@mdpi.com

www.mdpi.com/journal/materials